Exhibit 4.16 Summary of spreadsheet annuity or investment functions

Excel	Lotus 1-2-3[1]	Quattro Pro[2]
$PV(i, N, -A, -F, \text{Type})$	@PV(A, i, N)	@PVAL($i, N, -A, -F, \text{Type}$)
$PMT(i, N, -P, -F, \text{Type})$	@PMT(P, i, N)	@PAYMT($i, N, -P, -F, \text{Type}$)
$FV(i, N, -A, -P, \text{Type})$	@FV(A, i, N)	@FVAL($i, N, -A, -P, \text{Type}$)
$NPER(i, A, P, F, \text{Type})$	@TERM(A, i, F)$^{v2.0}$	@NPER(i, A, P, F, Type)
	@CTERM(i, F, P)$^{v2.0}$	
$RATE(N, A, P, F, \text{Type, guess})$	@RATE(F, P, N)$^{v2.0}$	@IRATE(N, A, P, F, Type)
= NPV(i, values)	@NPV(i, range)	@NPV(i, range)
= IRR(values, guess)	@IRR(guess, range)	@IRR(guess, range)

[1]No functions added in v3.0. In v4.0, functions with the names listed under Quattro Pro were added; however, the parameters were in a different order.
[2]Lotus 1-2-3 functions also work.

Exhibit 4.17 Converting factors to functions

ANSI Factor	Excel	Lotus 1-2-3[1]	Quattro Pro
$(P/F, i, N)$	PV($i, N, 0, -1$)		@PVAL($i, N, 0, -1$)
$(F/P, i, N)$	FV($i, N, 0, -1$)		@FVAL($i, N, 0, -1$)
$(P/A, i, N)$	PV($i, N, -1$)	@PV($1, i, N$)	@PV($1, i, N$)
$(F/A, i, N)$	FV($i, N, -1$)	@FV($1, i, N$)	@FV($1, i, N$)
$(A/P, i, N)$	PMT($i, N, -1$)	1/@PV($1, i, N$)	@PAYMT($i, N, -1$)
$(A/F, i, N)$	PMT($i, N, 0, -1$)	1/@FV($1, i, N$)	@PAYMT($i, N, 0, -1$)

[1]For versions 1.0, 2.0, and 3.0 of Lotus 1-2-3, the $(P/F, i, N)$ and $(F/P, i, N)$ factors are best found algebraically. For version 4.0, the functions listed for Quattro Pro can be used, although the parameter order is different.

Exhibit 16.1 Determined- vs. spent-to-date funds over a project's life

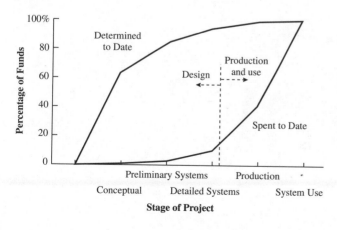

Exhibit 19.1 Three common objectives

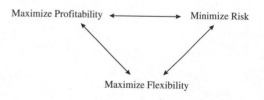

Chapter 16 Estimate accuracy

Estimate Type	Actual vs. Estimate
Order-of-magnitude	−30% to +50%
Budget	−15% to +30%
Definitive	−5% to +15%

ENGINEERING ECONOMY
APPLYING THEORY TO PRACTICE

IRWIN TITLES IN INDUSTRIAL ENGINEERING

Chase and Aquilano
Production and Operations Management: Manufacturing and Services
Seventh Edition

Duncan
Quality Control and Industrial Statistics
Fifth Edition

Eschenbach
Engineering Economy: Applying Theory to Practice
First Edition

Gitlow, Oppenheim, and Oppenheim
Quality Management: Tools and Methods for Improvement
Second Edition

Hill
Manufacturing Strategy: Text and Cases
Second Edition

Leenders, Fearon, and England
Purchasing and Materials Management
Ninth Edition

Nahmias
Production and Operations Analysis
Second Edition

Neter, Wasserman, and Kutner
Applied Linear Statistical Models
Third Edition

Neter, Wasserman, and Kutner
Applied Linear Regression Models
Second Edition

Niebel
Motion and Time Study
Ninth Edition

Vollman, Berry, and Whybark
Manufacturing Planning and Control Systems
Third Edition

OTHER IRWIN TITLES IN ENGINEERING

Bertoline, Wiebe, Miller, and Nasman
Engineering Graphics Communication
First Edition

Bertoline, Wiebe, Miller, and Nasman
Technical Graphics Communication
First Edition

Burgelman and Maidique
Strategic Management of Technology and Innovation
First Edition

Das, Kassimali, and Sami
Engineering Mechanics: Dynamics
First Edition

Das, Kassimali, and Sami
Engineering Mechanics: Statics
First Edition

Dhillon
Engineering Design: A Modern Approach
First Edition

Mills
Heat and Mass Transfer
First Edition

Mills
Basic Heat and Mass Transfer
First Edition

Mori
The New Experimental Design: Taguchi's Approach to Quality Engineering
First Edition

Rizzoni
Principles and Applications of Electrical Engineering
First Edition

Schaeffer
Material Science
First Edition

ENGINEERING ECONOMY
APPLYING THEORY TO PRACTICE

TED G. ESCHENBACH
University of Alaska Anchorage

IRWIN

Chicago • Bogotá • Boston • Buenos Aires • Caracas
London • Madrid • Mexico City • Sydney • Toronto

IRWIN

Concerned About Our Environment

In recognition of the fact that our company is a large end-user of fragile yet replenishable resources, we at IRWIN can assure you that every effort is made to meet or exceed Environmental Protection Agency (EPA) recommendations and requirements for a "greener" workplace.

To preserve these natural assets, a number of environmental policies, both companywide and department-specific, have been implemented. From the use of 50% recycled paper in our textbooks to the printing of promotional materials with recycled stock and soy inks to our office paper recycling program, we are committed to reducing waste and replacing environmentally unsafe products with safer alternatives.

Senior sponsoring editor: Richard T. Hercher, Jr.
Developmental editor: Carol L. Rose
Marketing manager: Brian Kibby
Project editor: Rita McMullen
Designer: Mercedes Santos
Graphics supervisor: Eurnice Harris
Compositor: Publication Services
Typeface: 10/12 Times Roman
Printer: R. R. Donnelley & Sons Company

Library of Congress Cataloging-in-Publication Data

Eschenbach, Ted.
    Engineering economy: applying theory to practice / Ted G.
Eschenbach.
        p.   cm.
    Includes index.
    ISBN 0-256-11441-2
    1. Engineering economy.   I. Title.
TA177.4.E833   1995
658.15—dc20   94–12751

Printed in the United States of America
   2   3   4   5   6   7   8   9   0   DO   1   0   9   8   7   6   5

This book is dedicated with my loving thanks for her decades of support and patience to my wife,
Chris Helen Matiukas

PREFACE

This text introduces the basic theory and application of engineering economy. It is suitable for a first course in engineering economy, for self-study by practicing engineers, and for reference during the practice of engineering. This text includes enough material so that instructors have some choices in approach for a first course at the undergraduate level. At the graduate level, instructors might add supplementary material, such as *Cases in Engineering Economy* (by Ted Eschenbach; Wiley, 1989). Examples from all major engineering disciplines are included, with both private- and public-sector applications.

KEY FEATURES OF THE TEXT

Studying what other texts have done well, surveying nearly 250 teachers of engineering economy, planning carefully, and listening to the comments of four sets of reviewers provide this text with clear advantages. The following points are the text's foundation:

1. Conceptually linking theory and practice.
2. Breadth and depth of coverage matched to introductory courses.
3. Emphasis on constrained project selection.
4. Appropriate pedagogical support for the students.
5. Modern computational tools.

1. Conceptually Linking Theory and Practice. Often the realism of engineering economy texts stops with the numbers plugged into examples and problems. This text presents the theory, and then it examines common practical violations of theoretical assumptions.

For example, most texts define *payback period* and demonstrate its inferiority to time value of money measures. This text also explains why payback continues to be used and why its use does not lead to bankruptcy. Similarly, this text, like most, explains the assumptions of repeated lives or salvage values that are needed for valid present worth comparisons. This text also discusses the impact of failing to satisfy the assumption of identical cost repetition while using equivalent annual techniques.

I believe that understanding how and why basic principles are applied is more important and interesting than fine theoretical nuances. In practice and in this text, ranking on the internal rate of return is used for constrained project selection, and present worth and equivalent annual cost are used for the selection of mutually exclusive alternatives. This text covers, but does not emphasize, integer programming models for capital budgeting, nor incremental internal rate of return and benefit/cost ratios for mutually exclusive alternatives.

Because students learn best when theory is clearly linked to real-world applications, the examples and homework problems are specifically chosen to illustrate practical realities and to include situations in which students can apply engineering economy concepts in their own lives.

2. Coverage Matched to Introductory Engineering Economy Courses.   Engineering economy relies on concepts and tools that are developed in accounting, finance, management science, probability, and statistics. Many topics have far more depth than can be included in an introductory course, and many engineering economy courses have no specific prerequisites. To best choose what to include here, I analyzed a list of needs established by a survey of nearly 250 faculty members in engineering economy.

For example, depreciation methods, income taxes, and the distinction between assets and income are covered, but other accounting topics are not. Similarly, the weighted cost of capital is discussed, but finance models for risk and stock prices are not. Basic models for multiple criteria are included, while utility theory is not. The application of expected value and decision trees is included from probability, and learning-curve models for cost estimation from statistics; however, general forecasting models are omitted.

The survey led to a text that has slightly more material than most instructors will cover in one semester. Thus, instructors have choices in the topics they will cover. To support the course work best, topics that were found to be "elective" were presented as independently as possible. For example, the text includes cost estimating, inflation, sensitivity analysis, multiple criteria evaluation, and economic decision trees, but each presentation is completely independent. Appendices of more advanced material are also included with several chapters.

3. Emphasis on Constrained Project Selection.   Most texts emphasize the choice between mutually exclusive alternatives with an externally defined discount rate. Capital budgeting is often relegated to the back of engineering economy texts, where the presentation focuses on theoretical models and advanced techniques that are rarely used.

However, the capital budgeting problem is often solved in the real world by ranking on internal rate of return. Discussing this constrained project selection problem in Chapter 9 provides a solid, intuitive foundation for selecting a discount rate. This is applied in Chapter 10 for the comparison of mutually exclusive alternatives.

The discussion of mutually exclusive alternatives focuses on the use of present worth and equivalent annual measures. Incremental internal rates of return and benefit/cost ratios are covered, but not as the technique of choice for this category of problem.

4. Pedagogical Support.   It is clear that students learn more when the structure of the presentation and the expectations are clear. Chapter objectives, key words and concepts, and lists of major points are used to facilitate student understanding.

Realistic examples can engage student interest, and problems drawn from student life can motivate high performance. These examples include factor-based solutions (the traditional basis of engineering economy) and newer spreadsheet-based solutions, which use computer power for more realistic models and more complete analysis. To support continuity of understanding and development of realistic problems, some examples and homework problems are carried through a sequence of chapters.

Exhibits are used liberally. Advances in computer power have made it possible and sometimes desirable to combine tables and graphics to form more powerful exhibits. A single numbering scheme for the exhibits is used to make it easier to find a referenced exhibit, whether table or figure.

5. Modern Computational Tools.   Spreadsheets enhance the student's analytical capabilities, allow the inclusion of more realistic problems, and better prepare students to use engineering economy after graduation. For these reasons spreadsheets are used in the examples, and they can be applied to the homework problems. The power of spreadsheets allows students to learn more and to address problems in realistic detail. Real-world engineering economy almost universally relies on spreadsheets as an analytical tool.

Spreadsheets can be included in course work with this text in several ways: (1) spreadsheets can be an integral part of the course, including instruction on their construction and use; (2) spreadsheets can be required or encouraged for homework without requiring their use in class; (3) students can use spreadsheets on their own to do homework. The spreadsheet instruction and setup sections are placed at the ends of chapters, so that they may be skipped if desired.

While spreadsheets are very useful, I have long been a proponent of understanding a problem and using hand calculations rather than relying on computers to solve it. That is why some examples and problems are designed to show that calculators are the easiest tool and the best choice. This text offers a variety of hand-calculation and computer problems that reflect what is expected of the practicing engineering economist. It also incorporates spreadsheets throughout the text for complex problems, especially those related to inflation, sensitivity, and taxes.

For access to spreadsheet information see page xiii.

ORGANIZATION OF THE TEXT

The overall flow of the material is shown in Exhibit P.1. The first three parts are included in most courses, but the last chapter in each part (Chapters 4, 8, and 11) can be easily skipped without loss of continuity.

Exhibit P1   Organization of the text
(Dotted material can be skipped)

```
┌──────────────────────────────────┐
│            Part One              │
│     Basic Concepts and Tools     │
│   1 Making Economic Decisions    │
│   2 The Time Value of Money      │
│   3 Equivalence—A Factor Approach │
├──────────────────────────────────┼──────────────────────────────────┐
│      4 Geometric Gradients       │           Part Two               │
│        and Spreadsheets          │       Analyzing a Project        │
│                                  │        5 Present Worth           │
│                                  │    6 Equivalent Annual Worth     │
│                                  │      7 The Rate of Return        │
│                                  ├──────────────────────────────────┼──────────────────────────────────┐
│                                  │ 8 Benefit/Cost Ratios and Other  │         Part Three               │
│                                  │            Measures              │    Comparing Projects            │
│                                  │                                  │      and Alternatives            │
│                                  │                                  │ 9 Constrained Project Selection  │
│                                  │                                  │ 10 Mutually Exclusive Choices    │
│                                  │                                  ├──────────────────────────────────┤
│                                  │                                  │  11 Replacement Analysis         │
├──────────────────────────────────┼──────────────────────────────────┴──────────────────────────────────┤
│          Part Four               │            Part Five             │
│ Enhancements for the Real World  │      Decision-Making Tools       │
│      12 Depreciation             │   16 Estimating Cash Flows       │
│      13 Income Tazes             │    17 Sensitivity Analysis       │
│      14 Public-Sector            │  18 Uncertainty and Probability  │
│   Engineering Economy            │    19 Multiple Objectives        │
│      15 Inflation                │                                  │
└──────────────────────────────────┴──────────────────────────────────┘
```

Part One presents concepts that form the fundamental basis for engineering economic calculations. Part Two reinforces this material for increasingly complex cash flows. Part Three compares alternatives—those that are constrained by a limited budget and those that are mutually exclusive. Part Four includes the effect of income taxes on the private sector, special concerns of the public sector, and inflation. Part Five explains tools that are needed to deal with the complexities of the real world.

THE INSTRUCTOR'S MANUAL

The instructor's manual includes solutions to end-of-chapter problems, suggested course outlines, and transparency masters for all exhibits.

The end-of-text tables are also available in pamphlet form in quantity from the publisher for those instructors who do closed-book testing. Similar pamphlets of tables for continuous compounding and distributed cash flows are also available.

ACKNOWLEDGMENTS

I have had an enormous amount of help in completing this text. The authors of the numerous other texts that I have used have provided fine examples and shaped my understanding of the subject. Similarly, students at the University of Alaska Anchorage, the University of Missouri–Rolla, and Merrimack College have endured classroom testing of drafts and patiently pointed out where further work was needed.

Over 200 faculty members responded to a six-page survey about their current texts, courses, and desired material for an introductory text. Their responses directed me to the best and worst aspects of existing texts and provided great insight into their courses. Their responses specifically guided me in the addition and deletion of topics in my outline.

Other faculty members have provided material, reviewed the manuscript, or tested it in class. They have provided many helpful comments and criticisms. I would like to thank the following individuals in particular:

James A. Alloway, Jr., *Syracuse University*

Daniel L. Babcock, *University of Missouri–Rolla*

Susan Burgess (deceased), *University of Missouri–Rolla*

John R. Canada, *North Carolina State University*

Barry Clemson, *Old Dominion University*

William J. Foley, *Rensselaer Polytechnic Institute*

Timothy J. Gallagher, *Colorado State University*

Carol S. Gattis, *University of Arkansas, Fayetteville*

Joseph E. Gust, Jr., *Northern Arizona University*

Kim Hazarvartian, *Merrimack College*

Donald P. Hendricks, *Iowa State University*

Ken Henkel, *California State University, Chico*

Leonard Hom, *California State University, Sacramento*

W. J. Kennedy, *Clemson University*

Robert G. Lundquist, *Ohio State University*

Richard W. Lyles, *Michigan State University*

Anthony K. Mason, *California Polytechnic State University at San Luis Obispo*

Paul R. McCright, *Kansas State University*

Nancy L. Mills, *University of Southern Colorado*

Murl Wayne Parker, *Mississippi State University*

Louis Plebani, *Lehigh University*

Jang W. Ra, *University of Alaska Anchorage*

Herbert P. Schroeder, *University of Alaska Anchorage*	David Veshosky, *Lafayette College*
Jack W. Schwalbe, *Florida Institute of Technology*	Ed Wheeler, *University of Tennessee, Martin*
Paul L. Shillings, *Montana State University*	Bob White, *Western Michigan University*
Sanford Thayer, *Colorado State University*	Henry Wiebe, *University of Missouri–Rolla*

Ed Wheeler of the University of Tennessee, Martin not only reviewed drafts of the manuscript, he also contributed sets of problems for every chapter. Because engineering economy and finance are closely related subjects, Tim Gallagher, professor of finance at Colorado State University, reviewed the manuscript. Jarad Golkar, then of Texas A&M University, independently solved the end-of-chapter problems for over half of the text. His "review" of the solutions manual has helped ensure a clean first edition.

I would like to thank Bill Stenquist, who as my initial editor at Irwin convinced me that the effort to write a new text was worthwhile and who then helped me shape the text to meet the needs of the market delineated by the survey. My first developmental editor, Max Effenson, got me started on the right track. My second developmental editor, Becky Johnson, vastly improved both my written product and my writing skills and managed to keep me both smiling and moving along. My second editor, Tom Casson, asked the tough questions that forced me to search out the remaining weak points. Pat Soberanis, in spite of the sleepless nights she caused me, is the best copy editor I have ever had. At the final stage, the team of Dick Hercher, Carol Rose, Rita McMullen, and Mark Malloy set a standard of professional competence, good humor, and tact that I endeavored to match.

I have had support at the University of Alaska Anchorage as well. I would like to thank my boss, Will Nelson, and my support staff of Pinky Miranda, Paulette Jennings, Mindy Nichols, Angie Damberg, and Jeannie Carpenter. They have typed manuscript, revised spreadsheets and graphics, and made numerous copies.

My deepest appreciation goes to my wife, Chris Matiukas, for her unflagging support and patience. I would also like to thank Andrew and Kelsey for dragging me away from my computer to play and for accepting the many times I had to say that I couldn't.

Ted Eschenbach
Anchorage, Alaska

SPREADSHEET SOFTWARE

Even though most of this text's problems can be solved with a handheld calculator and the tables in Appendix A and all of the theory can be understood without spreadsheets, I believe that using them enhances your understanding and provides a faster way to complete homework. Thus, I recommend that you use this course to improve or acquire spreadsheet mastery. That mastery will be useful for many other engineering applications that analyze and/or present numbers.

If you have any version of Excel™, or Quattro Pro™, or version 4 or later of Lotus 1-2-3™, the included financial functions can be used to quickly and easily solve most engineering economy problems. Current versions of all of these packages are sold at substantial academic discounts in university bookstores and through authorized educational resellers. I recommend taking advantage of this opportunity by:

1. Checking out your bookstore, since lower academic prices are available for volume purchases.

Then if necessary

2. Calling one of the following numbers.

Note: corporations do merge, change software distribution channels, or sell ownership of software packages, so the following information is subject to change. In November 1994, when this was written, PC Connection (800-800-0005) was authorized to sell all three packages to students who had faxed or mailed verification of their ID, and the prices were about $85. Other software distributors have or will develop the same capability.

For Microsoft Excel, authorized academic resellers include:

   Computer Discount Warehouse (CDW) 800-451-4239

   PC Connection 800-800-0005

In November 1994, the recommended academic price for a single copy was $100.

For Quattro Pro, by calling 800-321-3220 and providing your ZIP code, the authorized academic resellers closest to you can be identified. If none is nearby, there is provision for direct shipping with faxed verification of your student ID. In November 1994, the recommended academic price for a single copy was $99.

For Lotus 1-2-3, authorized academic resellers can be identified by calling 800-343-5414. In November 194, the recommended academic price for a single copy was $100.

CONTENTS IN BRIEF

CONTENTS

PART FOUR

ENHANCEMENTS FOR THE REAL WORLD   309

CHAPTER 14

Public-Sector Engineering Economy 370

APPENDIX

End-of-Period Compound Interest Tables   553

PART ONE   BASIC CONCEPTS AND TOOLS

One of the most important skills in engineering economy is that of structuring the problem and solving it. Data, objectives, and alternatives must be brought together in a logical process (Chapter 1) to ensure that the best projects are chosen.

Engineering projects involve different cash flows at different times; typically, the cost to build or buy occurs now and the benefits occur for the next 10 to 50 years. The basic principle underlying engineering economy is the time value of money—$1 now is worth more that $1 later. Compound interest is used to calculate equivalent cash flows (Chapter 2), so that cash flows at different times can be combined.

Because cash flows often occur in consistent patterns, engineering economy factors have been developed to find equivalent present worths, equivalent annual costs, etc. As shown in Chapter 3, these factors greatly simplify the task of economic evaluation.

When a computerized spreadsheet is available, engineering economy can be both easier to do and far more realistic in its modeling. As shown in Chapter 4, this is particularly true for cash flows that grow at a constant rate from year to year.

CHAPTER 1　MAKING ECONOMIC DECISIONS

The Situation and the Solution

Engineering decisions must consider money—both in the short term and in the long term. The choice between alternatives must balance economics, performance, esthetics, resources, etc.

The solution is to apply engineering economy as part of a well-structured decision-making process.

Chapter Objectives

After you have read and studied the sections of this chapter, you should be able to:

Section 1.1
Define engineering economy and provide example applications.

Section 1.2
State the principles of economic decision making.

Section 1.3
Structure and solve a problem.

Section 1.4
Identify problems where engineering economy is unnecessary, crucial, or a contributing factor.

Section 1.5
Identify economic and noneconomic criteria.

Key Words and Concepts

Time value of money   The principle that $1 today has more value than $1 some time in the future.

Engineering economy   Making engineering decisions by balancing expenses and revenues that occur now and in the future.

Sunk costs   Costs that have been incurred or dollars that have been spent or committed.

Stakeholders   Those who will be affected by the engineering project, including employees, customers, users, managers, owners, and stockholders.

Satisficing   Choosing a satisfactory alternative rather than searching for the best alternative.

1.1 WHAT IS ENGINEERING ECONOMY?

Engineering economy evaluates the monetary consequences of the products, projects, and processes that engineers design. Decisions requiring engineering economy are important because of capital investments that have significant costs and lives of 5 to 50 years. Often two design alternatives are compared, and the question is: "Which is cheaper in the long run?"

This question cannot be answered by comparing only physical efficiencies. Electric baseboard heating is nearly 100% efficient at converting input energy to heat, while a fuel oil furnace may have an efficiency as low as 45%. Nevertheless, fuel oil may be economically more efficient—if the cost of fuel oil is low enough. Economic efficiency is measured by dividing the dollar value of the outputs by the dollar value of the inputs.

In other cases, one alternative may be cheaper to build, and the other cheaper to operate. For example, a building without insulation is cheaper to build, but a building with thick insulation is cheaper to heat and to air-condition. Engineering design must answer the question, "How much insulation is the best?" Combining current building costs with future operating costs requires the tools of engineering economy, because $1 today is more valuable than $1 a year or 10 years later. The principle that $1 today is more valuable than $1 a year later is called the **time value of money.**

Engineering economy adjusts for the time value of money to balance current and future revenues and costs. Beginning with the next chapter, interest will be defined and used, but for now we need only recognize that $1 today has more value than $1 in a year. If you doubt that money has time value, why are you not eager to loan someone $1000 now to be repaid $1000 at the end of the year? Consider Example 1.1.

Engineering economy uses mathematical formulas to account for the **time value of money** and to balance current and future revenues and costs.

Example 1.1   Joe Miner's Student Loan

Joe Miner has been offered a student loan by an engineering firm in his hometown. He will borrow $5000 each year for 4 years. He is required to maintain a 3.0 GPA, major in some engineering discipline, and graduate within 5 years. Beginning 1 year after graduation and continuing for 3 more years, he is required to repay this loan at $5000 per year. Is this a good deal?

Solution
Yes! Since Joe repays each $5000 four or five years after he borrowed it, the time value of money says the payments are worth less than what was borrowed. Essentially, the engineering firm has given him the free use of $20,000 for 4 years. He is allowed to invest it in his education only. When he graduates, his skills and salary will be higher, and it will be easier for him to repay the loan.

The applications of engineering economy are many and varied. Examples include the choice between a concrete and a steel structure, between various insulation thicknesses, between possible loans for a car or for a robot, and between prices at which to sell a duplex, a firm, or a product. It can be applied by an engineer to size a pump or to buy a home. It can be applied by a design firm to analyze the purchase of engineering software. (Will the value of time saved or new capabilities achieved exceed the cost of adding the software?) It can be applied by a major corporation to analyze plans for a new manufacturing facility or a new research and development (R&D) thrust. In each case, as depicted in Exhibit 1.1,

Exhibit 1.1   Engineering economy balances current and future revenues and costs

engineering economy balances expenses and revenues that occur now and in the future.

The importance of money or engineering economy to engineering has long been recognized. To quote Arthur M. Wellington (1887):

> engineering . . . is the art of doing that well with one dollar which any bungler can do with two after a fashion.

1.2 PRINCIPLES FOR DECISION MAKING

Exhibit 1.2 highlights five principles that will be applied repeatedly in the economic evaluation of alternatives.

Exhibit 1.2 Decision-making principles

1. State consequences in a common measure.
2. Only differences between alternatives need to be considered, which implies ignoring sunk costs.
3. Separable decisions should be made separately.
4. Adopt a systems viewpoint.
5. Use a common planning horizon.

1. Common Measure.   The emphasis of engineering economy is on money's time value or, in other words, on converting dollars at different times into a common measure. This principle also implies that consequences such as time savings, reductions in accidents or pollution, and improvements in quality will be stated in dollar terms if at all possible.

2. Only Differences Matter.   This principle is used to simplify the economic evaluation of alternatives. Many engineering design problems focus on different alternatives for accomplishing the same end. For example, since each roof design ensures a dry interior, the value of a dry interior does not need to be determined. Instead, only differences in the cost of the roof are relevant. These differences include the original installation costs, annual maintenance costs, and how soon the roof must be replaced.

One crucial application of this principle is that **sunk costs** should be ignored. *Sunk costs* are defined as money that has been spent or irrevocably committed. Even if buying a robot, buying stock in Fly-by-Night Operators, or buying a particular computer was a mistake, the money spent on the original purchase is sunk. When considering alternatives, the robot's salvage value, the stock's current price, and the computer's resale value are the relevant concerns. Example 1.2 describes possible sources of pressure to erroneously consider sunk costs.

Sunk costs are spent and should be ignored.

Example 1.2   HiTek's Computer Workstations

HiTek Engineering purchased eight computer workstations from Turkey Technologies late last year. Last week the government announced a standard format for computer-based drawings and specifications. Unfortunately, none of the various software packages that currently meet that standard will run on the eight Turkeys. Last year the machines cost $250K, including software (K stands for thousands). Buying similar new machines would cost $190K today. If HiTek sells the Turkeys on the used-machine market, HiTek would receive only $80K. If HiTek buys new workstations, the eight Turkeys could be shifted to another use, which has a value of $90K.

Which of the following arguments on what to do with the old machines when new ones are purchased is true?

Argument 1.   HiTek must sell the Turkeys because it cannot justify using $190K worth of machines for a $90K use.

Argument 2.   HiTek must keep the Turkeys because it cannot afford to lose $110K (= $190K − $80K) by selling them.

Argument 3.   HiTek cannot afford to sell the Turkeys because then it would be admitting that it made a $250K mistake.

Argument 4.   Even though the alternative use is worth only $90K, HiTek should keep the Turkeys since they will be worth only $80K if sold.

Solution

The true argument is 4. It ignores the sunk costs of $250K and the costs of a separate decision, which is whether to spend $190K for new workstations.

Argument 1.   False, since the Turkeys are not worth $190K.

Argument 2.   False, since the only relevant cost is the $80K that could be received for sale of the workstations.

Argument 3.   False. The $250K is a sunk cost that cannot be recovered. Buying the workstations may or may not have been a mistake. How much value was received from having them for the past year? Was it possible to predict the forthcoming standard?

Even if a decision was a mistake, failing to admit and to correct a mistake is yet another mistake—often with larger consequences.

Even if this decision was a mistake, failing to admit and to correct a mistake is yet another mistake—often with larger consequences.

3. Separable Decisions Should Be Made Separately.   This principle avoids justifying a low-quality result for decision B by combining it with a high-quality result for decision A. Why not get a high-quality result for both A and B by examining them separately?

Applying this principle separates the financing and investing decisions for firms and public agencies. Firms and governments raise funds (finance) through a variety of mechanisms, including stocks, bonds, loans, user fees, and taxes. How they select projects to implement (invest) is a separable question. Choosing which investments to make requires the tools of engineering economy.

4. Adopt a Systems Viewpoint. For private-sector problems, the viewpoint should be that of the firm as a whole, not just a department, a plant, or a division. For the public sector, the viewpoint should be that of the agency and the public it serves and taxes, not that of one department.

For example, a firm's R&D group must consider potential revenues from new products to the entire firm, not just the interesting challenges the star researchers are eager to tackle. Similarly, a hospital's physical therapy unit might be well advised to purchase improved equipment, which would increase the unit's costs and cause its bottom line to suffer—if that improved equipment would allow patients to be discharged from other hospital units more quickly.

5. Use a Common Planning Horizon. To avoid introducing misleading differences into the comparison of alternatives, the same time period for project justification should be used for all alternatives. For example, a stainless-steel pump may have a longer life than a brass pump, but each alternative should be considered with the same time horizon for use of the pumps.

1.3 THE DECISION-MAKING PROCESS

The real world contains messes [Ackoff], not problems. A chaotic conglomeration of conflicting objectives, undefined constraints, and incomplete or contradictory data is a mess, not a problem. A well-structured decision-making process is required to turn a mess into a problem.

The steps shown in Exhibit 1.3 can be used to solve a mess. These steps are not unique to engineering economy. They can also be applied to engineering design and to managerial and personal decisions. In Example 1.3, they are applied to a simple mess in engineering design and economics.

Exhibit 1.3
Decision-making flowchart

1. Define problem
↓
2. Choose objective(s)
↓
3. Identify alternatives
↓
4. Evaluate consequences
↓
5. Select
↓
6. Implement
↓
7. Audit

Example 1.3   Charlene Selects a Pump

Charlene is a newly hired chemical engineer. She has been asked to select a pump to move 15,000 liters of slurry a day to the pollution treatment center. What process should she use to select the pump?

Solution

1. **Define Problem**.   Key problem characteristics include, but are not limited to: (1) slurry characteristics, such as particle size and corrosiveness; (2) pipe sizes and flow velocities for input and output; (3) any size, power, and location restrictions; (4) time horizon for use of the pump and how soon it is needed; (5) cost limits; (6) vendor selection process—purchase order, competitive bid, or request for proposal (RFP); and (7) how many pumps (identical, similar, or by the same vendor) are needed.

2. **Choose Objective(s)**.   Generally, cost over the pump's life and technical performance will be dominant, but other factors—such as vendor and pump reliability, similarity to current practice, maintainability, and flexibility for changed conditions—must also be considered.

3. **Identify Alternatives**.   If the pump will be purchased from a catalog or other vendor material, then Charlene must assemble and examine those materials from all appropriate vendors. If an RFP will be issued, then possible alternative specifications must be developed.

4. **Evaluate Consequences**.   The advantages and disadvantages of each alternative with respect to the objectives must be enumerated.

5. **Select**.   Charlene must select a pump based on costs and benefits adjusted for the time value of money. Technical performance objectives must be completely satisfied, so that any pump with deficiencies is eliminated from consideration.

6. **Implement**.   This is the pump's order and installation.

7. **Audit**.   This step evaluates the pump and the decision-making process after the pump is in use. The intent is to provide "lessons learned" to assist in future decisions.

The importance of each step can be shown by examples where an engineer has either erred or missed the opportunity to excel. The consequences to the firm, the public agency, the public, and the engineer can be severe. It is also useful to focus on how to do each step correctly.

1. Define Problem.   In Example 1.3, Charlene's problem was narrowly defined. This narrow definition precludes creative alternatives that might be outstanding solutions. For example, there might be another process design that would generate no slurry, or it might be better to dewater the slurry and use a conveyor to transport a sludge. If the problem is defined too broadly, then the new engineer may be left floundering; if it is defined too narrowly, then better solutions may be prohibited.

The key skills in problem definition seem to be in asking questions and correctly pulling together information from a variety of sources. In some areas, such as computer purchases, this step often leads to changes in procedures that remove the need to buy the computer. In other cases, completely new alternatives may be suggested.

2. Choose Objective(s). Often the wrong cost objective is chosen. Because of its immediacy, the cost to purchase or build may incorrectly be the only consideration. The correct objective is to consider the time value of money and all costs and benefits. This includes purchase, installation, overhaul, operation, and disposal costs, and the benefits that occur over the project's life.

At times the decision maker must consider factors in addition to the time value of money. The extra consideration is sometimes described by distinguishing between effectiveness and efficiency. *Effectiveness can be defined as doing the right things, while efficiency can be defined as doing things the right way.* Often effectiveness focuses on the firm's strategic direction, while efficiency is more of a tactical choice.

3. Identify Alternatives. The best alternative may be overlooked because the problem is defined too narrowly. Or the problem might be defined properly, but creativity is missing.

As many alternatives as possible should be suggested. Even if several have obvious defects, they may lead to better alternatives. Processes involving groups of individuals and specific techniques, such as brainstorming, are useful.

The potential payoff from innovative alternatives is often far greater than the benefits of more accurately analyzing existing alternatives. Choosing between two alternatives whose costs differ by 10% may require accurate estimates and detailed analysis—to achieve a maximum payoff of 10%. A new alternative that is 50% better can be justified and implemented with less detailed analysis, because it is clearly superior.

4. Evaluate Consequences. A common mistake when evaluating alternatives is a myopic focus on which consequences to include. A department considering the purchase of a computer might consider its needs alone, without considering whether other departments in the firm might use the computer (or might have surplus computers available).

For example, one dam on the Colorado River was built primarily to regulate the flow from one set of states to another set. It helped satisfy a legal requirement imposed on the first set of states. From a broader perspective, however, the increased evaporation and leakage into the canyon walls were important consequences that damaged the entire system.

At first, consequences may be evaluated in a cursory way, because this may be enough to eliminate obviously inferior alternatives. Then consequences for the better alternatives may be evaluated in more detail.

5. Select. Selecting the best alternative is closely linked to choosing the correct objective. The criteria used to select is based on the objective that has been identified. If the wrong objective, such as minimizing first cost, is used, then often the wrong selection will be made. Even if a correct objective is identified, such as minimizing life-cycle costs or maximizing present worth including benefits and costs, departmental infighting may result in second- or third-best selections.

It is unrealistic and inappropriate to expect estimates of costs and revenues to *always determine* the best alternative. Often the selection process will have to account for other factors, such as a corporate policy to minimize layoffs or a state highway agency's desire to have highway projects in every region.

6. Implement.   The selection process may determine the success or failure of an implementation. Capital projects require efforts from many individuals. If all are convinced that the best alternative has been selected, then all are likely to work to overcome any problems. On the other hand, if some believe that another alternative was better, they may not work as hard to make the implementation a success.

In some cases the users of engineering projects have not been consulted. Numerous projects in third-world countries have been designed and selected by engineers and bureaucrats from developed countries. In some cases the implementation fails because maintenance expertise is lacking. In other cases, implementation fails because the project violates local customs or power structures.

Stakeholders are the people affected by the engineering project. This includes employees, customers, users, managers, owners, and stockholders.

The key here seems to be involvement of **stakeholders** (those who will be affected by the engineering project) in the earlier steps.

7. Audit.   This step is often omitted. The pressure to move on to and complete the next project is substantial. Besides, comparing actual performance with predicted performance can be embarrassing, with few rewards. If showing that actual market growth lagged predicted growth by two years leads to punishment, then no one will want to audit the accuracy of the engineering economy studies that were used to make the decision.

The reward for auditing the engineering project is that audits are the solid foundation for learning. What should you do differently next time? Besides, if you know that someone will check your work later, you have an extra incentive to do it right now.

Summary of the Decision Process.   These seven steps provide a logical framework for making engineering economy decisions and solving other problems. Chapter problems will focus on the economic evaluation of alternatives, but the real world requires that you deal with the entire decision-making process.

1.4 THE ENVIRONMENT FOR DECISIONS

The model of Exhibit 1.3 is useful and effective. In practice it is modified to make it more useful and effective by adding feedback loops. This modification makes the process nonlinear, and it better recognizes how decisions are *and should be* made.

There are some key skills and attitudes that you must develop to maximize your effectiveness. These include developing high-quality communication skills, considering each decision as one in a continuing stream, and developing a bias for action over analysis.

Nonlinear Process.   Exhibit 1.3 outlined a linear flow from step to step. At times some of these steps are executed simultaneously rather than sequentially. Often there are feedback loops, and steps are repeated.

Example 1.4 expands upon the first several steps that were outlined in Example 1.3. This expansion includes several feedback loops in which information from a later stage is used to "redo" an earlier stage.

Example 1.4   Charlene Selects a Pump (Revisited)

What is a reasonable chronology for the early days of the project in Example 1.3?

Solution

Day 1:   Charlene is assigned the job by her boss and given a brief overview. Her boss states that a 10-year study period, an interest rate of 15%, and minimizing equivalent annual costs are specified for engineering economy studies. *(1) Some defining problem and (2) some choosing objectives.*

Day 1 (later):   The senior engineer assigned to the project provides Charlene with information on the technical specifications for the pump. This includes the name of the manufacturer for a similar pump that was purchased last year. *(1) More defining problem and (3) some identifying alternatives.*

Day 2:   Charlene calls the sales engineer for the pump manufacturer to get current pricing, performance, and specifications. While waiting for the information to be faxed, she visits the senior engineer to ask whether some piping and controls are part of her project. She is told they are not, and that final design specifications on those items should be available in a week. *(3) Some identifying alternatives and (1) some defining problem.*

Day 3:   Charlene calculates the requirements that the pump must fulfill and finds three possibilities within the faxed information. They differ substantially in their expected lives, so her calculation of equivalent annual costs must explicitly recognize this. *(3) Identifying and (4) evaluating alternatives.*

Day 4:   Charlene drafts a memo recommending a pump for this application. *(5) Selecting and (6) beginning implementation for her recommendation.*

Day 8:   Charlene is called in by her boss and shown a memo sent out by the maintenance department earlier this year. That memo severely criticizes the reliability of this manufacturer's pumps, and complains bitterly about the poor support that is provided. *Back to (3) identifying alternatives.*

Days 9–30:   The process continues.

Example 1.4 shows that no single step in the process is "make the decision." The decision is shaped by the entire process. The engineer often decides which alternatives reach the final stage of consideration. Eliminating an alternative is a decision and it may determine that only one alternative remains to be refined and selected. This may happen on Day 8 of a month-long process, so determining when a decision was made can be difficult.

Iterative Modeling and Spreadsheets.   Refining the data and recalculating the model are common feedback loops in the decision-making process. In fact, the first four steps of the decision-making process are often conducted as a loop, where each pass through the loop is an iteration. Each iteration adds more detail to the model.

This modeling philosophy builds simple models first and then adds detail *only where needed*. If a deadline is reached, then there is at least a simple model to use for decision making.

The iterative nature of modeling and decision making is one reason that spreadsheets are popular for solving engineering economy problems. A better-quality estimate of a project's cost, revenue, or life can easily be changed in the spreadsheet, and then the computer automatically updates all calculations. As explained in Chapter 17 on sensitivity analysis, the results of one iteration can be used to decide which data estimates should be refined for the next iteration.

Spreadsheets also allow the results of simple and complex models to be displayed graphically. Exhibit 1.4 illustrates the costs of the slurry pipeline vs. a sludge conveyor for Charlene's design problem. For volumes above 16,000 liters per day the sludge conveyor is less expensive, but its capacity is limited to less than 17,000 liters per day. For the expected volume of 15,000 liters per day as well as smaller volumes, the pipeline is more cost-effective.

Exhibit 1.4   How does annual cost depend on the volume/year?

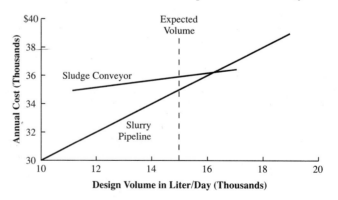

Importance of Communication Skills.   Engineering schools have asked employers on many occasions, "What needed skills and/or knowledge are our graduates missing?" The overwhelming reply relates to communication skills and the ability to work in groups or teams. Teamwork also depends on communication skills.

The decision-making process requires that data be provided by others. The emphasis is on asking the *right* questions of the *right* people and then listening to the answers. The selection and implementation steps depend on your ability to write and speak (at least before a small group). The audit step depends on your ability to ask the right questions and to listen.

The accrediting agency for engineering programs ensures that engineering graduates will have had courses in both written and oral communication. Most courses in the latter emphasize speaking and not listening; thus, Exhibit 1.5 emphasizes listening skills.

Exhibit 1.5 The skills of active listening

If the assignment is unclear or incomplete, do your questions of your boss or your professor lead to clarification? (Yes)

Do you state what the objectives or alternatives are and ask if you are correct, or do you listen after you ask questions to elicit others' views of objectives and alternatives? (The latter)

Suppose colleagues disagree with you on the objective, on an alternative's performance, or on the viewpoint you should adopt. Can you restate their viewpoint in your own words to their satisfaction? (Yes)

In the middle of your presentation to the board of directors, someone asks a question. Do you ignore it; label it as stupid; say, "I'll get to that"; or restate it in your own words before answering it? (Restate, then answer)

One Decision in a Continuing Stream.   With rare exceptions, each engineering project is only one of many implemented each year—and firms and agencies go on year after year. Consequently, the decision-making process might emphasize building a relationship between two departments, or the recommended alternative might focus on maximizing future possibilities rather than current returns.

If the manufacturing and engineering departments disagree over which specification should be adopted, it might be useful for engineering to simply accept manufacturing's proposal. Admitting that someone else might know more or might be a better judge than you can be humbling, but effective. It also makes your vehement arguments on the next issue more convincing. Building trust between units that must cooperate is an effective goal.

Introducing a new product or new process may not be cost-effective when considered in isolation. Rather, it may be justified by the knowledge that is acquired. That knowledge may determine whether or not future products or processes are considered and adopted. The knowledge may also be the foundation for the effective implementation of later alternatives.

Analysis vs. Action.   As shown in Exhibit 1.6, instructors and students are often concerned with the quality of analysis. Firms and agencies are concerned with the benefits of action. Once an alternative has been identified as the best, then the cost of delay and the cost of further analysis usually exceed the expected improvement in results.

Exhibit 1.6   Analysis paralysis vs. the benefits of action

Often, the best choice is obvious early in the analysis. For instance, a process analysis during the definition stage may validate improvements in operating procedures that remove the need for capital expenditures. This would short-circuit the decision-making process at the first step.

Since results come from implementation and not analysis, there are circumstances in which firms and agencies rely on weak data rather than waiting for more analysis. This can be frustrating for engineers who are trained to seek the best answer. Managers are often rewarded for acting, and punished for a failure to act. The emphasis on action over analysis implies that firms often seek satisfactory rather than optimal solutions. This behavior has been termed **satisficing.**

Satisficing is choosing a satisfactory alternative, rather than searching for the best alternative.

When further analysis would be done if time permitted, the best solutions place a premium on flexibility to accommodate learning and changed conditions. That way current recommendations can be improved as conditions evolve.

Private and Public Politics.   The role of interest groups or stakeholders in public decision making is obvious. Individuals, organized special-interest groups, political parties, agencies (city, state, and federal) with competing departments, citizens, and firms attempt to influence regulations, laws, and policies. Some regulations, laws, and policies mandate engineering projects to reduce pollution; others fund engineering projects to improve national competitiveness or national security; and still others select buildings, roads, and dams to be built.

Although less apparent, political activity is common in private firms as well. Individuals, departments, and divisions have their own agendas. HiTek's departments for new-product engineering and for manufacturing engineering may have different solutions for problems in product quality. HiTek's customers, suppliers, production department, and purchasing agents might have other solutions.

The conflicting interests of all stakeholders emphasize the importance of adopting a systems viewpoint. For the government, the system includes all stakeholders. For the private firm, the system is often defined based on the total firm's bottom line—both now and as influenced in the future by relationships with employees, vendors, customers, and government agencies.

These conflicting interests also increase the importance of communication skills within the decision-making process. You must communicate to gather data for the problem definition, objective selection, alternative identification, consequence evaluation, and audit steps. You must also convince others for the selection and implementation steps.

1.5 ENGINEERING ECONOMY'S AREA OF ACTION

Problems can be divided into three broad categories with respect to engineering economy:

1.   Some problems are too small to merit the application of engineering economy.
2.   Some problems' solutions are determined by engineering economy.
3.   Some problems are so big that engineering economy is only one factor in recommending a solution.

The general principle is that the scale of analysis needs to match the scale of the problem. Exhibit 1.7 illustrates that the larger the number of dollars, the longer the time period, and the smaller the organization, the more important it is that engineering economy be considered.

This standard of small, medium, and large depends on whether the problem is a personal, divisional, or corporate application. For most of us, spending $1000 to $5000 is a large expenditure that merits careful consideration. On the other hand, a corporate board of directors might only consider nonmonetary aspects for decisions involving $1 million or more.

Spreadsheets are shrinking the number of problems that are too small to warrant engineering economy. When a simple model can be built in five minutes or

Exhibit 1.7   Engineering economy's pyramid of inapplicability

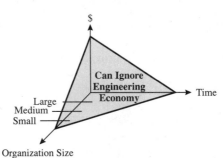

less, then it is harder to justify skipping the economic analysis. When new data can be substituted in seconds, economic analysis is easy to justify.

Problem Too Small to Merit Engineering Economy.   When you are grocery shopping and considering two package sizes, the time value of money is not significant. Instead, the money value of time, your available cash, or your storage space dominates the decision making. You simply compare unit prices and consider how fast the package will be used.

Similarly, an engineer may decide that a $10,000 pump choice should be made on the basis of the first acceptable alternative so that attention can be focused on the design of a $150,000 storage tank. If the number of hours for engineering design and economic analysis is limited, they should be expended where the payoff is highest.

Engineering Economy Dominates Decision Making.   The solution to many problems is dominated by money. *If money is to be spent or received over a year or more, then engineering economy is required to evaluate the alternatives.* If money is dominant, then engineering economy should dominate the decision-making process.

A very large number of engineering design problems fall in this category. Two designs achieve the same end, but which is more cost-effective?

A very large number of engineering design problems fall in this category. Two designs achieve the same end, but which is more cost-effective?

Engineering Economy Is One Factor in Decision Making.   Engineering economy influences many large decisions, but often uncertainty or multiple criteria dictate that other factors dominate. Engineering economy is still required; it is just not determinative.

When there are multiple criteria, the following distinction is relevant: From the perspective of engineering economy, the primary criterion is the time value of money (benefits measured in dollars); but from a broader perspective money may not be the most important criterion. For example, health and safety issues may not be stated quantitatively, yet they may still be the most important criteria for evaluating alternatives.

A secondary criterion is another quantifiable measure. Examples include rate of growth vs. steady rate of return, risk, lives saved/improved, decibel level, number of billboards torn down, market share, number of customers served, visitor days, and short-term cash flow situation.

Tertiary criteria are often termed **irreducibles.** These are not quantifiable, but are yes/no categories or subjectively estimated. Examples include entry into a new market, organizational image (for innovation, good treatment of personnel, or quality), admitting error, legal, ethical, corporate culture, and political factors.

Chapter 19 will discuss how quantifiable and irreducible criteria can be combined with the time value of money.

1.6 SUMMARY

Engineering economy is defined as considering the time value of money to decide which alternatives should be implemented. Examples from personal, corporate, and governmental situations have been used to show the broad area covered by engineering economy.

Correct economic analysis satisfies several principles: state consequences in a common measure, only consider differences between alternatives (ignore sunk costs), make separable decisions separately, adopt a systems viewpoint, and use a common planning horizon.

Correct economic analysis includes the following steps in the decision-making process: define the problem, choose the objective, identify the alternatives, evaluate their consequences, select, implement, and audit.

In reality, these steps must accommodate the environment in which decisions are made. (1) The decision-making process often includes parallel activities and feedback loops with repeated steps. (2) The modeling process is typically iterative, where the data is refined and the evaluation recalculated. This is one of the reasons for using spreadsheets. (3) Communication skills are overwhelmingly important. (4) Each decision is only one of many to be made. (5) There is an organizational emphasis on action rather than on analysis. (6) Politics are important in both public- and private-sector decision making.

Problems can be categorized as too small to justify engineering economic analysis, as likely to be dominated by engineering economy, and as large enough that engineering economy is one of several factors to be considered. Spreadsheets are shrinking the number of problems in the "too small" category.

REFERENCES

Ackoff, Russell L., *The Art of Problem Solving: Accompanied by Ackoff's Fables*, 1978, Wiley-Interscience.

VanGundy, Arthur B., *Techniques of Structured Problem Solving*, 1981, Van Nostrand Reinhold.

Wellington, Arthur M., *The Economic Theory of the Location of Railways*, 1887, Wiley.

PROBLEMS

1.1   In which of the following problems do you need engineering economy?
   a.   To compare a 4-year and a 5-year car loan.
   b.   To decide whether a new or a used car is cheaper.
   c.   To decide whether to pay your car insurance quarterly or annually.

1.2   An engineer for HiTek Manufacturing will analyze which of the following problems using engineering economy?
   a.   To compare a robot's and a person's cost per part produced.
   b.   To compare overhauling a forklift with buying a new one.
   c.   To compare the costs of leasing and buying a tractor trailer.
   d.   To decide between making and buying a part that requires some new equipment.

1.3   Which of the following is most suitable for engineering economic analysis? Why?
   a.   The purchase of a new automobile.
   b.   The selection of a major in college.
   c.   The lease vs. buy decision for a new milling machine.
   d.   The decision of whether to purchase collision insurance for your automobile.

1.4   Using Exhibit 1.2, analyze your choice of college or university.

1.5   Analyze (audit) a major decision that you have made in the last year. What steps in Exhibit 1.3 did you include? Which did you skip? Did you make a mistake in any step?

1.6   Analyze (audit) your choice of an academic major using the decision-making model in Exhibit 1.3.

1.7   Apply the steps of the decision-making process to your situation as you decide what to do after graduation.

1.8   George has just finished the cost analysis of a brass pump with and without an interior coating. The brass pump will last 3 years, but if an interior coating is applied, less energy will be needed for pumping. The interior coating costs $300, and the energy savings is $120 per year. George claims that the coating will save $60. Is George's work correct? If not, what has George done wrong?

1.9   The brass pump of Problem 1.8 would cost $5000 installed. Instead, George could use a stainless-steel pump that has been sitting in the maintenance shop for a year. The pump should last 3 years. The pump cost $15,000, but it was used for 2 years before being put on the shelf. The accountants say the pump is worth $6000. The maintenance supervisor says it will cost an extra $500 to reconfigure the pump for the new use and that he could sell it used for $3000. How much would George save or spend by using the stainless-steel pump?

1.10 You are reevaluating the electrical-generator choice that was made last year—when your boss had your job. The promised cost savings have not materialized, because the generator is too small. You have identified two alternatives. Alternative A is replacing the generator with one that is the right size and selling the old one. That will cost $50K. Alternative B is buying a small generator to use in addition to the one purchased last year. This will cost $60K. The two-generator solution has slightly higher maintenance costs, but it is more flexible and reliable. What are the primary, secondary, and irreducible criteria? Which alternative would you recommend and why?

1.11   Give an example in which you, your firm, or your city has made an incorrect decision because of sunk costs.

CHAPTER 2 THE TIME VALUE OF MONEY

The Situation and the Solution

A dollar today, a dollar a year from now, and a dollar 10 years from now differ in value. These dollars that occur at different times must be combined to evaluate engineering projects. How is the difference in value due to the time value of money accounted for?

The concept of compound interest adjusts for the time value of money to validly measure economic worth and cost.

Chapter Objectives

After you have read and studied the sections of this chapter, you should be able to:

Section 2.1
Define the interest rate for calculating the time value of money.

Section 2.2
Contrast simple and compound interest.

Section 2.3
Draw a cash flow diagram.

Section 2.4
Describe the concept of equivalence as applied to cash flow diagrams.

Section 2.5
Explain what equivalence does not include.

Section 2.6
Calculate nominal and effective interest rates.

Key Words and Concepts

Interest   The return on capital.

Capital   Invested money and resources.

Cash flow diagram   The pictorial description of when dollars are received and spent.

Equivalence   Occurs when different cash flows at different times are equal in economic value at a given interest rate.

Nominal interest rate   The rate per year without adjusting for the number of compounding periods.

Effective interest rate   The rate per year after adjusting for the number of compounding periods.

Present worth   A time 0 cash flow that is equivalent to one or more later cash flows.

2.1 WHAT IS INTEREST?

Interest is what you pay the bank for your car loan or your credit card. Interest is what the bank pays you for the money in your savings account. Interest is a rental fee for money; that is, a fee paid or a fee earned for the use of money.

Engineering economy generalizes this definition. **Interest** is the return on capital. **Capital**, in turn, is invested money and resources. Whoever owns the capital should expect a return on it from whomever uses it. For example, if the firm owns the capital and invests it in a project, then the project should *return* that capital plus interest as cost savings or added revenues.

Interest is the return on capital.

When you borrow from the bank, you pay interest. When you loan your money to the bank by depositing it, the bank pays you interest. When a firm invests in a project, that project should earn a return on the capital invested in the project.

This interest is typically expressed as an interest rate for a year. That rate equals the ratio of the interest amount and the capital amount. For example, if the interest rate is 5%, $5 of interest is paid when $100 is borrowed for a year. If the amount borrowed and the amount of interest paid are doubled, then the interest rate is still 5%, since $10/$200 = 5\%$.

When the bank owns the capital, a loan document will state the interest rate. When the owner of the capital buys shares in a firm, the interest is earned through dividends and increasing share values. The return to the shareholder is less certain, but it is still interest.

Similarly, when a firm or an individual invests money or resources in an engineering project, that project must earn a return on the capital invested in it. This return is earned by increasing future revenues and/or by decreasing future costs, as shown in Example 2.1. Interest is used to calculate the time value of money, and it is crucial to the practice of engineering. A history of engineering economy shows the parallel development of engineering and engineering economy [Lesser, 1969 and 1975].

Example 2.1   Interest Is Needed to Evaluate Engineering Projects

The engineering group of Baker Designs must decide whether to spend $90K (K for thousands) on a new project. This project will cost $5K per year for operations, and it will increase revenues by $20K annually. Both the costs and the revenues will continue for 10 years. Should the project be done?

Solution
The project costs $140K and earns $200K over the 10 years. To decide which is worth more, Baker Designs must include the time value of money. The $90K is spent immediately, and the revenues come later—as much as 10 years later. An interest rate must be used to compare the costs and the revenues. Does the $90K earn enough interest if invested in this project?

Interest Rates Vary.   The rate of interest that is appropriate depends on many factors, including risks, security, economic conditions, regulations, and time frame.

An unsecured loan made to an individual is a higher risk to the lender than a similar loan made to Microsoft Corporation. A mortgage secured by a house is a lower risk to the lender than a student loan secured only by a promise to pay. A loan made in a South American currency has a higher inflation risk than one made in U.S. dollars. Banks and credit unions are regulated by different agencies and may have different limits on the interest rates they can charge and pay. A loan to be repaid over 20 years is more vulnerable to inflation than one to be repaid over 5 years.

Some example rates are included in Exhibit 2.1. In principle, higher risks are matched with higher rates. Thus, the interest rate on an adjustable rate mortgage is initially lower than the rate on a fixed rate mortgage. The risk to the bank is lower with adjustable rates, since the bank can raise the interest rate on the adjustable mortgage to match inflation and interest-rate increases. These market interest rates change over time as inflation rates change and as the relative demand for and supply of borrowed money change. These changes can be rapid. For example, the typical rate for adjustable rate mortgages changed from 4.2% in December 1993 to 7% in May 1994. Inflation is detailed in Chapter 15.

Exhibit 2.1   Example interest rates

August 1991	May 1994	Type of Investment or Loan
6.75%	4.25%	Series EE savings bond
5.25%	3%	Daily passbook savings account
6.5%	4.25%	1-year certificate of deposit
8.5%	6.75%	Prime rate (loans to corporations)
7.25%	7%	Adjustable rate mortgage
9%	8.75%	Fixed rate mortgage
10.75%	7.5%	New car loan
12.25%	8.25%	Used car loan
16–18%	14–16%	Credit card account

Engineering projects also represent different risks and potential returns. Drilling an oil well, developing a new computer chip, and building a new municipal transit system are relatively risky. Opening a new gas station, redesigning a production line, and building a new municipal water reservoir are less risky. Firms and agencies can and do use different interest rates to evaluate projects in different classes of risk. Riskier projects are required to generate a higher rate of return.

From now through Chapter 17, a single interest rate for the time value of money is assumed. Chapter 9 discusses how this rate can be determined, but until then the interest rate will be specified in each example or problem.

2.2 SIMPLE VS. COMPOUND INTEREST

Compound, not simple, interest is used in engineering economy. To clarify the difference between compound and simple interest and to introduce the terminology that will be used throughout the text, the following variables are defined.

$i \equiv$ interest rate per year

$N \equiv$ number of years

$P \equiv$ initial deposit

$F \equiv$ future value after N years

Simple Interest.   With simple interest, the interest calculated for years 2, 3, ... is based on the initial deposit. There is no interest computed on the accrued interest. Thus in Equation 2.1 the interest rate is simply multiplied by the number of years. Simple interest is defined here, but it should not be used in engineering economy problems. Its principal use is in short-term loans, but compound interest is the norm even then.

$$F = P(1 + Ni) \tag{2.1}$$

Compound Interest.   The standard assumption is that interest is computed on the current balance which includes accrued interest that has not yet been paid. Savings accounts, loans, credit cards, and engineering economy rely on this assumption. In Equation 2.2, the number of years is a power on the factor that includes the interest rate.

$$\text{⚹} F = P(1 + i)^N \tag{2.2}$$

The computation of compound interest is illustrated in Example 2.2. Example 2.3 illustrates the difference between simple and compound interest. This difference increases as the interest rate and/or the number of years increases.

Example 2.2   Interest on a Savings Deposit

If $100 is deposited in a savings account, how much is in the account at the end of each year for 20 years, if the interest is deposited? The account earns 5% interest compounded annually.

Solution

The easiest way to compute each end-of-year value is to simply multiply the previous year's value by 1.05 (which is $1 + i$). Exhibit 2.2 includes the tabular results and a graph that shows the geometric nature of compound interest. Note that if simple interest were used, it would result in a linear increase in the amount in the bank.

Exhibit 2.2   Compound interest at 5% for 20 years

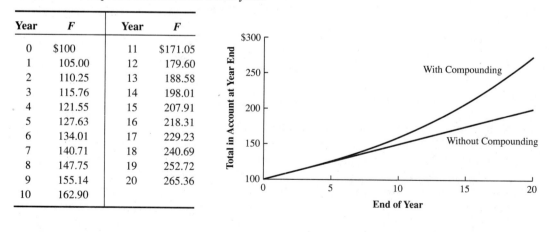

Year	F	Year	F
0	$100	11	$171.05
1	105.00	12	179.60
2	110.25	13	188.58
3	115.76	14	198.01
4	121.55	15	207.91
5	127.63	16	218.31
6	134.01	17	229.23
7	140.71	18	240.69
8	147.75	19	252.72
9	155.14	20	265.36
10	162.90		

Example 2.3 Comparing Simple and Compound Interest

How large is the difference between simple and compound interest if $N = 20$ years, $i = 5\%$, and $P = \$2$?

Solution

$$F_{\text{simple}} = \$2(1 + 20 \cdot .05) = \$4.00$$
$$F_{\text{compound}} = \$2 \cdot 1.05^{20} = \$5.31$$

Thus, the compounding increases the accrued interest by \$1.31. In percentage terms, the compound interest which totals \$3.31 is 65.5% larger than the \$2 which accrues with simple interest.

2.3 CASH FLOW DIAGRAMS

The pictorial description of when and how much money is spent or received is a **cash flow diagram.** This diagram summarizes an engineering project's economic aspects.

Cash flow diagrams depict the timing and amount of expenses and revenues for engineering projects.

As shown in Exhibit 2.3, the cash flow diagram has a horizontal axis that represents time. The vertical axis represents dollars, and arrows are used to represent the timing and the amount of receipts and expenses. The diagram's convention is that positive cash flows are receipts (those arrows point up), and negative cash flows are expenses (those arrows point down).

Exhibit 2.3 also illustrates that in some cases, the diagram can simply be reversed—depending on which viewpoint is taken. In Exhibit 2.3 the depositor and the bank both have the initial deposit and the final withdrawal as cash flows. But they have opposite perspectives on cash in and cash out.

Exhibit 2.3   Cash flow diagrams for bank and depositor

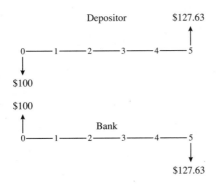

More complex cash flow diagrams involve more than two parties. For example, labor expenses are paid to employees, machine purchases to a manufacturer, and power purchases to a utility company, while revenues are received from customers. These complex diagrams cannot be reversed, as each party sees a different set of cash flows.

Categories of Cash Flows.   The expenses and receipts of engineering projects usually fall into one of the following categories. Costs and expenses are drawn as cash outflows (negative arrows), and receipts or values are drawn as cash inflows (positive arrows).

First cost ≡ expense to build or to buy and install

Operations and maintenance (O&M) ≡ annual expense, which often includes electricity, labor, minor repairs, etc.

Salvage value(*s*) ≡ receipt at project termination for sale or transfer of the equipment (can be a salvage cost)

Revenues ≡ annual receipts due to sale of products or services

Overhauls ≡ major capital expenditure that occurs partway through the life of the asset

Prepaid expenses ≡ annual expenses, such as leases and insurance payments, that must be paid in advance

Individual projects will often have specific categories of costs, revenues, or user benefits that will be added as needed. For example, annual operations and maintenance (O&M) expenses on an assembly line might be divided into direct labor, power, and other. Similarly, a public-sector dam project might have its annual benefits (revenues) divided into flood control, agricultural irrigation, and recreation.

Timing of Cash Flows.   When dealing with the time value of money, the assumption of when cash flows occur is clearly an issue. Cash flows could be assumed to occur at the beginning of the year, the middle of the year, the end of the year, or distributed throughout the year. In fact, tables of engineering economy factors have been constructed for all of these assumptions. This text follows the normal practice of engineering economy by assuming end-of-period cash flows for most cash flows.

This text will make the following assumptions:

End-of-period cash flows: Overhauls, salvage values, and annual receipts and expenses, such as O&M costs.

Time 0 or beginning of period 1: First cost.

Beginning-of-period cash flows: Prepaid expenses. Landlords and lessors insist that rent and lease payments be made in advance. Insurance companies insist that insurance be paid for in advance of the period covered. Most universities require that tuition be paid at the start, not the end, of the school term.

In Exhibit 2.4, all of the cash flows except the initial purchase cost are drawn as end-of-period cash flows. Exhibit 2.5 modifies the diagram for leasing rather than purchasing the heavy equipment.

Example 2.4   Ernie's Earthmoving Buys the Equipment

Ernie's Earthmoving is considering the purchase of a piece of heavy equipment. What is the cash flow diagram if the following cash flows are anticipated?

First cost	$120K
O&M cost	$30K per year
Overhaul cost	$35K in year 3
Salvage value	$40K after 5 years

Solution

The first three types of cash flows are costs, so they are drawn as negative arrows in Exhibit 2.4. The salvage value is a receipt, so it is drawn as a positive arrow.

Exhibit 2.4   Ernie's Earthmoving with purchased equipment

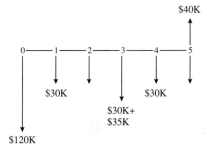

Example 2.5   Ernie's Earthmoving Leases the Equipment

Rather than purchasing the heavy equipment, as in Example 2.4, Ernie's Earthmoving is planning on leasing it. The lease payments will be $25K per year. What is the cash flow diagram?

Solution

The changes in the cash flow diagram shown in Exhibit 2.5 are three: (1) The lease payments are added at the beginning of years 1 through 5. (2) No first cost is incurred. (3) No salvage value is received, since Ernie's Earthmoving does not own the equipment.

Exhibit 2.5   Ernie's Earthmoving with leased equipment

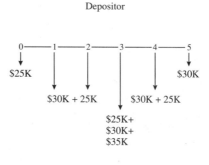

Do Not Simplify the Cash Flow Diagrams.   Often two or more cash flows occur in the same year, such as the overhaul and the year 3 O&M expense or the salvage value and the year 5 O&M expense. It seems that combining these into one total cash flow per year would simplify the cash flow diagram. It would. Don't combine. The translation of words into the diagram is clearer if the cash flows are not combined.

The purpose of the cash flow diagram is to ensure that all cash flows are included and that all are properly placed. For Examples 2.4 and 2.5, combined payments of $55K, $65K, and $90K are not as clear as separate entries of $25K, $30K, and $35K.

2.4 EQUIVALENCE FOR FOUR LOANS

Equivalence Defined.   Equivalence is adjusting for the time value of money. **Equivalence** means that different cash flows at different times are equal in economic value at a given interest rate. In Exhibit 2.3, the $100 at year 0 and the $127.63 at year 5 were equivalent at 5%.

Equivalence equates two or more dollar values that occur at different times.

Equivalence uses an interest rate or discount rate to adjust for the time value of money. Since $1 today and $1 a year from now have different values, this adjustment must occur before the cash flows can be added together.

One way to understand equivalence is through a loan. Money is borrowed, interest accrues on the unpaid balance, payments are made, and eventually the loan is repaid and the payments stop. The initial borrowing and the payments are equivalent at the loan's interest rate.

Exhibit 2.6 illustrates four ways to repay an initial loan of $1000 at 10% interest over 5 years. The business loan, bond, and consumer loan patterns are labeled according to their typical use. The constant principal payment pattern is rarely used, because the early payments are more burdensome than in the other patterns.

Exhibit 2.6   Cash flow diagram for four loans

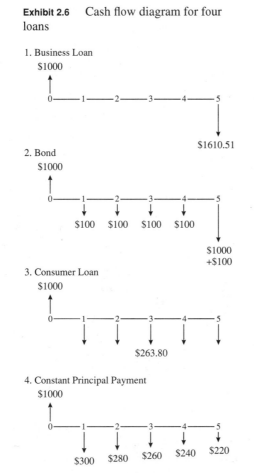

1. Business Loan
2. Bond
3. Consumer Loan
4. Constant Principal Payment

The business loan is repaid with a single, final payment of principal and interest. The bond is repaid with an interest payment each period, and a final payment of principal with interest in the last period. Consumer loans have the same payment each period, but the mix of principal and interest changes each period. Note that the cash flow diagram for the consumer loan illustrates that identical cash flows are sometimes labeled with a single value and the same length arrows.

Calculating the Interest for Each Year.   For each loan in Exhibit 2.6, the interest that accrues in year 1 is $100, which is 10% of the initial amount borrowed. Since the payment at the end of year 1 differs for each loan, the remaining amount owed is different, and so is the interest that accrues in year 2. In all cases the interest accruing in a year equals 10% of the amount owed at the year's beginning.

The calculation for the business loan uses the compound interest equation (Equation 2.2: $F = P(1 + i)^N$).

$$\$1000 \cdot 1.1^5 = \$1610.51$$

The interest on the bond is constant at $100. Since this is paid each year the amount owed remains constant, and so does the interest. In the last year the $1000 in principal is repaid along with $100 of interest.

Formulas for calculating the constant total payment used for the consumer loan will be presented in Chapter 3. Exhibit 2.7 confirms that at 10%, the payment of $263.80 is the right amount. Between years 1 and 2, the interest portion drops and the principal portion increases by $16.38. Between years 4 and 5, the interest portion drops and the principal portion increases by $21.80. Thus, the mix of principal and interest changes at an increasing rate as payments are made.

Exhibit 2.7   Annual repayment table for consumer loan

Year	Amount Owed	Interest	Principal Payment
1	$1000.00	$100.00	$163.80
2	836.20	83.62	180.18
3	656.02	65.60	198.20
4	457.82	45.78	218.02
5	239.80	23.98	239.82
6	−.02 (not 0 due to slight round-off error)		

The calculation for the constant principal payment loan is straightforward. Since five principal payments will be made to repay $1000, each is $200. The interest owed each year decreases by $20, which is the interest rate (10%) times the decrease in amount owed each year ($200).

Exhibit 2.8 summarizes the amount owed for each loan just before each year's payment (also shown) is made. The bar charts on the right divide the cash flows of each year into interest and principal.

Finding an Equivalent Present Worth.   The compound interest formula (Equation 2.2), $F = P(1 + i)^N$, can be applied to each loan payment. The payment occurs later so it is the F. Then a P is calculated for each payment by dividing F by $(1 + i)^N$. This P is the amount of initial borrowing that could be repaid by that F.

Exhibit 2.9 shows this calculation for the consumer loan. The payment is listed as $263.797 to avoid round-off error. The payments are divided by the entries in the interest rate column to compute the **present worth** (PW) of each payment. The final column, or PW, is the amount of initial borrowing that could be repaid by each payment. Each present worth amount is equivalent *at 10%* to a payment.

Each equivalent amount is called a present worth, because each is at the initial point of the loan, or time 0. If the present worths of all of the payments are summed, the total is $1000, which was the initial loan amount.

Exhibit 2.8 Four loan types: amount owed and paid at end of year

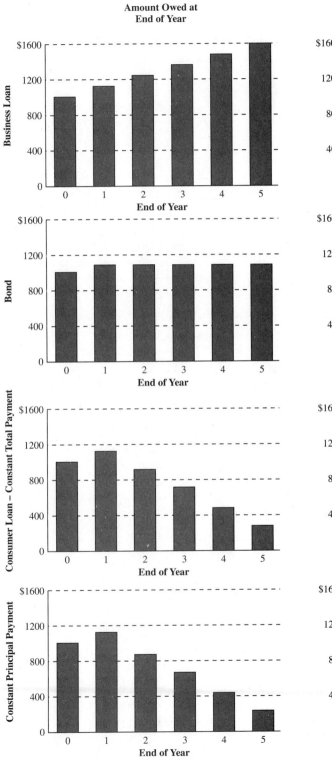

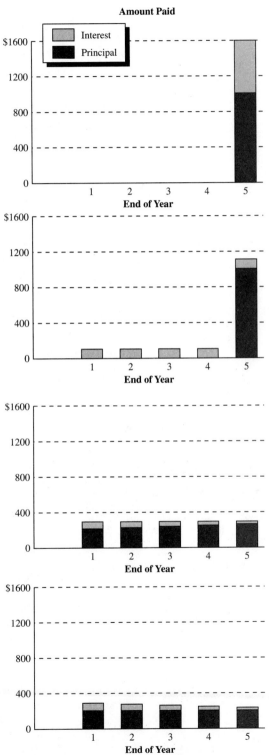

Exhibit 2.9   Present worth for consumer loan payments to repay $1000 at 10%

Year	Payment	$(1 + i)^{\text{Year}}$	Present Worth
1	$263.797	1.1000	$239.82
2	263.797	1.2100	218.01
3	263.797	1.3310	198.19
4	263.797	1.4641	180.18
5	263.797	1.6105	163.80
			Total $1000.00

The series of cash flows, or payments, are equivalent in total to a present worth of $1000 at 10%. Similar calculations for the other three loan patterns would show the same total present worth. Because all four loan patterns are equivalent to a present worth of $1000 at 10%, *they are also equivalent to each other at 10%.*

2.5 LIMITS ON EQUIVALENCE

Equivalence Depends on the Interest Rate.   The present worth of each set of loan payments can be calculated at other interest rates. For this we use the payments in years 1 through 5 that are shown in the cash flow diagrams of Exhibit 2.6. The method matches Exhibit 2.9, with a different interest rate used for the $(1 + i)^{\text{Year}}$ calculation.

As shown in Exhibit 2.10, the *only* interest rate where the four sets of loan payments have the same present worth is 10%. When two cash flows or two sets of cash flows are equivalent, that equivalence is at a particular interest rate. If the cash flows are compared at a different interest rate, they will not be equivalent.

Exhibit 2.10 shows that the interest rate changes the present worth of the business loan the most. This happens because the business loan has the maximum

Exhibit 2.10   Equivalence depends on i

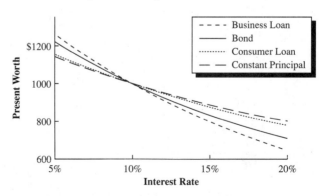

cash flow at the latest date. The pattern with the largest payments in the early years, the constant principal pattern, is the least affected.

Equivalence with Respect to Time Value of Money Only.   When two cash flow patterns are described as equivalent at 10%, this equivalence is limited to the time value of money. The four example loans apply to different situations, because the risks of default and how long the borrowed money is available are different and not equivalent.

A large firm can borrow funds with a single final payment of principal and interest because the lender is sure that the business can repay. When the federal government borrows money from you through a Series EE savings bond, the principal and interest are repaid in a single final payment. The firm and the government use all of the money for the loan's entire period instead of having to begin paying it back immediately.

Governments and firms issue bonds for which interest payments are made annually or semiannually and the principal is paid off in one lump sum at the end of the last period. Again, the lender who buys the bond is sure that payment will be made at the end of the last period.

Loans to individuals or consumers, and often to businesses, have level or constant payments. Almost all car loans, home mortgages, and in-store financing have this pattern, because the lender wants to be sure the loan is paid off. Suppose a home mortgage did not require monthly payments. How many of us would save the $1.47 million that would be required in year 30 to pay off a $100,000 loan (30-year term at 9%, compounded monthly)?

Consumer loans often require security or collateral. For example, most car loans are structured so that the car's value is always greater than the amount owed on the loan. If payments are missed, the lender has some protection from complete default.

Constant principal payments, with interest on the outstanding balance, lead to declining payments. The high payment required in the first year limits the total amount that can be borrowed. If this pattern were applied to home mortgages and car loans, many people would still be renting apartments and driving old cars.

In summary, equivalence is defined using the time value of money at a specified interest rate. Cash flows may not be equivalent to the firm in terms of risk or usefulness. As seen in Example 2.6, this also applies to individuals.

Example 2.6   Retirement—Play Now vs. Play Later

An engineering student is considering two plans for accumulating $1 million for retirement. In both cases, the money will be deposited in certificates of deposit that pay 8% interest, compounded annually. The "play now" plan requires no deposits for the first 9 years. Then, for the next 31 years, $8107 is deposited per year. The "play later" plan requires deposits of $7368 per year for the first 9 years, and nothing for the next 31 years. Are these plans equivalent?

Solution

Since both plans accumulate $1 million after 40 years, they are equivalent at 8%. *They are economically equivalent at 8%.* However, they are not the same from the perspective of the student. Which is better, "playing" for the first 9 years or for the next 31 years?

2.6 NOMINAL VS. EFFECTIVE INTEREST RATES

Compounding Periods Are *M* per Year.   Most loans and savings accounts compound the interest more than once per year. The annual rate might be stated as 12% (per year), with monthly compounding. An effective interest rate can be computed for comparisons that do not require that the compounding period be stated.

A nominal interest rate is an annual rate that must be compounded to calculate the **effective interest rate**.

The **nominal interest rate** is the annual rate that is not adjusted for the number of compounding periods. It is the 12% rate, which will be compounded monthly. The **effective interest rate** adjusts for the number of compounding periods. It is the rate that can be compared with other annual rates.

To show the relationship between the nominal and effective interest rates, define:

$r \equiv$ nominal interest rate per year

$M \equiv$ number of compounding periods per year

$i \equiv$ effective interest rate per year

The sequence is to convert the nominal rate to the interest rate per compounding period, r/M. That interest rate is then compounded for M periods, and set equal to $1 + i$. Mathematically, this is stated as:

$$1 + i = (1 + r/M)^M$$

Moving the 1 to the right-hand side allows us to state Equation 2.3 for the effective interest rate:

$$\ast \ i = (1 + r/M)^M - 1 \tag{2.3}$$

If 12 compounding periods per year are assumed for the nominal interest rate of 12% per year, then the effective interest rate of 12.68% is calculated as follows:

$$i = (1 + .12/12)^{12} - 1 = (1.01)^{12} - 1 = .1268 = 12.68\%$$

Exhibit 2.11 summarizes the effective interest rates for different compounding periods for a nominal interest rate of 12% per year. A nominal interest rate after compounding produces an effective interest rate. These effective interest rates can be compared without knowing the number of compounding periods. Note that *with only one compounding period, the nominal and effective interest rates are the same.* Note also that shifting from annual to semiannual compounding has about the same impact as shifting from semiannual to daily compounding.

Exhibit 2.11   Effective interest rates
$(r = 12\%)$

i	Number of Compounding Periods	
.12	1	annual
.1236	2	semiannual
.1255	4	quarterly
.1268	12	monthly
.1273	52	weekly
.1275	365	daily

Example 2.7 illustrates that effective interest rates can be used to compare nominal interest rates with different compounding periods. The federal government mandates that an annual percentage rate (APR) be stated for all loans. As shown in Example 2.8, this is the nominal interest rate. Example 2.9 illustrates finding a nominal interest rate given an effective interest rate.

Example 2.7   Comparing Effective Interest Rates

Which savings account pays interest at a higher effective rate? New York Savings (NYS) compounds interest daily at a nominal rate of $5\frac{1}{4}\%$. Montana Sky Bank (MSB) compounds interest quarterly at a nominal rate of $5\frac{3}{8}\%$.

Solution

$$i_{NYS} = (1 + .0525/365)^{365} - 1 = 5.390\%$$
$$i_{MSB} = (1 + .05375/4)^4 - 1 = 5.484\%$$

The MSB effective interest rate is higher.

Example 2.8   APR and Effective Credit Card Interest Rates

Many credit cards charge interest at $1\frac{1}{2}\%$ per month. What are the nominal and effective interest rates? Which one is the annual percentage rate (APR) that the federal government mandates be stated on loan agreements?

Solution
Since the interest rate is already stated as a rate per month, nothing needs to be divided by 12.

$$r = 12(.015) = .18 \text{ or } 18\%$$
$$i = 1.015^{12} - 1 = .1956 \text{ or } 19.56\%$$

The APR is the nominal interest rate of 18%.

Example 2.9   What Nominal Rate Equals an Effective Rate?

If interest is compounded monthly, what nominal interest rate will result in an effective interest rate of 9.3%?

Solution

The first step is to rearrange Equation 2.3 so that r is on the left-hand side, as follows:

$$1 + r/M = (1 + i)^{1/M} \quad \text{or} \quad r = M[(1 + i)^{1/M} - 1]$$

In this case, the result is

$$r = 12(1.093^{1/12} - 1) = 8.926\%.$$

Continuous Compounding.   It is possible to let the number of compounding periods approach infinity, which is called continuous compounding. Let $k = M/r$, then Equation 2.3 becomes

$$i = (1 + 1/k)^{kr} - 1 \quad \text{as } k \rightarrow \infty$$

The limit of the $(1 + 1/k)^k$ term is e, so the effective interest rate for continuous compounding is

$$\text{\Large $\ast$} \quad i = e^r - 1 \tag{2.4}$$

In the case of the 12% rate used for Exhibit 2.11, the effective interest rate is

$$i = e^{.12} - 1 = 12.7497\%.$$

For comparison, daily compounding yields an effective interest rate of 12.7475%.

Continuous compounding is used in the computation of interest for some savings accounts. In engineering economy its largest use is in conjunction with the cash flow assumptions that are discussed in Appendix 3A.

2.7 SUMMARY

Interest is defined as the rate of return on capital. Engineering projects typically require that money or capital be spent to build, buy, and assemble capital equipment. Interest is used to determine the time value of money, so that the costs can be compared with the benefits that may occur over the next 10 to 80 years.

Engineering economy relies on compound interest, not on simple interest. Simple interest ignores the time value of previous interest payments.

Cash flow diagrams have been used to describe the cash flows that often occur in engineering problems. These cash flows are first cost, annual operations and maintenance, salvage value, revenues, overhauls, and prepaid expenses.

Economic equivalence was illustrated using four common loan patterns. Each set of loan payments was equivalent to an initial borrowing of $1000 at an interest rate of 10%. In general, equivalence means that different sets of cash flows are equal in economic value at a given interest rate.

The economic equivalence of two sets of cash flows is at a specified interest rate and for the time value of money only. For example, two different loan payment schedules may represent different levels of risk that the loan will not be repaid.

Effective interest rates consider the number of compounding periods that are applied to a nominal interest rate. Effective interest rates can be used to compare loans or investments, as they are comparable even when the number of compounding periods is different. Effective interest rates are used throughout the text.

REFERENCES

Lesser, Jr., Arthur, "Engineering Economy in the United States in Retrospect—An Analysis," *The Engineering Economist,* Volume 14, Number 2, Winter 1969, pp. 109–116.

Lesser, Jr., Arthur, "Reminiscences of the Founder and Editor of the Engineering Economist," *The Engineering Economist,* Volume 20, Number 4, Summer 1975, pp. 314–318.

PROBLEMS

2.1 Define interest in your own words and provide an example.

2.2 What are the interest rates for the different kinds of savings accounts at your bank, credit union, or savings and loan? Why are they different?

2.3 What are the interest rates this week for the different kinds of loans at your bank, credit union, or savings and loan? What other differences between the loans are there (security, term, etc.)?

2.4 What are the interest rates at three different financial institutions in your area:
a. For similar new car loans?
b. For home mortgages?

2.5 Use the newspaper or a news magazine to find the interest rates on government securities with different terms. What are the interest rates on a savings bond, a 90-day note, and a 1-year note?

2.6 A 10-year loan is available from a relative at 10% simple interest. The amount borrowed will be $2000, and all interest and principal will be repaid at the end of year 10. How much more interest would be paid if the interest were compounded?   (*Answer:* $1187)

2.7 If $100 is deposited in a savings account that pays 6% annual interest, what amount has accumulated by the end of the eighth year? How much of this is interest and how much is principal?   (*Answer:* $159.4, $59.4, and $100)

2.8   Sam Boilermaker borrows $4000 from his parents for his final year of college. He agrees to repay it at 7% interest in one payment 3 years later. How much does he repay? How much of this is interest and how much is principal?

2.9   Joe Miner has $3000 in his savings account from an inheritance he received 10 years ago. If the savings account pays interest annually at a rate of 5%, how much was deposited?   (*Answer:* $1842)

2.10   Susan Cardinal deposited $500 in her savings account, and six years later the account has $600 in it. What compound rate of interest has Susan earned on her capital?   (*Answer:* 3.1%)

2.11   Draw the cash flow diagram for
        a.   Problem 2.7.
        b.   Problem 2.8.
        c.   Problem 2.9.
        d.   Problem 2.10.

2.12   Identify your major cash flows for the current school term as first costs, O&M expenses, salvage values, revenues, overhauls, or prepaid expenses. Using a week as the time period, draw the cash flow diagram.

2.13   A robot will be purchased for $18,000. It will cost $2500 per year to operate and maintain for 6 years. At the end of year 6 it will be donated to Metro U's School of Engineering. The donation will save $2000 in taxes. Draw the cash flow diagram.

2.14   The robot in Problem 2.13 requires an overhaul at the end of year 4 that will cost $5000. Draw the cash flow diagram.

2.15   A computer system will be leased for his last 2 years of school by Richard Husky. Disks, paper, and printer cartridges will cost him $120 per year. The lease payments are $80 per month, and there is a security deposit of $500. Draw a monthly cash flow diagram.

2.16   Grand Junction expects the cost of maintaining its sewer lines to be $20K in the first year after reconstruction and to increase by $2K each year thereafter until another major reconstruction after 8 years. Draw the cash flow diagram.

2.17   Mary Tipton is buying an auto that costs $12,000. She will pay $2000 immediately and the remaining $10,000 in four annual end-of-year principal payments of $2500 each. In addition to the $2500, she must pay 15% interest on the unpaid balance of the loan each year. Prepare a cash flow diagram.

2.18   Geske Sausage Company has just purchased a new meat-grinding machine. The machine's purchase price was $22,500. Geske made a 20% down payment and agreed to make 9 monthly principal payments of $2000 each. Geske also agreed to pay 1% interest on the unpaid principal each month. Prepare a cash flow diagram.

2.19   Draw the cash flow diagram if Strong Metalworking borrows $100,000, to be repaid in 5 years. The interest rate is 11%, and no payments are made before the end of year 5.

2.20   Draw the cash flow diagram if Strong Metalworking issues a $100,000 bond that pays 11% annual interest. The bond is repaid in 5 years.

2.21   Draw the cash flow diagram if Strong Metalworking repays a $100,000 loan with 5 annual payments of $27,057. Construct an interest table similar to Exhibit 2.7 to show that this repays the loan at an interest rate of 11%.

2.22   Draw the cash flow diagram if Strong Metalworking repays a $100,000 loan with 5 annual payments. Each payment includes one-fifth of the principal, plus interest on the unpaid balance. The interest rate is 11%.

2.23   Draw the cash flow diagram for City Utilities if it borrows $50,000, to be repaid over 10 years at an interest rate of 4% and:
   a.   One final payment is made.
   b.   Interest is paid annually, and the principal is paid at the end.
   c. $6164.55 is paid annually. What interest is paid in year 3?
   d. $5000 plus interest on the unpaid balance is paid each year.

2.24   Complete a table similar to Exhibit 2.9 for Exhibit 2.8's business loan.

2.25   Complete a table similar to Exhibit 2.9 for Exhibit 2.8's bond.

2.26   Complete a table similar to Exhibit 2.9 for Exhibit 2.8's constant principal payment, with interest on the outstanding balance.

2.27   Compute the present worth of the principal and interest payments in Problem 2.19 at an interest rate of
   a.   7%.   (Answer: −$120.1K)
   b.   11%.
   c.   15%.

2.28   Compute the present worth of the principal and interest payments in Problem 2.20 at an interest rate of
   a.   7%.   (Answer: −$116.4K)
   b.   11%.
   c.   15%.

2.29   Compute the present worth of the principal and interest payments in Problem 2.21 at an interest rate of
   a.   7%.   (Answer: −$110.94K)
   b.   11%.
   c.   15%.

2.30   Compute the present worth of the principal and interest payments in Problem 2.22 at an interest rate of
   a.   7%.
   b.   11%.
   c.   15%.

2.31   Which of the loan patterns in Problems 2.19–2.22 has its present worth most strongly affected by the interest rate? Which is least affected?
    a.   Graph the values for Problem 2.27.
    b.   Graph the values for Problem 2.28.
    c.   Graph the values for Problem 2.29.
    d.   Graph the values for Problem 2.30.

2.32   A certificate of deposit compounds interest annually at 10%. What amount has accumulated for retirement at age 65 if $10,000 is deposited at
    a.   Age 25?
    b.   Age 35?   (*Answer:* $174,494)
    c.   Age 45?
    d.   Age 55?   (*Answer:* $25,937)

2.33   Ramon has set a goal for his retirement fund at age 65 of $500K. His one deposit is $100K. What interest rate is required if the deposit is made at
    a.   Age 25?   (*Answer:* 4.11%)
    b.   Age 35?
    c.   Age 45?
    d.   Age 55?

2.34   If the nominal interest rate is 18%, what is the effective interest rate when interest is compounded
    a.   Annually?
    b.   Semiannually?   (*Answer:* 18.81%)
    c.   Quarterly?
    d.   Monthly?   (*Answer:* 19.56%)
    e.   Daily?
    f.   Continuously?

2.35   If the nominal interest rate is 24%, what is the effective interest rate when interest is compounded
    a.   Annually?
    b.   Semiannually?   (*Answer:* 25.44%)
    c.   Quarterly?
    d.   Monthly?
    e.   Daily?
    f.   Continuously?

2.36   If the nominal interest rate is 9%, what is the effective interest rate when interest is compounded
    a.   Annually?
    b.   Semiannually?   (*Answer:* 9.20%)
    c.   Quarterly?
    d.   Monthly?
    e.   Daily?
    f.   Continuously?

2.37   The following savings plans are offered at various banks in Skullbone, Tennessee. Which one is the best?
   a.   7% compounded annually
   b. $6\frac{7}{8}$% compounded semiannually
   c. $6\frac{15}{16}$% compounded quarterly
   d. $6\frac{3}{4}$% compounded monthly
   e. $6\frac{13}{16}$% compounded weekly
   f. $6\frac{31}{32}$% compounded daily (360 days)

2.38   If the effective rate equals 6.14% with monthly compounding, determine the nominal annual interest rate.   (*Answer: 6%*)

2.39   What monthly interest rate is equivalent to an effective annual interest rate of 12%?

2.40   The local loan shark will loan a person $1000 if the person agrees to repay the loan in two weeks with a payment of $1020. What is the nominal interest rate on this loan? The effective interest rate?

2.41   If the effective interest rate is 18%, what is the nominal interest rate when interest is compounded
   a.   Annually?
   b.   Semiannually?   (*Answer: 17.26%*)
   c.   Quarterly?
   d.   Monthly?
   e.   Daily?
   f.   Continuously?

2.42   Complete the following table for one nominal interest rate.

Compounding Period	Formula	Effective Interest Rate
Annual		
	$(1.025)^4 - 1$	
Monthly		
Continuously		

2.43   Compute a dollar-year measure of the amount owed for the four loan patterns in Exhibit 2.6. For example, the business loan has $1000 owed for the first year, for 1000 dollar-years. For the second year, $1100 is owed, so this year has 1100 dollar-years, for a cumulative total of 2100 dollar-years. Continue through 5 years. Then divide the sum of the interest paid over the 5 years by the total dollar-year measure of the amount owed. What is the result?

2.44   The *rule of 72's* states that the number of years for an investment to double is approximately 72 divided by the interest rate percentage. For example, about 12 years are required for an investment to double at 6% interest. Find the exact rate required to double an investment in 6, 8, 9, 12, 18, and 24 years.

2.45   A sum of money invested at 6% interest, compounded monthly, will double in how many months?   (*Answer:* 139 months)

2.46   At what nominal interest rate will money double in 6 years if interest is compounded quarterly? What is the effective interest rate?

2.47   Arrange the letters for the following interest rates from the least attractive loan to you to the most attractive loan (left to right). If two or more rates are the same, draw a circle around their letters in the sequence. (Credit for this problem goes to Sue Burgess.)
   a.   12% per year
   b.   12% per year, compounded monthly
   c.   6% per year, compounded semiannually
   d.   1% per month
   e.   1% per year
   f.   12% APR, compounded quarterly
   g.   3% per quarter

2.48   Many mortgages are computed using a 360-day year, and all months are assumed to have 30 days. Interest is computed as though all payments arrive on the first of the month, so that this is equivalent to monthly compounding. What is the effective interest rate if the nominal interest rate is 8.375%?

CHAPTER 3 EQUIVALENCE—A FACTOR APPROACH

The Situation and the Solution

The time value of money must be included in engineering decisions—quickly and easily. If long or complex calculations are required to find equivalent values, then engineering economy will not be applied wherever it should be.

A solution is to calculate equivalent values using the engineering economy factors that have been defined and standardized over the last 50 years.

Chapter Objectives

After you have read and studied the sections of this chapter, you should be able to:

Section 3.1
Define the variables and state the standard assumptions for calculating equivalence.

Section 3.2
Understand the notation of the tabulated engineering economy factors.

Section 3.3
Apply the factors to single-payment cash flows.

Section 3.4
Apply the factors to uniform-series cash flows.

Section 3.5
Combine the factors for more complex cash flows.

Section 3.6
Apply the factors for arithmetic gradients.

Appendix 3A
Use the factors for continuously flowing cash flows.

Key Words and Concepts

Engineering economy factors   The tabulated values to convert from one cash flow quantity ($P, F, A,$ or G) to another.

Annuity   A series of uniform cash flows that begins at the end of period 1 and continues through the end of period N.

Gradient   Constant period-by-period change in cash flows, which is an arithmetically increasing cash flow series.

Deferred annuity   A series of uniform cash flows that begins later than period 1 and continues through period N. There are no cash flows from period 1 to period D.

Prepaid expense   A cash flow that occurs at the beginning, not the end, of each period, such as leases and insurance.

End-of-period cash flow   Assumed for all cash flows except first costs and prepaid expenses.

3.1 DEFINITIONS AND ASSUMPTIONS

Chapter 2 introduced cash flow diagrams, compound interest, and equivalence. This chapter introduces the **engineering economy factors** that apply compound interest to calculate equivalent values for cash flows. These factors are tabulated and are often used instead of the formulas they represent. This text relies on these factors.

Engineering economy factors are the tabulated values for calculating equivalent values.

Like all formulas, those for the engineering economy factors are based on assumptions, and they require the definition of variables. This text uses the notation that has been developed by engineering economists for the American National Standards Institute [*The Engineering Economist*].

Assumptions.   The following assumptions are made for engineering economy factors:

1. Interest is compounded once per period.
2. Cash flows occur at the end of a period.
3. Time 0 is period 0 (beginning same as end) or the beginning of period 1.
4. All periods are the same length.

Cash flow diagrams often assume that each period is a year long. That assumption is not required. A period can be a year, a month, a quarter, etc. The only requirement is that both the interest rate and the number of periods be defined using the *same period length*. If the interest rate is 6% per year but periods are counted in quarters, then using a rate of 6% per quarter instead of 1 ½% per quarter in the formulas leads to nonsense.

Interest is computed or compounded at the end of each period. The cash flow diagram in Example 3.1 assumes that cash flows also occur at the end of each period. The first costs occur at the end of period 0, which is also the beginning of period 1. This is the stock assumption for engineering economy factors.

The timing of cash flows could be assumed to occur either continuously (Appendix 3A) or at the beginning (Section 3.5) or in the middle (Problem 3.42) of each period. However, it may be easiest to redefine the period's length. For a mid-year convention, simply use six-month periods. For continuously flowing cash flows, approximate with monthly periods.

Example 3.1   Cash Flows for a Heat Exchanger

Perfect Temp is a mechanical engineering design firm that specializes in heating and cooling systems for factories and warehouses. What is the client's cash flow diagram for the following costs and savings of an air-to-air heat exchanger?

First cost = $12,000
O&M cost = $2000 per year
Energy savings = $5000 per year
Heat-exchanger life = 10 years

Solution
As in Chapter 2, savings are shown as positive income (upward arrows) and expenses are shown as negative cash flows (downward arrows):

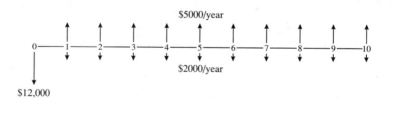

Definitions.   Four of the variables (P, F, i, and N) were defined in Chapter 2 on a yearly basis. Here they are defined more generally in terms of periods. Each period is usually, but not always, a year long. Definitions are also added for two more cash flow patterns (A and G). Exhibit 3.1 illustrates these definitions and the assumption of end-of-period cash flows.

$i \equiv$ interest rate per period.
$N \equiv$ number of periods (sometimes called study period or horizon).

Exhibit 3.1   Diagrams for *P, F, A,* and *G*

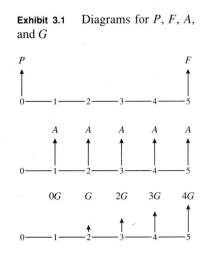

$P \equiv$ present cash flow or present value equivalent to a cash flow series.

$F \equiv$ future cash flow at the end of period *N*, or future worth at the end of period *N* equivalent to a cash flow series.

$A \equiv$ uniform periodic cash flow (often called an **annuity,** thus the letter *A*) at the end of every period from 1 to *N*. *A* is also used for a uniform constant amount equivalent to a cash flow series.

A is a uniform end-of-period amount for periods 1 through *N*, often called an **annuity** (for annual periods). *G* is an arithmetic **gradient**, which begins at 0 at the end of period 1 and increases to $(N-1)G$ at the end of period *N*.

$G \equiv$ gradient or constant period-by-period change in cash flows (so that cash flows form an arithmetically increasing series) from period 1 through period *N*. The amounts are 0 at end of period 1, *G* at the end of period 2, 2*G* at the end of period 3, etc.

3.2 TABLES OF ENGINEERING ECONOMY FACTORS

Factor Notation.   Before the advent of computers and electronic calculators, engineering economists needed tabulated conversion factors to speed their work. These factors are still often the fastest, easiest way to solve problems. The notation for these factors has been standardized with mnemonics that assist in their use. The variables, *i, N, P, F, A,* and *G*, are used as reminders of what each factor does.

The format of the engineering economy factors is *(X/Y, i%, N)*. The *X* and *Y* are chosen from the cash flow symbols *P, F, A,* and *G*. If *Y* is multiplied by the factor, then the equivalent value of *X* results (for *N* years at *i*%). Thus, it may be useful to think of the multiplication as clearing fractions, where the *Y*'s cancel and *X* is left. So to convert from a cash flow in year 10 (an *F*) to an equivalent present value (a *P*), the factor is *(P/F, i, 10)*.

$P = A(P/A, i, N)$ with cancellations through the *A*'s.

Appendix A consists of tables of these factors—one for each interest rate. Exhibit 3.2 is a simplified version of the 10% table that includes the columns for four factors: $(P/F, 10\%, N)$, $(F/P, 10\%, N)$, $(A/P, 10\%, N)$, and $(P/A, 10\%, N)$. If *N* is 20 years, the factor to convert to an *F* from a *P* is 6.727 and to an *A* from a *P* is .1175. Example 3.2 applies these factors.

Exhibit 3.2   Simplified table of factors for 10%

N	$(P/F, 10\%, N)$	$(F/P, 10\%, N)$	$(P/A, 10\%, N)$	$(A/P, 10\%, N)$
1	.9091	1.100	0.909	1.1000
2	.8264	1.210	1.736	.5762
3	.7513	1.331	2.487	.4021
4	.6830	1.464	3.170	.3155
5	.6209	1.611	3.791	.2638
10	.3855	2.594	6.145	.1627
15	.2394	4.177	7.606	.1315
20	.1486	6.727	8.514	.1175

Example 3.2   Finding Loan Payments for a $1000 Loan

Exhibit 2.6 presented ways to repay a loan of $1000 at 10% over 5 years. Use Exhibit 3.2 to find the payment amounts to repay the $1000 at 10% over 5 years for the business loan with a single final payment. Repeat for the consumer loan with five level payments.

Solution
In both cases, the initial borrowing of $1000 is a P. For the business loan, an F must be found. (In Chapter 2 we used Equation 2.2 to find that $F = 1610.51$.)

$$F = 1000(F/P, 10\%, 5) = 1000(1.611) = \$1611$$

For the consumer loan, five equal or uniform payments, which is an A, must be found.

$$A = 1000(A/P, 10\%, 5) = 1000(.2638) = \$263.8$$

Names of the Engineering Economy Factors.   It is easy and clear to rely on the letters for each cash flow type and the mnemonic factors. However, it is also useful to know the names of these factors. These are:

$(P/F, i, N)$ Present worth factor
$(F/P, i, N)$ Compound amount factor
$(P/A, i, N)$ Series present worth factor
$(A/P, i, N)$ Capital recovery factor
$(A/F, i, N)$ Sinking fund factor
$(F/A, i, N)$ Series compound amount factor

Most of these names are obvious, but the capital recovery factor and the sinking fund factor may need some explanation. If $10,000 is spent now to buy a machine, an (A/P) or capital recovery factor is used to calculate how much the machine needs to save or earn each year to recover the cost of the capital that was invested in it. This assumes that there is no salvage value.

A sinking fund was a savings account in which an annual deposit would be made to accumulate funds for the future replacement of machinery, such as a steam engine. Thus, if you knew the future cost of the machine, then you could calculate the annual deposit using a sinking fund or (A/F) factor.

Format for the Interest Rate.   Interest rates are stated as percentages or decimals, not integers. If the interest rate is 10%, or .1, it is harder to confuse it with the integer number of periods. Since i and N are not interchangeable, this small precaution is very important. For example:

$$(P/A, .1, 5) = 3.791 \neq (P/A, .05, 10) = 7.722$$

Interpolation.   The tables in Appendix A are comprehensive, but they do not include every possible interest rate or life. For example, there is no table for $i = 5.25\%$, nor for $i = 32\%$, and no row for $N = 37$. Finding factors for these values requires either interpolation or the formulas in Sections 3.3–3.6. Examples 3.3 and 3.4 illustrate interpolation.

To mentally double-check the accuracy of your interpolations, use two rules. First, the interpolated value should lie between the tabulated values. In Example 3.3, the value for $(F/P, .1, 7)$, 2.004, is between the values for $N = 5$ and 10. Second, the interpolated value should be closer to one of the tabulated values, and only occasionally will it be exactly in the middle. In Example 3.3, 7 is closer to 5 than to 10, so the interpolated value is closer to the tabulated value for $N = 5$ than to the one for $N = 10$.

Example 3.3   Interpolating for N

Using the values of Exhibit 3.2, interpolate to find $(F/P, .10, 7)$. How close to the exact value is this?

Solution
The two closest values are for years 5 and 10, $(F/P, .10, 5) = 1.611$ and $(F/P, .10, 10) = 2.594$. Rather than relying on a formula to ensure correct interpolation, it is better to sketch simple diagrams, such as Exhibit 3.3. Then it is possible to rely on similar triangles and simple geometric analogies.

Exhibit 3.3   Example interpolation for $(F/P, .1, 7)$

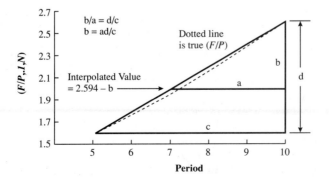

Thus, the interpolated value for

$$(F/P, .1, 7) = (F/P, .1, 10) - b$$
$$= 2.594 - (10 - 7)(2.594 - 1.611)/(10 - 5)$$
$$= 2.004$$

Using Equation 2.2, $(F/P, .1, 7) = 1.1^7 = 1.948717$. The exact and the interpolated values differ by 2.8%. This difference occurs because the linear interpolation does not exactly match the true nonlinear relationship. The value tabulated in Appendix A, 1.949, due to roundoff differs from the exact value by .001283 or .015%.

Example 3.4   Interpolating for an Interest Rate

Using the tables in Appendix A, interpolate for the value of $(P/F, 22\%, 10)$. How much does this differ from the results of the exact formula?

Solution
The two closest interest rates, 20% and 25%, have values of $(P/F, 20\%, 10) = .1615$ and $(P/F, 25\%, 10) = .1074$. Rather than relying on a formula to ensure the correct interpolation, it is better to sketch simple diagrams, such as Exhibit 3.4. Then it is possible to rely on similar triangles and simple geometric analogies.

Exhibit 3.4   Example interpolation for $(P/F, .22, 10)$

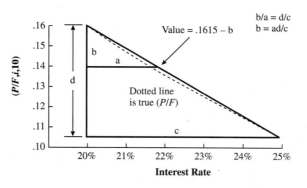

Thus, the interpolated value is

$$(P/F, .22, 10) = (P/F, .2, 10) - b$$
$$= .1615 - (.22 - .2)(.1615 - .1074)/(.25 - .2) = .1399$$

Using the inverse of Equation 2.2, $(P/F, .22, 10) = 1.22^{-10} = .1369$. The exact and the interpolated values differ by about 2%.

Formulas vs. Factors.   Factors are so useful that some engineering economists remember only Equation 2.2; that is, $F = P(1+i)^N$. To solve problems, they rely on knowing the symbols and using the basic tabulated factors. If the equations of

Sections 3.3, 3.4, and 3.6 are needed for an unusual interest rate or life, or if you want to program a formula into a calculator, then it is better to look up the formula rather than to rely on memory.

Often the main advantage of formulas is that calculating an exact value may be easier than interpolating for an approximate answer.

3.3 SINGLE-PAYMENT FACTORS (*P*s AND *F*s)

Formula Development. The basic formula relating the two single-payment quantities, P and F, was used in Chapter 2. Developed for compound interest (and also for population and demand), it is Equation 2.2, $F = P(1 + i)^N$. This formula is more completely developed in Exhibit 3.5. In each period, interest at rate i is earned, so that end-of-period quantities are $(1 + i)$ times the beginning-of-period quantities.

Exhibit 3.5   Development of $(F/P, i, N)$

Period	Value at Beginning-	Value at End-of-Period
1	P	$P(1 + i)$
2	$P(1 + i)$	$P(1 + i)(1 + i)$
3	$P(1 + i)^2$	$P(1 + i)^2(1 + i)$
...	...	...
...	...	...
N	$P(1 + i)^{N-1}$	$P(1 + i)^N$

Tabulated Factors. Equation 2.2 is the basis for two factors tabulated in Appendix A. These two factors, $(F/P, i, N)$ and $(P/F, i, N)$, are inverses. The first of these is called the compound amount factor, since it is compounding a present amount into the future with the factor $(1 + i)^N$. Equation 3.1 is basically the right-hand side of Equation 2.2.

$$(F/P, i, N) = (1 + i)^N \tag{3.1}$$

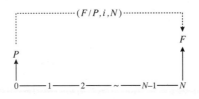

The second factor is the present worth factor. It discounts a single future value in period N to a present worth. This (Equation 3.2) is the inverse of Equation 3.1.

Exhibit 3.6 illustrates the application of these factors to a cash flow diagram. Example 3.5 applies the factors to a simple problem.

$$\ast\; (P/F, i, N) = (1 + i)^{-N} \tag{3.2}$$

Exhibit 3.6   Factors for P and F

Example 3.5   Which Loan?

NewTech has asked three banks for a loan of $15,000. As shown below, each bank has its own rate and compounding interval, but each loan would be paid back by a single, lump-sum payment of principal and interest at the end of 5 years. Which loan has the smallest payment? How much better than the next best is this?

First Federal Bank: Annual compounding at 10% per year

Second State Bank: Monthly compounding at 9% per year

City Credit Union: Quarterly compounding at 8% per year

Solution
The final payment for First Federal has 5 periods at 10% per period.

$$F = -15,000(F/P, 10\%, 5) = -15,000 \cdot 1.611 = -\$24,165$$

The final payment for Second State has 60 periods at .75% per period, since each period is a month long.

$$F = -15,000(F/P, .75\%, 60) = -15,000 \cdot 1.566 = -\$23,490$$

The final payment for City Credit has 20 periods at 2% per period, since each period is a quarter long.

$$F = -15,000(F/P, 2\%, 20) = -15,000 \cdot 1.486 = -\$22,290$$

The City Credit loan is the most attractive, and its final payment is $1200 lower than the Second State loan.

(P/F) and (F/P) as a Function of *i* and *N*.   Exhibit 3.7 summarizes the values of these factors for $i = 5\%$ and $i = 10\%$ as N is varied. Exhibit 3.8 summarizes the values of these factors for $N = 5$ and $N = 10$ as i is varied. These functions are nonlinear in both i and N, which is why a linear interpolation is only an approximation. The simplicity of Equations 2.2, 3.1, and 3.2 also implies that sometimes, as in Example 3.6, it is easier to calculate an interest rate using the formulas.

The limit for all of these single-payment factors is the same as *i* and/or *N* approaches 0. That limit is 1. As *i* and/or *N* approaches infinity, the present worth factors approach 0, and the future worth factors approach infinity.

Exhibit 3.7   Graph of (P/F) and (F/P) as a function of N, $i = 5\%$ and $i = 10\%$

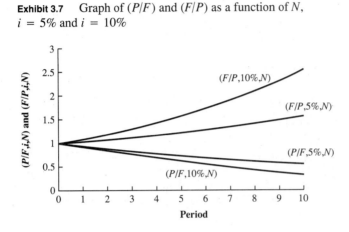

Exhibit 3.8   Graph of (P/F) and (F/P) as a function of i, $N = 5$ and $N = 10$

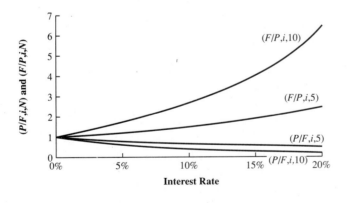

Example 3.6   The Value of a Dam

An earth-filled dam of the U.S. Department of Reclamation cost $850,000 to build in 1948. Today, in 1993, it is in mint condition and worth about $5 million. At what annual interest or growth rate has the value of the dam increased?

Solution

$$F = P(1 + i)^N$$

$$5{,}000{,}000 = 850{,}000(1 + i)^{45}$$

$$(1 + i)^{45} = 5{,}000{,}000/850{,}000 = 5.8824$$

$$(1 + i) = \sqrt[45]{5.8824} = 1.04016$$

$$i = 4.02\%$$

3.4 UNIFORM FLOWS

Formula Development.   This section develops the formula for converting from a uniform series (an A) to a present worth (a P). This formula is combined with the earlier $(F/P, i, N)$ formula to produce the $(F/A, i, N)$ formula.

A uniform series is identical to N single payments, where each single payment is the same and there is one payment at the end of each period. Thus, the $(P/A, i, N)$ factor is derived algebraically by summing N single-payment factors, $(P/F, i, 1)$ to $(P/F, i, N.)$ The derivation of Equation 3.3 is shown in Exhibit 3.9.

$$P = A[(1 + i)^N - 1]/[i(1 + i)^N] \qquad (3.3)$$

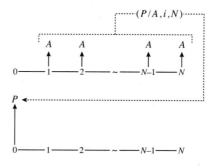

Exhibit 3.9   Derivation of $(P/A, i, N)$

$$P = A[\qquad 1/(1 + i)^1 + 1/(1 + i)^2 + \cdots + 1/(1 + i)^{N-1} + 1/(1 + i)^N]$$

$$(1 + i)P = A[1/(1 + i)^0 + 1/(1 + i)^1 + 1/(1 + i)^2 + \cdots + 1/(1 + i)^{N-1}]$$

$$(1 + i)P - P = A[1/(1 + i)^0 \qquad\qquad\qquad\qquad\qquad\qquad - 1/(1 + i)^N]$$

$$iP = A[1 - 1/(1 + i)^N]$$

$$= A[(1 + i)^N - 1]/(1 + i)^N$$

$$P = A[(1 + i)^N - 1]/[i(1 + i)^N]$$

The easiest way to derive Equation 3.4, which connects the uniform series, A, with the future single payment, F, is to multiply both sides of Equation 3.3 by $(1 + i)^N$.

$$(1 + i)^N \cdot P = A[(1 + i)^N - 1]/i$$
$$F = A[(1 + i)^N - 1]/i \qquad (3.4)$$

Tabulated Factors.   Equations 3.3 and 3.4 are each the basis for two factors tabulated in Appendix A. Both the series present worth factor, $(P/A, i, N)$, and the capital recovery factor, $(A/P, i, N)$, as shown respectively in Equations 3.5 and 3.6, are inverses based on Equation 3.3. The name of the capital recovery factor is based on the question, "How much must be saved or earned each period, A, to *recover the capital* cost of the initial investment, P?"

$$(P/A, i, N) = [(1 + i)^N - 1]/[i(1 + i)^N] \qquad (3.5)$$
$$(A/P, i, N) = 1/(P/A, i, N) = i(1 + i)^N/[(1 + i)^N - 1] \qquad (3.6)$$

The series compound amount factor, $(F/A, i, N)$, and the sinking fund factor, $(A/F, i, N)$, as shown respectively in Equations 3.7 and 3.8, are inverses based on Equation 3.4. The name of the sinking fund factor comes from an old approach to saving enough funds to replace capital equipment. Each period an amount, A, would be placed in a savings account (called a sinking fund) and then, after N years, a total of F would have accumulated through deposits and interest.

$$(F/A, i, N) = [(1 + i)^N - 1]/i \qquad (3.7)$$
$$(A/F, i, N) = 1/(F/A, i, N) = i/[(1 + i)^N - 1] \qquad (3.8)$$

Exhibit 3.10 illustrates these factors in cash flow diagrams, where both A and F are end-of-period cash flows. In Examples 3.7 and 3.8, these factors are applied to loan payments and to calculating the present worth of an investment.

Exhibit 3.10   Factors for A with P or F

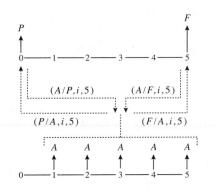

Example 3.7   Loan Payments

A loan of $8000 is to be repaid with uniform periodic payments at 12% over 5 years. What is the payment if payments are made annually? What is the payment if payments are made monthly? Why isn't the annual payment twelve times the monthly payment?

Solution

For annual payments, $i = 12\%$ and $N = 5$.

$$A = 8000(A/P, .12, 5) = 8000 \cdot .2774 = \$2219$$

For monthly payments, $i = 1\%$ and $N = 60$.

$$A = 8000(A/P, .01, 60) = 8000 \cdot .0222 = \$177.6$$

Twelve times the monthly payment is $2131, which is $88 less than the annual payment. This difference is due to a different cash flow pattern *and* a different interest rate. Twelve payments per year means the first payment is only one month away, not one year away. With monthly payments the "average" of 12 payments is in the middle of month 7. Since this is sooner than the end of the year, less interest accrues, and the monthly payment is lower than 1/12th of the annual payment.

The effective interest rates of the two loans differ as well. The annual loan has a rate of 12%, while the monthly payment loan has an effective interest rate of 12.68% (equals $1.01^{12} - 1$).

Example 3.8   Present Worth of Investment Returns

Northern Construction and Engineering plans to open an office in Duluth. Over the next 25 years the office is expected to have a positive cash flow of $250K per year. At an interest rate of 12%, what is the present worth equivalent of the expected cash flows?

Solution

$$PW = 250K(P/A, .12, 25) = 250K \cdot 7.843 = \$1961K$$

Calculating an Interest Rate.   Equations 3.3 and 3.4 cannot be solved for i unless N is 2 or 3. However, the equations can be solved numerically for specific values of i and N. The process is to:

1.   Write an equation, including the factor.
2.   Solve for the value of the factor.
3.   Find the i's that straddle the value of the factor.
4.   Interpolate.

Example 3.9 illustrates this process of finding the interest rate of a loan.

Example 3.9   Finding a Loan's Interest Rate

If $10,000 is borrowed and payments of $2000 are made each year for 9 years, then what is the interest rate?

Solution

This is equivalent to asking, "For what i does $(A/P, i, 9)$ equal .2?" First, draw the cash flow diagram.

Second, write the equation to calculate the nine end-of-year payments to repay a loan of $10,000, where i is an unknown. Set this equation equal to the given payment of $2000 and solve.

$$9 \text{ payments} = 10,000(A/P, i, 9)$$
$$2000 = 10,000(A/P, i, 9)$$
$$(A/P, i, 9) = .2$$

By paging through Appendix A, we find that the interest rate is between 13% and 14%. These capital recovery factors,

$$(A/P, .13, 9) = .1949 \quad \text{and} \quad (A/P, .14, 9) = .2022$$

straddle the value of .2.

Using Exhibit 3.11, we interpolate for the value of i. In particular,

$$i = .13 + (.2 - .1949)(.14 - .13)/(.2022 - .1949) = 13.699\%$$

or 13.7%. From a financial calculator, the exact value is found to be 13.704%.

Exhibit 3.11   Interpolating for i when $(A/P, i, 9) = .2$

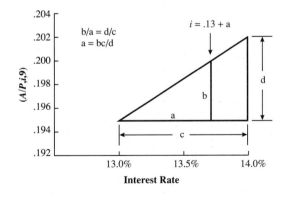

(A/P), (A/F), (P/A), and (F/A) vs. i and N.   Exhibit 3.12 shows how these factors vary with N. If N equals 1, then the capital recovery factor, $(A/P, i, 1)$, equals $(1 + i)$ and the series present worth factor, $(P/A, i, 1)$, equals $1/(1 + i)$.

As N approaches infinity, the (A/P) factors approach $1/i$ and the (P/A) factors approach i.

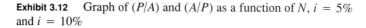

Exhibit 3.12   Graph of (P/A) and (A/P) as a function of N, $i = 5\%$ and $i = 10\%$

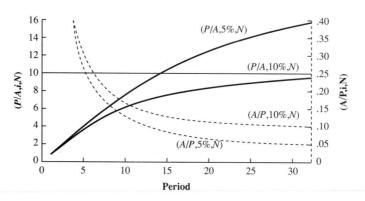

Exhibit 3.13 shows how these factors vary with i. If we let i equal 0%, we find that the series present worth factor, $(P/A, 0\%, N)$, equals N. Thus, $(P/A, 0\%, 5) = 5$ and $(P/A, 0\%, 10) = 10$. The capital recovery factor, $(A/P, 0\%, N)$, for $i = 0\%$ is $1/N$. Thus, $(A/P, 0\%, 5) = 1/5 = .2$ and $(A/P, 0\%, 10) = 1/10 = .1$. As i approaches infinity, the (P/A) and (A/P) factors approach 0 and infinity, respectively.

Exhibit 3.13   Graph of (P/A) and (A/P) as a function of i, $N = 5$ and $N = 10$

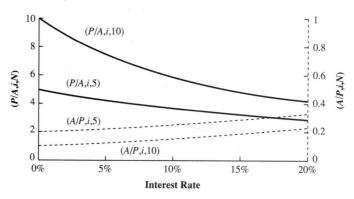

Clearly, Equations 3.7 and 3.8 are nonlinear in i and N. However, the quality of the linear interpolation is very high, as shown in Example 3.10.

Example 3.10   Interpolation Accuracy with (P/A) Factors

Compare exact and interpolated values for $(P/A, i, 10)$ for $i = 1.6\%$, 5.6%, 10.6%, and 20.6%.

Solution

The exact values are calculated with Equation 3.5 and are shown in Exhibit 3.14:

$$(P/A, i, 10) = [(1 + i)^{10} - 1]/[i(1 + i)^{10}]$$

Exhibit 3.14   Interpolation quality for $(P/A, i, 10)$

i	**Exact**	**Interpolated**	**% Difference**
1.6%	9.1735	9.1742	.007%
5.6%	7.5016	7.5048	.043%
10.6%	5.9893	5.9914	.035%
20.6%	4.1085	4.1175	.22%

The interpolated values for 1.6% are the most accurate (and, for 20.6%, the least accurate) for two reasons. First, Equation 3.5 is more nearly linear when i is close to 0. Second, the tabulated values are only .5% apart around $i = 1.6\%$, and they are 5% apart around $i = 20.6\%$. Notice that all of the accuracies are much higher than was the case for the interpolations with (P/F) factors.

The interpolated values in Exhibit 3.14 are calculated as detailed for $(P/A, 20.6\%, 10)$:

$$(P/A, 20.6\%, 10) \approx (P/A, 20\%, 10) - .006[(P/A, 20\%, 10) - (P/A, 25\%, 10)]/(.25 - .2)$$
$$\approx 4.192 - .006[4.192 - 3.571]/.05 = 4.1175$$

3.5 COMBINING FACTORS

Deferred Annuities.   A **deferred annuity** starts in a later period, not in period 1. Examples include revenues delayed by a multiyear construction period and maintenance costs that begin after the warranty expires.

To find the present worth of a deferred annuity, a pair of factors must be combined. Let the last period with no cash flow be period D, and let the last cash flow occur at the end of period N. Then Equation 3.9 describes the present worth. A_D is the periodic amount of the deferred annuity.

A **deferred annuity** is a series of uniform cash flows that begins later than period 1, that is in period $D + 1$, and continues through period N.

$$P = A_D(P/A, i, N-D)(P/F, i, D) \qquad (3.9)$$

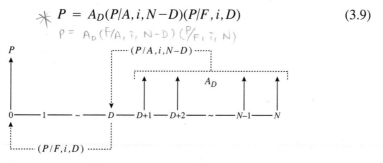

Example 3.11 applies this formula. Exhibit 3.15 depicts this in the cash flow diagram. A t' time series is defined for the 5-year annuity. Its P becomes the F for the 2-year (P/F) factor.

A common mistake is to use $(P/F, i, D+1)$, since the first cash flow of the deferred annuity occurs in year $D + 1$. This is wrong. The $(P/A, i, N-D)$ factor "moves" the annuity to a present equivalent at $t' = 0$ or $t = D$. This is clear when the factors are placed on the cash flow diagram, as in Exhibit 3.15.

You should imitate Exhibit 3.15 until using the factors correctly is automatic.

Example 3.11   Delayed Revenues

Revenues of $100 begin in year 3 and continue through year 7. If i is 6%, then what is the present worth of the cash flows?

Solution
Apply Equation 3.9 with $D = 2$ and $N = 7$. The number of cash flows is 5 (equals $N - D$).

$$P = 100(P/A, 6\%, 5)(P/F, 6\%, 2) = 100 \cdot 4.21236 \cdot .90000$$
$$= \$374.90$$

Exhibit 3.15   Combining factors for deferred annuities

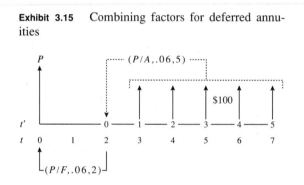

Equation 3.10 is another approach to deferred annuities. Here <u>one factor in-</u>cludes all N years and the other subtracts the D years that are not included. Example 3.12 applies Equation 3.10 to the data of Example 3.11, with the same result.

$$P = A_D[(P/A, i, N) - (P/A, i, D)] \tag{3.10}$$

Example 3.12   Delayed Revenues (revisited)

As in Example 3.11, revenues of $100 begin in year 3 and end in year 7. If the interest rate is 6%, then what is the present worth of the cash flows?

Solution
To apply Equation 3.10, $D = 2$ and $N = 7$.

$$P = 100[(P/A, 6\%, 7) - (P/A, 6\%, 2)]$$
$$= 100[5.58238 - 1.83334] = 100 \cdot 3.74899 = \$374.90$$

Prepaid Expenses and Other Beginning-of-Year Annuities.   Lease payments, rent, insurance premiums, tuition, and other prepaid expenses occur at the beginning of each period. In each case the service must be paid for, before the period of service begins. A beginning-of-period annuity is sometimes called an annuity due, while the standard end-of-period annuity is called an annuity in arrears or an an ordinary annuity.

> A **prepaid expense** is a cash flow that occurs at the beginning, not the end, of each period, such as lease and insurance payments.

To use tables that assume end-of-period cash flows for prepaid expenses, either Equation 3.11 or 3.12 must be used. Example 3.13 applies these formulas to a lease.

$$P = A_1 (P/A, i, N) \cdot (1 + i) \qquad (3.11)$$

$$(F/P, i, 1)$$

$$P = A_1 [1 + (P/A, i, N-1)] \qquad (3.12)$$

As shown in Exhibit 3.16, Equation 3.11 with the $(P/A, i, N)$ finds the equivalent value at $t = -1$, and then uses $(1 + i)$, which equals $(F/P, i, 1)$, to convert that to $t = 0$. Equation 3.12 relies on the first cash flow already occurring at $t = 0$, and then it calculates a present equivalent for the remaining $N - 1$ periods.

Exhibit 3.16   Combining factors for prepaid expenses

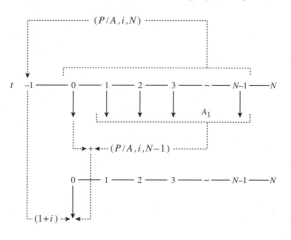

Example 3.13   Leasing a Crane

Calculate a present equivalent at 12% for the lease of a crane. The lease fee will be $40,000 per year for 5 years.

Solution
Using Equation 3.11,

$$P = -40,000(P/A, 12\%, 5)(1 + .12) = -40,000 \cdot 3.605 \cdot 1.12$$

$$= -\$161,500$$

Using Equation 3.12,

$$P = -40,000[1 + (P/A, 12\%, 4)] = -40,000[1 + 3.037]$$
$$= -\$161,500$$

Constructing Formulas from Cash Flow Diagrams.   If a cash flow diagram, such as in Example 3.14, is considered as a whole, it can be intimidating. But if it is considered in an organized step-by-step fashion, then the task is much simpler.

Begin with period 0 and consider the cash flows one at a time. Determine whether the cash flow is a single payment or part of an annuity, and then select the correct factor(s) for that cash flow. Then proceed to the next cash flow.

By proceeding from left to right, no cash flows will be overlooked. And by considering the cash flows one at a time, the factors are easier to choose.

Example 3.14   Example 2.4 Revisited

Calculate the present worth for the following cash flow diagram if the interest rate is 12%.

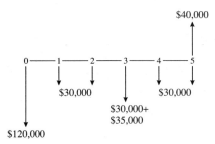

Solution
Begin with year 0, where the factor is 1, since P is at year 0.

$$P = -120,000$$

The year 1 cash flow is the first of a 5-year annuity, which uses $(P/A, 12\%, 5)$ to find the present equivalent.

$$P = -120,000 - 30,000(P/A, 12\%, 5)$$

The next cash flow is $35,000 in year 3, since the $30,000 in years 2 through 5 were included in the last factor. This $35,000 is a single payment at the end of year 3.

$$P = -120,000 - 30,000(P/A, 12\%, 5) - 35,000(P/F, 12\%, 3)$$

The only remaining cash flow is the positive $40,000 in year 5. This is also a single payment, but notice that its sign differs from the earlier cash flows.

$$P = -120,000 - 30,000(P/A, 12\%, 5) - 35,000(P/F, 12\%, 3) + 40,000(P/F, 12\%, 5)$$

Now that the equation is complete, it is time to plug in values from Appendix A.

$$P = -120{,}000 - 30{,}000 \cdot 3.605 - 35{,}000 \cdot .7118 + 40{,}000 \cdot .5674$$
$$= -120{,}000 - 108{,}150 - 24{,}913 + 22{,}696 = -\$230{,}367$$

Deriving One Factor's Formula from Another's.   The paired factors derived so far in this chapter have been inverses of each other. Specifically, $(P/F, i, N)$ equals $1/(F/P, i, N)$; $(P/A, i, N)$ equals $1/(A/P, i, N)$; and $(F/A, i, N)$ equals $1/(A/F, i, N)$.

There are other relationships, such as Equations 3.13 and 3.14, that can be verified algebraically or by substituting values. For example, let i equal 10% and N equal 20. Then to check Equation 3.13, $(A/F, 10\%, 20) = .0175$, and the (A/P) factor is .1 greater at .1175. To check Equation 3.14,

$$(F/A, 10\%, 20) = [(F/P, 10\%, 20) - 1]/.1 = (6.727 - 1)/.1 = 57.27$$

and the tabulated value is 57.275.

$$(A/F, i, N) = (A/P, i, N) - i \tag{3.13}$$

$$(F/A, i, N) = [(F/P, i, N) - 1]/i \tag{3.14}$$

Finally, there are two relationships, Equations 3.15 and 3.16, that can be justified intuitively, verified algrebraically (see Exhibit 3.9 for Equation 3.16), or graphed, as shown in Exhibit 3.17. The intuitive justification is that the increased present worth of adding another period to a cash flow series can be found by adding the present worth for that one period, N, to the present worth for periods 1 to $N - 1$.

$$(P/F, i, N) = (P/A, i, N) - (P/A, i, N-1) \tag{3.15}$$

$$(P/A, i, N) = (P/F, i, 1) + (P/F, i, 2) + \cdots + (P/F, i, N) \tag{3.16}$$

Exhibit 3.17   Linking formulas for (P/A) and (P/F) factors

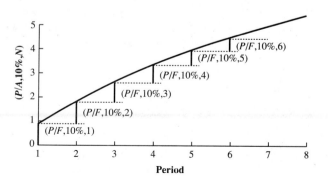

3.6 ARITHMETIC GRADIENTS

Definition.   A typical application of a gradient is thus: "Revenue is $4000 the first year, increasing by $1000 each year thereafter." The gradient is the $1000-per-year change in cash flows.

The gradients in Exhibits 3.1 and 3.18 have cash flows of zero at the end of the first period. The first nonzero cash flow occurs at the end of period 2, and an N-period gradient has $N - 1$ nonzero cash flows, which change by G per period.

A common mistake is to visualize a gradient series as $N - 1$ periods long. The gradient series is N periods long, but the first period has 0 cash flow. This assertion can be confirmed by examining the tables in Appendix A for any interest rate. The two factors at the right of the tables are the $(P/G, i, N)$ and $(A/G, i, N)$ factors. Both equal 0 for $N = 1$. Thus, the gradient and annuity series in Exhibit 3.18 both have a length of 5 or N periods.

Exhibit 3.18   Example gradient

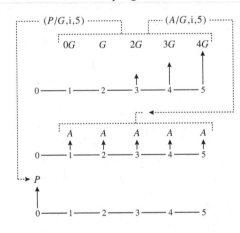

Using the Gradient Factors.   The gradient factors describe the change in a periodic cash flow—such as, increasing by $1000 per year. Thus, the factors are typically used, along with an annuity factor.

The first period's cash flow defines the annuity series, and the period-by-period change defines the gradient series. Since the annuity and the gradient can be positive or negative, there are four possible combinations. Exhibit 3.19 illustrates these in terms of cash flows:

Positive and increasing ($A > 0$ and $G > 0$)

Positive but decreasing ($A > 0$ and $G < 0$)

Negative but becoming less so ($A < 0$ and $G > 0$)

Negative and becoming more so ($A < 0$ and $G < 0$)

Exhibit 3.19   Gradient and annuity combinations

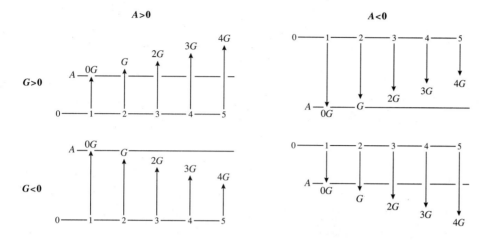

Examples 3.15 and 3.16 illustrate that solving gradients is easy, so long as the definitions are remembered. The first period's cash flow is the *A* for the annuity series, and the change each period is the *G* for the gradient series.

Example 3.15   Gradient for Positive and Increasing Cash Flows

Find the present worth of $4000 the first year, increasing by $1000 per year. The interest rate is 8% and *N* equals 5.

Solution

Exhibit 3.20   Gradient example

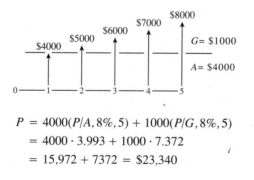

$$P = 4000(P/A, 8\%, 5) + 1000(P/G, 8\%, 5)$$
$$= 4000 \cdot 3.993 + 1000 \cdot 7.372$$
$$= 15,972 + 7372 = \$23,340$$

Example 3.16   Loan with Constant Principal Payment

Find the present worth, from the bank's perspective, of the constant principal payment loan in Exhibit 2.6. The interest rate was 10%, *N* was 5, and payments began at $300 and declined to $220.

Solution

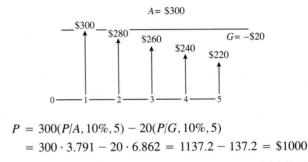

5xhibit 3.21   Gradient example for constant principal payment loan

$$P = 300(P/A, 10\%, 5) - 20(P/G, 10\%, 5)$$
$$= 300 \cdot 3.791 - 20 \cdot 6.862 = 1137.2 - 137.2 = \$1000$$

The present worth of the payments equals the amount that was initially borrowed. Note that if the diagram is drawn from the viewpoint of the borrower, then $A = -\$300$ and $G = +\$20$.

Gradient Formulas.   The formulas for the arithmetic gradient can be derived just as the annuity formulas were. (See Problem 3.52.) They are as shown in Equations 3.17 through 3.21.

$$(A/G, i, N) = 1/i - N/[(1 + i)^N - 1] \qquad (3.17)$$
$$= [(1 + i)^N - iN - 1]/[i(1 + i)^N - i] \qquad (3.17')$$
$$(P/G, i, N) = 1/i^2 - N/[i(1 + i)^N] - 1/[i^2(1 + i)^N] \qquad (3.18)$$
$$= [(1 + i)^N - iN - 1]/[i^2(1 + i)^N] \qquad (3.18')$$

These formulas can be derived by beginning with

$$(F/G, i, N) = (F/A, i, N-1) + (F/A, i, N-2) + \cdots + (F/A, i, 1)$$

which, through algebra, can be shown to lead to Equation 3.19:

$$(F/G, i, N) = [(1 + i)^N - 1]/i^2 - N/i \qquad (3.19)$$

The gradient formula can also be restated, based on the previously derived factors:

$$(A/G, i, N) = (N/i)[1/N - (A/F, i, N)]$$
$$= 1/i - N(A/F, i, N)/i \qquad (3.20)$$
$$(P/G, i, N) = (1/i)[(P/A, i, N) - N(P/F, i, N)] \qquad (3.21)$$

3.7 SUMMARY

In this chapter, the basic cash flow elements (P, F, A, i, N, and G) were defined. The standard assumptions are that P occurs at period 0 or the beginning of period

1. F occurs at the end of period N. A and G occur at the ends of periods 1 through N. Furthermore, G has a zero cash flow at the end of period 1.

These elements are combined into the engineering economy factors: $(P/F, i, N)$, $(F/P, i, N)$, $(P/A, i, N)$, $(A/P, i, N)$, $(A/F, i, N)$, and $(F/A, i, N)$. The periods for i and N must be the same, and the proper notation for i is percentages or decimals—5% or .05, not 5. Interpolation for values of factors, where i or N is not tabulated, is best done using similar triangles.

The single-payment factors, $(P/F, i, N)$ and $(F/P, i, N)$, were applied to the problem of finding an interest rate. Representative values of the factors were graphed and the limits over i and N were discussed.

The formulas for the annuity factors—$(P/A, i, N)$, $(A/P, i, N)$, $(A/F, i, N)$, and $(F/A, i, N)$—were developed. Representative values of the factors were graphed and the limits over i and N were discussed. Interpolation for an unknown i was presented.

Deferred annuities, such as annual repair costs after a warranty period, and beginning-of-period annuities, such as lease payments, were developed. Cash flow equations that combine (P/A) and (P/F) factors were linked to the cash flow diagrams. Formulas that permit the calculation of one factor from another were presented.

Two factors for the gradient series, $(A/G, i, N)$ and $(P/G, i, N)$, were developed. These series assume a $0 cash flow at the end of the first period, because the gradient series is usually combined with an annual series. The formulas for the gradient factors were given.

Appendix 3A will describe factors for distributed cash flows with continuous compounding. This will include example tables.

REFERENCE

"Report of the Engineering Economy Subcommittee (Z94.5) of the ANSI Z94 Standards Committee on Industrial Engineering Terminology," *The Engineering Economist,* Volume 33, Number 2, Winter 1988, pp. 145–151. Draft (Z94.7) expected 1995.

PROBLEMS

3.1 $(P/F, 8\%, 43)$ is not tabulated.
   a. Find the value using the appropriate formula.
   b. Find the value by interpolating between $N = 40$ and $N = 45$.
   c. How far off is the interpolation?   (*Answer:* absolute error of .00095, relative error of 2.6%)

3.2 $(P/F, 10\%, 37)$ is not tabulated.
   a. Find the value using the appropriate formula.
   b. Find the value by interpolating between $N = 35$ and $N = 40$.
   c. How far off is the interpolation?

3.3 $(P/F, 8.6\%, 10)$ is not tabulated.
   a. Find the value using the appropriate formula.
   b. Find the value by interpolating between 8% and 10%.

  c. How far off is the interpolation? (*Answer:* absolute error of .00017, relative error of .38%)

3.4 $(P/F, 12.3\%, 20)$ is not tabulated.
  a. Find the value using the appropriate formula.
  b. Find the value by interpolating between 12% and 13%.
  c. Find the value by interpolating between 10% and 15%.
  d. How much better is the *tighter* interpolation?

3.5 HiTek Manufacturing has agreed to license an industrial process to another firm. That firm will pay $5 million when it licenses the process, which may happen at any time during the next 10 years. The firm's interest rate is 8%. What is the present worth if the license is issued at the end of year 3? Of year 5? Of year 10? (*Answer:* $PW_5 = \$3.40M$)

3.6 Graph the answer for Problem 3.5 for years 1 to 10.

3.7 If you can make 6% interest on your money, how much is $1000 paid to you 6 years in the future worth to you now? Assume annual compounding.

3.8 Johnny B. Good wishes to have $3000 after 3 years in an account that draws 6% nominal interest compounded monthly. How much must he deposit each month, starting in one month?

3.9 Paul Randolph, a first-round NFL draft selection, signs a contract that pays him $750,000 for signing. If Paul's agent invests the total sum at 12% nominal interest compounded monthly, how much will Paul have at the end of 2 years? (*Answer:* $952,500)

3.10 A company in Texas will "buy" inheritances before they are actually received. If you will receive $20,000 when a rich uncle dies, how much would you accept now for this future inheritance? Money is worth 8% to you, and you "guesstimate" that your uncle will live 6 more years. (*Answer:* $12.60K)

3.11 George expects to inherit $500,000 someday from his grandparents. Assuming that it happens in 10 years, graph the present worth as a function of the interest rate. Include rates of 0%, 5%, 10%, 15%, and 20%. (George does not know what interest rate he should use.)

3.12 George (of Problem 3.11) does not really know when the inheritance will come. Assume 10 years and a 10% interest rate. Consider the present worth again.
  a. Which has more effect: halving the life or halving the interest rate?
  b. Which has more effect: doubling the life or doubling the interest rate?

3.13 What is the monthly payment for a 3-year new car loan when the nominal interest rate is 12%? After rebates, down payments, and other charges, the amount borrowed was $10,500. (*Answer:* $348.6)

3.14 A young engineer wishes to buy a house, but can only afford monthly payments of $500. Thirty-year loans are available at 12% interest with monthly payments. If she can make a $5000 down payment, what is the most expensive house she can afford?

3.15 Using a credit card, Tim Settles has just purchased a stereo system for $975. If he makes payments of $45 per month and the interest rate is 18% compounded monthly, how long will it take to pay off the loan?

3.16 Murphy Morsels just bought a new chocolate-chip-making machine for $25,000. Murphy's old machine was taken as a trade and was valued at $5000. Also, Murphy received a 5% discount on the new machine's price after the trade-in allowance. Murphy plans on making monthly payments for 3 years with 12% nominal interest. What is the monthly payment?

3.17 Nichols Electric expects to replace a $500,000 piece of equipment in 4 years. It plans to accumulate the necessary funds by making equal monthly deposits into a bank fund that earns $\frac{3}{4}$% per month. What is the monthly deposit? (*Answer:* $8700)

3.18 To buy a new automobile, Brian Williams takes out a loan of $12,000 at 6% interest, with monthly payments over 5 years. How much is each monthly payment? How much interest is paid over the loan's life?

3.19 Bob Jones wants to buy a new home, which costs $87,500. Bob is planning to make a 10% down payment. If Bob can obtain a loan at 9% with monthly payments, how much will each payment be if the loan is paid off over 30 years? Ignoring the time value of money, how much money could Bob save if the loan was for 20 years at 9% interest and a 15% down payment?

3.20 Cranford Cranberry Bogs just purchased new water-whipping equipment for $100,000. Cranford agreed to repay the loan over a period of 10 years with semiannual payments. The loan's interest rate is 8%. Determine the payment amount and the total finance charge. (*Answer:* total finance charge = $47.2K)

3.21 Bob and Mary Johnson are expecting their first child. They have decided to deposit $1000 into a savings account that pays 6% interest compounded annually on the day the child is born. They will then deposit $1000 on each birthday through the child's 18th birthday. How much money will be in the account on the child's 19th birthday to finance a college education? (*Answer:* $35,786)

3.22 HiTek Manufacturing is purchasing a large tract of land for a new facility. Its loan is for $3 million and it will be paid off in 15 equal annual payments. If the interest rate is 10%, how much is each payment?

3.23 $(P/A, .064, 10)$ is not tabulated.
 a. Find the value using the appropriate formula.
 b. Find the value by interpolation.
 c. How far off is the interpolation?

3.24 $(P/A, i, 10)$ is 6.503. What is i?

3.25   E. Z. Money, who owns a chain of nightclubs, needs to borrow $20,000 to remodel the local dive. Ima Shark offers to loan the $20,000 to Money for 24 months. Money must make monthly payments of $1118. Determine the loan's (a) monthly interest rate and (b) effective annual interest rate.

3.26   A new product will generate net revenues of $600,000 per year. The interest rate is 10%.
   a.   What is the present worth of the revenue stream if the life is 6 years? 12 years?
   b.   Graph the present worth for lives of 1 to 20 years.

3.27   A lottery pays the winner $1 million, in 20 equal payments of $50,000. The payments are received at the end of the year, and the winner's interest rate is 12%. What is the present worth of the winnings?

3.28   Graph the answer for Problem 3.27 for interest rates ranging from 0% to 20%.

3.29   For the new product in Problem 3.26, the interest rate is also uncertain. Assume the interest rate is 10% and the life is 6 years.
   a.   Which has more effect: halving the life or doubling the interest rate?
   b.   Which has more effect: doubling the life or halving the interest rate?
   c.   Do your answers to parts a and b change if the life is 20 years rather than 6 years?

3.30   What is the present worth of the following cash flows?
   a. $1400 per year for 8 years at 9%
   b. $1400 per year starting in year 4 and continuing through year 11 at 9%

3.31   The first lottery payment in Problem 3.27 will not be received until the end of the third year. What is the present worth if the interest rate is 12%?

3.32   Determine the present value of 10 monthly lease payments of $1000 if the nominal interest rate is 9%.   (Answer: $9672)

3.33   Creekside Engineering will pay $24,000 per year for errors and omissions liability insurance. What is the present worth of the coverage for 10 years? (Firms similar to Creekside average one claim every 10 years.) The interest rate is 9%.

3.34   Creekside Engineering pays $3000 per month to lease its office space. If the nominal annual interest rate is 12%, what is the present worth of 1 year's lease payments?

3.35   How can the tables of single-payment factors be used to compute $(P/F, i, 200)$? $(P/F, i, 150)$?

3.36   How can the tables be used to compute $(P/A, i, 200)$? $(P/A, i, 150)$?

3.37   Matuk Construction has purchased some earthmoving equipment that comes with a 2-year warranty. Repair costs are expected to be $1500 per year, beginning in year 3 and continuing through the equipment's 10-year life. What is the PW of the repair expenses if the interest rate is 9%?   (*Answer:* −$6988)

3.38 MicroTech expects to receive royalties of $750K per year on a patent. The royalties begin at the end of year 4 and continue through year 11. If the interest rate is 8%, what is the PW of the royalties?

3.39 You want to bank enough money to pay for 4 years of college at $15,000 per year for your child. The savings account will pay an effective rate of 6% per year. The first annual payment for tuition and for room and board is made on your child's 18th birthday. If you deposit the money on your child's 3rd birthday, how much must you deposit?

3.40 Helen pays rent of $400 per month for the academic year (September through May). She is going to Australia for the summer and won't be working. How much must she set aside in her savings account on June 1 to pay her rent for the next year? The account pays 6% interest with monthly compounding.

3.41 How much difference is there when the assumption changes from beginning- to end-of-period payments in
a.   Problem 3.33?
b.   Problem 3.34?

3.42 Equation 3.11 multiplied the end-of-period factor by $(1 + i)$ to calculate a value for beginning-of-period cash flows. For a middle-of-period cash flow, the adjustment is $\sqrt{(1 + i)}$. Apply this principle to calculate the present worth of the annual payroll of HiTek Manufacturing over the next 10 years. Employees are paid every two weeks, so on average they are paid in the middle of the year, not at the end. The annual payroll is $650,000. The interest rate is 12%.

3.43 Prove the following relationships algebraically:
a. $(A/F, i, N) = (A/P, i, N) - i$
b. $(P/F, i, N) = (P/A, i, N) - (P/A, i, N-1)$
c. $(P/A, i, N) = (P/F, i, 1) + (P/F, i, 2) + \cdots + (P/F, i, N)$
d. $(F/A, i, N) = [(F/P, i, N) - 1]/i$

3.44 What is the present worth of the following cash flows?
a. $2800 per year for 9 years at 8%
b. $2800 per year starting in year 4 and continuing through year 12 at 8%
c. $2800 for the first year and decreasing by $150 for each of the next eight years at 8%

3.45 What is the present worth (at 7%) of maintenance costs that start at $3500 per year and then climb by $825 per year over a 10-year life? (*Answer:* −$47,450)

3.46 Determine the maximum amount that could be loaned at 8% interest, if it is repaid as follows:

Year	1	2	3	4	5
Payment	$200	400	600	800	1000

3.47   What is the present worth (at 8%) of an inheritance that begins at $8000 per year at the end of year 1 and then falls by $500 per year over its 16-year life?   (*Answer:* $44,676)

3.48   What is the present worth (at 12%) of maintenance costs that start at $4500 per year and then climb by $225 per year when the warranty expires. Since the warranty expires at the end of year 3, the mainte-nance costs for year 4 are $4725. The machine has a 10-year life.

3.49   Determine the present value at 9% interest of the following cash flow:

Year	1	2	3	4	5–10
Payment	$100	100	150	175	200

3.50   Determine the present worth, at 12% interest, of the maintenance costs listed below.

Year	Cost
1–4	$10,000
5–8	15,000
9–10	18,000

3.51   Given the following cash flows, determine the equivalent uniform cash flow at 6% interest.

Year	1	2	3	4	5
Payment	$200	300	400	500	600

3.52   Derive Equation 3.19 for the arithmetic gradient.

3.53   Show that Equation 3.20 is always positive for $i > 0$.

3.54   Show that Equations 3.17 and 3.18 equal 0 when N is 1.

3.55   Find the limit for Equation 3.18 as N approaches infinity.

3.56   If the following two cash flows have the same present worth at i, which is preferred at $2i$?

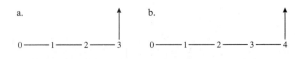

3.57 If the following two cash flows have the same present worth at i, which is preferred at $2i$?

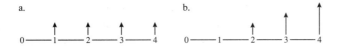

a.

b.

APPENDIX 3A: CONTINUOUS FLOW AND CONTINUOUS COMPOUNDING

Definitions. The factors discussed in this appendix assume that cash flows are continuous or distributed throughout a period, rather than occurring in one lump sum at the period's end. For example, when a highway improvement or a new computer system saves time for the users, that time is spread throughout the year. The time, and thus the savings, does not occur all at once on December 31. It is usually more accurate to assume that the benefits and costs of engineering projects are distributed (1) over a year, rather than (2) in one lump sum at the year's end.

The change in notation that identifies continuous cash flows is the addition of a bar ($^-$) above the A and the F. The specific dollar amount is the number of dollars per period. Thus, $\overline{F}$ = \$100 says that \$100 is distributed over the Nth period; that is, from the end of period $N - 1$ to the end of period N. Similarly, $\overline{A}$ = \$200 says that \$200 *per period* is continuous from time 0 to the end of period N.

Both the nominal and the effective interest rates, r and i, are used. Section 2.6 developed Equation 2.4, $e^r - 1 = i$, for converting a continuously compounded nominal rate of interest, r, to an effective interest rate, i. Continuous compounding at rate r must be assumed, so that the distributed or continuously flowing cash flow "earns interest" as it occurs. Thus, the formulas for distributed cash flows use r.

The tables for continuous flow factors must be for an effective interest rate, i, since the effective interest rates must match the rates for the other end-of-period cash flows and tables. Thus, the rates r are chosen so that the effective interest rates, i, are 2%, 5%, etc.

Exhibit 3A.1 Cash flow diagrams for continuous flow

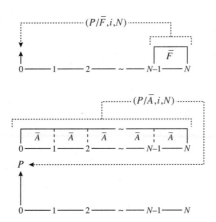

Tabulated Factors.   The factors for $\overline{A}$ and $\overline{F}$ do not convert between them. Instead, the factors convert continuous cash flows to present worths at time 0, as shown in Exhibit 3A.1. Since we already have tables for end-of-period P's, F's, and A's, Exhibit 3A.2 for $\overline{A}$ and Exhibit 3A.3 for $\overline{F}$ are sufficient to use the distributed cash flow model.

Note: the values of r are chosen so that the effective interest rates match the rates in Appendix 1. The formula is $r = \ln(1 + i)$, since $(1 + i) = e^r$. This ensures that the same assumption about the time value of money is made for end-of-period cash flows and for continuous cash flows. The effective interest rates are the same.

Example 3A.1   Distributed Highway Benefits

The benefits for a highway project are $1.2 million (or M) per year in time savings. During year 3, the benefits will be tripled because of planned construction on another highway. The regional transportation authority mandates an interest rate of 4%, and the project has an expected life of 50 years. What is the present worth of the benefits?

Solution

$$P = 1.2\text{M}(P/\overline{A}, 4\%, 50) + 2.4\text{M}(P/\overline{F}, 4\%, 3)$$
$$= 1.2\text{M}(21.91) + 2.4\text{M}(.9067) = \$28.47\text{M}$$

Example 3A.2   Distributed Construction Costs

The construction of the highway project in Example 3A.1 will take a year. The $35M in costs are distributed evenly over the year before the project opens (opening day is time 0). What is the present worth at time 0 of the construction costs?

Solution

$$P = 35\text{M}(P/\overline{F}, 4\%, 1)(F/P, 4\%, 1)$$
$$= 35\text{M}(.9806)(1.04) = \$35.69\text{M}$$

Formulas.   The formulas for the tabulated values are Equations 3A.1 and 3A.2. For derivation of these and other formulas for continuous cash flows, see [Park and Sharp-Bette]. Different cash flow patterns can be included, such as P and several gradients (linear, exponential increase, exponential decay, and growth curves).

$$(P/\overline{F}, i, N) = (e^r - 1)/(re^{rN}) \qquad (3\text{A}.1)$$

$$(P/\overline{A}, i, N) = (e^{rN} - 1)/(re^{rN}) \qquad (3\text{A}.2)$$

For developing theory, these continuous formulas are far easier to use than the discrete end-of-period formulas. The reasons are that e^r is far easier to integrate and manipulate than is $1 + i$, and that the formulas for $\overline{A}$ and $\overline{F}$ are so similar.

Exhibit 3.A2   Present worth of cash flow distributed from end of period $N - 1$ to end of period N, $P/\overline{F}$

i	2%	4%	6%	8%	10%	15%	20%	25%	50%
r	1.98%	3.92%	5.83%	7.70%	9.53%	13.98%	18.23%	22.31%	40.55%
N									
1	.9902	.9806	.9714	.9625	.9538	.9333	.9141	.8963	.8221
2	.9707	.9429	.9164	.8912	.8671	.8115	.7618	.7170	.5481
3	.9517	.9067	.8646	.8252	.7883	.7057	.6348	.5736	.3654
4	.9331	.8718	.8156	.7641	.7166	.6136	.5290	.4589	.2436
5	.9148	.8383	.7695	.7075	.6515	.5336	.4408	.3671	.1624
6	.8968	.8060	.7259	.6551	.5922	.4640	.3674	.2937	.1083
7	.8792	.7750	.6848	.6065	.5384	.4035	.3061	.2350	.0722
8	.8620	.7452	.6461	.5616	.4895	.3508	.2551	.1880	.0481
9	.8451	.7165	.6095	.5200	.4450	.3051	.2126	.1504	.0321
10	.8285	.6890	.5750	.4815	.4045	.2653	.1772	.1203	.0214
11	.8123	.6625	.5424	.4458	.3677	.2307	.1476	.0962	.0143
12	.7964	.6370	.5117	.4128	.3343	.2006	.1230	.0770	.0095
13	.7807	.6125	.4828	.3822	.3039	.1744	.1025	.0616	.0063
14	.7654	.5889	.4554	.3539	.2763	.1517	.0854	.0493	.0042
15	.7504	.5663	.4297	.3277	.2512	.1319	.0712	.0394	.0028
16	.7357	.5445	.4053	.3034	.2283	.1147	.0593	.0315	.0019
17	.7213	.5236	.3824	.2809	.2076	.0997	.0494	.0252	.0013
18	.7071	.5034	.3608	.2601	.1887	.0867	.0412	.0202	.0008
19	.6933	.4841	.3403	.2409	.1716	.0754	.0343	.0161	.0006
20	.6797	.4655	.3211	.2230	.1560	.0656	.0286	.0129	.0004
21	.6664	.4476	.3029	.2065	.1418	.0570	.0238	.0103	.0002
22	.6533	.4303	.2857	.1912	.1289	.0496	.0199	.0083	.0002
23	.6405	.4138	.2696	.1770	.1172	.0431	.0166	.0066	.0001
24	.6279	.3979	.2543	.1639	.1065	.0375	.0138	.0053	.0001
25	.6156	.3826	.2399	.1518	.0968	.0326	.0115	.0042	.0000
30	.5576	.3144	.1793	.1033	.0601	.0162	.0046	.0014	.0000
35	.5050	.2585	.1340	.0703	.0373	.0081	.0019	.0005	.0000
40	.4574	.2124	.1001	.0478	.0232	.0040	.0007	.0001	.0000
45	.4143	.1746	.0748	.0326	.0144	.0020	.0003	.0000	.0000
50	.3752	.1435	.0559	.0222	.0089	.0010	.0001	.0000	.0000
60	.3078	.0969	.0312	.0103	.0034	.0002	.0000	.0000	.0000
70	.2525	.0655	.0174	.0048	.0013	.0001	.0000	.0000	.0000
80	.2072	.0442	.0097	.0022	.0005	.0000	.0000	.0000	.0000
90	.1699	.0299	.0054	.0010	.0002	.0000	.0000	.0000	.0000
100	.1394	.0202	.0030	.0005	.0001	.0000	.0000	.0000	.0000

Exhibit 3.A3   Present worth of cash flows distributed from 0 to N at $\overline{A}$ per period $P/\overline{A}$

i	2%	4%	6%	8%	10%	15%	20%	25%	50%
r	**1.98%**	**3.92%**	**5.83%**	**7.70%**	**9.53%**	**13.98%**	**18.23%**	**22.31%**	**40.55%**
N									
1	.990	.981	.971	.962	.954	.933	.914	.896	.822
2	1.961	1.924	1.888	1.854	1.821	1.745	1.676	1.613	1.370
3	2.913	2.830	2.752	2.679	2.609	2.450	2.311	2.187	1.736
4	3.846	3.702	3.568	3.443	3.326	3.064	2.840	2.646	1.979
5	4.760	4.540	4.338	4.150	3.977	3.598	3.281	3.013	2.142
6	5.657	5.346	5.063	4.805	4.570	4.062	3.648	3.307	2.250
7	6.536	6.121	5.748	5.412	5.108	4.465	3.954	3.542	2.322
8	7.398	6.867	6.394	5.974	5.597	4.816	4.209	3.730	2.370
9	8.244	7.583	7.004	6.494	6.042	5.121	4.422	3.880	2.402
10	9.072	8.272	7.579	6.975	6.447	5.386	4.599	4.000	2.424
11	9.884	8.935	8.121	7.421	6.815	5.617	4.747	4.096	2.438
12	10.68	9.572	8.633	7.834	7.149	5.818	4.870	4.173	2.447
13	11.46	10.18	9.116	8.216	7.453	5.992	4.972	4.235	2.454
14	12.23	10.77	9.571	8.570	7.729	6.144	5.058	4.284	2.458
15	12.98	11.34	10.00	8.897	7.980	6.276	5.129	4.324	2.461
16	13.71	11.88	10.41	9.201	8.209	6.390	5.188	4.355	2.463
17	14.43	12.41	10.79	9.482	8.416	6.490	5.238	4.381	2.464
18	15.14	12.91	11.15	9.742	8.605	6.577	5.279	4.401	2.465
19	15.83	13.39	11.49	9.983	8.777	6.652	5.313	4.417	2.465
20	16.51	13.86	11.81	10.21	8.932	6.718	5.342	4.430	2.466
21	17.18	14.31	12.11	10.41	9.074	6.775	5.366	4.440	2.466
22	17.83	14.74	12.40	10.60	9.203	6.824	5.385	4.448	2.466
23	18.47	15.15	12.67	10.78	9.320	6.868	5.402	4.455	2.466
24	19.10	15.55	12.92	10.94	9.427	6.905	5.416	4.460	2.466
25	19.72	15.93	13.16	11.10	9.524	6.938	5.427	4.464	2.466
30	22.62	17.64	14.17	11.70	9.891	7.047	5.462	4.476	2.466
35	25.25	19.04	14.93	12.11	10.12	7.101	5.476	4.480	2.466
40	27.63	20.19	15.49	12.40	10.26	7.128	5.481	4.481	2.466
45	29.78	21.13	15.92	12.59	10.35	7.142	5.483	4.481	2.466
50	31.74	21.91	16.23	12.72	10.40	7.148	5.484	4.481	2.466
60	35.11	23.07	16.64	12.87	10.46	7.153	5.485	4.481	2.466
70	37.87	23.86	16.87	12.93	10.48	7.155	5.485	4.481	2.466
80	40.14	24.39	17.00	12.97	10.49	7.155	5.485	4.481	2.466
90	42.00	24.75	17.07	12.98	10.49	7.155	5.485	4.481	2.466
100	43.53	24.99	17.11	12.99	10.49	7.155	5.485	4.481	2.466

Example 3A.3   Deriving a Distributed Cash Flow Formula

Derive the formula for $(F/\overline{A}, i, N)$.

Solution
Using Equation 2.4, $F = Pe^{rN}$ with continuous compounding. Now using Equation 3A.2, we know that

$$(P/\overline{A}, i, N) = (e^{rN} - 1)/(re^{rN})$$

so that

$$P = \overline{A}(e^{rN} - 1)/(re^{rN})$$

We then substitute the last expression for P in Equation 2.4.

$$F = [\overline{A}(e^{rN} - 1)/(re^{rN})]e^{rN}$$

Canceling the e^{rN} terms leaves

$$(F/\overline{A}, i, N) = (e^{rN} - 1)/r$$

Continuous vs. End-of-Period Assumptions.   In most engineering situations the distributed cash flow factors are more realistic than the end-of-period factors. However, the end-of-period factors were developed first, and their use has become accepted practice.

Historically, the reason may have been that end-of-period flows were more suitable for banking, which relies heavily on the time value of money. Another possible reason is that managers, who were not engineers, were uncomfortable with using e.

Today, spreadsheets include formulas for the end-of-period factors, but not for the distributed cash flow factors. The widespread use of spreadsheets ensures that the domination of end-of-period assumptions will continue.

This domination is unfortunate, because the factors are equally easy to use when tabulated. When formulas are used, continuous cash flows require fewer keystrokes, and it is easier to use e than $(1 + i)$ on a modern calculator or computer.

More importantly, the costs of building engineering projects and the benefits received from their use are distributed through time. December 31 is not a magic day on which all of the year's benefits and costs are suddenly realized. Payments to employees may most accurately be modeled as occurring at the end of two-week periods, but the distributed cash flow model with yearly periods will be very close to that ideal—much closer than assuming a single payment at the end of a year.

Distributed cash flows are often closer to reality than end-of-year cash flows; however, end-of-month or end-of-pay-period models may be even more accurate. The added advantage of using shorter periods is that the cash flows in each sub-period need not be the same. For example, construction costs or recreational benefits in the summer months might greatly exceed those in the winter months.

Example 3A.4 shows how much difference the assumption of distributed cash flows can make.

Example 3A.4   Distributed vs. End-of-Period Cash Flows

How much difference does the assumption of continuous cash flows make for the present worth of a cash flow of $500?

a. Let $i = 4\%$, and consider at $N = 1$ and $N = 20$.

b. Let $i = 8\%$, and consider at $N = 1$ and $N = 20$.

Solution

These differences are due to the fact "on average" distributed cash flows occur at the middle rather than the end of the period.

a. For $N = 1$, the present worths of the $500 are

$$P_F = 500(P/F, 4\%, 1) = 500(.9615) = \$480.75$$
$$P_{\overline{F}} = 500(P/\overline{F}, 4\%, 1) = 500(.9806) = \$490.30$$

For $N = 20$, the present worths of the $500 are

$$P_F = 500(P/F, 4\%, 20) = 500(.4564) = \$228.19$$
$$P_{\overline{F}} = 500(P/\overline{F}, 4\%, 20) = 500(.4655) = \$232.75$$

In both cases, the distributed cash flow is worth 1.99% more.

b. For $N = 1$, the present worths of the $500 are

$$P_F = 500(P/F, 8\%, 1) = 500(.9259) = \$462.95$$
$$P_{\overline{F}} = 500(P/\overline{F}, 8\%, 1) = 500(.9625) = \$481.25$$

For $N = 20$ the present worths of the $500 are

$$P_F = 500(P/F, 8\%, 20) = 500(.2145) = \$107.25$$
$$P_{\overline{F}} = 500(P/\overline{F}, 8\%, 20) = 500(.2230) = \$111.50$$

In both $i = 8\%$ cases, the distributed cash flow is worth 3.95% more.

Thus, the distributed cash flow assumption makes more of a difference for single-payment cash flows when the interest rate is larger, and the difference is not affected by the number of periods.

REFERENCE

Park, Chan S, and Gunter P. Sharp-Bette, *Advanced Engineering Economics*, 1990, Wiley.

PROBLEMS

3A.1   Added insulation will save HiTek Manufacturing $150,000 per year for the next 20 years. Assume that the savings are continuous throughout the year, and that $i = 6\%$. What is the present worth of the savings?   (*Answer:* $1.77M)

3A.2   Construction costs for a large coal-fired generating facility will be spread over a 3-year period. Costs for the first year are $25M, for the second $150M, and for the third, $90M. The costs are evenly distributed within each year, and the interest rate is 10%. What is the present worth of the construction costs?

3A.3   Construction costs for a petroleum refinery will be spread over a 4-year period. Costs are $10M; $50M; $80M; and $50M for years 1 to 4. Costs are evenly distributed within each year, and i is 15%. What is the present worth?

3A.4   Time savings for travelers using Big City's airport would be worth $2.5 million per year for 25 years if roadway and parking improvements are made. During the 2 years of construction, delays will "cost" $1.4 million. If the interest rate for the 27-year period is 6% and cash flows are evenly distributed within each year, what is the present worth?   (*Answer:* $26.64M)

3A.5   Redo Example 3A.4 for $500 per year, using $\overline{A}$ rather than $\overline{F}$.

3A.6   Define a $\overline{P}$ that begins at $t = 0$ and is distributed over period 1.
a.   What is the formula for $(F/\overline{P}, i, N)$?
b.   Why is this not the inverse of $(P/\overline{F}, i, N)$?

3A.7   HiTek Manufacturing has an annual payroll cost of $36 million. What is the present worth of 1 year's payroll, assuming (1) end-of-year cash flows, (2) continuous cash flows, and (3) end-of-month cash flows?
a.   The interest rate is 8%.
b.   The interest rate is 20%.

3A.8   PetroChem spends $120M per year on crude oil. What is the present worth of 1 year's purchases, assuming (1) end-of-year cash flows, (2) evenly distributed cash flows, (3) end-of-month cash flows, and (4) end-of-week cash flows?
a.   The interest rate is 10%.
b.   The interest rate is 25%.

3A.9   Redo Problem 3A.8 for 10 years' worth of purchases at $120M per year. Are the results consistent with Problem 3A.8?

CHAPTER 4 GEOMETRIC GRADIENTS AND SPREADSHEETS

The Situation and the Solution

The last chapter covered single, uniform, and linearly changing cash flows, but many cash flows do not fit these patterns. For example, population growth and product sales are often modeled as increasing at a constant *rate* rather than by a constant amount. Cash flows that change at a constant *rate* are called geometric gradients.

The solution is to model and evaluate these cash flows using computerized spreadsheets or a tabular approach. Appendix 4A describes a more mathematical, formula-based approach to geometric gradients.

Chapter Objectives

After you have read and studied the sections of this chapter, you should be able to:

Section 4.1
Contrast arithmetic and geometric models of changing cash flows and use the formula for a geometric gradient.

Section 4.2
Define four common sources of geometric gradients.

Section 4.3
Set up problems with multiple geometric gradients.

Section 4.4
Apply manual tabular methods to geometric gradients.

Section 4.5
Define the terms associated with computerized spreadsheets.

Section 4.6
Create computerized spreadsheets using the COPY command and absolute and relative addressing.

Section 4.7
Demonstrate the power of computerized spreadsheets with financial functions and graphics.

Appendix 4A
Calculate an equivalent discount rate and use it with the engineering economy factors introduced in Chapter 3.

Key Words and Concepts

Geometric gradients   Cash flows that change at a constant rate, g.

Inflation   A decrease in the value of the dollar (or other monetary unit); or, equivalently, increases in the prices of purchased items.

Cell   The basic element of a computerized spreadsheet. It corresponds to an entry in a table made up of columns and rows.

Absolute address   Column and/or row reference that is fixed so it does not change when the cell is copied to another location.

Relative address   Column and/or row reference that is not fixed at that column or row.

Amortization schedule   List of the principal paid, interest paid, and remaining balance for each period of a loan.

4.1 GEOMETRIC VS. ARITHMETIC GRADIENTS

Chapter 3 presented several factors for finding equivalent worths of common cash flow patterns. These patterns included single payments at time 0 or time N, uniform payments at the end of every period from 1 to N, and an arithmetic gradient that increased from a base of 0 at the end of period 1. Example factors include $(F/P, i, N)$, $(P/A, i, N)$, and $(A/G, i, N)$.

The real world often has cash flows that follow a different pattern—one that has a constant *rate* of change. This type of cash flow is a **geometric gradient** (also known as an *escalating series*). A salary that increases 6% per year is increasing geometrically, just as a market demand that decreases 15% per year is decreasing geometrically.

A **geometric gradient** is a cash flow that increases or decreases at a *constant rate.*

To develop these examples further, consider an engineer whose starting salary is $30,000 per year, with an expected increase of 6% per year for the next 15 years.

Six percent of $30,000 is $1800, and an arithmetic gradient of $1800 leads to a final salary of $55,200. However, as shown on the *geometric* curve of Exhibit 4.1, a constant 6% rate of increase leads to a final salary of $67,827. Thus, a geometric gradient leads to a final salary that is 23% larger than that of an arithmetic gradient.

Exhibit 4.1   Geometric and arithmetic salary increases

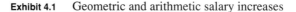
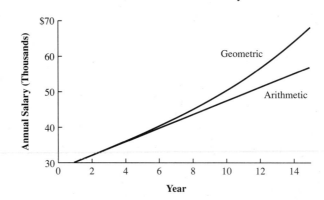

For negative gradients, we have the example of a $500,000 market share that is decreasing at 15% per year. If the initial drop of $75,000 is modeled arithmetically, then the market share is $50,000 by the end of year 7. But if modeled geometrically, the year 7 demand is $188,570 (see Exhibit 4.2).

Exhibit 4.2   Geometric and arithmetic market share

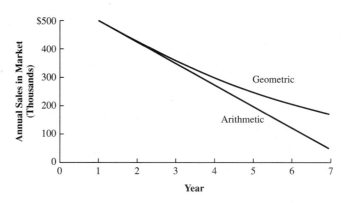

These differences between arithmetic and geometric gradients are greater for larger rates of change and longer time periods. For a short time interval at a low percentage change, it may not matter which model we use. In other cases, the geometric model will be more accurate, and the simpler arithmetic model might be misleading.

The Basic Geometric Formula.   To calculate the values that are graphed in Exhibits 4.1 and 4.2, we rely on the basic compound interest equation. The only change from the $(F/P, i, N)$ factor is that the value at the end of the first year is the base. Thus, the $30,000 salary or $500,000 market share is assumed to occur *at the end of the first year,* rather than at time 0.

The formula for the value in year t is Equation 4.1, where g is the rate of the geometric gradient. This formula is easy to use, but restating it recursively makes the constant rate of change even clearer (see Equation 4.2). This recursive format is easier to use in manual and computerized tables.

$$F_t = P(1 + g)^{(t-1)} \qquad (4.1)$$

$$F_t = F_{t-1}(1 + g) \qquad (4.2)$$

For example, applying the recursive model to the engineer's salary leads to $F_1 = \$30,000$, $F_2 = F_1 \cdot 1.06 = \$31,800$, and $F_3 = F_2 \cdot 1.06 = \$33,708$. The raise for year 2 is $1800, and for year 3 it is $1908. This also illustrates why we need geometric gradients. When the raise for the second year is given, it will be based on many factors (the engineer's performance, the firm's profitability, the rate of inflation, and the current market for that engineering specialty). However, it will almost certainly be thought of by both the firm and the engineer as a percentage increase over the current salary. This is a geometric gradient, and *it is compound growth.*

Growth at a constant rate is often the best model of future product sales, of traffic on a new highway, or of the population served by a public project. Example 4.1 illustrates this growth.

Example 4.1   Sales of Personal Computers

Given the following data, calculate the average annual rate of increase in the number of personal computers sold in the U.S. between 1980 and 1991.

Personal Computer Sales

Year	Sales (Millions)	Growth over Previous Year
1981	1.11[1]	NA
1982	3.53[1]	218.0%
1983	6.90[1]	95.5
1984	7.61[1]	10.3
1985	6.75[1]	−11.3
1986	7.04[1]	4.3
1987	8.34[1]	18.5
1988	9.50[1]	13.9
1989		−.8
1990	9.34[2]	−.8
1991	12.01[2]	28.6

[1] 1992 *Statistical Abstract of the United States,* Table 1247.
[2] 1990 and 1991 *Current Industrial Reports,* Department of Commerce. Note series shift means measurement difference is possible cause of drop in volume.

Solution
This cannot be solved by averaging the percentages in the third column. That wrong answer would be 37.6%. Rather, the formula for a geometric rate of increase must be used: $F_t = P(1 + g)^{(t-1)}$. In this case, $t = 11$, and $t - 1$ is a 10-year period (*recalling that the first value is assumed to be at the end of year one*), and the unknown is g.

$$(1 + g)^{10} = \text{Sales}_{1991}/\text{Sales}_{1981} = 12.01/1.11$$

$$g = 10.82^{1/10} - 1 = 26.9\%.$$

4.2 CHANGING VOLUME AND OTHER GEOMETRIC GRADIENTS

The Four Geometric Gradients.   Future costs and revenues are often estimated as changing at the same *rate* as in the past. Common examples include the cost of supplying more power to a growing city, the labor cost to run an assembly line, and the cost to repay a loan during an inflationary period. Respectively, these are geometric gradients for the amount of power, the labor's unit cost, and the dollar's value. To compare two cash flows, we have to find present or annual worths using compound interest—a fourth geometric gradient.

This chapter covers two of these geometric gradients—those for changes in volume and for compound interest.

$g \equiv$ rate of change in the volume of the item

$i \equiv$ interest rate for money's time value

Exhibit 4.2 illustrated a constant rate of change in market demand. Example 4.1 illustrated, for personal computers, an average rate of increase in the volume sold. For projecting into the future, a constant rate of change is the normal assumption.

The two geometric gradients linked to inflation—those for an item's price and for the value of a dollar—are detailed in Chapter 15, with a brief overview here. The salary increase for the engineer in Exhibit 4.1 is an example of increases in the dollars per unit; in this case, dollars per year of work. To calculate the purchasing power of the engineer's salary, an estimate of inflation is also needed.

Inflation is a decrease in the value of the dollar (or other monetary unit); or, equivalently, it is increases in the prices of purchased items.

Inflation most often refers to the economy as a whole, but it also applies to the price of a single item. If the price of an item is inflating, then more dollars (pesos, yen, etc.) will be needed to buy it in the future than now. If an economy is inflationary, then the prices of most or all items are increasing. Deflation is rarer, but it does occur; this term applies when prices are falling rather than increasing, as with inflation.

Mathematical Model.   Obviously, a single problem could involve all four sources of geometric gradients. However, until Chapter 15 inflation will be handled by assuming that all cost estimates are stated in inflation-adjusted terms, rather than including geometric gradients for an item's price and the buying power of a dollar.

The remaining two geometric gradients (for volume changes and for the time value of money) can be expressed mathematically as shown in Exhibit 4.3. Notice that Exhibit 4.3 is based on Equation 4.1, since it does not include years 3 to $t-1$. For many problems, every year will be included, and it may be easier to use the recursive formula, Equation 4.2.

Exhibit 4.3 Mathematical model of volume change and present worth

Period	Base Amount	g % Volume Change	i PW Factor	Present Worth (PW)
1	100	1.0	$1/(1+i)^1$	$100/(1+i)$
2	100	$1+g$	$\dfrac{1}{(1+i)^2}$	$\dfrac{100(1+g)}{(1+i)^2}$
t	100	$(1+g)^{(t-1)}$	$\dfrac{1}{(1+i)^t}$	$\dfrac{100(1+g)^{(t-1)}}{(1+i)^t}$

Inflation Assumptions.   When the inflation rate is 5%, most goods cost 5% more each year. Even though a particular item might have a different inflation rate, the best possible forecast may still be 5% per year. In this case, costs are typically stated and used as inflation-adjusted values.

For example, if a salvage value is stated as 10% of the first cost, or if the operating cost of a chemical plant is forecast to be $15M per year for 10 years, then these costs have been adjusted for inflation; the values can be used without further adjustment in our analysis. We need only ensure that the interest rate is for a real rate of return and that the interest rate does not include an allowance for inflation. This text's approach matches a common practice of estimating costs in constant-value terms.

Suppose those same operating costs are described as $15M the first year, with yearly increases of 3%. Then the reason for the 3% rate of increase must be examined. If it is because the equipment is operating at a 3% greater volume each year or because the equipment needs 3% more maintenance each year as it gets older, then maintenance has a volume-related geometric gradient (covered in this chapter). If the 3% increase is due to inflation, then the more complex analysis presented in Chapter 15 is needed to consider both the inflation rate in the operating costs and the inflation rate in the economy.

4.3 EXAMPLES OF GEOMETRIC GRADIENTS

Problems can have one or more geometric gradients for volume changes. The production volume may increase by 10% per year, but productivity improvements of 6% per year may keep total labor-hour increases to 4% per year. Meanwhile, conservation measures may cause energy use to fall by 2% per year.

Example 4.2 illustrates the method for finding the present worth (PW) for a single volume change. Example 4.3 reflects multiple volume changes in the context of finding the PW of a new product. Example 4.4 illustrates finding the equivalent annual worth (EAW) of a new public park.

Example 4.2   The Value of Energy Conservation

Susan works in the electrical and mechanical engineering department of MidWestTech. She is evaluating whether a more efficient motor should be installed on one of the assembly lines. The line is producing (and operating) 8% more each year. The more efficient motor has no salvage value after a life of 5 years; it costs $3000 more installed; and it will save $800 the first year. MidWestTech uses an i of 10%.

Solution
The first step is to calculate the cash flows. At year 0, the $3000 cost is incurred. In year 1, there are savings of $800. These savings increase by 8% per year for the next 4 years. Using Equation 4.2, the savings in year 2, F_2, equal $800 · 1.08 or $864. Similarly, the savings in year 3, F_3, equal $864 · 1.08 or $933. The values are summarized in the following cash flow diagram:

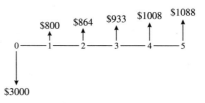

The calculations of the PW are summarized in Exhibit 4.4. This exhibit shows that the energy savings increase at 8% per year as the production volume increases. However, the exhibit does use four gradient multiplications for the volume and five (P/F) factors to find the PW.

Exhibit 4.4   PW of geometrically increasing energy savings

Year	Cost or Energy Savings at 8%	PW Factor at 10%	PW
0	− $3000.00	1.0000	− $3000.00
1	800.00	.9091	727.27
2	864.00	.8264	714.05
3	933.12	.7513	701.07
4	1007.77	.6830	688.32
5	1088.39	.6209	675.81
		Total	$506.51

Example 4.3   MidWestTech's New Product

George works in the sales engineering department of MidWestTech. He is evaluating a proposed new product. Final design and assembly-line tooling and setup will cost $150,000. Sales are expected to begin at 10,000 units the first year with 12% annual growth for 5 years. The selling price will be $6 per unit. The production cost will begin at $2 per unit the first year, with a 2% annual decline due to productivity increases. The product should have a life of 5 years. MidWestTech uses an i of 10%. What is the product's PW?

Solution
In this case, two calculations of geometric growth must be made along with the geometric PW factors. The first growth factor is at 12% per year for the number of units produced. For example, Volume$_3$ (the volume for year 3) shown in Exhibit 4.5 equals $1.12 \cdot$ Volume$_2$ = 12,544. The revenues for each year are $6 times the volume.

The calculation of the production costs is more complicated. A column for the cost per unit could be added to Exhibit 4.5; then that column multiplied by the volume would be the annual cost. Instead, the 2% annual decline in unit production costs is combined with the 12% annual increase in volume. In this case, $1 + g$ equals $(1.12)(.98)$. Rather than solving for the value of g ($g = 9.76\%$), it is easier to simply use the 1.12 and .98 in any formulas. Thus, Costs$_2$ in Exhibit 4.5 equals Costs$_1 \cdot 1.12 \cdot .98 = -\$21,952$. The cash flow for years 1 to 5 is simply the sum of the sales and the costs. Exhibit 4.5 summarizes the calculations of the PW.

Exhibit 4.5   PW of multiple geometric gradients

Year	Volume (+12%/year)	Sales ($6/unit)	Costs (−2%/year)	Cash Flow	PW Factor at 10%	PW
0				−$150,000	1.0000	−$150,000
1	10,000	$60,000	−$20,000	40,000	.9091	36,364
2	11,200	67,200	−21,952	45,248	.8264	37,395
3	12,544	75,264	−24,095	51,169	.7513	38,444
4	14,049	84,296	−26,446	57,850	.6830	39,512
5	15,735	94,411	−29,027	65,384	.6209	40,598
					Total	$42,313

Example 4.4   Annual Worth for a Park with Geometrically Increasing Use

MetroCity is growing rapidly and expanding its freeway network. It expects to use a 10-acre parcel of land for a highway connection 5 years from now. Meanwhile, the parcel could be developed as a set of playing fields with parking. The spring soccer leagues and the fall football leagues all need additional room.

Find the PW and the equivalent annuity if MetroCity uses an i of 6%. The development cost is $40,000, and the annual operating cost is $3,000. The soccer fields would serve 1500 people in the first year, with annual increases of 20%. The football fields would serve 2000 people in the first year, with annual increases of 3%. Assume each user receives a benefit of $3.

Solution

The two sets of users for the playing fields must be calculated separately, as shown in Exhibit 4.6. For example, Soccer$_3$ equals Soccer$_2 \cdot 1.2 = 1800 \cdot 1.2 = 2160$. The number of soccer and football users are added together for each year, multiplied by $3, and combined with the $3,000 operating cost to calculate the annual benefit or cash flow from having the fields. For example, the cash flow for the first year is $(3500 \cdot \$3) - \$3,000$ or $7,500.

The total PW is multiplied by $(A/P, 6\%, 5)$ to find the equivalent annuity of $411 (equals .2374 $\cdot$ $1732).

Exhibit 4.6   MetroCity's playing fields with multiple geometric gradients

Year	Soccer (+20%/year)	Football (+3%/year)	Cash and Benefits	PW Factor at 6%	PW
0			−$40,000	1.0000	−$40,000
1	1500	2000	7500	.9434	7075
2	1800	2060	8580	.8900	7636
3	2160	2122	9845	.8396	8266
4	2592	2185	11,332	.7921	8976
5	3110	2251	13,084	.7473	9777
				Total	$1732

4.4 SPREADSHEET TABLES BY HAND

The calculations in Examples 4.2–4.4 were tabulated using a format based on Exhibit 4.3. Such tabulations will be referred to as spreadsheets, whether tabulated by hand or by computer.

These spreadsheets were constructed by repeatedly using the formulas for geometric increases and decreases. The disadvantages of these spreadsheets are that the formulas are not easily displayed and that numerous calculations are required. The advantage is that we can see all of the factors, even for problems with a large number of years.

In these examples, the base amount has been constant, but any other pattern that fits reality can be used. For example, a manufacturer might have a jump in output when a new facility is opened or another shift of workers added. Also, rates of volume changes are assumed to be constant. Again, the tables are easy to modify to fit a product whose market grows at 20% for 3 years and then at 10% for 6 years, as in Example 4.5.

Example 4.5   Northern Machine Tools

Jose works as a design engineer for Northern Machine Tools. He is evaluating whether a new product should go through final design and into production. The cost of completing and introducing the product will be \$225K. Revenues will be \$30K higher than production costs for the first year. This net revenue will increase at 20% per year for 3 years and then at 10% per year for the last 6 years of its life. If Northern Machine Tools uses an i of 12%, should the product be developed?

Solution

The first step is to estimate the cash flows. The values are summarized in the following cash flow diagram, and the calculations of the PW are summarized in Exhibit 4.7. This exhibit clearly shows the $1 + g$ term for the geometric gradient of each year. In each case, the net revenue for year t is found by multiplying the $1 + g_t$ term by the net revenue for year $t - 1$.

Specifically, at year 0, the \$225K cost is incurred. In year 1, there are net revenues of \$30K. These net revenues increase by 20% per year for the next 3 years (years 2 through 4). For example,

$$\text{Revenue}_4 = \text{Revenue}_3 \cdot 1.2 = 43{,}200 \cdot 1.2 = \$51{,}840$$

The net revenues increase by 10% per year for the next 6 years (years 5 through 10). For example,

$$\text{Revenue}_5 = \text{Revenue}_4 \cdot 1.1 = 51{,}840 \cdot 1.1 = \$57{,}024$$

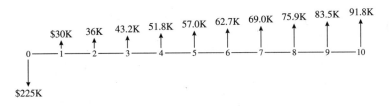

Exhibit 4.7   Northern Machine Tools

Year	Factor for Volume Change, g_t	Cost (Year 0) or Net Revenues (Years 1–10) at 20% and 10%	PW Factor at 12%	PW
0		−\$225,000	1.0000	−\$225,000
1	1.00	30,000	.8929	26,786
2	1.20	36,000	.7972	28,699
3	1.20	43,200	.7118	30,749
4	1.20	51,840	.6355	32,945
5	1.10	57,024	.5674	32,357
6	1.10	62,726	.5066	31,779
7	1.10	68,999	.4523	31,212
8	1.10	75,899	.4039	30,654
9	1.10	83,489	.3606	30,107
10	1.10	91,838	.3220	29,569
			Total	\$79,857

One important advantage of these cash flow spreadsheet models is the flexibility to *accurately* model the expected reality. The disadvantage of doing spreadsheet models by hand is that accuracy and flexibility are achieved by using simple algebraic relationships, rather than the tabulated factors, such as $(A/P, i, N)$. These simple algebraic relationships require massive numbers of calculations. As a result, it is only practical to do relatively small problems by hand.

4.5 COMPUTERIZED SPREADSHEETS

Computerized spreadsheets are available nearly everywhere, and they can easily be applied to models of geometric gradients. All of the arithmetic and algebra required by the tabular approach to geometric gradients can be done instantly by a computer. In fact, spreadsheets were originally developed to analyze the financial data of businesses, and they are often credited for fueling the explosive growth in demand for desktop computing. It is also worth noting that the simple calculations in Exhibits 4.4–4.7 can be done as mathematical formulas within the tables feature in common word processing programs.

The general structure of computerized spreadsheets mimics the tables developed in the last section, but formulas and menus for command management are also shown on the computer screen. Computerized spreadsheets bring two key distinctions: (1) the computer can automatically update the rest of the table when one entry is changed, and (2) table entries can be displayed as formulas and/or results. To display a table entry as a formula, either the cursor is moved to that entry, the spreadsheet is printed in formula format, or (for some packages, such as Microsoft Excel™) the entire spreadsheet can be displayed as formulas.

The Elements of a Computerized Spreadsheet.   A spreadsheet's two-dimensional table labels the columns in alphabetical order: A to Z, AA to AZ, BA to BZ, etc. The rows are numbered from 1 to 100 or from 1 to 100,000, depending on the capabilities of the spreadsheet package. Thus a **cell** of the spreadsheet is specified by its column letter and row number. For example, B7 is the 2nd column in the 7th row, and AA6 is the 27th column in the 6th row. This cell can contain a label, a numerical value, or a formula.

A label is any cell in which the contents should be treated as a word or words. Arithmetic cannot be performed on labels. Labels are used for row and column headings, as well as for any explanatory notes. In a spreadsheet model that is used as a personal phone book, for instance, row labels would be people's names. Also, a phone number that was entered as 333-0200 must be specified as a label; otherwise, 200 would be subtracted from 333, and a result of 133 would be displayed. Excel treats 2 + 4 or 333-0200 as a label, unless the formula begins with an equal sign ($= 2 + 4$).

Formulas can include many functions—trigonometric, financial, statistical, etc.—that are part of a spreadsheet package (and others that can be defined by the user). These formulas can use numbers, such as 1.07, or they can reference other cells, such as B7.

> A **cell** is the basic element of a computerized spreadsheet. It corresponds to an entry in a table made up of columns and rows. Cells are referenced by a letter (column) and a number (row). A cell contains a label, a numerical value, or a formula.

Defining Variables in a Spreadsheet. A basic principle of spreadsheet modeling is to use variables in your models. If your interest rate is 7%, define i as a variable whose value is .07. Then, instead of using 1.07 in a formula, use $1 + i$. That allows you to change the value of i, and to instantly recompute every value in your spreadsheet. When you are solving real problems, this is essential, as it allows you to make better decisions. Basically, your data will not be exact, and you must try other possible values. This is called sensitivity analysis (see Chapter 17). For homework problems you can use the same spreadsheet, with only small changes from one problem to the next.

To use i as a variable, choose a cell such as B7 to contain the value of i. Then, in any formula that uses i, specify the cell (B7) rather than the value of i (7%). To change i, only the value in B7 is changed; the rest of the spreadsheet is automatically recalculated.

Examples 4.6, 4.7, and 4.8 formulate Examples 4.2, 4.3, and 4.5 as spreadsheets, which are shown in Exhibits 4.8, 4.9, and 4.10, respectively. Notice that these exhibits repeat parts of the spreadsheet to break the process into steps that can be individually explained.

In Exhibit 4.8, step 1 is to label the spreadsheet in cell A1 and to build a data block. Step 2 creates the cash flows. Step 3 adds the PW. Exhibits 4.9 and 4.10 follow the same order, except that steps 1 and 2 from Exhibit 4.8 are combined into one for space reasons. Both exhibits also include extra explanatory notes about the formulas used. The next section discusses the commands that are used to build these models.

Explaining your spreadsheet formulas can be done by (1) converting the cell with the formula to a label, (2) copying that label to an adjacent cell, and (3) converting the original formula back into a formula from a label. These explanatory notes make it easier for you to remember what you've done and easier for others to find any errors.

Example 4.6   **Spreadsheet for Energy Conservation (Based on Example 4.2)**

Susan of MidWestTech is evaluating a more efficient motor for an assembly line that is operating 8% more each year. The more efficient motor has no salvage value after 5 years; it costs $3000 more installed, and it will save $800 the first year. Build a spreadsheet to find the PW at MidWestTech's i of 10%.

Solution
The first step is to define a data block for the 8% geometric gradient, the $3000 first cost, the $800 year 1 savings, and the 10% interest. The number of years, 5, determines the number of rows and is not part of the data block.

The second step is to calculate the annual savings, the present worth factors (PWF), and the total PW. Notice that the annual Savings$_t$ equals $1.08 \cdot$ Savings$_{t-1}$, since g, the geometric gradient, is 8%. The PWF are found in the same way, but the geometric gradient factor is $(1+i)^{-t}$. These are exactly the same calculations as presented previously, but now they are explained using labels within the spreadsheet.

Exhibit 4.8   Spreadsheet for geometrically increasing energy savings

	A	B	C	D	E	F
1	Exhibit 4.8 Spreadsheet for Example 4.2 & 4.6					
2						
3	Step 1: Create data block					
4	800	Base annual savings			-3000	First cost
5	8%	g			10%	interest rate
3	Step 2: Calculate cash flows					
4	800	Base annual savings			-3000	First cost
5	8%	g			10%	interest rate
6		Cash				
7	Year	Flow				
8	0	-3000	=$E$4			
9	1	800	=$A$4			
10	2	864	=B9*(1+$A$5)			
11	3	933	=B10*(1+$A$5)			
12	4	1008	=B11*(1+$A$5)			
13	5	1088	=B12*(1+$A$5)			
3	Step 3: Adds the PW factors & computes the PW					
4	800	Base annual savings			-3000	First cost
5	8%	g			10%	interest rate
6		Cash	PWF@			
7	Year	Flow	10%	PW		
8	0	-3000	1.0000	-3000		
9	1	800	0.9091	727	=B9*C9	
10	2	864	0.8264	714		
11	3	933	0.7513	701		
12	4	1008	0.6830	688		
13	5	1088	0.6209	676		
14				507	= @ SUM(D8..D13)	

Example 4.7   Spreadsheet for MidWestTech (Based on Example 4.3)

George's data for a proposed new product included a first cost of $150K, sales that begin at 10,000 units with 12% annual growth, and a selling price of $6 per unit. The production cost begins at $2 per unit, with a 2% annual decline due to productivity increases over a 5-year life. MidWestTech uses an i of 10%. Use a spreadsheet to calculate the product's PW.

Solution
In this case, there are seven different data items that will be treated as variables. The eighth variable, number of years, is used to determine the number of rows in the spreadsheet. This spreadsheet calculates the number of units for each year, and then calculates the revenue and the production costs.

In this case, the recursive formula for geometric gradients is used to calculate the volume in each year; that is, the volume for year $t - 1$ is multiplied by 1.12. If a column for the cost per unit had been created, then a similar approach could be applied to the costs. However, it is easier to use $(1 - .02)^{t-1}$ instead. Either approach could be used for the $(P/F, i, N)$ factors, although 1.1^{-N} is used here. Note that Exhibit 4.9 uses lines to connect example cells from each column to explanatory formulas.

Exhibit 4.9 Spreadsheet for new product's multiple geometric gradients

	A	B	C	D	E	F	G
1	Exhibit 4.9 Spreadsheet for Examples 4.3 & 4.7						
2							
3	Step 1: Create data block						
4	10000	Base volume			-150000	First cost	
5	12%	growth rate			$2	production cost	
6	$6	initial selling price			-2%	g	
7	10%	interest rate					
8	Year	Volume					
9	0	0					
10	1	10000	=$A$4				
11	2	11200	=B10*(1+$A$5)				
12	3	12544	=B11*(1+$A$5)				
13	4	14049	=B12*(1+$A$5)				
14	5	15735	=B13*(1+$A$5)				

	A	B	C	D	E	F	G
3	Step 2: Create cash flows						
4	10000	Base volume			-150000	First cost	
5	12%	growth rate			$2	production cost	
6	$6	initial selling price			-2%	g	
7	10%	interest rate			Cash		
8	Year	Volume	Sales	Costs	Flow	(P/F,i,N)	PW
9	0	0		-150000	-150000	1.0000	-150000
10	1	10000	60000	-20000	40000	0.9091	36364
11	2	11200	67200	-21952	45248	0.8264	37395
12	3	12544	75264	-24095	51169	0.7513	38444
13	4	14049	84296	-26446	57850	0.6830	39512
14	5	15735	94411	-29027	65384	0.6209	40598

42313

=C14+D14

= @ SUM(G9..G14)

=-E5*B14*(1+$E$6)^(A14-1)

=B14*$A$6

=1/(1+$A$7)^A14

Example 4.8 **Spreadsheet for Northern Machine Tools (Based on Example 4.5)**

Jose of Northern Machine Tools is evaluating a new product. The cost of completing and introducing the product will be $225K. Revenues will be $30K higher than production costs in year 1, and will increase by 20% per year for 3 years and then by 10% per year for 6 years. Use a spreadsheet to find the product's PW at Northern Machine Tools' i of 12%.

Solution
The first step is to build a data block for the five variables. These are the $225K first cost, the $30K in initial net revenues, the 20% geometric gradient for years 2–4, the 10% geometric gradient for years 5–10, and the 12% interest rate. The years will be used to define the rows, rather than being defined in the data block.

The calculations of the PW are summarized in Exhibit 4.10. This exhibit clearly shows the $1 + g$ term for the geometric gradient of each year. In each case the net revenue for year t is found by multiplying the $1 + g_t$ term by the net revenue for year $t - 1$. As shown by cell C11, which equals C10 · B11, the recursive Equation 4.2 is used to build the geometric gradient. Example 4A.3 (in Appendix 4A) illustrates the difficulties in using the nonrecursive Equation 4.1.

Exhibit 4.10   Spreadsheet for Northern Machine Tools

	A	B	C	D	E	F
1	Exhibit 4.10 Spreadsheet for Examples 4.5 & 4.8					
2						
3	Step 1: Create data block and cash flows					
4	30000	Base revenue			-225000	First cost
5	20%	growth rate 1			12%	interest rate
6	10%	growth rate 2				
7		Volume	Cash			
8	Year	Factor	Flow			
9	0		-225000	=$E$4		
10	1	1	30000	=$A$4		
11	2	1.2	36000	=C10*B11		
12	3	1.2	43200		=1+$A$5	
13	4	1.2	51840			
14	5	1.1	57024		=1+$A$6	
15	6	1.1	62726			
16	7	1.1	68999			
17	8	1.1	75899			
18	9	1.1	83489			
19	10	1.1	91838			
3	Step 2: Calculate PW					
4	30000	Base revenue			-225000	First cost
5	20%	growth rate 1			12%	interest rate
6	10%	growth rate 2				
7		Volume	Cash			
8	Year	Factor	Flow	PWF	PW	
9	0		-225000	1.0000	-225000	=C9*D9
10	1	1	30000	0.8929	26786	=D9/(1+$E$5)
11	2	1.2	36000	0.7972	28699	
12	3	1.2	43200	0.7118	30749	
13	4	1.2	51840	0.6355	32945	
14	5	1.1	57024	0.5674	32357	
15	6	1.1	62726	0.5066	31779	
16	7	1.1	68999	0.4523	31212	
17	8	1.1	75899	0.4039	30654	
18	9	1.1	83489	0.3606	30107	
19	10	1.1	91838	0.3220	29569	
20					79857	= @ SUM(E9..E19)

4.6 CREATING A SPREADSHEET MODEL

To explain how computerized spreadsheet models are created, this text will present Lotus 1-2-3®, Quattro Pro™, and Excel™ commands. This discussion focuses on the basics, since updated versions are released frequently. The financial functions of all three packages will be discussed, but the menu tree of only Lotus 1-2-3 for DOS will be presented in examples. The more intuitive, mouse-driven menu trees in versions for Macintosh™ computers and the Windows™ interface need less explanation.

Entering Labels and Formulas, and Using Basic Commands. A display from Lotus 1-2-3 is shown in Exhibit 4.11. The cursor is located at cell B5, so that the cell, column label, and row label are highlighted on the computer screen. At the top of the screen, the contents of that cell (a formula) are shown. Below those cell contents are two rows of commands—the menu bar. The menu bar is displayed because a slash (/) has been entered. The top line of the menu bar indicates the current choices, while the second line indicates the lower level of choices linked to the command that is highlighted on the first line. A choice on the first line of the menu bar can be made by clicking the mouse button, moving the highlight with the arrow keys and pressing the RETURN key, or entering the first letter of the choice. For example, /f accesses the File menu, and /c the COPY command.

Exhibit 4.11 Lotus 1-2-3 display

```
B5:  (C2) U [W10] 30000                                                    MENU
Worksheet    Range  Copy   Move   File   Print  Graph  Data  System  Add-In  Quit    —— Menu on
Global       Insert Delete Column Erase  Titles Window Status Page    Learn ⌐         second
                                                                                      line

          A           B          C          D          E         F
  1  INCOME STATEMENT 1989:  Goodwin's Sports Supply
  2                                                                          ⌐ Submenu or
  3                 Q1         Q2         Q3         Q4        YTD              description
  4  ─────────────────────────────────────────────────────────────           on third line
  5  Net Sales    $30,000.00 $38,000.00 $32,000.00 $51,000.00 $151,000.00
  6  ─────────────────────────────────────────────────────────────
  7
  8  Operating Expenses:
  9   Payroll      6,000.00   7,600.00   6,400.00  10,200.00   30,200.00
 10   Utilities    4,500.00   5,700.00   4,800.00   7,650.00   22,650.00
 11   Rent         2,000.00   2,000.00   2,000.00   2,000.00    8,000.00
 12   Ads          2,400.00   3,040.00   2,560.00   4,080.00   12,080.00
 13   COG Sold    10,500.00  13,300.00  11,200.00  17,850.00   52,850.00
 14  ─────────────────────────────────────────────────────────────
 15  Tot Op Exp   25,400.00  31,640.00  26,960.00  41,780.00  125,780.00
 16  ─────────────────────────────────────────────────────────────
 17  Op Income    $4,600.00  $6,360.00  $5,040.00  $9,220.00  $25,220.00
 18
 19
 20
TUTOR9X.WK1                      UNDO
```

For Lotus 1-2-3 and Quattro Pro, numbers, letters, cell references, and formulas can simply be typed in, then entered in the current cell by pressing RETURN or by using an arrow key to move the cursor. The first exception is a label, such as a phone number, that begins with a number or an arithmetic operator (+, −, etc.). This requires that the first character entered be a label signal (″, ′, or ^). The second exception is a formula that begins with a letter, as in a cell reference. This requires that the first character be an operator, which is usually a plus or a minus sign. For Excel, any cell that contains more than a simple number is treated as a label unless it is a formula, which must begin with an equals sign, such as $=3*4^2$ or $=B7*A3$.

Several commands are used to format the output or to modify the spreadsheet. Some of these are introduced in Exhibit 4.12, but the most important is the COPY command, which is described in the next subsection and presented in Exhibits 4.13–15.

Exhibit 4.12   Basic commands

Formatting the display

/wc	Changes the column width (note that labels flow into the next cell on the right—if it is empty)
/rl	Changes whether a label is left (′) or right (″) justified, or centered (^)
/rf	Changes how a numerical value is displayed (number of decimal places; is it a percentage?, etc.)

Changing the range

Example ranges include: A1..A5 (first five rows of first column); C2 (a single cell); and A1..D5 (first five rows of first four columns).

Esc	"Unanchors" the current cursor
.	"Anchors" the current cursor

Editing the spreadsheet

/wi	Inserts one or more blank rows or columns
/wd	Deletes one or more blank rows or columns
/m	Moves a range of values to a new location

Miscellaneous commands

/g	Opens the submenu for creating graphs
/f	Opens the submenu for retrieving and saving files
/p	Opens the submenu for printing a block of values
/r n	Gives a cell address a name (e.g., B7 = Rate)
F1	Help
F2	Edit the current cell
F4	Toggle between relative and absolute addresses
F10	Display current graph

COPY Command.   The COPY command is singled out to provide an example of a computerized spreadsheet's power, to emphasize the key command that makes spreadsheet models easy to build, and to explain the sometimes confusing topic of relative and absolute addressing.

If the range of cells to be copied contains only labels, numbers, and functions, then the COPY command is easy to use and understand. As shown in Exhibit 4.13, the result is an exact copy at a new location. However, if cell addresses are part of the range being copied, then the process is more complicated.

Exhibit 4.13   Lotus 1-2-3 example of COPY command without cell addresses

```
         A          B          C          D          E
1  Exhibit 4.13 Using the Lotus COPY Command
2
3  Step 1: create material to be copied
4     Label A    <= cell contains a label
5      123       <= cell contains the numerical value 12
6     16.4446    <= cell contains @ EXP(2.8)

   Step 2: complete the COPY command
   start the command                /c
   enter from range             A4..16
   enter to range               D7..D9
     then cell contents are copied


         A          B          C          D          E
1  Exhibit 4.13 Using the Lotus COPY Command
2
3  Step 1: create material to be copied
4     Label A    <= cell contains a label
5      123       <= cell contains the numerical value 12
6     16.4446    <= cell contains @ EXP(2.8)
7                                      Label A
8                                       123
9                                      16.4446
```

An absolute address is denoted by adding dollar signs ($) before the row and/or column (e.g., $A$7). When an **absolute address** is copied, the row and/or column that is fixed by the $ sign (fixed means it does not change) is copied exactly. Thus, $A$7 is completely fixed, $A7 fixes the column, and A$7 fixes the row. One common use for absolute addresses is any data-block entry, such as the interest rate.

In contrast, a **relative address** is best interpreted as directions from one cell to another. For example, suppose cell C4 contains +A1*B1. Cell A1 is three rows higher and two columns to the left of cell C4, so the formula is really: "contents of (3 up, 2 to the left) · contents of (3 up, 1 to the left)." When a cell containing a relative address is copied to a new location, it is these directions that are copied to determine any new relative addresses.

An **absolute address** is a column and/or row for a referenced cell that is fixed so that it does not change when the cell is copied to another location.

A **relative address** is a column and/or row for a referenced cell that is not fixed at that column or row.

To create a table of present worth factors, an absolute address is used for the interest rate and a relative address for the number of years. The formulas for $(P/F, i, N)$, $(P/A, i, N)$, and $(P/G, i, N)$ are entered in the cells for the first year and then copied for all of the other years (see Exhibit 4.14).

Exhibit 4.14   Copying using relative and absolute addresses for present value factors

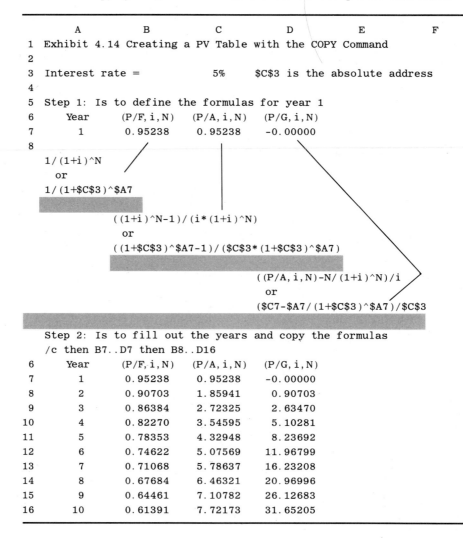

	A	B	C	D	E	F
1	Exhibit 4.14 Creating a PV Table with the COPY Command					
2						
3	Interest rate =		5%	$C$3 is the absolute address		
4						
5	Step 1: Is to define the formulas for year 1					
6	Year	(P/F,i,N)	(P/A,i,N)	(P/G,i,N)		
7	1	0.95238	0.95238	-0.00000		

1/(1+i)^N
  or
1/(1+$C$3)^$A7

((1+i)^N-1)/(i*(1+i)^N)
  or
((1+$C$3)^$A7-1)/($C$3*(1+$C$3)^$A7)

((P/A,i,N)-N/(1+i)^N)/i
  or
($C7-$A7/(1+$C$3)^$A7)/$C$3

Step 2: Is to fill out the years and copy the formulas
/c then B7..D7 then B8..D16

	Year	(P/F,i,N)	(P/A,i,N)	(P/G,i,N)
6	Year	(P/F,i,N)	(P/A,i,N)	(P/G,i,N)
7	1	0.95238	0.95238	-0.00000
8	2	0.90703	1.85941	0.90703
9	3	0.86384	2.72325	2.63470
10	4	0.82270	3.54595	5.10281
11	5	0.78353	4.32948	8.23692
12	6	0.74622	5.07569	11.96799
13	7	0.71068	5.78637	16.23208
14	8	0.67684	6.46321	20.96996
15	9	0.64461	7.10782	26.12683
16	10	0.61391	7.72173	31.65205

An **amortization schedule** divides each scheduled payment into principal and interest and includes the outstanding balance for each period.

Similarly, to calculate a loan repayment schedule, or **amortization schedule,** as in Exhibit 4.15, the first two payment years are created and then the second is copied for the remaining years.

Exhibit 4.15   Copying using relative and absolute addresses for a loan repayment schedule

	A	B	C	D	E	F
1	Exhibit 4.15 Loan Repayment Schedule Using the COPY Command					
2						
3	Interest rate =		5%			
4	Amount borrowed =		$10,000			
5	Number of periods =		10			
6	Annual payment =		$1,295.05	=$10,000*(A/P,.05,10)		
7				= @ PMT($C$4,$C$3,$C$5)		
8						
9	Step 1: Is to define the formulas for years 1 and 2					
10		Beginning	Interest	Principal	Ending	
11	Year	Balance	for Year	for Year	Balance	
12	1	10000.00	500.00	795.00	9204.95	
13	2	9204.95				
14						
15	+$C$4	+E12	+$C$3*B12	+$C$6-C12	+B12-D12	
16						

Step 2: Is to fill out the years and copy the formulas
/c then C12..E12 then C13..E13
/c then B13..E13 then B14..E21

	Year	Beginning Balance	Interest for Year	Principal for Year	Ending Balance
10		Beginning	Interest	Principal	Ending
11	Year	Balance	for Year	for Year	Balance
12	1	10000.00	500.00	795.05	9204.95
13	2	9204.95	460.25	834.80	8370.16
14	3	8370.16	418.51	876.54	7493.62
15	4	7493.62	374.68	920.36	6573.25
16	5	6573.25	328.66	966.38	5606.87
17	6	5606.87	280.34	1014.70	4592.17
18	7	4592.17	229.61	1065.44	3526.73
19	8	3526.73	176.34	1118.71	2408.02
20	9	2408.02	120.40	1174.64	1233.38
21	10	1233.38	61.67	1233.38	0.00

4.7 USING SPREADSHEETS FOR ECONOMIC ANALYSIS

Spreadsheets have been introduced with geometric gradients because of the associated heavy computational load. Obviously, spreadsheets can be used for other kinds of cash flows. In fact, there are many who claim that spreadsheets are essential to the practice of engineering economy.

The main advantages of spreadsheets are:

1. They make it easy to construct year-by-year cash flow tables.
2. They allow easy "what-if" calculations for different circumstances.

3. The available graphics are useful for analysis and for convincing presentations.
4. They include annuity functions that are more powerful than the tabulated engineering economy factors.

Spreadsheet Annuity Functions.   In the tabulated engineering economy factors that are in Appendix A, i is the table; N is the row; and then two of P, F, A, and G define a column. The corresponding spreadsheet annuity functions for P, F, and A are shown in Exhibit 4.16. (See Problem 4.32 for G functions [Eschenbach].) Exhibit 4.17 substitutes 0's and 1's in the functions of Exhibit 4.16 to state the equivalents of the engineering economy factors.

Exhibit 4.16   Summary of spreadsheet annuity or investment functions

Excel	Lotus 1-2-3[1]	Quattro Pro[2]
$PV(i,N,-A,-F,\text{Type})$	$@PV(A,i,N)$	$@PVAL(i,N,-A,-F,\text{Type})$
$PMT(i,N,-P,-F,\text{Type})$	$@PMT(P,i,N)$	$@PAYMT(i,N,-P,-F,\text{Type})$
$FV(i,N,-A,-P,\text{Type})$	$@FV(A,i,N)$	$@FVAL(i,N,-A,-P,\text{Type})$
$NPER(i,A,P,F,\text{Type})$	$@TERM(A,i,F)^{v2.0}$	$@NPER(i,A,P,F,\text{Type})$
	$@CTERM(i,F,P)^{v2.0}$	
$RATE(N,A,P,F,\text{Type},\text{guess})$	$@RATE(F,P,N)^{v2.0}$	$@IRATE(N,A,P,F,\text{Type})$

[1]No functions added in v3.0. In v4.0, functions with the names listed under Quattro Pro were added; however, the parameters were in a different order.
[2]Lotus 1-2-3 functions also work.

Exhibit 4.17   Converting factors to functions

ANSI Factor	Excel	Lotus 1-2-3[1]	Quattro Pro
$(P/F,i,N)$	$PV(i,N,0,-1)$	[2]	$@PVAL(i,N,0,-1)$
$(F/P,i,N)$	$FV(i,N,0,-1)$	[3]	$@FVAL(i,N,0,-1)$
$(P/A,i,N)$	$PV(i,N,-1)$	$@PV(1,i,N)$	$@PV(1,i,N)^{4}$
$(F/A,i,N)$	$FV(i,N,-1)$	$@FV(1,i,N)$	$@FV(1,i,N)^{5}$
$(A/P,i,N)$	$PMT(i,N,-1)$	$1/@PV(1,i,N)$	$@PAYMT(i,N,-1)$
$(A/F,i,N)$	$PMT(i,N,0,-1)$	$1/@FV(1,i,N)$	$@PAYMT(i,N,0,-1)$

[1]For versions 1.0, 2.0, and 3.0 of Lotus 1-2-3, the $(P/F,i,N)$ and $(F/P,i,N)$ factors are best found algebraically. For version 4.0, the functions listed for Quattro Pro can be used, although the parameter order is different.
[2] $(1+i)^{-N}$ rather than $@PV(1,i,N) - @PV(1,i,N-1)$
[3] $(1+i)^{N}$ rather than $1/[@PV(1,i,N) - @PV(1,i,N-1)]$
[4]Can also use $@PVAL(i,N,-1)$
[5]Can also use $@FVAL(i,N,-1)$

The functions in Exhibit 4.16 have existed from early versions of Excel and Quattro Pro (through v5.0 as of mid-1994). Because Lotus 1-2-3 provided these function in later versions only, Exhibit 4.16 details their evolution. This detail is provided since many campuses and students do not have the latest software release. Also the earlier releases are powerful enough for the homework in this text, and those earlier releases run on less expensive hardware.

Exhibits 4.16 and 4.17 show that the annuity functions for Excel and Quattro Pro are considerably more powerful than those for versions 1, 2, and 3 of Lotus 1-2-3. The important differences include the ability:

1. To calculate the internal rate of return with an A.
2. To convert between P and F.
3. To include all three of P, A, *and* F as nonzero values, rather than assuming that one equals 0.
4. To switch between a default end-of-period assumption and a beginning-of-period assumption, by specifying a *Type,* which is an optional argument to pick beginning-of-period rather than the default end-of-period.

These more powerful functions have the same sign conventions and the same power as the functions built into many business calculators (such as those made by Hewlett-Packard). The sign convention is to calculate the value that leads to a present worth of 0. Thus, the present worth or present value PV (Excel) or @PV (Quattro Pro) of $200 per period for 10 periods is negative.

Spreadsheet Block Functions.   Cash flows can be specified period by period. These cash flows are analyzed by block functions that identify the row or column entries for which a present worth or an internal rate of return should be calculated. In Excel's documentation, these cells are called *values,* and the functions are NPV(i,values) and IRR(values,guess). In the other two packages the term *range* is used, and the general format is @NPV(i,range of cells) or @IRR(i,range of cells), as illustrated below.

For cash flows involving only P, F, and A, this block approach seems to be inferior to the annuity functions. However, this is a conceptually easy approach for more complicated cash flows, such as the geometric gradients presented in this chapter. Suppose the years (row 1) and the cash flows (row 2) are specified in columns B through E.

	A	B	C	D	E
1	Year	1	2	3	4
2	**Cash flow**	$6000	8000	9000	5000

If an interest rate of .08 is assumed, then the present worth of the cash flows can be calculated as follows: Excel's function is NPV(.08, B2:E2). For Lotus and Quattro Pro the function is @NPV(.08, B2..E2).

If the investment cost were inserted as period 0, then the internal rate of return could be calculated using IRR(values,guess) or @IRR(guess, range). It is worth emphasizing that the two common block functions for present worth and internal rate of return make different assumptions about the range of years included. @NPV(i, range) and NPV(i, values) assume year 0 is *not* included, while @IRR(guess, range) and IRR(values, guess) assume year 0 is included. These functions require equally spaced cash flows. The cash flows for periods 1 to N are assumed to be end-of-period flows.

For example, consider Exhibit 4.18, which replaces the columns C and D of Exhibit 4.8 with a formula calculation of net present value NPV. Not only does this simplify the calculations, this function lets us create a small table to analyze what-if the interest rate is different.

Exhibit 4.18   Example of a financial function (based on Exhibit 4.8)

	A	B	C	D	E	F
1	Exhibit 4.18 Spreadsheet for Examples 4.2 & 4.6					
2						
3	Step 1: Create data block and cash flows					
4	800	Base annual savings			-3000	First cost
5	8%	g			10%	interest rate
6		Cash				
7	Year	Flow				
8	0	-3000	=$E$4			
9	1	800	=$A$4			
10	2	864	=B9*(1+$A$5)			
11	3	933	=B10*(1+$A$5)			
12	4	1008	=B11*(1+$A$5)			
13	5	1088	=B12*(1+$A$5)			

	A	B	C	D	E	F
3	Step 2: Uses the annuity function to find the PW					
4	800	Base annual savings			-3000	First cost
5	8%	g			10%	interest rate
6		Cash				
7	Year	Flow		i	PW	
8	0	-3000		0%	1693	=$B$8+
9	1	800		5%	1034	@ NPV (D8, $B$9..$B$13)
10	2	864		10%	507	
11	3	933		15%	80	
12	4	1008		20%	-270	
13	5	1088				
14		506.51	=$B$8+ @ NPV ($E$5, $B$9..$B$13)			

Creating Graphs.   As an example of how easy it is to create graphs, Exhibit 4.19 contains the commands to create the graph of the what-if data on the interest rate in Exhibit 4.18.

Exhibit 4.19   Example graph, with spreadsheet commands to produce it

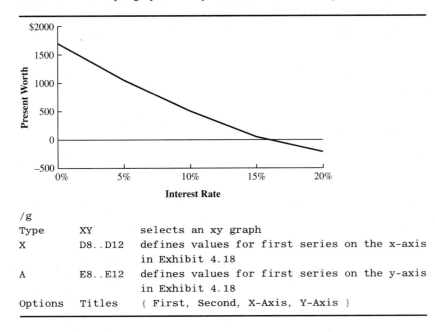

```
/g
Type       XY          selects an xy graph
X          D8..D12     defines values for first series on the x-axis
                       in Exhibit 4.18
A          E8..E12     defines values for first series on the y-axis
                       in Exhibit 4.18
Options    Titles      { First, Second, X-Axis, Y-Axis }
```

4.8 SUMMARY

This chapter has focused on spreadsheets and geometric gradients, which are cash flows modeled by a constant rate of change. Geometric gradients due to changes in a physical volume, such as the number of items manufactured and sold each year or the amount of electicity generated each year, and due to the time value of money were covered. Geometric gradients due to inflation are covered in Chapter 15.

Whether a particular problem has one or more geometric gradients, and whether they influence one or several sources of cash flows (energy costs, labor costs, loan repayments, etc.), geometric gradients are arithmetically cumbersome. Thus, this chapter has introduced spreadsheets, a tabular approach that is most efficient when computerized.

Computerized spreadsheets are useful for many engineering economy problems. First, they permit flexible models whose cash flows are estimated for each year. Second, they can be updated easily to analyze new estimated values for the data. Third, their graphs are useful in the approval process for new projects.

Simple spreadsheets can be calculated by hand, but this approach is more work than the equations and factors introduced in Chapters 2 and 3, so it is only occasionally useful. However, the power of computerized spreadsheets makes many problems easier to solve. That power is almost required for after-tax problems that use the current tax code (Chapter 13), and it is essential for others, such as geometric gradients (this chapter) and sensitivity analysis (Chapter 17).

REFERENCE

Eschenbach, Ted G., "Technical Note: Using Spreadsheet Functions to Compute Arithmetic Gradients," *The Engineering Economist,* Volume 39, Number 3, Spring 1994, pp. 275–280.

PROBLEMS

4.1   Sam receives a starting salary offer of $30,000 for year 1. If he expects a 4% raise each year, what is his salary for year 10? Year 20? Year 30? Year 40?   (*Answer:* $F_{30} = \$93,560$)

4.2   Assume that Sam (of Problem 4.1) receives an annual raise of $1200. Compared with Problem 4.1, how much less is his salary in year 10? Year 20? Year 30? Year 40?

4.3   Net revenues at an older manufacturing plant will be $2M for this year. The profitability of the work will decrease 15% per year for 5 years, until the assembly plant will be closed (at the end of year 6). If the firm's interest rate is 10%, calculate the PW of the revenue stream.

4.4   What is the present worth of cash flows that begin at $10,000 and increase at 8% per year, when N is 4 years? The interest rate is 6%. (*Answer:* $38,817)

4.5   What is the present worth of cash flows that begin at $30,000 and decrease by 15% per year ($N = 6$ years)? The interest rate is 10%.

4.6   The ABC Block Company anticipates receiving $4000 next year from its investments, with increases of 5% per year. N equals 5 years. If ABC's interest rate is 8%, determine the present worth of the cash flows.

4.7   Felix Jones, a recent engineering graduate, expects a starting salary of $35,000 per year. His future employer has averaged 5% per year in salary increases for the last several years. What is the PW and equivalent annuity for Felix's salary over the next 5 years? Felix uses an interest rate of 6%.   (*Answer:* PW = $162,009)

4.8   The cost of garbage pickup in Green Valley is $4.5 million in year 1. Estimate the cost each year for the first 5 years. The population are increasing at 6% and the tons per resident are increasing at a rate of 1%.

4.9   If the residents of Green Valley (of Problem 4.8) reduce their individual trash generation by 2% per year, how much does the cost in each year decrease?   (*Answer:* Year 3 savings = $.30M)

4.10   Redo Problem 4.8 for $N = 10$. For $N = 25$.

4.11   Redo Problem 4.9 for $N = 10$. For $N = 25$.

4.12   Calculate and print out an amortization schedule for a used car loan. The nominal interest rate is 12% per year, compounded monthly. Payments are made monthly for 3 years. The original loan is for $11,000. (*Answer:* Monthly payment = $365.36)

4.13   Calculate and print out an amortization schedule for a new car loan. The nominal interest rate is 9% per year, compounded monthly. Payments are made monthly for 5 years. The original loan is for $17,000.

4.14   For the used car loan in Problem 4.12, graph the monthly payment:
   a.   As a function of the interest rate (from 5% to 15%).
   b.   As a function of the number of payments (from 24 to 48).

4.15   For the new car loan in Problem 4.13, graph the monthly payment:
   a.   As a function of the interest rate (from 4% to 14%).
   b.   As a function of the number of payments (from 36 to 84).

4.16   Your beginning salary is $30,000. You deposit 10% at the end of each year in a savings account that earns 6% interest. Your salary increases by 5% per year. What value does your savings account show after 40 years?   (*Answer:* $973,719)

4.17   Your beginning salary is $30,000. You deposit a fixed percentage at the end of each year in a savings account that earns 6% interest. Your salary increases by 5% per year. What percentage must you deposit to have $1M after 40 years?

4.18   The market volume for widgets is increasing by 15% per year from current profits of $200,000. Investing in a design change will allow the profit per widget to stay steady; otherwise, it will drop 3% per year. The interest rate is 10%. What is the present worth of the savings over the next 5 years? Ten years?

4.19   Bob lost his job and had to move back in with his mother. She agreed to let Bob have his old room back on the condition that he pay her $1000 rent per year, and an additional $1000 every other year to pay for her biannual jaunt to Florida. Since he is down on his luck, she will allow him to pay his rent at the end of the year. If Bob's interest rate is 15%, how much is the present cost for a 5-year contract? (Trips are in years 2 and 4.)   (*Answer:* $4680)

4.20   City University's enrollment is 12,298 full-time-equivalent students (at 12 credit hours each), with 3% annual increases. The university's cost per credit hour is now $105, which is increasing by 8% per year. State funds are decreasing by 4% per year. State funds currently pay half of the costs for City U., while tuition pays the rest. What is tuition per credit hour each year for the next 5 years?

4.21   A 30-year mortgage for $120,000 has been issued. The interest rate is 10% and payments are made monthly. Print out an amortization schedule.

4.22   A 30-year mortgage for $120,000 has been issued. The interest rate is 10% and payments are made monthly. Calculate the monthly payment for principal and interest. If an extra $50 principal payment is made at the end of month 1, and this extra amount increases by 1% per month, how soon is the mortgage paid off? (Hint: Construct an amortization schedule.)

4.23 For Problem 4.22, graph the number of years until payoff:
  a.   For an initial extra payment of $20 to $100.
  b.   For a geometric gradient of 0% to 3%.

4.24 Specialty Chemicals has a new plastic formulation that should have a market life of 10 years. First costs will be $15M. Initial raw materials costs will be $4.3M per year with a constant 3% rate of increase. Production costs for labor, energy, and facility maintenance are $1.8M per year initially, with a 2% rate of increase as the facility ages. If revenue is constant at $11M per year, what is the PW at Specialty's interest rate of 10%?   (*Answer:* $11.06M)

4.25 In Problem 4.24, how fast would revenue have to fall (or increase) for the PW to be $0?

4.26 A homeowner is considering an upgrade from a fuel-oil-based furnace to a natural gas unit. The investment in the fixed equipment, such as a new boiler, will be $2500 installed. The cost of the natural gas will average $60 per month over the year, instead of the $145 per month that the fuel oil costs. If the interest rate is 9% per year, how long will it take to recover the initial investment?

4.27 John earns a salary of $40,000 per year, and he expects to receive increases at a rate of 4% per year for the next 30 years. He is purchasing a home for $80,000 at 10.5% for 30 years (under a special veterans' preference loan, with 0% down). He expects the home to appreciate at a rate of 6% per year. He will also save 10% of his gross salary in savings certificates that earn 5% per year. Assume that his payments and deposits are made annually. What is the value of each of John's two investments at the end of the 30-year period?

4.28 In Problem 4.27, how much does the value of each investment change if the appreciation rate, rate of salary increases, fraction of salary saved, or savings account interest rate is halved? What is the relative importance of the changes?

4.29 An arithmetic gradient is to be used to approximate a geometric gradient for an engineer's salary, which is $35,000 initially. Both models assume the same raise for the second year. What is the maximum annual raise and corresponding rate of increase that result in less than a 10% difference between the two final values at the end of 5 years?

4.30 Redo Problem 4.29 for 10, 15, 20, 30, and 50 years.

4.31 Using the data developed in Problems 4.29 and 4.30, construct a graph of the maximum rate of increase vs. the life. Label the areas above and below the graphed line as "Within 10%" and "Approximation fails," respectively.

4.32 Construct a spreadsheet function for a geometric gradient that relies on existing spreadsheet annuity functions (see Equations 3.20 and 3.21):
  a.   For $(A/G, i, N)$.
  b.   For $(P/G, i, N)$.

APPENDIX 4A: FORMULAS BASED ON THE EQUIVALENT DISCOUNT RATE

This chapter introduced computerized spreadsheets for creating detailed, yet flexible models of cash flows. In particular, these models were used for problems with geometric gradients—cash flows that change at a constant rate. This appendix develops formulas and an equivalent discount rate that can sometimes replace spreadsheets. This could be described as using finesse to replace brute force.

The two approaches described here are most useful (1) if financial calculators rather than computerized spreadsheets are used, or (2) for "what-if" analysis, where manipulating equations is often easier than manipulating tables.

The first approach is the formula for a new factor: $(P/A, g, i, N)$. This formula (Equation 4A.1) is not tabulated, but it can be calculated for positive or negative geometric gradients [*The Engineering Economist*].

The second approach will be generalized in the inflation chapter (Chapter 15) for more than two geometric gradients for the same cash flow. This approach relies on the following: the factors for geometric gradients for volume change and the time value of money (and inflation) have the same form of $(1 + \text{rate})^{\text{year}}$. Thus, for some problems, these rates can be combined into an equivalent discount rate.

Present Worth Formula for a Single Geometric Gradient.   Algebra can be used to develop Equations 4A.1 and 4A.2, just as the engineering economy factors were developed using algebra in Chapter 3 (see Problem 4A.7). This formula defines g as the constant rate of geometric change for a uniform periodic cash flow; that is, the A_1 at the end of the first period. Thus, as defined earlier in this chapter, the cash flow is $(1 + g)A_1$ at the end of period 2, and $(1 + g)^{t-1} \cdot A_1$ for period t.

For $i \neq g$,

$$(P/A, g, i, N) = [1 - (1 + g)^N (1 + i)^{-N}]/(i - g) \qquad (4A.1)$$

For $i = g$,

$$(P/A, g, i, N) = N/(1 + i) \qquad (4A.2)$$

Example 4A.1 applies this formula to the data in Example 4.2. As shown in Example 4A.2, if g is greater than i then the present worth (PW) factor is still positive, since both the denominator and the numerator are negative. If g is less than i then $(P/A, g, i, N) < N$, as is $(P/A, i, N)$. If g is only slightly greater than i, then $(P/A, g, i, N)$ is close to N. Whether $(P/A, g, i, N)$ is smaller or greater than N depends on the relative importance of the difference between the two rates and the one-period delay before g starts. If $g > i$, then $(P/A, g, i, N)$ will generally exceed N. Note that an easy check for this formula is to let $g = 0$ (see Problem 4A.6).

While easy to use in Examples 4A.1 and 4A.2, this formula cannot easily be used for problems like Example 4.3 (where two gradients apply to the same cash flow). It can be applied to problems like Example 4.5, with different gradient rates over time, but, as shown in Example 4A.3, it is necessary to separately calculate the cash flows for the year when the gradient rates change.

Example 4A.1   The Value of Energy Conservation (Based on Example 4.2)

Susan works in the electrical and mechanical engineering department of MidWestTech. She is evaluating whether a more efficient motor should be installed on one of the assembly lines. The line is producing (and operating) 8% more each year. The more efficient motor has no salvage value after a life of 5 years, costs $3000 more installed, and will save $800 the first year. What is the PW if MidWestTech uses an i of 10%?

Solution
In this case, $A_1 = \$800, N = 5, g = 8\%$, and $i = 10\%$.

$$(P/A, 8\%, 10\%, 5) = [1 - (1 + .08)^5(1 + .1)^{-5}]/(.1 - .08)$$
$$= (1 - .9123)/.02$$
$$= 4.3831$$

Since $g < i$, a check on our calculations shows that 4.3831 is less than $N = 5$.

$$PW = -3000 + 800 \cdot 4.3831 = \$506.5$$

Note: This answer matches that of Example 4.2.

Example 4A.2   High Geometric Gradients

Susan (of Example 4A.1) can install the same more-efficient motor in another assembly line. This other line is producing (and operating) 25% more volume each year. The more efficient motor has no salvage value after a life of 5 years, costs $3000 more installed, and will save $800 the first year. What is the PW if MidWestTech uses an i of 10%?

Solution
In this case $A_1 = \$800, N = 5, g = 25\%$, and $i = 10\%$.

$$(P/A, 25\%, 10\%, 5) = [1 - (1 + .25)^5(1 + .1)^{-5}]/(.1 - .25)$$
$$= (1 - 1.8949)/(-.15)$$
$$= 5.9660$$

Since g is much greater than i, a check on our calculations shows that 5.9660 exceeds $N = 5$.

$$PW = -3000 + 800 \cdot 5.9660 = \$1773$$

Example 4A.3   Northern Machine Tools (Based on Example 4.5)

Jose is evaluating whether a new product should go through final design and into production, which will cost $225K. Revenues will be $30K higher than production costs for the first year. This net revenue will increase at 20% per year for 3 years and then at 10% per year for the last 6 years of its life. If Northern Machine Tools uses an i of 12%, should the product be developed?

Solution
The first step is to calculate the net revenue for year 5 (Revenue$_5$), which is the first or base year for the second gradient. This includes a 10% geometric gradient for year 5.

$$Revenue_5 = 30K \cdot 1.2^3 \cdot 1.1 = \$57,024$$

The second step is to draw the cash flow diagram. Note that only the first cost of $225K and two values for the revenue stream are shown ($30K and $57,024). The other revenue cash flows do not need to be calculated.

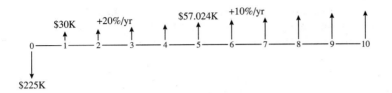

The third step is to calculate the two (P/A) factors for the geometric gradients (20% for 4 years and 10% for 6 years). Remember that both geometric gradients are 0 for their first year.

$$(P/A, 20\%, 12\%, 4) = [1 - (1 + .2)^4(1 + .12)^{-4}]/(.12 - .2)$$
$$= (1 - 1.3178)/(-.08)$$
$$= 3.9726$$
$$(P/A, 10\%, 12\%, 6) = [1 - (1 + .1)^6(1 + .12)^{-6}]/(.12 - .1)$$
$$= (1 - .8975)/.02$$
$$= 5.1236$$

$$\text{PW} = -225K + 30K(P/A, 20\%, 12\%, 4) + 57.0K(P/A, 10\%, 12\%, 6) \cdot (P/F, 12\%, 4)$$
$$= -225K + 30K \cdot 3.9726 + 57.024K \cdot 5.1236 \cdot .6355$$
$$= \$79,851$$

Within the limits of round-off error, this matches the $79,857 that was calculated in Example 4.5.

Using Equivalent Discount Rates for Geometric Gradients. Geometric gradients assume a constant rate of change, g, applied to a base amount. This assumption permits development of an equivalent discount rate, x, which is a single interest rate, that replaces both g and i. This interest rate is usually not an integer, so it is most useful with financial calculators and spreadsheets.

As above, A_1 is defined as the base periodic cash flow at the end of the first period. To find the PW, Equations 4A.3 and 4A.4 are developed from the following equation:

$$\text{PW} = A_1[1/(1+i) + (1+g)^1/(1+i)^2 + \cdots + (1+g)^{N-1}/(1+i)N]$$
$$= A_1[(1+g)/(1+i) + (1+g)^2/(1+i)^2 + \cdots + (1+g)^N/(1+i)^N]/(1+g)$$
$$= \frac{A_1}{(1+g)} \sum_{t=1}^{N} \frac{(1+g)^t}{(1+i)^t}$$

If $g < i$, let $(1 + x) = (1 + i)/(1 + g)$

$$\text{PW} = [A_1/(1 + g)](P/A, x, N) \qquad (4A.3)$$

If $g = i$,

$$PW = [A_1/(1 + g)]N \qquad (4A.4)$$

If $g > i$, where volume is increasing rapidly, then the x as calculated in Equation 4A.3 would be negative. We have no tables for negative interest rates. However, defining x inversely and forming a summation for a future worth calculation leads to a similar result. First, the limits of the summation are reduced by 1 by pulling $(1 + g)/(1 + i)$ outside of the summation. Then

$$PW = \frac{A_1}{(1 + i)} \sum_{t=0}^{N-1} \frac{(1 + g)^t}{(1 + i)^t}$$

If $g > i$, let $(1 + x) = (1 + g)/(1 + i)$.

$$PW = \frac{A_1}{(1 + i)} \sum_{t=0}^{N-1} (1 + x)^t$$

$$= [A_1/(1 + i)](F/A, x, N) \qquad (4A.5)$$

Note 1:   We are using a future worth factor to calculate a PW.

Note 2:   If $g < i$, we divide A_1 by $1 + g$, but if $g > i$, we divide A_1 by $1 + i$.

It may be less obvious than looking at the first few rows of a spreadsheet, but these equations make the same assumptions as the tables that were developed earlier. The cash flow in year 1 is the base value and all later cash flows are derived by multiplying that base value by a series of geometric gradients.

Example 4A.4   The Value of Energy Conservation (Based on Examples 4.2 and 4A.1)

Susan works in the electrical and mechanical engineering department of MidWestTech. She is evaluating whether a more efficient motor should be installed on one of the assembly lines. The line is producing (and operating) 8% more each year. The more efficient motor has no salvage value after a life of 5 years, costs $3000 more installed, and will save $800 the first year. What is the PW if MidWestTech uses an i of 10%?

Solution
In this case $A_1 = \$800, N = 5, g = 8\%,$ and $i = 10\%$. Since $g < i$, then

$$x = (1 + i)/(1 + g) - 1 = 1.1/1.08 - 1 = 1.85\%$$

$$(P/A, 1.85\%, 5) = 4.7338$$

$$PW = -3000 + 800 \cdot 4.7338/1.08 = \$506.5$$

Note: This answer matches those of Examples 4.2 and 4A.1.

Example 4A.5   High Geometric Gradients (Based on Example 4A.2)

Susan (of Example 4A.1) can install the same more-efficient motor in another assembly-line. This other line is producing (and operating) 25% more each year. The more efficient motor has no salvage value after a life of 5 years, costs $3000 extra installed, and will save $800 the first year. What is the PW if MidWestTech uses an i of 10%?

Solution
In this case $A_1 = \$800, N = 5, g = 25\%$, and $i = 10\%$. Since $g > i$, then

$$x = (1 + g)/(1 + i) - 1$$
$$= 1.25/1.1 - 1$$
$$= 13.64\%$$
$$(F/A, 13.64\%, 5) = 6.5626$$
$$PW = -3000 + 800 \cdot 6.5626/1.1 = \$1772.8$$

which matches the answer to Example 4A.2.

REFERENCE

"Report of the Engineering Economy Subcommittee (Z94.5) of the ANSI Z94 Standards Committee on Industrial Engineering Terminology," *The Engineering Economist,* Volume 33, Number 2, Winter 1988, pp. 145–151. Draft (Z94.7) expected 1995.

PROBLEMS

4A.1   Smallville is suffering annual losses of taxable properties and property values of 1% each. Even so, Smallville must maintain its tax collections at a constant value of $3.2 million. What is the required rate of increase in the tax rate? Note: While Smallville uses a rate of 6% for the time value of money, that rate is irrelevant to this problem.

4A.2   Redo Problem 4.3.
    a.   Using Equation 4A.1 or 4A.2.
    b.   Using Equation 4A.3, 4A.4, or 4A.5.

4A.3   Redo Problem 4.6.
    a.   Using Equation 4A.1 or 4A.2.
    b.   Using Equation 4A.3, 4A.4, or 4A.5.

4A.4   Redo Problem 4.7.
    a.   Using Equation 4A.1 or 4A.2.
    b.   Using Equation 4A.3, 4A.4, or 4A.5.

4A.5   Redo Problem 4.18 using Equation 4A.1 or 4A.2.

4A.6   Check Equation 4A.1 by setting $g = 0$ and comparing the result with the equation for $(P/A, i, N)$.

4A.7   Derive Equation 4A.1 for $(P/A, g, i, N)$.

PART TWO ANALYZING A PROJECT

INTRODUCTION TO SELECTION OF A MEASURE

When correctly applied, all of the following evaluation methods lead to the same result. However, the audience, principal cash flows, or situation may imply that one measure is preferred.

PW—present worth. An example is: "How much should we pay for these patent rights?" This measure is commonly preferred for purchasing decisions, for problems where first costs are particularly important, and for problems where benefits and costs are included. (Chapter 5)

EAC—equivalent annual cost. An example is: "Is it worth $20,000 per year for the added reliability of a new machine?" This measure is commonly preferred for the comparison of known costs vs. annual benefits that can only be approximated, for the evaluation of operating improvements, and for presentation to operating personnel such as foremen. (Chapter 6)

IRR—internal rate of return. Examples include: "Which loan is better?" and "Which capital projects should be done if the total budget equals $1.5M?" This measure is commonly preferred for evaluation of loans and financing, for presentation to top management, and for comparing dissimilar projects that are competing for limited funds. (Chapter 7)

B/C—benefit/cost ratio. An example is: "Should this dam be built?" Government uses this measure, although some firms use an analogous present worth index. (Chapter 8)

CHAPTER 5 PRESENT WORTH

The Situation and the Solution

Equivalence can be applied to cash flows that occur at different times, but which equivalent measure should be calculated and how should it be evaluated?

Often the solution is to calculate a time 0 equivalent value or a present worth.

Chapter Objectives

After you have read and studied the sections of this chapter, you should be able to:

Section 5.1
Define present worth, state its standard assumptions, and use it to measure the economic attractiveness of a project or an alternative.

Section 5.2
Apply present worth to a variety of examples.

Section 5.3
Calculate present worth using a rollback procedure.

Section 5.4
Include and evaluate the significance of salvage values and salvage costs.

Section 5.5
Calculate capitalized cost for perpetual life.

Section 5.6
Compare the advantages of different approaches to building projects in stages.

Section 5.7
Include the cost of underutilized capacity in a present worth model.

Section 5.8
Use spreadsheets for more accurate models built while shorter periods.

Section 5.9
Use spreadsheets for more accurate models that detail irregular cash flows.

Key Words and Concepts

Present worth (PW)   The equivalent value at time 0 of a set of cash flows.

Capitalized cost   The present worth of annual costs that are assumed to occur in perpetuity.

5.1 THE PRESENT WORTH MEASURE

Is PW > 0?   The present worth (PW) measure is easy to understand, easy to use, and matched to our intuitive understanding of money. The very name, present worth, conjures an image of current value. More formally, **present worth** is the value at time 0 that is equivalent to the cash flow series of a proposed project or alternative.

> **Present worth (PW)** is the equivalent value at time 0 of a set of cash flows.

Present worth is easy to use because of the sign convention for cash flows. Cash flows that are revenues are greater than 0, and cash flows that are costs are less than 0, so the standard for a desirable PW is PW > 0. (PW = 0 represents economic indifference and alternatives while a PW < 0 should be avoided if possible.)

Present worth is an attractive economic measure because we have an intuitive feel for the result's meaning. We have an image of what $100, $1000, or $1,000,000 is worth right now. A similar intuitive feel for future dollars is difficult to develop, as illustrated in Example 5.1.

Example 5.1   Booker Lee's Bonus

Booker Lee expects a $10,000 bonus when the product his team is developing is marketed in 4 years. He will invest in a money market fund that earns 8% per year. He plans to use the money as a down payment on a house, or to start a business, or for traveling during retirement. Assume these occur 10, 20, and 40 years, respectively, after he receives the bonus. At 8%, what equivalent values can be calculated? Which one is most meaningful? Least meaningful?

Solution
The four appropriate equivalent values are at time 0 and at the ends of years 14, 24, and 44. These are calculated as follows:

$$P_0 = 10{,}000(P/F, 8\%, 4) = 10{,}000 \cdot .7350 = \$7350$$
$$F_{14} = 10{,}000(F/P, 8\%, 10) = 10{,}000 \cdot 2.159 = \$21{,}590$$
$$F_{24} = 10{,}000(F/P, 8\%, 20) = 10{,}000 \cdot 4.661 = \$46{,}610$$
$$F_{44} = 10{,}000(F/P, 8\%, 40) = 10{,}000 \cdot 21.725 = \$217{,}250$$

Of these, the time 0 value (PW of \$7350) is the most meaningful. The year 44 value of \$217,250 is the least meaningful, simply because it is the furthest in time from "now." Because we understand what \$7350 will buy now, the PW is the one that we should use.

Because each of these values is equivalent to \$10,000 at the end of year 4, they are also equivalent to each other.

Standard Assumptions.   The present worth measure is commonly applied by making the following assumptions:

1.   Cash flows occur at the end of the period, except for first costs and pre-payments like insurance and leases.

2.   Cash flows are known, certain values. Known, certain values are *deterministic* values.

3.   The interest rate, i, is given.

4.   The problem's horizon or study period, N, is given.

These assumptions imply that you can state a problem as a cash flow diagram plus an interest rate. That is, the project or alternative can be completely described by specifying which cash flows occur at the beginning and the end of each period, along with an interest rate used for calculating equivalent values.

5.2 EXAMPLES OF WHEN TO USE PRESENT WORTH

Because it is easy to understand, the PW measure is often effective. This is especially true for: (1) <u>setting a price to buy or sell a project or an alternative</u>; (2) <u>eval-</u>

uating an investment or project where the price to invest or the first cost is given; and (3) calculating an equivalent value for an irregular series of cash flows.

Example 5.2 sets a price for buying a bond; Example 5.3 sets a price for buying a patent. Example 5.4 evaluates a project using PW. Example 5.5 calculates an equivalent PW for an irregular series of cash flows.

Example 5.2 Buying a Bond

A 15-year municipal bond was issued 5 years ago. It pays 8% annual interest on its face value of $15,000. At the end of 15 years, the face value is paid. If your time value of money, or i, is 12%, what price should you offer for the bond?

Solution
The first 5 years are past, and there are 10 more annual payments. The price you should offer is the PW of the cash flows that will be received if the bond is purchased. The annual interest is 8% of $15,000, or $1200. Thus, the cash flows are $1200 at the ends of years 5 through 15 and an additional $15,000 in year 15. In the cash flow diagram the years are labeled as the number of years from now.

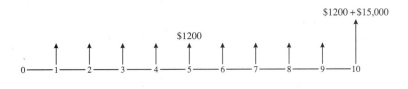

$$PW = 1200(P/A, 12\%, 10) + 15,000(P/F, 12\%, 10)$$
$$= 1200 \cdot 5.650 + 15,000 \cdot .3220 = \$11,610$$

The $11,610 is the discounted price; that is, the PW at 12% of the cash flows from the $15,000 bond. The $3390 discount raises the investment's rate of return from 8% for the face value to 12% on an investment of $11,610.

The calculation in Example 5.2 is done routinely when bonds are bought and sold during their life. Bonds are issued at a face value and a face interest rate, and the face value is received when the bonds mature. A cash flow diagram for the remaining interest payments and the final face value is used with a current interest rate to calculate a price. However, as shown in Exhibit 5.1, interest rates do vary over the 10 to 50 years between when the bonds are issued and when they mature.

Exhibit 5.1 Interest rates for industrial bonds

Example 5.3 Buying a Patent

MoreTech Drilling may buy a patent with 14 years of life left. If MoreTech spends $1.5M to implement the technology, it expects net revenues of $650,000 per year for the patent's life. If MoreTech's discount rate for money's time value is 10%, what is the maximum price that MoreTech can pay for the patent?

Solution
This maximum amount is the present worth of the cash flows to be paid and received.

$$PW = -1500K + 650K(P/A, 10\%, 14)$$

$$= -1500K + 650K \cdot 7.367 = 3289K \text{ or } \$3.3M$$

Example 5.4 MoreTech Drilling's Patent (Revisited)

Suppose that a price of $3M has been set for the patent in Example 5.3. In addition, a more accurate model of net revenues is that they begin at $300,000 per year and increase by $100,000 per year for the 14 years. At a discount rate of 10%, is this an attractive project?

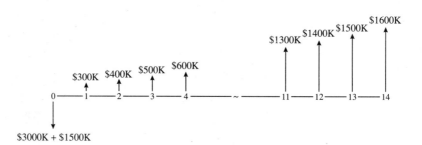

Solution
This is easily evaluated by computing the project's present worth at 10%. The cash flows
at time 0 total $4.5M ($3M purchase cost plus $1.5M for implementation).

$$PW = -4500K + 300K(P/A, .1, 14) + 100K(P/G, .1, 14)$$

$$= -4500K + 300K \cdot 7.367 + 100K \cdot 36.800$$

$$= 1390K = \$1.4M$$

Since the PW is positive at 10%, this is an attractive project.

Example 5.5 Voice Recognition for Construction Inspection

A large construction job will take 4 years to complete. The costs of the many required
inspections can be reduced by purchasing a voice recognition system to act as the "front
end" for a word processor. Then inspectors can be trained to record their comments and
revise the written output of an automated form.

The firm's interest rate is 12%. What is the PW of the cash flows for the voice recognition system? The first cost for purchase and training is $100K. Savings are estimated at
$30K, $40K, $65K, and $35K for the 4 years of construction.

Solution
The first step is to draw the cash flow diagram.

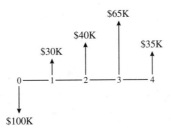

Calculating the PW must be done by considering each cash flow individually, because
there is no uniform or gradient series.

$$PW = -100K + 30K(P/F, .12, 1) + 40K(P/F, .12, 2)$$
$$+ 65K(P/F, .12, 3) + 35K(P/F, .12, 4)$$

$$= -100K + 30K \cdot .8929 + 40K \cdot .7972 + 65K \cdot .7118 + 35K \cdot .6355$$

$$= \$27.2K$$

This is an attractive investment.

5.3 ROLLING BACK IRREGULAR CASH FLOWS FOR PW CALCULATIONS

In Example 5.5, where an irregular set of cash flows was evaluated, it was necessary to look up and use a different (P/F) factor for each period. For a problem of 10, 15, or 20 periods, the chance of an arithmetic error is unacceptably high. Another approach is needed for irregular cash flows to be evaluated without a computer.

Rather than looking up N (P/F) factors, it is much easier to divide N times by $(1 + i)$. Starting with the last period, the single-period (P/F) factor, $(1 + i)^{-1}$, is used repeatedly to "roll back" to the PW. The rollback value for any period is the next period's rollback value divided by $(1 + i)$ plus that period's cash flow. Example 5.6 uses this technique for the data in Example 5.5. Example 5.7 considers the economics of graduate study in engineering.

More generally, intermediate rollback values for periods t through N, V_t, are calculated at the ends of each period. The formula for the value from t to N is Equation 5.1. V_N equals the cash flow (CF) in period N, which is CF_N. The PW is the value from periods 0 to N.

$$\text{\Large ✳} \quad V_t = V_{t+1}/(1 + i) + CF_t \tag{5.1}$$

This approach is simplified by storing $1 + i$ in the calculator's memory and then recalling it. This calculation is so easy that it can be done in some word processing packages using the math functions included as a feature for constructing tables.

Example 5.6   Voice Recognition (Revisited) (Based on Example 5.5)

Calculate the PW at 12% of the cash flows for the computerized voice recognition system. Use a rollback approach. The cash flows are $-\$100K$, $\$30K$, $\$40K$, $\$65K$, and $\$35K$.

Solution
Because there is no uniform or gradient series, the rollback approach is easier than using the tabulated factors.

$$V_4 = 35K$$
$$V_3 = 35K/1.12 + 65K = 96.25K$$
$$V_2 = 96.25K/1.12 + 40K = 125.94K$$
$$V_1 = 125.94K/1.12 + 30K = 142.44K$$
$$V_0 = 142.44K/1.12 - 100K = 27.18K \text{ or } \$27.2K$$

This final value is the PW of the cash flows from periods 0 to N, and it is the same as the value calculated in Example 5.5.

Example 5.7   Payoff from Graduate School

Considering the data in Exhibit 5.2, what is the PW of completing an M.S. degree in engineering? Assume a horizon of 40 years, an i of 5%, and that the degree can be completed in 1 year. Assume that a graduate assistant's income is $12,000, and that a tuition waiver is included.

Exhibit 5.2   Salary vs. number of years since graduation

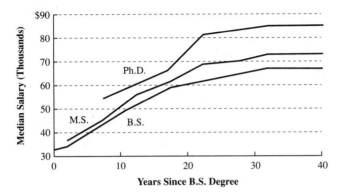

	Annual Salary Difference between	
Years since B.S.	**B.S. and M.S**	**M.S. and Ph.D.**
1–4	$2407	(too little data)
5–9	2308	$ 8322
10–14	4297	4203
15–19	2470	5400
20–24	6600	12,700
25–29	5771	13,200
30+	6270	11,980

Solution

The salaries increase, but not with an arithmetic gradient nor a geometric gradient. Because there is no pattern, the economic model must treat each cash flow individually. To show the order of calculations, Exhibit 5.3 is constructed with the years from 0 to 40 in reverse order. In fact, the cash flows for the salary differences are entered from the bottom up. The assumed salary for year 1 is $33,816 if employed as an engineer and $12,000 if employed as a graduate assistant, for a net loss of $21,816 for going to graduate school. Then the $2407 difference for the first 4 years since receiving the B.S. are entered for years 2 to 5. All other salary differences apply to 5-year blocks.

The first rollback value, V_{40}, equals the final cash flow of $6270. Exhibit 5.3 illustrates that it is easier to divide by 1.05 forty times than to look up 40 (P/F) factors. For example, V_{39} equals V_{40} rolled back one period, plus the cash flow for period 39, which is $V_{39} = 6270/1.05 + 6270 = \$12,241$.

The last rollback value, $V_0 = \$41.1K$, is the PW. This is positive, so graduate school is economically justified at $i = 5\%$.

Exhibit 5.3   PW calculation for M.S. in engineering

Year	Salary difference between M.S. and B.S.	V_t
40	$6270	$6270
39	6270	12,241
38	6270	17,929
37	6270	23,345
36	6270	28,503
35	6270	33,416
34	6270	38,095
33	6270	42,551
32	6270	46,794
31	6270	50,836
30	5771	54,186
29	5771	57,377
28	5771	60,416
27	5771	63,310
26	5771	66,066
25	6600	69,520
24	6600	72,810
23	6600	75,942
22	6600	78,926
21	6600	81,768
20	2470	80,344
19	2470	78,988
18	2470	77,697
17	2470	76,467
16	2470	75,296
15	4297	76,007
14	4297	76,685
13	4297	77,330
12	4297	77,945
11	4297	78,530
10	2308	77,099
9	2308	75,735
8	2308	74,437
7	2308	73,200
6	2308	72,022
5	2407	71,000
4	2407	70,026
3	2407	69,098
2	2407	68,215
1	−21,816	43,151
0	0	41,096

5.4 SALVAGE VALUES

Section 2.3 defined salvage value (S) as a net receipt in period N for the sale or transfer of equipment (which can be a salvage cost). Theoretically, salvage values are simple. The salvage value is simply the amount received minus the cost of removal. Calculating its PW is even simpler, since a single $(P/F, i, N)$ factor will do the job.

For most projects, including salvage values in the calculation of PW is easy. Equipment that has been used for 5 or 10 years will usually have a salvage value that is 20% or less of its first cost. Even if there are no removal costs, this is usually a small part of a project's PW (see Example 5.8).

The difficulties with salvage value arise on projects whose positive or negative termination value is large. When this is true, the value is also likely to be uncertain. Examples with positive salvage values include land or a business's goodwill (the economic value of established customer relationships and general reputation).

Examples with negative salvage values or salvage costs often focus on environmental reclamation. For example, there are substantial salvage costs associated with restoring an open-pit coal mine or decommissioning a nuclear power facility. In a specific example, when the Trans-Alaska Pipeline is finally shut down, plugging buried sections of pipe and removing above-ground sections, pump stations, and processing facilities is expected to cost about $2B.

Nuclear power facilities, as shown in Example 5.9, provide another use of salvage value—representing the value of costs to manage nuclear waste that occur after the problem's horizon or study period. Similar examples of salvage value can be used for the value of an improved market position to a private firm, or for the residual value of infrastructure improvements beyond a 20-year horizon that may have been used to evaluate a public project.

Example 5.8   How Important Is a Typical Salvage Value?

Assume that the salvage value is 20% of the $1000 first cost for a 5-year life and 10% of the first cost for a 10-year life. If $i = 7\%$, compare the PW of the salvage values with the first cost.

Solution

$$PW_S = 1000 \cdot .2(P/F, 7\%, 5) = .2 \cdot .7130 \cdot 1000 = \$142.6$$

$$PW_S = 1000 \cdot .1(P/F, 7\%, 10) = .1 \cdot .5083 \cdot 1000 = \$50.83$$

While these two values differ by nearly a factor of three, even the larger is less than 15% of the first cost. Thus, if the life of a piece of equipment is even moderately long, the PW of the salvage value should be relatively small.

Example 5.9   Decommissioning a Nuclear Facility

Nuclear power facilities must be rebuilt or shut down about 30 to 40 years after they begin operating. The shutdown in August 1989 of the 330 mW Ft. St. Vrain power plant in Colorado provides an example of salvage costs. These costs are linked to partially dismantling the facility to remove high-level radioactive material, to closing the facility for about 50 years, to the final decommissioning of the facility, and to caring for the hazardous wastes generated during operation and dismantling. The initial dismantling cost is a typical salvage value at the end of the study period. The cost of caring for the waste is really the present value of costs over a centuries-long period. These costs are estimated to exceed $200M after 20 years of use for a facility that cost $1 billion [Barrett].

In the U.S., about $180 billion has been spent on nuclear power plant construction. The total cost to shut down these facilities is estimated to be $50 billion to $100 billion [*Energy Journal*]. Using an interest rate of 5% and an operating period of 30 years, what is the PW at project start-up of the salvage costs?

Solution

The total cost to shut down the facility is the value at the end of year 30 of the costs into the future. Using the total figures of $50B to $100B (with B meaning billion), the PW of these costs is

$$PW = -50B(P/F, 5\%, 30) = -50B \cdot .2314 = -\$11.57B$$

or

$$PW = -100B(P/F, 5\%, 30) = -100B \cdot .2314 = -\$23.1B$$

5.5 CAPITALIZED COST AND PERPETUAL LIFE

Chapter 3 mentioned that the limit of $(P/A, i, N)$ as N approaches infinity is $1/i$. Assuming that N equals infinity is assuming perpetual life. Tunnels, roadbeds, rights-of-way, earthworks, and dams often have lives that are very close to perpetual in an economic sense. If the annual costs of these facilities are assumed to be perpetual, then the corresponding PW is defined to be the **capitalized cost.** Note that accountants commonly use a different definition of capitalized cost, which is costs to be depreciated (see Chapter 12).

* **Capitalized cost** is the present worth of annual costs that are assumed to occur in perpetuity.

This capitalized cost, P, equals A/i. The relationship can be derived mathematically using formulas and limits, or it can be derived intuitively, using as an example a savings account that lasts forever. If amount P is in the account, then the annual interest is $A = i \cdot P$. That exact amount may be withdrawn each year without changing the problem for later years. If less than $i \cdot P$ is withdrawn each year, then capital accumulates; if more than $i \cdot P$ is withdrawn, then eventually the principal is all gone. Equation 5.2 states this in terms of the factors, and Example 5.10 calculates a capitalized cost. Equation 5.3, for finding the capitalized cost of a perpetual gradient, is stated without proof and then applied in Example 5.11.

$$\ast \quad (P/A, i, \infty) = 1/i \qquad (5.2)$$

$$\ast (P/G, i, \infty) = 1/i^2 \qquad (5.3)$$

Example 5.10   Capitalized Cost of an Endowed Scholarship

Creekside Engineering wants to establish an engineering scholarship in honor of its founder, John Freeborn. The endowment will be placed in a savings account by the university, and the interest income will be available for the scholarship. The account's interest rate is 6%, and the scholarship amount is $3000 per year. What is the required endowment?

Solution

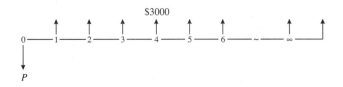

The perpetual life assumption is appropriate here, and the required endowment is the capitalized cost of the scholarship.

$$\text{PW} = 3000(P/A, i, \infty) = 3000/i = 3000/.06 = \$50\text{K}$$

Example 5.11   Perpetual Maintenance Gradient

A state historic park has maintenance costs of $40K for the first year. Due to increased use, these costs increase by $2K per year. Assuming an interest rate of 5% and perpetual life, what is the capitalized cost?

Solution

$$\begin{aligned}
\text{PW} &= -40\text{K}(P/A, i, \infty) - 2\text{K}(P/G, i, \infty) \\
&= -40\text{K}/i - 2\text{K}/i^2 = -40\text{K}/.05 - 2\text{K}/.0025 \\
&= -1600\text{K} = -\$1.6\text{M}
\end{aligned}$$

5.6 STAGED PROJECTS

There are many projects for which some alternatives involve building in stages. Often these alternatives can be divided into three sets, based on what is built now vs. what will be built later. Capacity can be matched (1) to current needs, with later additional projects; (2) to future needs; or (3) to current needs, with some provision for future expansion.

Meeting only current requirements requires that future needs be met by future projects. On the other hand, matching capacity to future needs means building excess capacity that will only be fully used later. Between these extremes are alternatives that include the ability to more easily modify or add onto the structure later. On a building, this could mean designing the foundation and the structure for additional floors. For a manufacturing plant, this could mean building a large structure with space for five assembly lines, but installing only three now. For a hydroelectric dam, this could mean installing two of the four turbines that the powerhouse is designed to hold. Example 5.12 illustrates these three possibilities.

Example 5.12   Construction of a School—Staged or Not?

Metro University must choose from among three plans for a new classroom building. (1) Spending $21M now will meet the needs of the next 10 years. (2) Spending $13M now will meet the needs of the next 6 years. After 6 years, $12M would be spent on another building to meet the needs of the last 4 years. (3) For $15M now, the new building can be designed for expansion later at a cost of $7M. If Metro U uses an interest rate of 5%, find the PW of each alternative.

Solution

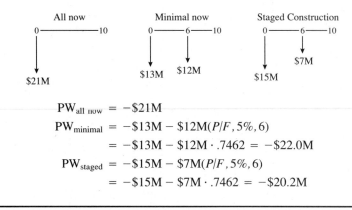

$$PW_{all\ now} = -\$21M$$
$$PW_{minimal} = -\$13M - \$12M(P/F, 5\%, 6)$$
$$= -\$13M - \$12M \cdot .7462 = -\$22.0M$$
$$PW_{staged} = -\$15M - \$7M(P/F, 5\%, 6)$$
$$= -\$15M - \$7M \cdot .7462 = -\$20.2M$$

Example 5.12 illustrated finding the PW of the construction costs for the three approaches. However, each strategic approach has advantages and disadvantages that cannot be incorporated into the cash flow diagram. Exhibit 5.4 summarizes these.

Exhibit 5.4   Advantages and disadvantages of staged projects

	Current Needs (Minimal Now)	Foundation for Later Expansion (Staged)	All at Once
Advantages	Lowest cost now Maximum flexibility for design of later facility	Flexible timing, so can be matched to actual growth in need Can utilize the most cost-effective investments for later expansion	Fully utilizes economies of scale No disruption for later construction
Disadvantages	Difficulty in integrating original and later buildings Incurs largest costs for later construction	Limited flexibility in size Later expansion will cause some disruption for original facility	If needs change, capacity may be unneeded or poorly suited

In general, alternatives that meet current requirements tend to minimize current costs and debt loads, but maximize the cost of later expansion—especially when disruption of activities during later construction is important. Expanding later as a separate project provides flexibility in the size and timing of the additional capacity. The later project can be built when it is needed and sized appropriately, but it is harder to mesh that capacity with what was built earlier.

Alternatives that emphasize excess capacity reduce the costs of building that capacity because there are many economies of scale in construction. This minimizes later costs and total costs (PW at $i = 0\%$) due to economies of scale and minimal disruption of activities. Since the capacity is already built, however, there is no flexibility in the size of the capacity increment, nor in its timing. In fact, there is even the risk that the excess capacity will never be needed.

Laying the foundation for future expansion has the advantages and disadvantages of compromise. For example, the timing of an expansion is flexible, but the expansion's exact size may be heavily constrained by current decisions.

5.7 COST OF UNDERUTILIZED CAPACITY

One difficulty in evaluating projects or alternatives is that firms and agencies may have excess capacity due to changed plans or changed situations. Once-busy warehouses or schools may now be only half full. As long as that excess capacity is not used, it has no value. Yet if a marginal alternative or project uses that excess capacity, then that capacity will not be available for more attractive alternatives or projects.

Common sense suggests that if a warehouse costs $85 per square foot to build, then each project that uses warehouse space should include a cost of $85 per square foot. Suppose that no current project proposal is justifiable if the full $85 is used, but several alternatives can contribute $20 to $60 per square foot. If no alternative is accepted, then the warehouse keeps its excess capacity, and does not earn even that $20 to $60 per square foot. The capacity is unused for longer periods—perhaps forever. If a standard of less than $85 per square foot is used, then the excess capacity may be assigned to marginal projects before a "good" project can be approved. If a price below $85 is used, it must at least cover additional operating costs for the facility. In a warehouse this would include additional staffing, heating, insurance, etc.

The following are rough guidelines for the maximum period of cheap pricing. The period should be (1) less than 5 years (long for dynamic industries, but the other two criteria will automatically reduce it), (2) less than one-third of the life of the alternative receiving the cheap pricing, and (3) less than the time until better alternatives could start. (Guidelines 1 and 2 have appeared in Reul and in Grant, Ireson, and Leavenworth.) Example 5.13 illustrates these guidelines.

Example 5.13   Charging for Unused Computer Capacity

MoreTech Drilling has had a downturn in business, causing it to lay off one of its six engineers. One of its computer workstations has been idle for 3 months. A proposal has come forward to use the workstation for process control in the plant.

MoreTech Drilling has identified the cost to buy and operate this workstation at $55 per hour. If the machine is idle, the cost is $40 per hour. If the machine is used for process control, it will be worn out in 9 years. How should this proposal be evaluated?

Solution
The guidelines suggest that the "cheap" rate of $40 per hour should be limited to the smallest of (1) 5 years; (2) one-third of 9 years, or 3 years; or (3) the predicted time for returning to six engineers. After that point, the process control should be able to repay a charge of $55 per hour to be worthwhile.

These guidelines are not universally applicable. For example, many school systems rent out classrooms, gyms, pools, and assembly halls for use outside of school hours—in the evenings, on weekends, and during the summer. Often there is no possibility of an alternative use by the school system, and any fee above the cost of setting up and cleaning the room may be much-needed revenue.

5.8 SPREADSHEETS AND SHORTER PERIODS

One advantage of using spreadsheets to calculate present worths is that greater complexity or detail can be included efficiently. This allows more accurate models with shorter time periods. The same situation can be modeled using 10 years, 120 months, or even 260 biweekly pay periods. Accuracy can be enhanced by estimating the cash flows for each time period, rather than assuming that years 1 through 10 will be the same. The more accurate model is easier to evaluate if a spreadsheet is used.

Spreadsheets can easily use monthly rather than yearly periods, and any interest rate is entered exactly, so no interpolation is needed. The resulting economic models are more accurate because the assumption of end-of-period cash flows is not the best approximation of most real-world problems.

For example, if you are modeling the repayment schedule for a car loan or a house mortgage, it is more accurate to use monthly periods rather than an annual approximation. Similarly, firms may build more accurate models using 26 biweekly pay periods rather than an annual approximation. These more accurate models can include energy costs that vary by the season, employment levels that vary by the month, and construction equipment rentals that vary by the week.

Example 5.14 models the construction period for a project. The more accurate assumption of monthly cash flows increases the absolute value of the PW by more than 5%. This increase happens because the cash flows at the ends of months 1 to 11 occur sooner than the cash flows at the end of year 1 (month 12). The solution to Example 5.14 also includes formulas for solving the problem, but as the problem grows in size and complexity this method becomes more and more cumbersome.

Example 5.14 AAA Construction

AAA Construction is bidding on a project whose costs are divided into $20,000 for start-up, $120,000 for the first year, and $180,000 for the second year. If the interest rate is 1% per month, or 12.68% per year, what is the PW with annual compounding? With monthly compounding?

Solution

Exhibit 5.5 illustrates the spreadsheet solution for this problem, with the assumption that costs are distributed evenly throughout the year.

Exhibit 5.5 Monthly vs. annual periods

	A	B	C	D	E
1	Exbt 5.5	1.00%	i		12.68%
2		($282,375)	PW		($268,256)
3	Month	Arith.		Year	Arith.
4	0	($20,000)		0	($20,000)
5	1	($10,000)		1	($120,000)
6	2	($10,000)		2	($180,000)
7	3	($10,000)			
8	4	($10,000)	@NPV(B1,B5..B28)+B4		
9	5	($10,000)			
10	6	($10,000)			
11	7	($10,000)			
12	8	($10,000)			
13	9	($10,000)	= $E$5/12		
14	10	($10,000)			
15	11	($10,000)			
16	12	($10,000)			
17	13	($15,000)			
18	14	($15,000)			
19	15	($15,000)			
20	16	($15,000)	= $E$6/12		
21	17	($15,000)			
22	18	($15,000)			
23	19	($15,000)			
24	20	($15,000)			
25	21	($15,000)			
26	22	($15,000)			
27	23	($15,000)			
28	24	($15,000)			

Since the costs are uniform a factor solution is also reasonable.

$$PW_{annual} = -20K - 120K/1.1268 - 180K/1.1268^2 = -\$268.3K$$

$$PW_{monthly} = -20K - 10K(P/A, 1\%, 12) - 15K(P/A, 1\%, 12)(P/F, 1\%, 12)$$

$$= -\$282.4K$$

Why do these two values differ by more than $14,000? The interest rates are equivalent, so that is not the answer. The answer involves different assumptions about *when* cash flows occur. $10K at the ends of months 1 through 12 is not the same as $120K at the end of year 1 (month 12).

5.9 SPREADSHEETS AND MORE EXACT MODELS

The factor approach to engineering economy problems works well so long as most cash flows are part of uniform or gradient series. If a more exact model is produced and if that model involves different cash flows in most periods, then a spreadsheet is the best approach. If 20 factors must be looked up in the tables, then the danger of a mistake is substantial.

Many problems cannot be analyzed easily using factors for uniform flows, arithmetic gradients, and single payments. For example, production from oil fields and most mineral resources declines over time. Demand for manufactured products typically starts low, increases rapidly for a period of months to years, and then declines more slowly. Demand for municipal services such as sewer, water, and roads grows irregularly as subdivisions and major industrial facilities are built. If a spreadsheet is not available for problems with this level of detail, then the rollback approach explained in Section 5.3 should be used. This divides repeatedly by $(1 + i)$ rather than using $N(P/F)$ factors.

Example 5.15 converts Example 5.14's data into a monthly pattern, with heavier summer construction activities (month 1 is June). Example 5.16 calculates the PW for a new product, where the demand first increases and then declines more slowly. The unit price is first stable, and then it declines.

Example 5.15   AAA Construction (Revisited)

Assume that the project is initiated June 1 (time 0), that the maximum monthly cost is $14K for the first year and $21K for the second year, and that the minimums are $6K and $9K. The detailed monthly costs are shown in Exhibit 5.6. What is the PW for the monthly model?

Solution
The PW calculated in Exhibit 5.6 is not significantly different from the monthly PW calculated in Example 5.14; however, the pattern is quite different. This pattern may be crucial in determining the project's required financing and the cost of that financing.

Exhibit 5.6   Detailed time-varying cash flow estimates

	A	B	C	D	E
1	Exbt 5.6	1.00%	i		12.68%
2		($283,970)	PW		($268,256)
3	Month	Arith.		Year	Arith.
4	0	($20,000)		0	($20,000)
5	1	($14,000)		1	($120,000)
6	2	($14,000)		2	($180,000)
7	3	($14,000)			
8	4	($12,000)			
9	5	($10,000)			
10	6	($8,000)			
11	7	($6,000)			
12	8	($6,000)			
13	9	($6,000)			
14	10	($8,000)			
15	11	($10,000)			
16	12	($12,000)	-120000 = @SUM(B5..B16)		
17	13	($21,000)			
18	14	($21,000)			
19	15	($21,000)			
20	16	($18,000)			
21	17	($15,000)			
22	18	($12,000)			
23	19	($9,000)			
24	20	($9,000)			
25	21	($9,000)			
26	22	($12,000)			
27	23	($15,000)			
28	24	($18,000)	-180000 = @SUM(B17..28)		

Example 5.16   PW for Hi-Tech's New Product

Hi-Tech Industries has a new product whose sales are expected to be 1.2, 3.5, 7, 6, 5, 4, and 3 million units per year over the next 7 years. Production, distribution, and overhead costs are stable at $120 per unit. The price will be $200 per unit for the first 3 years, then $180, $160, $140, and $130 for the next 4 years. The remaining R&D and production costs are $400M. If i is 15%, what is the PW of the new product?

Solution
It is easiest to calculate the yearly net revenue per unit before building the spreadsheet. Those values are $80, $80, $80, $60, $40, $20, and $10. The rest of the calculations are shown in Exhibit 5.7. Note that it would be easy to include costs per unit that varied with time and production volume.

Exhibit 5.7 PW of a new product with growing and declining demand

	A	B	C	D	E
1	Exbt. 5.7				
2	($400)	P		i= 15%	
3					
4			Net	Cash	PW at
5	Year	Sales	Revenue	Flow	15%
6	0	0.0		-400	($400)
7	1	1.2	$80	$96	$83
8	2	3.5	$80	$280	$212
9	3	7.0	$80	$560	$368
10	4	6.0	$60	$360	$206
11	5	5.0	$40	$200	$99
12	6	4.0	$20	$80	$35
13	7	3.0	$10	$30	$11
14	D6+ @NPV(D2,D7..D13) =			$615	
15			@SUM(E6..E13)=	$615	

5.10 SUMMARY

This chapter defined and applied the PW measure to problems of setting a price for a patent or a bond and of evaluating private- and public-sector projects. This included salvage values, capitalized cost of perpetual annuities, staged projects, and underutilized capacity. Finally, spreadsheets have been used to improve model accuracy by using more and shorter periods and by using cash flow estimates that change each period.

The standard is to accept projects whose PW is positive at the specified interest rate. Cash flows are assumed to be known and certain, and to occur at the end of the period, except for first costs and prepayments such as leases or insurance. Computations are usually made using engineering economy factors, but if each cash flow is different, then a spreadsheet or rollback procedure is recommended.

PWs were calculated to compare staged projects where the current project (1) meets only current needs (minimal now), (2) meets future and current needs (all now), and (3) includes some provision for later expansion (staged). Other advantages and disadvantages were also summarized.

Guidelines were developed for pricing underutilized capacity. These guidelines are that "cheap" pricing should always exceed the added operating costs, and that the maximum period should be less than the smallest of (1) 5 years, (2) one-third of the project's life, or (3) the starting point for better alternatives.

Spreadsheets were used to build more accurate models that use monthly or bi-weekly periods rather than years. These models can also accommodate estimates for each year or period rather than assuming that cash flows are uniform series or arithmetic gradients.

REFERENCES

Barrett, Amy, "The Big Turnoff," *Financial World,* Volume 160, Number 15, July 23, 1991, pp. 30–32.

Energy Journal, Volume 12, special issue, decommissioning articles, 1991.

Grant, Eugene L., W. Grant Ireson, and Richard S. Leavenworth, *Principles of Engineering Economy,* 7th ed., 1982, Wiley, p. 385.

Reul, Ray I., "Profitability Index for Investments," *Harvard Business Review,* Volume 35, Number 4, July–August 1957, p. 122.

PROBLEMS

5.1 George, a graduating civil engineering student, is offered a job at a remote construction project overseas. If George is still with the project when it is completed at the end of year 5, he will receive a bonus of $100,000. His discount rate is 12%. What is the present worth of the bonus?   (*Answer:* $56,740)

5.2 Graph the answer for Problem 5.1 for discount rates ranging from 0% to 20%.

5.3 An R&D project for a new product has already required the expenditure of $200K. Spending another $100K for a license on another firm's patent will complete the project. Net returns from the product will be $20K per year for 10 years. If the firm's i is 10%, what is the PW for continuing? Should the firm continue?

5.4 The annual income from an apartment house is $20,000. The annual expenses are estimated to be $2000. If the apartment house can be sold for $100,000 at the end of 10 years, how much can you afford to pay for it if the time value of money is 10%?

5.5 Hollingsworth Machine Group needs a new precision grinding wheel. If 12% is the acceptable return for Hollingsworth, determine the present worth of one of the best alternatives.

Cost	$80,000
Useful life	5 years
Annual cost	$12,000
Salvage value	$10,000

(*Answer:* −$117.6K)

5.6   The Burger Barn is evaluating the Quick Fry french fryer, which has an initial cost of $6000. The new equipment will allow Burger Barn to save on costs and hopefully sell more fries. The annual expected net income increase is $1380. The machine has a salvage value of $250 at the end of its 5-year useful life. Determine the present worth at an interest rate of 5%.

5.7   We-Clean-U will receive $32,000 each year for 15 years from the sale of its newest soap, Rub-A-Dub-Dub. The initial investment is $150,000. Manufacturing and selling the soap will cost $7350 per year. Determine the investment's present worth if interest is 12%.

5.8   A new drill press has a first cost of $10,000. The net annual income is $1500 the first year, which decreases by $250 each year thereafter. After 6 years, the press can be sold for $500. What is the press's present worth at an interest rate of 12%?

5.9   MoreTech Drilling is considering the sale of its Alaskan subsidiary to another company. The subsidiary is expected to produce profits of $800,000 next year, which drops by $50,000 per year for the following 14 years. If MoreTech's interest rate is 12%, what price should MoreTech charge for the subsidiary?   (*Answer:* PW = $3.75M)

5.10   What is the PW (at $i = 10\%$) of the following electric power generating project? There is a $180,000 overhaul cost in year 8. The first cost is $420,000, and the annual operating cost is $30,000. Revenue from power sales is expected to be $100,000 per year. The facility will have a salvage value of $75,000 at the end of year 15.   (*Answer:* PW = $46.4K)

5.11   A project has a first cost of $10,000, net annual benefits of $2000, and a salvage value after 10 years of $3000. The project will be replaced identically at the end of 10 years, and again at the end of 20 years. What is the present worth of the entire 30 years of service if the interest rate is 10%?

5.12   Pass-Time Pool Hall may buy 10 of the finest slate pool tables available. Each table will cost $12,000 and have an unknown residual value. Pass-Time's management is certain that business will increase, and that the hall will be able to host a number of tournaments. The expected increase in net revenues is $22,000 per year for the next 8 years. Determine *each* table's residual value if the PW is to equal 0. The cost of money for Pass-Time is 10%.

5.13   What is the PW (at $i = 5\%$) of SuperTool's new test equipment? The development cost is $1.2M. Net revenues will begin at $300,000 for the first 2 years and then decline at $50,000 per year. SuperTool will terminate sales when net revenues decline to 0.

5.14   A 10-year, 9% bond, issued by Cheap Motors Manufacturing on January 1, 1990, matures on December 31, 1999. If investors require a 12% return, how much should they be willing to pay on January 1,

1994? (Face value of the bond is $10,000 and interest is paid annually.) (*Answer:* $8766)

5.15 A 10-year, $10,000 bond will be issued at an interest rate of 8%. Interest is paid annually. Interest rates have fallen since the planning for this bond was completed, and interest rates are now 7%. At what price should this bond sell?

5.16 A 10-year, $20,000 bond was issued at a nominal interest rate of 8% with semiannual compounding. Just after the fourth interest payment, the bond will be sold. Assume that an effective interest rate of 10¼% will apply, and calculate the price of the bond. (*Answer:* PW = $17,832)

5.17 The head coach at State University has received an offer to buy out his contract. The offer is $175,000 now and a semiannual annuity of $75,000 for the next 3 years. If money is worth 6.1% to the coach, what is the buyout worth?

5.18 Determine the purchase price of a new home if it was financed with a 10% down payment and payments of $539.59 per month for 30 years at 6% interest. (*Answer:* $100K)

5.19 Determine the down payment that must be made on an automobile that costs $20,000 if payments are limited to $375 per month for 48 months at 9% interest.

5.20 Firerock Tire and Rubber Company can invest $2.6M to receive the cash flows shown in the table. If money costs Firerock 8%, determine the investment's present worth.

Year	Cash Flow	Year	Cash Flow
1	$125K	5	$1000K
2	250K	6	750K
3	500K	7	500K
4	750K	8	250K

(*Answer:* $258K)

5.21 Determine the PW of the following cash flows at 6% interest.

Year	0	1	2	3	4	5	6
Cash Flow	−$14K	−8K	13K	22K	−14K	−4K	12K

5.22 What is the PW of the following cash flows at an interest rate of 8%?

Year	0	1	2	3	4	5	6	7	8	9
Cash Flow	−$40K	3K	5K	7K	9K	11K	13K	11K	9K	7K

(*Answer:* PW = 9.70K = $9700)

5.23   What is the PW of the following cash flows at an interest rate of 6%?

Year	0	1	2	3	4	5	6	7
Cash Flow	−$40K	8K	9K	10K	11K	12K	10K	8K

5.24   The Ding Bell Import Company requires a return of 15% on all projects. If Ding is planning an overseas development project with the following cash flows, what is the project's present worth?

Year	0	1	2	3	4	5	6	7
Net Cash	0	−$100K	−40K	−20K	30K	80K	100K	30K

5.25   Recalculate the present worth of an M.S. degree using the data from Example 5.5. Assume that graduate school takes 2 years to complete.

5.26   Calculate the PW of a Ph.D. degree over a B.S. degree using the data from Example 5.7. Assume that graduate school takes 5 years to complete.

5.27   Graph the PW vs. the interest rate for the following equipment. (Use $i = 4\%$, 6%, 8%, and 10%.) What conclusions can you draw concerning the PW and the interest rate at which the PW is evaluated.

First cost	$10,000
Net savings per year	$ 2500
Useful life	5 years

5.28   A common standard of significant change is 10%. A project's first cost is $20,000, the expected life is 15 years, and the interest rate is 8%.
   a.   How large must the salvage value be so that its PW is 10% of the first cost?
   b.   If the salvage value is $4000, how short must the life be so that the salvage value's PW is 10% of the first cost?
   c.   If the salvage value is $4000, how low must the interest rate be so that the salvage value's PW is 10% of the first cost?

5.29   The industrial engineering department for Invade Air Fresheners has found that a new packing machine should save $45,000 for each of the next 8 years. The machine will require a major overhaul at the end of 5 years costing $12,000. The machine's expected salvage value after 8 years is 7.5% of its original cost. If money is worth 5%, determine the amount Invade should be willing to pay for the machine.

5.30   Western Coal Inc. is considering an open-pit mine. The rights to mine on public land are expected to cost $25M, and development of the infrastructure will cost another $12M. The mine should produce a net revenue of $8M per year for 20 years. When the mine is closed, $15M will be spent for reclamation. If Western Coal uses an interest rate of 15%, what is the PW of this mine?

5.31   Western Coal is considering a modification of its plans for the open-pit mine described in Problem 5.30. Adding a coal-slurry pipeline (its life is also 20 years) will increase the first cost by $4M, increase annual net revenues by $750K, and increase closure costs by $1M. If the interest rate is 15%, what is the PW of the slurry pipeline?   (*Answer:* PW = $633.2K)

5.32   MidCentral Light and Power is planning a nuclear facility. Expected first costs are $1.2B, and the annual net revenue is estimated to be $75M. The life is 40 years, and the interest rate is 7%. Shut-down costs are estimated to be $400M, and the cost of caring for the hazardous wastes is $10M per year for 100 years after the facility is shut down. What is the PW of the facility?

5.33   Baker and Sons specializes in land development and construction management. It is considering whether to develop a research park near a major engineering school. The land will cost $20M, and the design and construction of the infrastructure will cost $25M. The plan is to build a new building each year for 15 years at an average cost of $10M. Operating costs for the buildings will be paid by the tenants. In year 20 the research park will be sold for $250M (land, buildings, and goodwill). Tenants will pay 15% of the cost of the buildings each year in rent. If Baker and Sons uses an interest rate of 8%, what is the PW of the research park? (Assume 1 year for infrastructure construction, and 1 year for each building.)

5.34   The Moo-Cow Dairy may purchase new milking equipment for its milking barn. The equipment costs $135,000. The annual savings due to increased production and decreased labor costs will be $25,000. The equipment requires a minor overhaul every 2 years that costs $5000. At the end of the sixth year, a major overhaul costing $15,000 will be required. The equipment is expected to last 10 years with scheduled overhauls. No salvage value is anticipated. If money is worth 7% to the dairy, determine the equipment's PW. (*Answer:* $19.56K)

5.35   National Motors must build a bridge to access land for its manufacturing plant expansion. If made of normal steel, the bridge would initially cost $30,000 and it should last 15 years. Maintenance (cleaning and painting) will cost $1000 per year. If a more corrosion-resistant steel is used, the annual maintenance cost will be only $100 per year (with the same life). If the firm's cost of money is 12%, what is the maximum amount that should be spent on the corrosion-resistant bridge?

5.36   Fishermen's Bay used to have a thriving fishing industry, but the boats got bigger and now few fit in the harbor's protected area. The city council would like to attract the boats back, and it believes a new breakwater and docking area would do the job. The city engineer's estimated first cost is $800,000, and the annual expenses are estimated at $75,000. The planning department estimates annual benefits to be $150,000. The city uses a 5% interest rate and a study period of 20 years. How large does the salvage or residual value have to be for the project to have a PW > 0?

5.37   Lazy Links Country Club has been more successful than expected in the 3 years since it opened. The owners believe that this growth will continue for 10 years before leveling off. The pool, racquet, and golf facilities (both indoor and out) can be expanded during each sport's off-season. However, the locker rooms, sports shop, and dining facilities are used year-round. These facilities are approaching capacity (4 years sooner than projected), so an expansion is required.

If all of the work is done immediately, it will cost $4.5M to build and $.6M per year to maintain. If capacity is matched to the 5-year forecast, then the capital cost will be $2M now and $5M in 5 years. (These cost estimates include a "cost of disruptions.") For the staged alternative, maintenance costs are estimated to be $.2M and $.6M before and after expansion. If the owners' interest rate is 12%, what is the PW of each option over the next 5 years? Will costs after 5 years make a difference?

5.38   National Motors is planning a new manufacturing facility to support a three-wheeled car called the Trio. Once the car is introduced, demand is expected to grow 40% per year. Much of this growth can be handled by adding shifts of workers, but clearly the manufacturing facilities will need to be expanded at some point. Find the PW for the following options. Assume that the interest rate is 12% and that expansion will be required in year 6.

	Build All Now	Build Now with Expansion Plans	Build Now without Expansion Plans
First cost	$200M	$120M	$100M
Expansion	$0	$120M	$180M

5.39   Assume mortgage payments of $1000 per month for 30 years and an interest rate of 1% per month. What initial principal will these figures repay? (That is, find the PW.) If annual payments of $12,000 are assumed with an interest rate of 12.68% per year, what initial principal will be repaid? Why do these differ?

CHAPTER 6 EQUIVALENT ANNUAL WORTH

The Situation and the Solution

Decision makers often think in terms of annual expenses, annual receipts, and annual budgets—not present worths. In other cases, the benefits are not studied and only the costs of alternatives are compared.

The solution is to state equivalence in yearly terms—that is, as an equivalent annual worth or an equivalent annual cost.

Chapter Objectives

After you have read and studied the sections of this chapter, you should be able to:

Section 6.1
Define equivalent annual worth (EAW) and explain when that measure is preferred.

Section 6.2
Explain the assumptions and sign conventions of EAW and equivalent annual cost (EAC).

Section 6.3
Understand typical examples of EAC use.

Section 6.4
Find the EAC of an overhaul or a deferred annuity.

Section 6.5
Find the EAC with salvage values and working capital.

Section 6.6
Calculate EACs for perpetual life, including perpetual gradients.

Section 6.7
Calculate EAWs if some costs occur periodically, such as a road that is restriped every 3 years.

Section 6.8
Analyze loan repayments and balances due using spreadsheets.

Key Words and Concepts

Equivalent annual worth (EAW)   The uniform dollar amounts at the ends of periods 1 through N that are equivalent to a project's cash flows.

Equivalent annual cost (EAC)   The uniform dollar amounts at the ends of periods 1 through N that are equivalent to the *negative* of a project's cash flows. For EAC, costs are positive.

Deferred annuity   A uniform cash flow that first occurs at the end of period $D + 1$ rather than at the end of period 1. It continues until the end of period N.

Working capital   Money required to implement a project that is recovered at the project's end.

Principle of repeated renewals   Implies that if cash flows occur every T periods and if this event repeats for N years, where N is an integer multiple of T, then the EAW or EAC over T periods equals the EAW or EAC over N periods.

6.1 THE EQUIVALENT ANNUAL WORTH MEASURE

Equivalent annual worth (EAW) is the value per period. More rigorously: EAW represents N identical cash flow equivalents occurring at the ends of periods 1 through N. A desirable project or alternative has an EAW > 0, just as a desirable project or alternative has a PW > 0. In fact, the EAW equals the PW$(A/P, i, N)$.

Equivalent annual worth is the uniform dollar amounts at the ends of periods 1 through N that are equivalent to a project's cash flows.

The "equivalent annual" approach is the most convenient method for comparing project evaluations with an annual budget. For example, equipment to automate a manual process is best evaluated by comparing an equivalent annual amount for automation with the annual labor savings.

Equivalent annual is often the best approach because saying that this project will cost or save $3500 per year makes intuitive sense, even to people who have never heard of engineering economy.

Chapter 10 will show that equivalent annual measures are particularly convenient when alternatives with different length lives are compared. For example, a stainless steel design may be considered precisely because it will last longer than a brass design. The question to be answered is: "Does the longer life of the stainless steel design *make up* for its higher first cost?" Or: "Which is cheaper over its life, stainless steel or brass?"

6.2 ASSUMPTIONS AND SIGN CONVENTIONS

Equivalent annual measures are commonly applied using exactly the same assumptions used with present worth measures. As noted in Section 5.1, these are:

1. Cash flows occur at time 0 and at the ends of periods 1 through N.
2. Cash flows are known; the horizon, N, is known; and there is no uncertainty.
3. The interest rate, i, is known.

There are two sign conventions that are used with equivalent annual measures. Equivalent annual worth (EAW) uses the same sign convention as does present worth (PW): positive or incoming cash flows have positive signs.

Equivalent annual cost is the uniform dollar amounts at the ends of periods 1 through N that are equivalent to the *negative* of a project's cash flows.

The second convention is that of **equivalent annual costs** (EAC): negative cash flows have positive signs. Logically, a *cost* when it is positive represents a negative cash flow.

In either case, within a problem *be consistent*. Do not mix EAC and EAW. For each problem, pick EAW or EAC and use it for all terms.

The EAC convention exists because many problems consider only costs. The question is, Which is the most cost-effective solution? For example, a civil engineer may design a roof; an electrical engineer may design a power transmission system; a mechanical engineer may design a heating and ventilation system; and an industrial engineer may design an assembly line.

In each case, the design engineer searches for the most cost-effective solution, often without even quantifying the value of solving the problem. There is no quantified value of having the roof, transmission system, heating and ventilation system, or assembly line. There is only the cost to build each. If all numbers are costs, it is easier if all numbers have a positive sign. Thus, the EAC sign convention and the EAW sign convention will both be used. Example 6.1 illustrates the EAC convention.

Other terminology that is sometimes used for EAW and EAC includes EUAW and EUAC, which stand for equivalent uniform annual worth and cost, respectively. Both EAWs and EACs are annuities. Even though all of these terms have the word *annual* as part of their names, they are used to describe uniform amounts for each month, each quarter, or each of another time period. In engineering economy, however, the year or annual period is most common.

Example 6.1 Measuring the Cost of a New Pump

George is a recent chemical engineering graduate. One of his first jobs is to calculate a large pump's cost using the following data. The first cost will be $54,000; the annual operating costs for energy and maintenance will be $18,000; the expected life is 10 years; the expected salvage value is 0; and the firm's interest rate is 9%. What is the EAC?

Solution

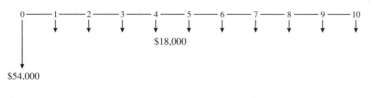

$$EAC = 54K(A/P, 9\%, 10) + 18K$$

$$= 54K \cdot .1558 + 18K = \$26.41K$$

6.3 EXAMPLES OF ANNUAL EVALUATIONS

Example 6.2 illustrates the use of EAW and EAC to evaluate automating a process, and Example 6.3 illustrates the use of EAC to evaluate a road improvement.

Example 6.2 The EAW of a Robot

Jose is evaluating the payoff for using two robots in place of one person on each shift at an assembly station. The robots together cost \$135,000, and they are expected to have a life of 10 years, with a salvage value of 0. This life is based on economic obsolescence, not on wearing out. Their maintenance and operating costs are expected to be \$5000 per year per shift operated. Each person per shift costs Jose's company \$25,000 per year. If the firm's interest rate is 12%, what is the EAW for the robots, assuming two shifts for the 10-year period? One shift? Three shifts?

Solution
The cash flow diagram is drawn, with the number-of-shifts assumption shown as N_{Shift}.

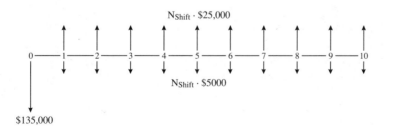

Substituting 1, 2, or 3 for the number of shifts leads to the EAWs:

$$EAW_1 = -135K(A/P, .12, 10) + 1(25K - 5K)$$

$$= -135K \cdot .1770 + 20K = -3.9K = -\$3900$$

$$EAW_2 = -135K(A/P, .12, 10) + 2(25K - 5K)$$
$$= -135K \cdot .1770 + 40K = 16.1K = \$16,000$$
$$EAW_3 = -135K(A/P, .12, 10) + 3 \cdot (25K - 5K)$$
$$= -135K \cdot .1770 + 60K = 36.1K = \$36,000$$

This problem could also be analyzed by using the EAC for buying the two robots:

$$EAC_{buy} = 135K(A/P, .12, 10) = 135K \cdot .1770 = \$23,895$$

This number could then be compared with the $20,000 per year per shift that the robots would save.

Example 6.3   Cost for Highway Improvements

For 40 years Wishbone Hill has been notorious for the accidents that have occurred there. Railings, signs, wider shoulders, etc., have only managed to keep the increasing level of traffic from increasing the rate of accidents. The highway department uses an interest rate of 6% and a study period of 40 years. What is the EAC for building another two-lane road to separate the traffic by direction?

Purchasing the right-of-way is expected to cost $15M and construction is expected to cost $60M. Snowplowing, painting, repair of guardrails, and periodic pavement work are expected to cost $3M per year.

Solution
The right-of-way and the construction costs occur at time 0, while the $3M occurs every year.

$$EAC = (15M + 60M) \cdot (A/P, 6\%, 40) + 3M = 75M \cdot .0665 + 3M = \$7.99M$$

6.4 FINDING THE EAC OF IRREGULAR CASH FLOWS

EAC of a Single Interior Cash Flow.   There are often single cash flows that occur neither at time 0, nor at the end of period N. These "interior" cash flows may be due to equipment overhauls or staged construction. To calculate an equivalent annual amount, first calculate an equivalent P at period 0 or F at period N; then the EAW or EAC for each of the N periods can be found.

Exhibit 6.1 illustrates the calculations defined in Equations 6.1 (using P) and 6.2 (using F). Example 6.4 applies these formulas and shows the error in not calculating P or F first. Let F_t equal the cash flow in year t.

$$EAC = F_t \cdot (P/F, i, t)(A/P, i, N) \qquad (6.1)$$

or

$$EAC = F_t \cdot (F/P, i, N-t)(A/F, i, N) \qquad (6.2)$$

Exhibit 6.1 EAC for a Single, Interior Cash Flow

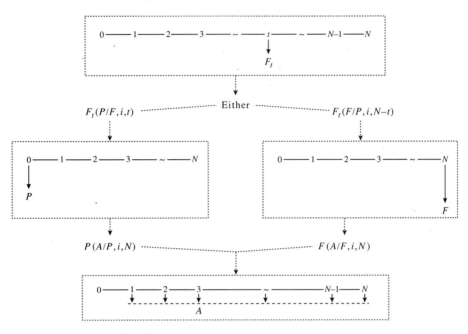

Example 6.4 EAC for an Overhaul Cost

Find the EAC for a 5-year life of a $680 overhaul at the end of year 3. Let $i = 8\%$.

Solution

Applying Equation 6.1 converts the overhaul cost to time 0 and then to an annual amount, which results in the following:

$$\text{EAC} = 680(P/F, .08, 3)(A/P, .08, 5)$$

$$= 680 \cdot .7938 \cdot .2505 = \$135.2$$

Applying Equation 6.2 results in the same answer, but it converts the overhaul cost to a future cost at the end of year 5, and then to an annual amount,

$$\text{EAC} = 680(F/P, .08, 2)(A/F, .08, 5)$$

$$= 680 \cdot 1.166 \cdot .1705 = \$135.2$$

As shown in Exhibit 6.2, this EAC which occurs for years 1–5 does not match either of the two incorrect approaches which are sometimes used. A common mistake is to *spread* the overhaul costs just over the years before it (years 1–3 here) or over the years after it (years 4 and 5 here). These respectively find an equivalent annual amount for years 1–3 and 4–5, rather than for years 1–5 as required.

$$\text{EAC}_{1-3} = 680(A/F, .08, 3) = 680 \cdot .3080 = \$209.4 \text{ and years 4 and 5} = \$0$$

$$\text{EAC}_{4-5} = 680(A/P, .08, 2) = 680 \cdot .5608 = \$381.3 \text{ and years 1–3} = \$0$$

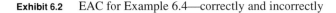

Exhibit 6.2   EAC for Example 6.4—correctly and incorrectly

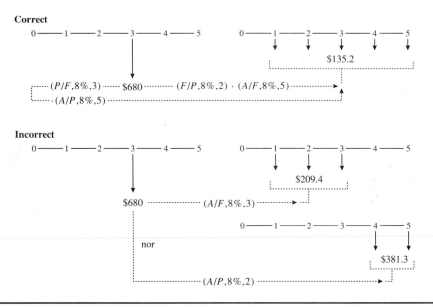

A **deferred annuity** is
a uniform cash flow in
periods $D + 1$ to N.

Deferred Annuities to Regular Annuities.   The difficulty with a **deferred annuity** is that the uniform cash flows occur in periods $D + 1$ to N, but not at the ends of the first D periods. Example 6.5 illustrates a common situation for a deferred annuity, in which a warranty covers all repair costs for an initial period of 1, 2, or more years. The warranty period equals D periods.

Another example of a deferred annuity would be project revenues that are $300K per year once a 4-year development period is completed. Thus, the deferred annuity would start in year 5, and D would equal 4.

Exhibit 6.3 illustrates an application of Equation 6.3, which converts a deferred annuity to a regular annuity. A formula with three factors that relied on Ps rather than Fs could easily be developed by finding a worth at the end of period D, then the PW, and then the annuity (see Problem 6.14).

$$\text{EAC} = A_D(F/A, i, N-D) \cdot (A/F, i, N) \tag{6.3}$$

Exhibit 6.3   EAC for a deferred annuity

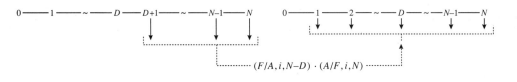

Example 6.5   EAC for Repair Costs after Warranty

North Central Power & Light has just purchased a generating unit that comes with a 2-year warranty. Once the warranty expires, repair costs are expected to average $3500 per year.

Find the EAC for the repair costs over the 15-year life of the generating unit. The interest rate is 7%.

Solution
This is a deferred annuity, since the warranty covers repair costs during the initial 2 years. Repair costs are incurred only for the last 13 years. Note that even this short 2-year warranty lowers the EAC for repairs to 20% below the $3500 per year.

$$\text{EAC} = 3500(F/A, 7\%, 13) \cdot (A/F, 7\%, 15)$$

$$= 3500 \cdot 20.141 \cdot .0398 = \$2805$$

Exhibit 6.4   EAW for Example 6.5—repair costs deferred by a warranty

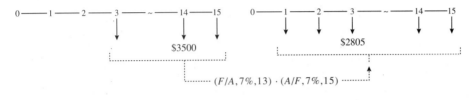

6.5 EAC FORMULAS FOR SALVAGE VALUES AND WORKING CAPITAL

Capital Recovery with Salvage Values.   A project's capital costs are its first cost and its salvage value, S. The EAC of these two cash flows is the capital recovery cost. If there is no salvage value, then only the $(A/P, i, N)$, or capital recovery factor, is needed.

There are two equivalent formulas for computing the capital recovery cost. The first, as shown in Exhibit 6.5, is a natural one; that is, the initial cash flow, P, and the final cash flow or salvage value, S, are multiplied by (A/P) and (A/F) factors, respectively. Note that Exhibit 6.5 is drawn as though the salvage value is positive; however, Equations 6.4 and 6.5 are correct even for negative salvage values $(S < 0)$, such as for a salvage or demolition cost.

$$\text{EAC} = P(A/P, i, N) - S(A/F, i, N) \tag{6.4}$$

$$\text{EAC} = (P - S)(A/P, i, N) + S \cdot i \tag{6.5}$$

Exhibit 6.5   EAC for capital recovery

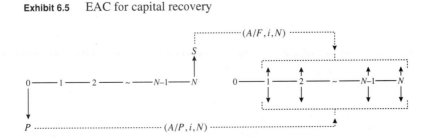

The second formula can be visualized as the drop in the asset's value $(P - S)$ that must be recovered over its life, plus the interest on the capital that is "tied up" in the asset during its life. The two formulas can be shown to be equivalent by setting them equal to each other, or by substituting $(A/P, i, N) = (A/F, i, N) + i$, or by example, as in Example 6.6.

Working Capital.   If an item's salvage value is assumed to equal its first cost, then Equation 6.5 is particularly useful, since its first term goes to 0. Two common examples are a project's land and its working capital. The value of land after adjusting for inflation is often assumed to be constant. Also, the U.S. income tax code does not allow land to be depreciated.

Working capital is money that is required at a project's beginning and that is recovered at the project's end.

Working capital, the money required for operations, is best explained by example. A manufacturing firm pays for its supplies, energy, labor, and purchased parts with working capital, which is recovered 6 weeks or 6 months later, when customers pay for the goods that have been manufactured and delivered. Working capital is even required by engineering design firms whose principal cost is employee labor. These employees must be paid every 2 weeks, but the firm may not receive payment from the client until 6 weeks after the job is complete when all accounts are cleared. Thus, a firm or a project requires working capital to pay for costs that will be recovered once the firm's product or service is paid for.

Working capital is important in practice, but it is often mistakenly omitted by engineers who are unfamiliar with accounting and finance. It is also often omitted from texts on engineering economy [Weaver]. In accounting and finance a definition of working capital that is based on the firm's current assets (cash + inventory + receivables + prepaid expenses) is used. However, for engineering economy a definition that is linked to individual projects is needed.

Example 6.6   EAC for HeavyMetal's Working Capital

HeavyMetal Recovery specializes in removing transuranic elements from hazardous waste streams. It is considering a large expansion of an existing facility in eastern Washington. The expansion requires $3M in working capital. If HeavyMetal's interest rate is 12%, find the EAC for its required working capital.

Solution
The working capital requirement is best found using Equation 6.5, since working capital's salvage value equals its first cost.

$$\text{EAC}_{\text{working capital}} = (3M - 3M) \cdot (A/P, .12, 30) + 3M \cdot .12$$
$$= 0 + 360K = \$360K$$

Using Equation 6.4 leads to the same answer:

$$\text{EAC}_{\text{working capital}} = 3M(A/P, .12, 30) - 3M(A/F, .12, 30)$$
$$= 3M \cdot .12414 - 3M \cdot .00414$$
$$= 372.42K - 12.42K = \$360K$$

6.6 PERPETUAL LIFE

Assumption and Formulas.   Chapter 3 mentioned that the limit of $(A/P, i, N)$ as N approaches infinity is i. Assuming that N equals infinity is assuming perpetual life. Thus, assuming perpetual life allows substituting i for a table lookup or for the more complicated $(A/P, i, N)$ formula. This assumption was very useful before calculators and computers existed, and it can still provide a very close approximation. Tunnels, roadbeds, rights-of-way, earthworks, and dams often have lives that are very close to perpetual in an economic sense. Capitalized cost was defined in Chapter 5 to support calculations for these long-lived assets.

Rather than deriving the relationship as a mathematical exercise in limits, it can be derived intuitively from a savings account perspective, as shown in Chapter 5. If you have amount P in the bank, then the annual interest is $A = i \cdot P$. That exact amount may be withdrawn from the bank each year without changing the problem for later years. If less than $i \cdot P$ is withdrawn each year, capital accumulates; if more than $i \cdot P$ is withdrawn, then eventually the principal is all gone. Equations 6.6 and 6.7 state this in terms of the factors (as did Equations 5.2 and 5.3), and Example 6.7 applies this assumption.

$$\text{\Large ✳}\quad (A/P, i, \infty) = i \qquad\qquad (6.6)$$

$$\text{\Large ✳}\quad (P/A, i, \infty) = 1/i \qquad\qquad (6.7)$$

Example 6.7 EAC for a Right-of-Way

The Department of Highways will be widening a two-lane road into a four-lane freeway this summer. The project requires the purchase of $22M worth of land for the additional two lanes. If the department uses an interest rate of 6%, what is the EAC of the right-of-way?

Solution
Rights-of-way do not wear out, and roads tend to be used for centuries. Thus, the perpetual life assumption seems particularly appropriate here:

$$\text{EAC} = i \cdot P = .06 \cdot 22M = \$1.32M$$

N vs. Infinity.   Using infinity as an approximation of N can be a good approximation for the right combination of i and N. The higher i is, the lower N can be for a good approximation. Exhibit 6.6 illustrates how the $(A/P, i, N)$ factor approaches its limit of i for $i = 5\%$, 10%, and 20%. Exhibit 6.7 illustrates the difference between i and $(A/P, i, N)$, if perpetual life is assumed as an approximation.

Exhibit 6.6 $(A/P, i, N)$ vs. N (approaching i)

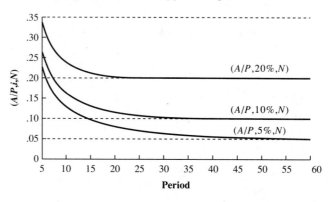

Exhibit 6.7   Approximation error if perpetual i is used rather than $(A/P, i, N)$

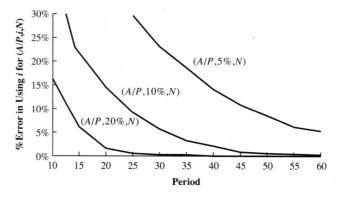

Arithmetic Gradients and Perpetual Life.   As shown in Exhibit 6.8, perpetual life for a gradient can be viewed as a series of conversions from uniform amounts of G to single payments. The first G begins at period 2's end and continues in perpetuity. Dividing by i converts this to a single payment at the end of period 1.

The second G begins at period 3's end and continues in perpetuity. Dividing by i converts this to a single payment at the end of period 2. This continues for periods 4, 5, etc. Each single payment is the same, and the first one is in period 1, so the series of single payments is an A. The result is Equation 6.8.

$$(A/G, i, \infty) = 1/i \tag{6.8}$$

Multiplying Equation 6.8 by $(P/A, i, \infty) = 1/i$ leads to Equation 6.9. Example 6.8 illustrates one potential application.

$$(P/G, i, \infty) = 1/i^2 \tag{6.9}$$

Exhibit 6.8 Perpetual gradient to perpetual uniform series

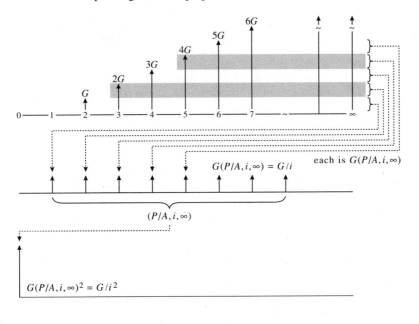

Example 6.8 Perpetual Tourist Gradient

The National Monuments Commission is considering the imposition of viewing fees for several monuments. At one monument, these fees would bring in $800K for the first year, and this should increase by $50K per year as the number of tourists increases. If the government uses an interest rate of 10% and perpetual life is assumed, what is the EAW for the income stream?

Solution

$$\text{EAW} = 800K + 50K(A/G, i, \infty) = 800K + 50K/.1 = \$1.3M$$

Note that the *capitalized* present worth of this is 1.3M/i, or $13M.

Exhibit 6.9 shows that $(P/G, i, N)$ gradient factors have a larger approximation error if i^2 is used, since they approach their limits somewhat more slowly than do $(A/P, i, N)$ factors. Note that the curves for $(A/P, i, N)$ in Exhibit 6.7 match the curves that would represent $(A/G, i, N)$, since both are $1/i$.

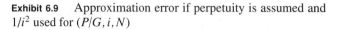

Exhibit 6.9   Approximation error if perpetuity is assumed and $1/i^2$ used for $(P/G, i, N)$

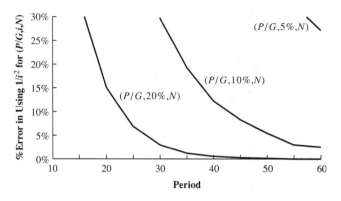

6.7 REPEATED RENEWALS

Often in a major project there are some costs that are incurred periodically, but not every year. For example, a road is resurfaced every 8 years over a 40-year life; a building is painted every 5 years over a 30-year life; or an assembly line is reconfigured every 2 years over a 20-year life. However, all of these costs are incurred over the life of the road, the building, or the assembly line. Furthermore, the same life must be used for computing the total cost of all components.

A more complex example would be purchasing a machine that must be maintained, overhauled, and sold for a salvage value over T years. What is the annual cost of 6 of these machines purchased one after another every T years over a period of $N = 6T$ years?

EAW and EAC provide a substantial shortcut for problems in which repeated renewals occur. For the shortcut to work, the repetition should be consistent. In other words, the repeated cost must be constant and the interval must be constant. Then the **principle of repeated renewals** is that the EAW or EAC for one machine over T years equals the EAW or EAC for six successive machines over $6T$ years.

If cash flows occur every T periods and if this event repeats for N years, where N is an integer multiple of T, then the **principle of repeated renewals** implies that the EAW or EAC over T periods equals the EAW or EAC over N periods.

Repetition for Every Subperiod.   The easiest case has a repeated cost every T years, the first cost occurs at time 0 or at year T, and the study period, N, is an integer multiple of T. In this case the EAC or EAW over the T-year period exactly matches the EAC or EAW over the N-year period. In Example 6.9, the repeated cash flow occurs at time 0, so the EAC for T or N periods is computed using $(A/P, i, T)$. If the repeated cash flow occurred at year T, then the factor would be $(A/F, i, T)$.

If there are multiple items in a large project, then the EAC can be found for each and they can simply be added together, as in Example 6.10.

Example 6.9 Replacing Pumps in a Sewage Treatment Plant

Crystal Lake Village is building a sewage treatment facility that has a 35-year design life. This facility includes a pump that costs $13,000 and that will last 7 years. The pump has no salvage value, and its energy costs are included with other energy costs. If $i = 8\%$, find the EAC for the series of five pumps.

Solution
If you were asked to find the first pump's cost over its 7-year life, the answer would be obvious:

$$EAC = 13K(A/P, 8\%, 7) = 13K \cdot .1921 = \$2497$$

This is also the cost for the second pump over its 7-year life, which is years 8 through 14. In fact, $2497 is the EAC for all five pumps over each of their 7-year lives. Together, these EACs jointly cover the 35-year period. Since $2497 is uniform over all 35 periods, this is the EAC for the full 35-year life.

$$EAC_{1-35} = \$2497$$

Example 6.10 Three Sisters Construction

George graduated in civil engineering 10 years ago and his mother and her two sisters are ready to let him open a branch office of their firm in another state. Their pattern is to own only one grader, one Caterpillar tractor (cat), one loader, and two tractor trailers for hauling at any time. If more machinery is needed, they hire it, just as they hire all dump trucks or belly dumps.

Use an interest rate of 10% and a study period of 30 years to find the EAC of the owned equipment.

	Grader	Cat	Loader	Tractor Trailer (each)
First cost	$195K	$365K	$270K	$140K
O&M	$28K	$40K	$50K	$18K
Salvage value	$40K	$90K	$40K	$21K
Life	6	5	10	15

Solution
The easiest solution is to compute EACs over 6, 5, 10, and 15 years, respectively, for the grader, cat, loader, and each tractor trailer. Over the 30 years, six graders will be purchased one after another. Since each one has the same EAC of $67.6K, that cost is constant over the 30 years. Similarly, the cost of purchasing five cats successively will cost $122K per year over 30 years.

Since 30 years is a multiple of each life, the total EAC equals the sum of two tractor trailers and one of each of the other pieces of equipment. In fact, 30 years is the least common multiple of the four lives, so it is the shortest horizon in which this approach will work.

$$EAC_{grader} = 195K(A/P, .1, 6) + 28K - 40K(A/F, .1, 6)$$
$$= 195K \cdot .2296 + 28K - 40K \cdot .1296 = \$67.59K$$

$$EAC_{cat} = 365K(A/P, .1, 5) + 40K - 90K(A/F, .1, 5)$$
$$= 365K \cdot .2638 + 40K - 90K \cdot .1638 = \$121.55K$$

$$EAC_{loader} = 270K(A/P, .1, 10) + 50K - 40K(A/F, .1, 10)$$
$$= 270K \cdot .1627 + 50K - 40K \cdot .0627 = \$91.42K$$

$$EAC_{truck} = 140K(A/P, .1, 15) + 18K - 21K(A/F, .1, 15)$$
$$= 140K \cdot .1315 + 18K - 21K \cdot .0315 = 35.75K$$

Adding these together, including two trucks, gives a total of $352K.

Capitalized Cost.   One application of repeated renewals is in problems with an assumption of perpetual life. Obviously, infinity is a multiple of any shorter life. Thus, the EAC is found for each item in the problem, then the total EAC is computed. The capitalized cost is found using $P = A/i$, as shown in Example 6.11.

Example 6.11   Capitalized Cost for a Dam

A hydroelectric dam has a design life of 90 years. The dam will cost $275M to build. The electric turbines will last 30 years and cost $45M. The interest rate is 8% and all salvage values are 0. What is the EAC? What is the capitalized cost?

Solution

The EAC is found by using the 90-year study period, and then the capitalized cost is found by assuming that this EAC will continue forever.

$$EAC = 275M(A/P, 8\%, 90) + 45M(A/P, 8\%, 30)$$
$$= 275M \cdot .0801 + 45M \cdot .0888 = \$26.02M$$
$$\text{Capitalized cost} = 26.02M/.08 = \$325.3M$$

Repeated Renewals with Neither an Initial nor a Final Cash Flow.   In the previous examples, each T-year cycle was repeated throughout. However, a repainting or repaving cost will occur neither at time 0 nor at time N. You do not repave as part of the initial construction, nor just as the project is terminated.

But the technique of repeated renewals can still be applied by partitioning the first cost among the different series (see Example 6.12a). Another similar, and mathematically equivalent, approach is to create two balancing cash flows that are opposite in sign in year N and to partition one (see Example 6.12b). A third approach is demonstrated in Example 6.12c.

Example 6.12a   Building Maintenance Using (A/P)s

Find the EAC for a building that has a life of 90 years, a first cost of $1.2M, and no salvage value. The building will be repainted every 5 years for $80K, reshingled every 15 years for $225K, and replumbed and rewired every 30 years for $750K. Find the EAC, if $i = 10\%$.

Solution

The three periodic repairs occur neither at time 0, nor at year 90. The $1.2M in first cost can, however, be partitioned into four separate cash flow series, as shown in Exhibit 6.10. These are, respectively, $80K, $225K, $750K, and $145K. The first three are chosen to match the repair series, and the fourth is chosen so that the total time 0 cash flow is still $1.2M. (Note that this last cash flow can be whichever sign is required.)

Exhibit 6.10 Cash flow diagrams for repeated maintenance using (A/P)s

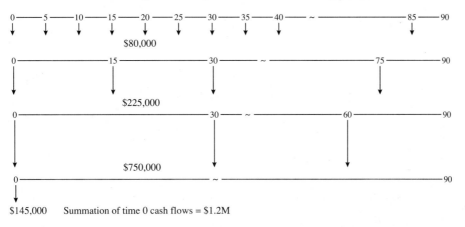

$145,000   Summation of time 0 cash flows = $1.2M

If the four cash flow diagrams in Exhibit 6.10 are added together, then the original cash flows result. Thus, the EAC of 25 different cash flows can be found by the following simple equation:

$$EAC = 80K(A/P, .1, 5) + 225K(A/P, .1, 15) + 750K(A/P, .1, 30) + 145K(A/P, .1, 90)$$

$$= 80K \cdot .2638 + 225K \cdot .1315 + 750K \cdot .1061 + 145K \cdot .1000$$

$$= \$144.8K$$

Example 6.12b   Building Maintenance Using (A/F)s

Resolve Example 6.12a without partitioning the initial cash flow. Assume instead that the cost of repainting, reshingling, etc., occurs in the Tth period for each repetition.

Solution

The solution is to compute the EACs using (A/F) factors. However, in year 90 (since the project is being terminated), none of the costs are really incurred. The solution, shown in Exhibit 6.11, is to create two cash flows of $1055K in year 90. The negative cash flow is partitioned among the three repair series so that each has a "cash flow" in year 90 that yields the following equation. The positive cash flow is equated to a 90-year annuity.

$$EAC = 1200K(A/P, .1, 90) + 80K(A/F, .1, 5)$$

$$+ 225K(A/F, .1, 15) + 750K(A/F, .1, 30) - 1055K(A/F, .1, 90)$$

$$= 120,023 + 13,104 + 7082 + 4559 - 20$$

$$= \$144,748$$

Exhibit 6.11   Cash flow diagrams for repeated maintenance using (A/F)s

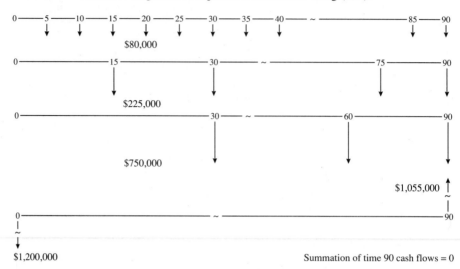

Example 6.12c   Building Maintenance without Balancing Cash Flows

Resolve Example 6.12a without partitioning the cash flow at time 0 or creating balancing cash flows in year 90.

Solution
Repainting happens at the end of years 5, 10, ..., 85. If an $(A/F, .1, 5)$ factor were applied, an 85-year series would be created. A PW can be computed, and then a 90-year EAC computed:

$$\text{EAC} = 80\text{K}(A/F, .1, 5)(P/A, .1, 85)(A/P, .1, 90) = \$13,102$$

The reshingling happens at the end of years 15, 30, ..., 75. The calculation for the reshingling is:

$$\text{EAC} = 225\text{K}(A/F, .1, 15)(P/A, .1, 75)(A/P, .1, 90) = \$7077$$

The calculation for the rewiring and replumbing is:

$$\text{EAC} = 750\text{K}(A/F, .1, 30)(P/A, .1, 60)(A/P, .1, 90) = \$4545$$

The calculation for the initial construction is:

$$\text{EAC} = \$1200\text{K}(A/P, .1, 90) = \$120,023$$

The total EAC is:

$$\text{EAC} = 13,102 + 7077 + 4545 + 120,023 = \$144,747$$

6.8 SPREADSHEETS AND ANALYZING LOAN REPAYMENTS

One of the most important periodic cash flows that firms, agencies, and individual engineers incur is payments on loans and bonds. Usually, these payments are constant for each period. Spreadsheets are not required to solve the following problems, but they provide a powerful means of confirming that the factor-based approaches are correct. Three problems will be analyzed:

1. Finding the balance due on a loan.

2. Shortening the time to pay off a loan by making extra payments.

3. Deciding how a payment is split between principal and interest.

In Chapters 2, 3, and 4, some schedules of loan repayment or amortization were calculated. Here, the goal is to answer similar questions without having to calculate complete amortization tables.

Finding the Balance Due.   To answer the question of how much is still owed after 12 monthly payments are made on a 3-year car loan, an amortization table with nine more rows than are shown below could be constructed. In this case, the initial loan was for $6800, and the nominal interest rate is 12% compounded monthly. Each of the 36 payments equals $225.86, or 6800(A/P,.01,36). The interest each month is 1% of the month's initial balance. The principal payment for each month is $225.86 minus the interest payment.

Month	Beginning Balance	Interest	Principal Payment	Ending Balance
1	$6800.00	$68.00	$157.86	$6642.14
2	6642.14	66.42	159.44	6482.70
3	6482.70	64.83	161.03	6321.67

An easier approach is to recognize that after any payment is made, the remaining balance is the PW of the remaining payments. This is true because interest is paid in full after each payment, and later payments are simply based on the balance due. For the above example, the amount owed after 12 payments, F_{12}, is:

$$F_{12} = (P/A, i, N_{\text{remaining}}) \cdot \text{payment}$$

$$= (P/A, 1\%, 24)225.86 = 225.86 \cdot 21.240 = \$4797.27$$

This approach can be applied within a spreadsheet to create a simple table of amount owed by month, without creating any other columns. Exhibit 6.12 illustrates this, where the key function is @ PV(payment, $i, N-t$). Exhibit 6.12 also includes a simple amortization table as a check.

Exhibit 6.12   Finding the balance due on a loan

	A	B	C	D	E
1	$6,800	initial amount			
2	36	N, monthly periods			
3	1%	i, monthly interest rate			
4	$225.86	PMT, monthly payment			
5			Beginning		Remaining
6	t	@ PV(PMT,i,N-t)	Bal.	interest	Bal.
7	1	6642.14	6800.00	68.00	6642.14
8	2	6482.71	6642.14	66.42	6482.71
9	3	6321.68	6482.71	64.83	6321.68
10	4	6159.04	6321.68	63.22	6159.04
11	5	5994.77	6159.04	61.59	5994.77
12	6	5828.86	5994.77	59.95	5828.86
13	7	5661.29	5828.66	58.29	5661.29
14	8	5492.05	5661.29	56.61	5492.05
15	9	5321.11	5492.05	54.92	5321.11
16	10	5148.46	5321.11	53.21	5148.46
17	11	4974.09	5148.46	51.48	4974.09
18	12	4797.97	4974.09	49.74	4797.97

@ PV($A$4, $A$3, $A$2-A18)   +E17   +C18*$A$3   +C18+D18-$A$4

Shortening the Time to Payoff by Increasing Payments.   Corporations and individuals sometimes find that paying off debt can be a good investment. For example, consider an individual with a 30-year mortgage for $100,000 at a nominal interest rate of 9%. Since payments are monthly, there are 360 periods and the interest rate per period is .75%. The monthly payment is

$$100K(A/P, .0075, 360) = 100K \cdot .00805 = \$805$$

Extra payments are earning an effective interest rate of $9.38\% = 1.0075^{12} - 1$, which may be the highest safe rate available for investments. If you have outstanding balances on your credit cards, paying those off is an even better investment.

However, one common question is: How much sooner will the debt be paid off? Until the debt is paid off, any early payments are essentially locked up. Those early payments reduce the balance owed, but in terms of the payment schedule, the extra early payments merely replace payments at the end of the schedule. Until the debt is paid off, then, the same payment amount is owed each month.

Spreadsheets are convenient for this problem because loans are often made at fractional interest rates. For example, an auto loan might be made at a nominal rate of 13% with monthly compounding; that is, 1.08333% per month. Solving for the number of remaining periods often requires linear interpolation, and with noninteger interest rates, a two-variable interpolation is required. As shown in Example 6.13, spreadsheets are easier.

Example 6.13   110% Payments on a 30-Year Mortgage

Ilene has decided to celebrate earning her P.E. license by buying a condo. Her bank has told her that she can qualify for a $100,000, 30-year mortgage at the current nominal interest rate of 9 ¼ %. The monthly payment at that rate is $822.68. She believes she can afford to pay 10% more than that. If she makes the extra payment each month, how soon will her mortgage be paid off?

Solution

Mathematically, the problem is simple. At time 0 the $100,000 is borrowed, and for periods 1 through N, payments of $904.95 are made. At an interest rate of .0925/12 or .0077083, what is N? Stated using a cash flow equation, it is:

$$0 = 100,000(A/P, .0077083, N) - 904.95$$

This can be solved easily with a financial calculator (the answer is 249 payments, or 20 to 21 years).

Exhibit 6.13 shows an easy-to-construct spreadsheet that has solved the problem for 10%, 20%, and 30% extra payments. The formula at the bottom of the exhibit could be copied into all relevant cells. This formula simply computes the monthly payment by

Exhibit 6.13   Spreadsheet for accelerated payments

	A	B	C	D
1	$100,000	initial amount		
2	360	N, monthly periods		
3	9.25%	nominal annual interest rate		
4	0.77%	i, monthly interest rate		
5	$823	PMT, monthly payment = @ PMT(A1,A4,A2)		
6				
7	Guess	Total Paid Off for Each Payment Level		
8	for N	110%	120%	130%
9	240	98807		
10	241	98949		
11	246	99644		
12	248	99915		
13	249	100049		
14				
15	200		100498	
16	199		100285	
17	198		100071	
18				
19	170			101135
20	167			100258
21	166			99961

@ PV($A$5*E$8,$A$4,$A21) =

multiplying the required payment by 110%, 120%, or 130%. The PW of the guessed number of payments is then computed.

A trial-and-error substitution of Ns is used. For example, at 110% the first guess is 20 years, or 240 periods. The total amount repaid comes to less than $100K, so the number of payments is increased to 249, where the required amount has been repaid. (Note: in Chapter 10, the SOLVE FOR function available in many spreadsheets is described. The "solving for" method is used instead of the trial-and-error search for the number of periods that is illustrated here.)

How Much Goes to Interest? How Much Goes to Principal?   The split of any payment between principal and interest can be solved in one of three ways. First, and most laboriously, the amortization table can be constructed; this answers the question for all periods. Second, the split for period t can be constructed by calculating the balance owed at the end of period $t - 1$. Then the interest for period t is that balance multiplied by the interest rate.

The third approach uses functions such as those available in Quattro Pro or Excel. To find the interest and principal portions of the payment in period t, the Quattro Pro functions are @IPAYMT and @PPAYMT, respectively. For simple problems, both functions have the following format: @IPAYMT$(i, t, N, -P)$. Both functions have optional arguments that permit adding a balloon payment (an F) and changing from end-of-period payments to beginning-of-period payments. The Excel functions (IPMT and PPMT) have the same arguments. The Lotus 1-2-3 functions added in version 4 have the same names as Quattro Pro's, but the parameters appear in a different order. Exhibit 6.14 illustrates this approach using Quattro Pro functions.

6.9 SUMMARY

This chapter defined and applied equivalent annual measures, which occur uniformly at the ends of periods 1 through N. Equivalent annuities were calculated for irregularly placed overhaul expenses, deferred annuities, and working capital.

For equivalent annual worth (EAW), the standard sign convention applies, and projects with $EAW > 0$ are attractive. For equivalent annual cost (EAC) the sign convention is reversed, and projects with the lowest possible EAC are attractive.

To find the EAC of an overhaul or other expense that occurs between periods 0 and N, first a present or future value is calculated, and then the EAC is calculated. Similarly, to find the EAW for a deferred annuity, first a future value and then the EAW is calculated.

Calculating the EAC for working capital and real estate is simplified, since the salvage value is usually assumed to equal the first cost. Another simplification is assuming perpetual life, which also allows for calculation of the capitalized cost.

The use of EACs with periodic or repeated renewals greatly simplifies the calculation of economic worth for projects with periodic repainting, overhauls, etc. Similarly, the use of spreadsheets speeds calculations for different loan repayment approaches.

Exhibit 6.14 Using spreadsheet functions to compute the interest/principal split on a payment

	A	B	C	D
1	$100,000	initial amount		
2	240	N, monthly periods		
3	9.25%	nominal annual interest rate		
4	0.77%	i, monthly interest rate		
5	$915.87	PMT, monthly payment = @ PMT(A1,A4,A2)		
6				
7	period	Principal	Interest	Total
8	12	157.82	758.05	915.87
9	24	173.05	742.82	
10	36	189.75	726.11	
11	48	208.07	707.80	
12	60	228.15	687.71	
13	72	250.17	665.69	
14	84	274.32	641.54	
15	96	300.80	615.07	
16	108	329.84	586.03	
17	120	361.67	554.19	
18	132	396.58	519.28	
19	144	434.86	481.00	
20	156	476.84	439.03	
21	168	522.86	393.00	
22	180	573.33	342.53	
23	192	628.67	287.19	
24	204	689.35	226.51	
25	216	755.89	159.97	
26	228	828.86	87.01	
27	240	908.86	7.01	

@ IPAYMT($A$4,$A27,$A$2,-$A$1)

@ PPAYMT($A$4,$A27,$A$2,-$A$1)

REFERENCE

Weaver, James B., "Analysis of Textbooks on Capital Investment Appraisal," AACE Technical Monograph Series, EA-1, 1990.

PROBLEMS

6.1 What is the EAW (at 8%) of an inheritance that begins at $8000 per year and then falls by $500 per year over its 15-year life? (*Answer:* $5203)

6.2 What is the EAC (at 7%) of maintenance costs that start at $3500 per year and then climb by $825 per year over a 10-year life?

6.3   The ABC Company may buy a heat exchanger for $80,000 installed. The exchanger will save $20,000 per year over an 8-year life. If ABC's cost of money is 8%, determine the EAW.   (*Answer:* $6080)

6.4   Burke Brake Shoes may buy a new grinding machine. The best alternative costs $10,500. The operating cost for it is $6500 per year. The expected savings from the machine are $9600 per year. The machine has no salvage value after 5 years. Determine the machine's EAW at 5% interest.

6.5   Determine the first cost of a machine that has an EAC of $4100 and annual operating costs of $2012. The machine is expected to last 6 years and money is worth 7%.   (*Answer:* $9952)

6.6   Joe Smith is buying a new car for $15,000. Maintenance the first year is expected to be $100. In each year for the next 5 years, the costs are expected to increase by $75 per year, so that in the last year (year 6), the maintenance cost will be $475. If money is worth 10%, determine the EAC for the auto.

6.7   The Shout Soap Company's annual revenues will be $320,000 for 15 years from its new soap, Rub-A-Dub-Dub. The soap requires an initial investment of $900,000. The annual expenses of manufacturing and selling the soap equal $73,500. Determine the EAW if the interest rate is 15%.   (*Answer:* $92,600)

6.8   The Sandwich Company may buy a new piece of equipment for $25,000. The equipment's useful life is 4 years and its salvage value is $5000. The company's net annual earnings should increase by $8000 for 4 years. Determine the EAW if Sandwich's cost of money is 10%.

6.9   A certain machine costs $30,000. Expected revenues are $2500 per quarter for the next 6 years. The quarterly operating cost is $500. The machine can be sold for $1000 at the end of the 6 years. If interest is 8% compounded quarterly, determine the equivalent quarterly worth and the EAW.   (*Answer:* EAW = $1842)

6.10   Creekside Consulting and Construction (CCC) is considering the purchase of a minicomputer that costs $45,000. The estimated salvage value for the computer at the end of 6 years is $6000. Find the computer's EAC, if CCC uses an interest rate of 15% for economic evaluation.

6.11   A new roof will be needed for the civil engineering labs at the 30-year point of the labs' 50-year life. If the new roof will cost $140,000, what is the EAC of the new roof at $i = 5\%$?   (*Answer:* $1774)

6.12   There is a glut of office space in OilTown, and one potential landlord is offering 3 initial months free to new tenants who sign a 3-year lease. Oilfield Engineering's current offices are slated for demolition, and their potential lease with this landlord is for $4000 per month. If their interest rate is 1% per month, what is the equivalent monthly cost for a 3-year lease? What is a comparable monthly lease cost?

6.13 MicroMotors is evaluating a new line of products that will require several years of R&D. New lab equipment that costs $200,000 is needed, and salaries and other expenses will cost $300,000 per year for 4 years. Revenues from the new products begin at the end of year 5, and they are $250,000 per year higher than expenses. If MicroMotors uses an interest rate of 10% and a time horizon of 20 years, what is the EAW of this product line?

6.14 Develop an equation for deferred annuities that is similar to Equation 6.3 but that uses Ps rather than Fs.

6.15 Find the EAC for a process that costs $35,000 to install and $2750 per year to operate. At the end of its 9-year life, the process has an $8000 salvage value. In year 5 a major overhaul is required at a cost of $18,000. The firm's discount rate is 8%. *(Answer:* $9674)

6.16 A project will cost $250,000 to start, $80,000 per year for operations, and $45,000 for working capital. The annual income is $170,000 per year for the 9 years the project will operate. Find the project's EAW at an interest rate of 10%.

6.17 A mining venture has a first cost of $6M, annual operating costs of $2M, annual revenues of $3.2M, working capital requirements of $1.5M, a life of 10 years, and a final reclamation cost of $4M. If i is 15%, what is the EAW?

6.18 Does the annual cost of working capital depend on the project's time horizon?

6.19 A small town's sewer system costs $1M installed. With proper maintenance, it should last indefinitely. Twenty percent of the cost will be for the pumping system, which must be replaced every 20 years. Sixty percent of the cost will be for the tiles and pipes, which will be replaced every 30 years. Annual maintenance costs are estimated to be $3000 per year. Right-of-way and earthworks account for the rest of the cost. If the town uses 12% as its cost of money, determine the EAC. *(Answer:* $128.3K)

6.20 The right-of-way and the roadbed for a new highway will cost $4.5 million. The pavement will cost $2.5 million, and it has a much shorter life of 20 years. If the discount rate is 9%, find the project's capitalized cost.

6.21 Bryson R. Tinfoot IV wants to endow a position in perpetuity for a chaired professor at Midstate University. Such a position requires $125,000 per year. If the funds earn interest at 7%, what must he deposit now?

6.22 The State Parks Department is considering the purchase of a marsh to form a bird sanctuary. Bird watching is valued at $2 per hour; the state's interest rate is 5%; and perpetual life is assumed. Viewing hours are estimated at 40,000 for the first year, with increases of 4000 per year. What is the EAW for the sanctuary? What fraction of the EAW is due to the expected increases?

6.23   By letting N approach infinity, show that the following are true:
a.   Equation 6.6 for $(A/P, i, \infty)$
b.   Equation 6.8 for $(A/G, i, \infty)$
c.   Equation 6.9 for $(P/G, i, \infty)$

6.24   What is the EAC of a road that costs \$450,000 to build and \$12,000 per year for street cleaning? The road is restriped every 2 years for \$28,000 and repaved every 10 years for \$290,000. Assume that the road will last 60 years and that the interest rate is 8%.   (*Answer:* \$81.58K)

6.25   A proposed steel bridge has a life of 50 years. The initial cost is \$350,000, and the annual maintenance cost is \$12,000. The bridge deck will be resurfaced every 10 years (in years 10, 20, 30, and 40) for \$90,000, and anticorrosion paint will be applied every 5 years (in years 5, 10, 15, ..., 45) for \$22,000. If the interest rate is 5%, what is the EAC?

6.26   A golf complex has a first cost of \$21M, annual O&M costs of \$1.7M, salvage value in year 100 of \$6M, clubhouse renovation every 20 years costing \$3M, and reseeding every 5 years starting in year 5 costing \$.4M. Find the EAC for a 100-year horizon if $i = 8\%$.

6.27   A proposed tunnel has a first cost of \$800M, annual O&M costs of \$14M, a salvage value in year 100 of \$225M, repaving every 10 years at \$45M, and repainting lines and mending railings every 5 years at \$1.2M (note that this cost is included in the repaving cost for the years where repaving is done). Find the EAC ($i = 5\%$) for a 100-year horizon.   (*Answer:* \$57.92M)

6.28   A new airport expansion will cost \$850 million. Of the total, \$550 million is for land acquisition and major earthworks, which will last as long as the airport is used, and \$200 million is for terminal buildings, which will last 50 years. These buildings will cost \$25 million per year in O&M. The last \$100 million will be spent on runways. These will also last forever, with a major repaving every 20 years at a cost of \$40 million. What is the expansion's EAC at $i = .08$ over a perpetual life?

6.29   Determine the EAW of the following cash flow at 12% interest:

Year	0	1	2	3	4	5	6
Cash flow	−\$18K	5K	6K	3K	7K	4K	3K

6.30   Brown Boxboard is considering investing in new automated production equipment. What is the EAC of each alternative, if i is 6%?

	Stamp-It-Fast	Stamp-Quicker	Fast-Press
First cost	$75,000	$60,000	$55,000
Annual benefit	7500	6000	6500
Annual cost	3000	2000	1500
Salvage value	13,000	9000	6000
Life (years)	8	8	8

6.31 A new automobile costs $28,000. The car's value will decrease 15% the first year and 10% each year thereafter. The maintenance costs are expected to be $200 the first year, with an annual increase of $75. If an interest rate of 6% is used, determine the EAC if the car is sold after 5 years.

6.32 A new car is purchased for $10,000 with a 0% down, 9% interest rate loan. The loan's length is 4 years. After making 30 monthly payments, the owner desires to pay off the loan's remaining balance. How much is owed?   (*Answer:* $4178)

6.33 For an $85,000 mortgage with a 30-year term and a 12% nominal interest rate, what is the monthly payment? After the first year of payments, what is the outstanding balance? After the first 10 years of payments, what is the outstanding balance?

6.34 For the mortgage in Problem 6.33, how much interest is paid in month 25? How much principal?

6.35 How much extra do you have to pay each month for the loan in Problem 6.33 if you want to pay it off in 20 years?   (*Answer:* $30.88)

6.36 A 30-year mortgage for $140,000 is issued at a 9% nominal interest rate. What is the monthly payment? How long does it take to pay off the mortgage if $1500 per month is paid? If $2000 per month is paid? If double payments are made?

6.37 A construction firm owns a concrete batching plant on which there is a $150,000 mortgage. The mortgage's interest rate is 7% and it is to be paid off in 25 equal annual payments. After the 13th payment, the firm refinances the balance at 6%, to be paid off in 35 equal annual payments. What will be the firm's new annual payments?

6.38 Whistling Widgets (WW) may establish a line of credit at First Local Bank. The line of credit would permit WW to meet its short-term cash needs on a routine basis, rather than having to process loan requests each time. A typical loan under either the old or the new basis is paid back 6 months later in a lump sum of principal and interest. Over the last 5 years, WW has had such loans outstanding about half the time.

However, a $100,000 line of credit requires that WW continuously maintain at least $20,000 in an account that pays 6% compounded monthly. If invested internally, WW would earn a nominal 18% on this required deposit. What equivalent annual loan-processing cost is equivalent to the cost of the required deposit?

6.39   In evaluating projects, NewTech's engineers use a discount rate of
15% for a before-tax analysis. One year ago, a robotic transfer machine was installed at a cost of $38,000. At the time, a 10-year life
was estimated, but the machine has had a down-time rate of 28%,
which is unacceptably high. A $12,000 upgrade should fix the prob-
lem, or a labor-intensive process costing $3500 in direct labor per year
can be substituted. The plant estimates indirect plant expenses at 60%
of direct labor costs, and it allocates front-office overhead at 45% of
plant expenses (direct and indirect). The robot has a value in other
uses of $15,000. What is the EAC for upgrading? For switching to the
labor-intensive process?

CHAPTER 7

RATE OF RETURN

The Situation and the Solution

A project's return may have to be compared with the financing cost. Firms, government agencies, and individuals must compute the cost of financing projects.

The solution is to calculate the internal rate of return for a project or for a financing source. That rate can be compared with an interest rate for the time value of money.

Chapter Objectives

After you have read and studied the sections of this chapter, you should be able to:

Section 7.1
Define the internal rate of return (IRR).

Section 7.2
State the assumptions for the IRR.

Section 7.3
Calculate the IRR for simple problems.

Section 7.4
Calculate the IRR for a variety of loans and leases.

Section 7.5
Use spreadsheets to calculate the IRR.

Section 7.6

Identify problems where a meaningful IRR may not exist because of multiple sign changes in the cash flow diagram.

Section 7.7

Use the project balances over time as a sufficient condition for a meaningful IRR when multiple sign changes exist.

Key Words and Concepts

Internal rate of return (IRR)   The interest rate that makes both the PW and EAW equal 0.

Multiple sign changes   Occur when there are two or more switches between positive and negative cash flows in a sequence of cash flows.

Loan   A cash flow sequence in which the initial cash flows are positive and there is a single sign change to a series of negative cash flows, which "repay" the initial cash flows.

Investment   A cash flow sequence in which the initial cash flows are negative and there is a single sign change to a series of positive cash flows, which are the return on the initial cash flows.

Evaluating the IRR   If a loan's IRR is $\leq i$, then the loan is attractive. An attractive investment's IRR is $\geq i$.

Investment functions   Spreadsheet functions that calculate the IRR, given one each of *P, A, F,* and *N* (also called *annuity* functions).

Block functions   Spreadsheet functions that calculate the IRR, given the cash flow in every period.

Project balances over time   For a loan these are the amounts owed at the end of each period. For an investment, these are the unrecovered investment balances.

7.1 THE INTERNAL RATE OF RETURN

The **internal rate of return (IRR)** is based only on a project's cash flows, which is the basis for the *internal*. As the interest rate that makes the PW equal 0, it is the project's rate of return. Since the PW equals 0, all other equivalent measures calculated at the IRR, such as the EAW, are also equal to 0.

The **internal rate of return (IRR)** is the interest rate that makes both the PW and EAW equal 0.

For the IRR to exist, both costs and benefits must be defined. If only a project's benefits are defined, then the PW and EAW are positive for all interest rates. If there are only costs, then the PW and EAW are negative. Only if there are costs *and* benefits can the PW and the EAW equal 0 for some interest rate.

Example 7.1's IRR of 5.24% is easy to calculate, and it is a good measure of investment quality. We have an immediate, intuitive feel for how good the investment is. Other measures, such as PW, require an interest rate, a calculated

PW, and some feel for the relative scale of the initial investment and the PW. Investments of $20K and $2M, each with a PW of $4500, are *not* equally attractive.

Example 7.1   Interest Rate on an Investment

Suppose that Ralph invests $300 in a Series E savings bond that is cashed in 10 years later for $500. What is the IRR for this investment?

Solution
The first step is to draw the simple cash flow diagram.

As shown in the cash flow diagram, only a P and an F are involved. The unknown i that equates the two cash flows can be found by using Equation 3.2.

$$PW = 0 = -300 + 500/(1 + i)^{10}$$

$$i = \sqrt[10]{(500/300)} - 1 = \sqrt[10]{1.667} - 1 = 5.24\%$$

The internal rate of return method is the most convenient approach if you are not sure what interest rate should be used for the time value of money (especially if different levels of risk are an issue). As shown in Chapter 9, the IRR is very important in deciding what interest rate is used to evaluate projects.

The IRR is also used to evaluate different ways of raising funds. The cost of a loan, bond, lease, etc., is the interest rate charged over its life. Firms, agencies, and individuals will economically prefer the financing source with the lowest cost. Why borrow at 18% if you can borrow at 8%?

The chief disadvantage of the IRR approach has been removed by the general use of spreadsheets for economic analysis. Without a financial calculator or a spreadsheet, some problems require many computations and several interpolations to find an accurate answer.

Multiple sign changes occur when there are two or more switches between positive and negative cash flows in a sequence of cash flows.

The second disadvantage of the IRR sometimes occurs when the sequence of cash flows has **multiple sign changes.** For example, a problem with three negative cash flows, then four positive cash flows, and then two negative cash flows has two sign changes between positive and negative cash flows. As a result, it may not have a unique solution in i for PW = 0.

There are a variety of *external* or *modified* rates of return that have been proposed and used, but only the basic IRR is covered here.

7.2 ASSUMPTIONS

Our first assumptions match the ones made with present worth and equivalent annual measures. Those were:

1. Cash flows occur at time 0 and at the ends of periods 1 through N.
2. Cash flows are known; the horizon, N, is known; and there is no uncertainty.
3. The interest rate, i, is known.

To allow calculation of the IRR, two other assumptions are made. Item 4 is a necessary condition; without it the IRR cannot exist. Item 5 is a sufficient condition. Thus, if it is true, the IRR must exist; but a unique IRR can exist even if item 5 is not true.

4. Both positive and negative cash flows must be part of the cash flow equation.
5. The sequence of cash flows should have only one sign change between positive and negative cash flows.

Exhibit 7.1 illustrates three possibilities. All three satisfy assumptions 1–4, but only (a) and (b) satisfy assumption 5. The first two illustrate the common cash patterns of loans and investments. The third pattern is one with multiple sign changes.

Exhibit 7.1   Loans, investments, and multiple sign changes

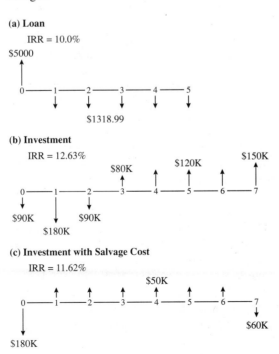

(a) Loan
IRR = 10.0%

(b) Investment
IRR = 12.63%

(c) Investment with Salvage Cost
IRR = 11.62%

A **loan** is a cash flow sequence in which the initial cash flows are positive and there is a single sign change to a series of negative cash flows, which "repay" the initial cash flow.

Loans.   Exhibit 7.1(a) corresponds to a standard **loan**, where money is first borrowed and then repaid. The initial positive cash flow might be cash or it might be an object. For example, it might be a car, where the dealer accepts the promise to repay the loan from the buyer. After the buyer's initial positive cash flow, there is one sign change to a series of loan payments, which are negative cash flows. In fact, cash flow sequences in which the positive cash flows come first are often referred to as loans, whether or not they involve a bank or other money institution.

Investments.   Exhibit 7.1(b) corresponds to an investment, in which the first costs are spread over a 3-year construction period. After the initial set of negative cash flows, there is a single sign change to a series of positive cash flows, which "repay" the initial investment. In fact, cash flow sequences in which the negative cash flows come first are often referred to as **investments.** Most engineering projects are investments. The initial costs are the negative cash flows to pay for designing and building the engineered product—facility, dam, etc. Then sales of the product, use of the building, electricity generated, etc. are the positive cash flows or benefits that justify the initial investment. From a bank's perspective, the *loans* to individuals and firms are really *investments* by the bank, which return future cash flows to the bank.

An **investment** is a cash flow sequence in which the initial cash flows are negative and there is a single sign change to a series of positive cash flows, which are the return on the initial cash flows.

Multiple Sign Changes.   Exhibit 7.1(c) corresponds to an investment project in which there is a salvage cost, such as might be required for environmental restoration. This is an example of a cash flow diagram that does not satisfy assumption 5 for calculation of the IRR. There are two sign changes in this sequence of cash flows. The initial negative cash flow of $180K is followed by a series of six positive cash flows, and then there is a final negative cash flow—for a total of two sign changes.

This cash flow diagram has a unique IRR of 11.62% even though it has multiple sign changes. This example shows that the condition of a single sign change is not necessary to calculate the IRR. However, assumption 5 is sufficient for the IRR to be defined (see also Section 7.6).

Reinvestment Assumption.   The interest rate at which an investment is evaluated is the assumed reinvestment rate, or the rate for the time value of money. This rate is specifically used in the computation of the PW, EAW, or EAC. For the IRR method, the reinvestment rate is the standard of comparison for a good loan or a good investment [Lohmann].

Applying the IRR Measure.   If the five assumptions are satisfied, then the IRR can be calculated. For investments such as Exhibit 7.1(b), the higher the IRR, the more attractive the project. For loans such as Exhibit 7.1(a), the lower the IRR, the more attractive the loan.

To **evaluate the IRR,** if an investment's IRR is $\geq$ *i*, then the investment is attractive. An attractive loan's IRR is $\leq$ *i*.

In both cases, the IRR is compared with the given interest rate for the time value of money. For an investment, if the IRR is greater than or equal to the interest rate, then the investment is attractive (see Example 7.2). For a loan, if the IRR is less than or equal to the interest rate, then the loan is attractive (see Example 7.3).

Example 7.2   Golden Valley Manufacturing

By replacing its assembly line conveyor with an asynchronous conveyor, Golden Valley Manufacturing will save $50,000 per year in rework, inspection, and labor costs. The asynchronous conveyor will cost $275,000, and its life is 15 years (when its salvage value equals 0). If Golden Valley uses a 10% interest rate, should the asynchronous conveyor be installed?

Solution

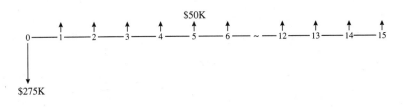

The PW equation is

$$PW = 0 = -275K + 50K(P/A, IRR, 15)$$

$$(P/A, IRR, 15) = 275K/50K = 5.500$$

By thumbing through the tables, we find that the IRR is between 16% and 17%. The next step is to interpolate for the IRR.

$$(P/A, .16, 15) = 5.575 \quad \text{and} \quad (P/A, .17, 15) = 5.324$$

So,

$$IRR = .16 + .01(5.575 - 5.500)/(5.575 - 5.324) = .1629 \text{ or } 16.3\%$$

This is an investment, and the IRR exceeds 10%, so the asynchronous conveyor should be installed. If knowing that the IRR exceeded 10% were enough, then the interpolation step could be skipped.

Example 7.3   IRR for a Simple Loan

Suppose that $5000 is borrowed at time 0, to be paid back in 5 equal annual payments of $1200. What is the IRR?

Solution

Writing an equation for the PW and setting it equal to 0 yields the following:

$$0 = 5K - 1200(P/A, \text{IRR}, 5)$$

$$(P/A, \text{IRR}, 5) = 4.1667$$

By thumbing through the tables, we find that the IRR is between 6% and 7%. The next step is to interpolate for the IRR.

$$(P/A, .06, 5) = 4.212 \quad \text{and} \quad (P/A, .07, 5) = 4.100$$

So,

$$\text{IRR} = .06 + .01(4.212 - 4.1667)/(4.212 - 4.100) = .064 \text{ or } 6.4\%$$

Mortgages typically have interest rates of 8% or more; rates for car loans are higher; and credit card rates are higher yet. This is an attractive loan at a relatively low interest rate.

7.3 FINDING THE IRR

If only two cash flows exist, the IRR problem can be solved using $F = P(1 + i)^N$, as in Example 7.1. However, if the problem involves even a uniform series (an A), then the equation is too complicated to solve directly and table lookup is required. Examples 7.2 and 7.3 illustrated the table lookup and interpolation procedures. If two or more factors are required to set the PW equal to 0, then an iterative solution or a spreadsheet is required, as in Examples 7.4 and 7.5.

Exhibit 7.2 returns to the loan and investment examples of Exhibit 7.1. It shows the graphs of PW vs. interest rate that occur for the assumptions given in Section 7.2. All three diagrams have a single $i \geq 0$, where the present worth is 0. The graph in Exhibit 7.2(b) is used to simplify the calculations shown in Example 7.4.

As the interest rate increases, the relative importance of later cash flows decreases. The higher the interest rate, the more heavily later cash flows are discounted in the equivalence calculations. For example, in Exhibit 7.2(a) the loan payments of $1318.99 per year are far less important at an interest rate of 20% than at 5%. Thus, the PW of the loan increases with i, as shown in Exhibit 7.2(a).

Similarly, if the investment in Exhibit 7.2(b) is evaluated, a higher interest rate decreases the value of the positive cash flows that come later. The PW of an investment decreases with i, as shown in Exhibit 7.2(b).

As the interest rate approaches infinity, the PW approaches the value of the period 0 cash flow. The limit of the curve for the loan in Exhibit 7.2(a) is $5000, while the limit for the investment with salvage cost in Exhibit 7.2(c) is –$180K. The limit for the construction example shown in Exhibit 7.2(b) is –$90K. However, because there are construction costs in periods 2 and 3, which would be ignored by an infinite interest rate, the PW at $i = 100\%$ is –$178K, which is lower than the –$90K limit as i approaches infinity, and the limit is approached from below.

Exhibit 7.2   Graphs of PW vs. *i* for loans and investments

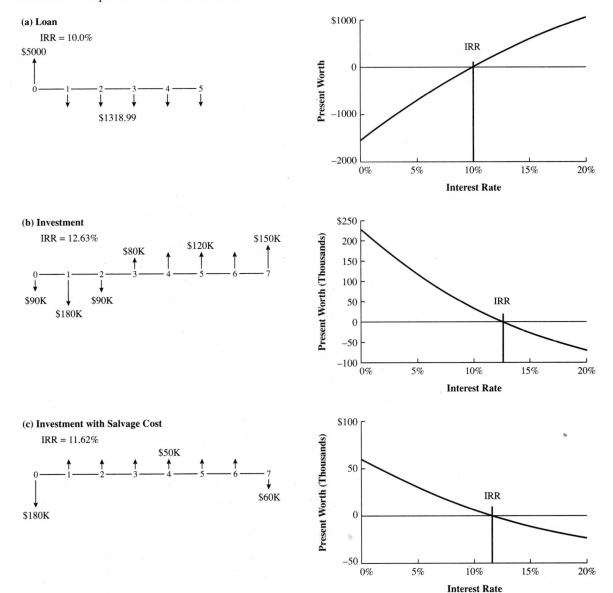

(a) Loan

IRR = 10.0%

(b) Investment

IRR = 12.63%

(c) Investment with Salvage Cost

IRR = 11.62%

Example 7.4   IRR for the Investment in Exhibits 7.1(b) and 7.2(b)

The information in Exhibit 7.2(b) can save us several steps because the IRR is known to be between 12% and 13%. Calculate the value of the IRR.

Solution
First the PWs at 12% and 13% are calculated, then the interpolation is done.

$$PW_{12\%} = -90K - 180K/1.12 - 90K/1.12^2$$
$$+ [80K + 120K(P/A, .12, 3)]/1.12^3 + 150K/1.12^7$$
$$= 7.482K = \$7482$$
$$PW_{13\%} = -90K - 180K/1.13 - 90K/1.13^2$$
$$+ [80K + 120K(P/A, .13, 3)]/1.13^3 + 150K/1.13^7$$
$$= -4.204K = -\$4204$$
$$IRR = .12 + .01 \cdot 7482/(7482 + 4204) = .1264 = 12.6\%$$

Hints and Shortcuts for Finding the IRR.   Without the graphical hint that the IRR for Example 7.4 was between 12% and 13%, the calculation of many PWs might have been required. The following hints can speed the process, as shown in Example 7.5.

1. Letting i approach infinity implies that the PW equals the period 0 cash flow. The value is useless for interpolation, but the sign can quickly define the search interval.

2. If PW_{i_1} and PW_{i_2} are opposite in sign, then the IRR is between i_1 and i_2. Thus, we could guess at a starting point, such as 10%, and calculate $PW_{10\%}$. If $PW_{10\%}$ and PW_∞ have different signs, then the IRR exceeds 10%.

3. If $PW_{10\%}$ and PW_∞ have the same sign, then the IRR is less than 10%. The next rate to try is 0%, since it is an easy rate to evaluate (the PW equals the sum of the cash flows). If there is a uniform series, $(P/A, 0\%, N) = N$ can be used.

4. Once the $PW_{0\%}$ and $PW_{10\%}$ values have been calculated, then rough interpolating or extrapolating can be used to estimate the next interest rate to try.
    For example, if $PW_{0\%}$ is 500 and $PW_{10\%}$ is −1000, then the IRR is about 3% to 4%—a third of the distance between 0% and 10%. Similarly, if $PW_{0\%}$ is 500 and $PW_{10\%}$ is 100, then the IRR is about 12% to 13%—a quarter of 10% higher than 10%.

5. The final interpolation should be between the two closest tabulated values (not between 0% and 10% or 10% and 20%).

6. For complicated cash flows, it may save time to use a single factor series as an approximation, as shown in Example 7.6. This will not find the IRR, but it may identify an interval for the initial calculations.

Example 7.5   Subsidized Student Loan

Golden Valley Manufacturing sponsors a student loan program for the children of employees. No interest is charged until graduation, and then the interest rate is 5%. Assume that George borrows $8000 per year at the beginnings of years 1 through 4 and that he graduates at the end of year 4. If equal annual payments are made at the ends of years 5 through 9, what is the internal rate of return for George's loan? Is this attractive to George?

Solution
The first step is to calculate the 5 equal annual payments. When he graduates, he owes $32,000 because no interest has been charged. The first payment is due one year after graduation, so each payment is

$$A = \$32{,}000(A/P, 5\%, 5) = 32K \cdot .2310 = \$7392$$

The cash flow series for the subsidized loan is

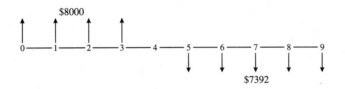

Solving for the IRR for this loan is perhaps most easily done by using $t = 4$ as the equivalence point. The first two interest rates tried are 0% (because it is easy) and 3% (because the subsidized rate will be below the 5% that is charged after graduation). The equivalent value, F_4, at the end of year 4 is a function of i:

$$F_4(i) = 8000(F/A, i, 4)(F/P, i, 1) - 7392(P/A, i, 5)$$

Setting this equal to 0 gives the equation where i = IRR. It is slightly simpler to use $(1 + i)$ rather than the 1-year (F/P) factor:

$$0 = 8000(F/A, \text{IRR}, 4)(1 + \text{IRR}) - 7392(P/A, \text{IRR}, 5)$$

$$F_4(0\%) = 8000 \cdot 4 - 7392 \cdot 5 = -\$4960$$

$$F_4(3\%) = 8000 \cdot 4.184 \cdot 1.03 - 7392 \cdot 4.580 = \$620.8$$

Since F_4 has opposite signs for 0% and 3%, there is a value of i between 0% and 3% that is the IRR. Because the value for 3% is closer to 0, the IRR will be closer to 3%. Try 2% next:

$$F_4(2\%) = 8000 \cdot 4.122 \cdot 1.02 - 7392 \cdot 4.713 = -\$1203.0$$

$F_4(2\%)$ is negative and $F_4(3\%)$ is positive, and interpolating between 2% and 3% leads to

$$\text{IRR} = 2\% + 1\%(1203/(1203 + 620.8))$$

$$= 2.6596\% \text{ or } 2.7\%.$$

The exact value is IRR = 2.66%. This rate is quite low, and it makes the loan look like a good choice.

Example 7.6   Using an Approximate Cash Flow Series

Find an initial approximation of the IRR for the following cash flows:

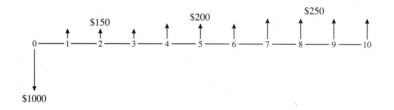

Solution

The average positive cash flow is about $200, so an approximate interest rate can be found using

$$PW = 0 = -1000 + 200(P/A, i, 10)$$

$$(P/A, i, 10) = 5 \Rightarrow i \approx 15+\%$$

Since the cash flow series has smaller returns at the beginning and larger cash flows at the end, this is likely to slightly overstate the IRR. The next step is calculating the PW at $i = 14\%$. Since $PW_\infty = -1000$, if $PW_{14\%}$ is greater than 0, then the IRR is between 14% and infinity. If $PW_{14\%}$ is less than 0, then the IRR is below 14%. (If you calculate the actual IRR, you should get 13.86%.)

7.4 LOANS AND LEASES

Interest rates or internal rates of return on level-payment loans are conceptually and numerically clear. The same principles apply to other ways of borrowing money. Often a problem's most difficult part is summarizing the words of a proposed loan in a cash flow diagram. In some cases, as shown in Examples 7.7 and 7.8, two cash flow diagrams are combined to show the difference between the choices.

Example 7.7 illustrates the impact of a cash discount, which is often available from sellers who also offer financing. Problem 7.22 is another example, where a car dealer offers a low interest rate or a rebate. Example 7.8 illustrates that a lease is another way of borrowing money. In a lease, someone else's money is used to purchase equipment that you use. Example 7.9 illustrates finding the IRR for a bond that is being sold at a discount.

Example 7.7   Interest Rate on a Loan with a Cash Discount

Golden Valley Manufacturing is considering the purchase of an adjacent property so that the warehouse can be expanded. If financed through the seller, the property's price is $40K, with 20% down and the balance due in 9 annual payments at 12%. The seller will accept 10% less if cash is paid. Golden Valley does not have $36K in cash, but it may be able to borrow it from another source. What is the IRR for the loan offered by the seller?

Solution
One choice is to pay $36K in cash. The other choice is to pay $8K down and 9 annual payments computed at 12% on a principal of $32K or

$$32K(A/P, .12, 9) = 32K \cdot .1877 = \$6006$$

annually. The cash flow diagrams are:

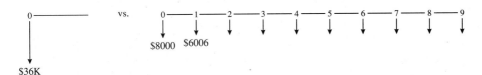

This choice between (1) $36K now and (2) $8K now and $6006 annually for 9 years can be stated in PW terms, as follows:

$$-36K = -8K - 6006(P/A, \text{IRR}, 9)$$

The true amount borrowed is the difference between the $8K down payment and the $36K cash price, or $28K. This is the amount that Golden Valley does not pay now. Collecting the terms, the PW equation is

$$0 = 28K - 6006(P/A, \text{IRR}, 9)$$

This can also be shown in a cash flow diagram, which is the difference of the two above.

In either case, the final equation for finding the IRR is:

$$\text{PW} = 0 = 28K - 6006(P/A, \text{IRR}, 9)$$
$$(P/A, \text{IRR}, 9) = 4.6622 \Rightarrow \text{IRR} = 15.66\%$$

on a borrowed amount of $28K.

Example 7.8 Leasing a Computer

Creekside Consulting is comparing the purchase of a computer network with leasing it. The purchase price is $36,000, or it can be leased for $3000 per month. In either case, Creekside expects to use the network for 3 years, when it plans on an upgrade. If Creekside buys the network, its salvage value is $5000 after 3 years. What is the IRR or cost of the lease?

Solution
The first step is to draw the diagrams for the two choices.

Setting the PWs of these two diagrams equal to each other leads to:

$$-36K + 5K(P/F, i, 36) = -3K - 3K(P/A, i, 35)$$

$$0 = 33K - 3K(P/A, i, 35) - 5K(P/F, i, 36)$$

In other words, if Creekside leases the computer, it avoids the expenditure of $33,000 at time 0. It incurs a cost of $3000 at the ends of months 1 through 35, and it gives up the salvage value of $5000 at the end of month 36. As shown in Section 7.5, this equation can be solved with $N = 36$ for both $A = -\$3K$ and $F = -\$2K$, which is easier for calculators and spreadsheets. The cash flow diagram for this statement (or the combined equation above) is:

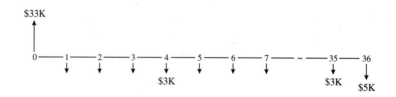

In either case, the final PW equation for the monthly interest rate is:

$$0 = 33K - 3K(P/A, i, 35) - 5K(P/F, i, 36)$$

Ignoring the extra $2000 in month 36 allows a quick calculation of the approximate monthly rate:

$$0 = 33K - 3K(P/A, i, 36) \quad \text{or} \quad (P/A, i, 36) = 11 \quad \text{so } i \approx 8+\%$$

Trying 9% leads to

$$PW_{9\%} = 33K - 3K \cdot 10.567 - 5K/1.09^{36} = \$1054$$

Since PW_∞ equals $33K, the next value to try is 8%:

$$PW_{8\%} = 33K - 3K \cdot 11.655 - 5K/1.08^{36} = -\$2278$$

Now, to interpolate for the monthly rate:

$$i = .08 + .01 \cdot 2278/(2278 + 1054)$$

$$= 8.68\% \text{ per month}$$

After converting this to an effective annual rate, we have the IRR:

$$IRR = 1.0868^{12} - 1 = 71.6\%$$

This extremely high interest rate suggests that this 3-year lease is a bad way to borrow money. Note that some computer rental rates are set to recover the cost in as few as 10 months, rather than the 12 months used here. In other cases, the monthly rental fee is a more reasonable 3% to 5% of the first cost.

Example 7.9 Calculate the IRR for a Discounted Bond

You own some municipal bonds that were issued at 8% with semiannual interest payments. They were to be repaid in 20 years. Calculate the effective interest rate for a $10K bond that you can now sell for $8200. The bond matures in 14 years.

Solution
The semiannual interest payments are 4% of the bond's face value, or $400. Selling the bond for $8200 brings in that amount immediately. However, then you forgo the $400 every 6 months and $10K after 14 years. There are 28 six-month periods remaining. The following cash flow diagram emphasizes that selling the bond is a way to borrow funds, since it is a loan with an initial positive cash flow followed by negative cash flows for the forgone interest and principal payments.

The solution uses a subscript on the IRR to emphasize that it is not an annual rate. Setting the PW equal to 0 yields:

$$0 = 8200 - 400(P/A, IRR_6, 28) - 10K(P/F, IRR_6, 28)$$

Because the bond is being sold at a discount, the interest rate will be higher than 4% every 6 months. Try 5% first:

$$P_{5\%} = 8200 - 400(P/A, 5\%, 28) - 10K(P/F, 5\%, 28)$$

$$= 8200 - 400 \cdot 14.898 - 10K \cdot .2551 = -\$310.2$$

The negative cash flows are greater than \$8200, so $P_{0\%}$ is negative. Because the PWs at 0% and 5% have the same sign, the IRR is greater than 5%. Try 6% next:

$$P_{6\%} = 8200 - 400(P/A, 6\%, 28) - 10K(P/F, 6\%, 28)$$
$$= 8200 - 400 \cdot 13.406 - 10K \cdot .1956 = \$881.6$$

Interpolating for the 6-month interest rate yields:

$$IRR_6 = .05 + .01 \cdot 310.2/(310.2 + 881.6) = 5.26\%$$

Calculating the effective annual interest rate yields:

$$IRR = 1.0526^2 - 1 = 10.8\%$$

(The exact IRR is 5.240% for 6 months, which yields an effective annual rate of 10.754%.)

7.5 SPREADSHEETS AND THE IRR

Investment or **annuity functions** account for one P, one A, and one F, all over N periods. **Block functions** identify each cash flow.

The arithmetic required to calculate the IRR is an obvious reason to prefer the use of spreadsheets to calculations using the tables of factors. As described in Chapter 4, spreadsheets include two kinds of functions designed to calculate IRRs. There are **investment** or **annuity functions,** which are used for cash flows made up of one each of P, A, and F, all over N periods; and there are **block functions,** which identify each cash flow individually.

Investment Functions.   The investment functions (or annuity functions) are very similar to the engineering economy factors used so far in this chapter. In fact, Chapter 4 summarized how to substitute 1 for the P, F, and/or A value of the investment function to calculate the engineering economy factors. For problems with at most one P, one A, and/or one F, these investment functions are the easiest way to calculate the internal rate of return.

However, some spreadsheets, such as Lotus 1-2-3 versions 2 and 3, are limited to @RATE(F, P, N) or the equivalent syntax. This function will solve for the IRR only for problems limited to P, F, and N, such as Example 7.1. Other spreadsheets have functions equivalent to those of Quattro Pro, Excel, and Lotus 1-2-3 version 4, which are, respectively, @IRATE(N, A, P, F, Type), RATE(N, A, P, F, Type), and @IRATE (N, A, P, Type, F). This far more capable function includes A; thus, it can be used for a wide variety of loans.

For Example 7.7, the Quattro Pro and Excel functions, with $P = 28K$, $A = -6006$, and $N = 9$, would be:

@IRATE(9,6006,−28000)   or   RATE(9,6006,−28000)

For Example 7.8: $P = 33K$, $A = -3K$, $F = -2K$, and $N = 36$. (If $F = -5K$ is used, then the A series has $N = 35$.) To calculate the monthly interest rate, using Quattro Pro or Excel use:

@IRATE(36,−3000,33000,−2000)   or   RATE(36,−3000,33000,−2000)

For Example 7.9: $P = 8200$, $A = -400$, $F = -10K$, and $N = 28$. To calculate the semiannual interest rate, use:

@IRATE(28,−400,8200,−10000)   or   RATE(28,−400,8200,−10000)

Example 7.10 applies this function to a loan with a balloon payment. The optional Type argument is used for beginning-of-period assumptions (Type = 1). Example 7.11 illustrates that these functions work even when the IRR is negative. Unfortunately, investments with negative IRRs are sometimes required by government regulations. Or, as in Example 7.11, it could be the worst of several possible outcomes of an investment.

Example 7.10   Loan with a Balloon Payment

A business is purchased for $250K. Payments of $50K are required at the ends of years 1 through 4, and a balloon payment of $150K is required at the end of year 5. What is the interest rate for this loan?

Solution
In this case, $N = 5$, which is used both for A and F. Thus, $A = -50$K and $F = -100$K. F is not -150K, however, because –50K in year 5 is allocated to the uniform series. The initial cash flow, P, for the loan is the $250K. The spreadsheet function for Quattro Pro and Excel can be entered, with cash flow stated in dollars or in $1000s, as follows:

@IRATE(5,−50000,250000,−100000)   or   RATE(5,−50,250,−100)

The calculation yields IRR = 10.21%.

Example 7.11   Investment with a Negative IRR

An investment of $10K in new machinery is required. After 5 years, the machinery will have no salvage value. Depending on how many shifts the machinery is used, it will save either $1250, $2500, or $3750 per year. Calculate the IRRs for one-, two-, and three-shift operations.

Solution
In each case, $P = -10$K and $N = 5$, while $A = 1250, 2500,$ or 3750:

@IRATE(5,1250,−10000)   or   RATE(5,1250,−10000)

returns −.1387, or –13.87% for the IRR.

@IRATE(5,2500,−10000)   or   RATE(5,2500,−10000)

returns .0793, or 7.93% for the IRR.

@IRATE(5,3750,−10000)   or   RATE(5,3750,−10000)

returns .2541, or 25.41% for the IRR.

If there is only one shift of operations, then the negative IRR implies that over 5 years, less than the original cost of $10K is saved. In fact, only $5 \cdot \$1250$, or $7.5K, is saved.

Block Functions. Most spreadsheets have a function that is equivalent to the Lotus 1-2-3 or Quattro Pro @IRR(guess,block). The guess argument is simply a starting point in the search for the IRR, and it is optional in some spreadsheets. The block argument is the range of $N + 1$ cells that contain the cash flows for periods 0 to N. Cash flows that are 0 must be explicitly entered as 0. This is the function that would be used to solve for the IRR in Examples 7.5 and 7.6.

For Example 7.5, the cash flows would be entered in 10 cells in the following order:

$$8000, 8000, 8000, 8000, 0, -7392, -7392, -7392, -7392, -7392$$

For Example 7.6, the cash flows would be entered in 11 cells in the following order:

$$-1000, 150, 150, 150, 200, 200, 200, 250, 250, 250, 250$$

The only disadvantage of the @IRR function is that each cash flow must be explicitly stated. It is not possible to define the revenues as an A for years 1 to 15, with an overhaul in year 5, a first cost, and a salvage value. Instead, 16 or $N + 1$ values must be specified. Example 7.12 illustrates that the @IRR function can be used even if there are multiple sign changes in the cash flow diagram.

Another advantage of spreadsheets is that it is easy to construct graphs of PW vs. i. This is done using another block function: @NPV(i,block). However, as noted in Chapter 4, the @NPV function's block is N cells for periods 1 to N. Time 0 is not included. Exhibit 7.3 is the spreadsheet that was used to calculate the IRRs in Exhibit 7.1 and produce the graphs of PW vs. i in Exhibit 7.2.

Example 7.12   IRR for an Investment with a Reclamation Cost (Based on Exhibit 7.1(c))

Find the IRR for the following cash flow series:

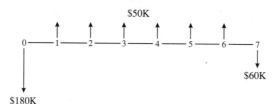

Solution

The simplest spreadsheet would be as follows:

	A	B	C	D	E	F	G	H
1	−180000	50000	50000	50000	50000	50000	50000	−60000
2	11.62%	cell A2 contains @IRR(0,A1..H1)						

Exhibit 7.3 Spreadsheet for computing IRRs and PW vs. *i*

	A	B	C	D
1		Exhibit 7.3		
2	Spreadsheet for Exhibits 7.1 & 7.2			
3	Year	a. Loan	b. Invest	c. INVw/SC
4	0	5000	-90000	-180000
5	1	-1318.99	-180000	50000
6	2	-1318.99	-90000	50000
7	3	-1318.99	80000	50000
8	4	-1318.99	120000	50000
9	5	-1318.99	120000	50000
10	6		120000	50000
11	7		150000	-60000
12	IRR	10.00%	12.63%	11.62%
13		@ IRR(0,B4..B11)		
14				
15		+B$4+ @ NPV($A17,B$5..B$11)		
16	i	NPVa	NPVb	NPVc
17	0%	-1595	230000	60000
18	2%	-1217	179100	47838
19	4%	-872	134866	36512
20	6%	-556	96335	25963
21	8%	-266	62697	16135
22	10%	-0	33271	6974
23	12%	245	7482	-1571
24	14%	472	-15159	-9545
25	16%	681	-35068	-16993
26	18%	875	-52600	-23955
27	20%	1055	-68058	-30469

7.6 MULTIPLE SIGN CHANGES

Cash flow diagrams, as in Example 7.12, may not have a unique IRR because the cash flow pattern has more than one sign change. In Example 7.12, there are two sign changes. The first is between periods 0 and 1 and the second is between periods 6 and 7. In this case, only one root of the equation is greater than 0. This root of 11.62% is an IRR, where the root of –43.25% does not seem to be meaningful.

The possibility of difficulties in calculating the IRR is a commonly cited reason for not relying on the IRR. However, the cases where it is most likely to occur can readily be identified. These include problems in mineral extraction, environmental restoration, and staged construction.

Mineral Extraction. The classic example of problems with calculating the IRR is mineral extraction. For example, additional oil wells may be added to an oil field; the effect is to recover the oil sooner more than to increase the field's total recovery. Exhibit 7.4 illustrates an example (note that the 8-year time span is compressed from the normal horizon). The initial cost of the wells is $4M, which is more than the additional $3M in recovered oil due to the new wells. However, the main effect of the additional wells is to shift $4.5M worth of production from years 6, 7, and 8 to earlier years. If the well is justified, one reason is because the oil is recovered sooner. Note that the additional recovery corresponds to an investment, and the shifting of recovery to earlier years corresponds to a loan (positive cash flow now and negative later). Thus, the oil wells are neither an investment nor a loan; they are a combination of both.

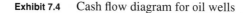

Exhibit 7.4   Cash flow diagram for oil wells

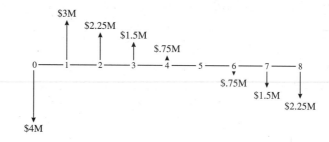

As shown in Exhibit 7.5, this oil well example has two interest rates where the PW equals 0. However, neither can be usefully interpreted as an IRR. In addition, this cash flow diagram cannot be interpreted as a loan nor as an investment. Even if there were a unique IRR, the project is not a loan where a low rate is good nor is the project an investment where a high rate is good.

Exhibit 7.5   Graph of PW vs. i for oil wells

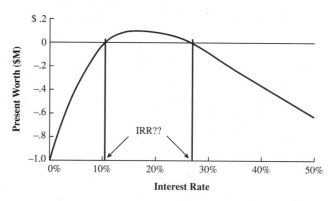

In fact, even though the project has a positive PW at 25%, it may not be appropriate to conclude that such a project is attractive even when evaluated at that given interest rate. Small uncertainties in the interest rate might lead to a different conclusion. Also, as shown in Exhibit 7.6, when multiple roots exist, small uncertainties in the cash flows can substantially change the calculated interest rates.

Environmental Restoration.   Example 7.12, which is based on Exhibit 7.1(c), can be modified to illustrate one problem that can occur when multiple sign changes cause multiple roots for the PW equation. Suppose the first cost in Example 7.12 is changed to $100K; the 6 years' worth of $50K receipts are the same; and the cost of environmental restoration is changed to $180K or greater. Exhibit 7.6 graphs PW vs. *i* for four different costs of environmental restoration. Not only can there be multiple roots, but "small" changes in the restoration cost can produce significant changes in the roots. If the cost of restoration exceeds $238.6K, then there are no roots to the equation PW = 0.

Exhibit 7.6   Graphs of PW vs. *i* for differing restoration costs

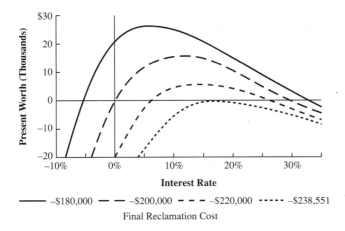

Staged Construction or Expansion.   Example 7.13 illustrates staged construction, which typically has three sign changes. There is an initial construction period, which is followed by a profitable period. Then additional capital costs are incurred to expand facilities. This is followed by another profitable period. If there are reclamation costs, then the problem ends with a fourth sign change.

Example 7.13   Staged Expansion of a Project

Expansion of a proposed project is expected at the end of year 5. This will lead to a substantial loss in year 5, but it will allow the business to operate through the project's horizon in year 10. Can an interest rate be calculated? Does this project appear to be attractive?

The project's first cost is $100K, and the expansion cost is $75K. Net revenues begin at $30K in year 1, and they increase by $1.5K per year.

Solution
The first step is to draw the cash flow diagram.

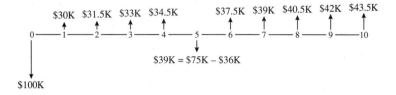

Using the @IRR function for this block of cash flows identifies an IRR of 24.19%. Section 7.7 describes one way to show that this root is unique. Since this project is an investment and 24.19% is a high IRR, the project is attractive.

Summary of Multiple Sign Change Consequences.   Having multiple sign changes violates the sufficient condition for a unique IRR, which is a single sign change. The outcomes may include a unique root, which is an IRR and which is useful. The outcomes may also include double roots, no roots, positive and negative roots, and extreme sensitivity.

In general, due to Descartes Rule of Signs, the number of roots cannot exceed the number of sign changes. Thus, a single sign change implies a unique root. However, we have seen that unique roots may exist even in the presence of multiple sign changes.

7.7 PROJECT BALANCES OVER TIME

This section describes a sufficient test for the uniqueness of the calculated root to PW = 0, and the root's meaningfulness as an IRR. The approach described here as the **project balances over time** (PB_t) is sometimes described as the unrecovered loan or investment balance [Bussey and Eschenbach, and Thuesen and Fabrycky]. The project balances over time are calculated using a root of the PW = 0 equation. If all values of PB_t are less than or equal to 0, then the root is an IRR for an investment. If all values of PB_t are greater than or equal to 0, then the root is an IRR for a loan.

This can be described mathematically by Equation 7.1, where PB_0 equals the cash flow in period 0 (CF_0). The name of this function comes from calculating for each period the cumulative value of all cash flows through that period.

$$\ast \quad PB_t = PB_{t-1}(1 + IRR) + CF_t \qquad (7.1)$$

The project balances over time for a loan are the amount owed for each period. For an investment, they are the unrecovered investment balances.

Exhibit 7.7   Staged expansion of an investment (based on Example 7.13)

Year	0	1	2	3	4	5	6	7	8	9	10
Cash flow	−$100K	30K	31.5K	33K	34.5K	−39K	37.5K	39K	40.5K	42K	43.5K
PB_t	−$100	−94.2	−85.5	−73.2	−56.4	−109	−97.9	−82.6	−62.0	−35.0	0

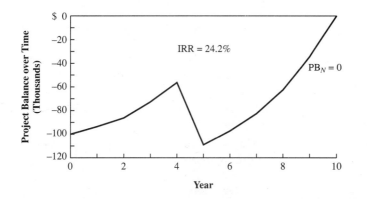

Exhibit 7.7 applies this equation to Example 7.13. All of the PB_t values are less than or equal to 0, so this is an investment. Furthermore, the 24.19% root is the unique IRR.

The graph of the project balances over time can have both positive and negative values, if the cash flow diagram has multiple sign changes. The cash flow diagram is then described as a mixed investment or mixed loan and more advanced techniques are required. If a mixed investment or loan has a unique IRR (see Problem 7.19c), then the IRR may still be useful for decision making [Eschenbach], but it is beyond the scope of this text.

7.8 SUMMARY

The chapter has discussed the definition, calculation, and use of the internal rate of return (IRR). It has focused on problems with the same assumptions as the chapters on present worth and equivalent annual worth/cost. Those assumptions are known cash flows and a known time horizon. The interest rate for the time value of money is also assumed to be known, but it is applied through comparison with the calculated IRR. This IRR is the interest rate at which the PW or EAW equals 0.

Most problems correspond to evaluating loans or investments. In these cases there is only one sign change in the cash flow diagram. This is a sufficient, but not a necessary, condition for the IRR to be defined. A low IRR is good for a loan, while a high IRR is good for an investment.

Calculating the IRR is done by calculating the value of the cash flow equation for different interest rates. Once negative and positive values have been found, it is possible to interpolate for the IRR. These calculations are automated through the financial functions of a spreadsheet. Some packages include investment functions that can solve for IRR given P, A, F, and N. All packages have the equivalent of an @IRR function that can be used with a block of cells that contain the cash flows.

If a cash flow diagram has multiple sign changes, then the project balances over time (PB_t) can be calculated. If the PB_t are less than or equal to 0 for all t, then the project is an investment. If the values are always nonnegative, then a loan is being evaluated. This is a sufficient, but not a necessary, condition for the IRR to be meaningful.

REFERENCES

Bussey, Lynn E., and Ted G. Eschenbach, *The Economic Analysis of Industrial Projects,* 2nd ed., Prentice-Hall, 1992.

Eschenbach, Ted G., "Multiple Roots and the Subscription/Membership Problem," *The Engineering Economist,* Volume 29, Number 3, Spring 1984, pp. 216–223.

Lohmann, Jack R., "The IRR, NPV and the Fallacy of the Reinvestment Rate Assumptions," *The Engineering Economist,* Volume 33, Number 4, Summer 1988, pp. 303–330.

Thuesen, Gerald J., and Wolter J. Fabrycky, *Engineering Economy,* 8th ed., Prentice-Hall, 1993.

PROBLEMS

7.1 Burke Brake Shoes may buy a new milling machine. The industrial engineer has asked for your recommendation on the best alternative. What is the IRR? If Burke requires a return of 8%, what do you recommend?

First cost	$15,500
Operating cost per year	$ 7300
Savings	$11,600
Useful life	5 years

(*Answer:* 12%)

7.2 One share of Festival Cruise Line stock initially sells for $25. Each year the company pays dividends of $.75. After holding the stock for 3 years the investor sells it for $30. What is the IRR for this investment?

7.3 Given the cost and benefits shown below, determine the IRR.

Year	0	1	2	3	4	5
CF_t	−$3000	400	600	800	1000	1200

(*Answer:* 8.77%)

7.4   A lot on the outskirts of town costs $35,000. The annual property taxes are calculated at 3.5% of the assessed value of $28,000. Assume that the lot will not be reassessed for the next 4 years (property taxes remain constant) and that the value increases by 20% before it is sold at the end of 4 years. What is the IRR?

7.5   What is the rate of return for the following project? It costs $300,000 to start and returns $45,000 the first year, with a $5000 increase each year thereafter. The project lasts 10 years.   (*Answer:* 15.8%)

7.6   The Jafar Jewel Mining Company has just issued a 0% coupon bond priced at $423. If the bond is bought today and in 2 years is sold for $502.50, determine the rate of return.

7.7   A bond has semiannual interest payments at a nominal annual rate of 12%. It has a life of 10 years and a face value of $5000. If it is currently selling for $4270, what is the effective interest rate?   (*Answer:* 15.4%)

7.8   A 6.5% coupon bond of the E Z Money Corporation now sells for $876. The bond's par or face value is $1000, and it is due in 9 years. Determine the yield to maturity (IRR) of this bond. (Assume that interest is paid annually.)

7.9   An 8% bond issued by Holiday Shipping sells for $679.40 on January 1, 1995. The bond matures on December 31, 2000. Assume that the bond is held until its maturity, and determine the IRR. The bond's face value is $1000, and interest is paid semiannually.   (*Answer:* 16.3%)

7.10 A 9% coupon bond of the ABC Company with a face value of $1000 is selling for $1057.53. If interest is paid annually and the bond matures in 8 years, determine the yield to maturity (the IRR).

7.11 Acme Bond Trading receives a 10% commission on all bond transactions. If an 8% coupon bond selling for $933 matures in 8 years and is bought from Acme, what is the rate of return? The bond's face value is $1000, and it pays interest annually.

7.12 NewTech has a new product that has incurred R&D expenses of $375,000. An additional $500,000 is needed if the product is to proceed. Initial sales will be $50,000 per month, and they will increase by $5000 per month. The market is moving fast, and the product will be closed out in 2 years. What is the new product's IRR for deciding whether or not to proceed?

7.13 Disneyworld's 4-day pass costs $140 and a 1-day pass is $50. A family visits Disneyworld every 3 years for 2 days each time. If the family buys the 4-day pass and uses 2 days on this visit and 2 days on the next visit, what IRR is earned?   (*Answer:* 35.7%)

7.14 Muddy Fields Earthmoving can buy a bulldozer for $30,000. If lease payments are $1000 per month for 36 months, what is the IRR for the lease? Lease payments occur at the beginning of each month. If the bulldozer is purchased, then it would have a salvage value of $5000 after 3 years. The operating costs can be ignored because they are the same for both leasing and buying.

7.15   My auto insurance can be paid in installments. If paid annually, it costs $450. If paid in installments, 40% is paid at month 0, and 15% at the ends of months 2, 4, 6, and 8. A delayed payment plan fee of $10 is added to all 5 installment payments. If the installment plan is chosen, what is the effective annual interest rate being paid on the "loan"?   (*Answer:* 53.7%)

7.16   Fred is a sales engineer who conducts much of his business at an exclusive country club. If his dues are paid annually, they are $3600. If paid quarterly, the payments are $1000 each. If Fred's company pays the dues annually, what effective annual interest rate is it earning?

7.17   Whistling Widgets can pay the fire insurance premiums on its factory either quarterly or annually. If paid annually, the premium is $3600. If paid quarterly, there is a $50 fee added to each $900 payment. If Whistling Widgets pays annually, what effective annual interest rate is it earning?

7.18   I purchased stock in Metal Stampings at $50 per share 2 years ago. The stock is now selling for $70 per share. Metal Stampings has paid annual dividends of $3 per year for the last 10 years and seems likely to continue doing so. I just received this year's dividend payment. If I expect the stock to be worth $90 per share in 3 years, what is the IRR for keeping the stock? Should I keep or sell the stock if my interest rate is 15%?

7.19   *The Engineering Economist* is a quarterly journal that costs $20 for 1 year, $38 for 2 years, or $56 for 3 years.
    a.   What is the IRR for subscribing for 2 years at once rather than 1 year at a time?
    b.   What is the IRR for subscribing for 3 years rather than 1 year at a time?

7.20   A 10-year, $10,000 bond was issued at a nominal interest rate of 8% compounded semiannually. Last year I bought it for $8580, just after the previous owner had received the fourth interest payment. I just received my second interest payment. If I sell it now, the price would be set to yield a nominal 10% rate of return. I believe that interest rates will drop, and that I can sell it in a year for $10,000. What is my expected IRR for the year?

7.21   You want to borrow $300,000 to start a business. The loan's initial balance is increased by 2% of the loan's face value for loan origination fees, mortgage insurance, credit checks, title searches, etc. Your annual payments are calculated using an initial balance of $306,000, an interest rate of 9%, and a 20-year term. The bank requires you to refinance the loan in 5 years by paying it off with a balloon payment. What interest rate are you paying?

7.22 A car dealer offers a $1500 rebate if you pay cash (or borrow the money elsewhere) for a $9500 car you are considering. Or the dealer offers .75% financing with monthly payments for 3 years. The dealer also requires a 10% down payment. What is the effective annual rate for the dealer's car loan?

7.23 A bank is offering a foreclosed condominium for $35,000. Its advertising on this property features a 7% mortgage (assume annual payments) with a 20-year term and *no* down payment. The bank charges 2 points (2%) for loan origination and closing costs. These fees are added to the initial balance of the loan. Nearly identical condominiums are selling for $33,000. What is the annual payment and what is the true rate for the loan?

7.24 A $2500 computer system can be leased for $79 per month for 3 years. After 3 years it can be purchased for $750. This is also the salvage value if the system was purchased originally. What is the effective annual rate for leasing the computer?

7.25 Briefly discuss the relationship between present worth and the calculation of rate of return.

7.26 Identify the two roots for the PW equation of the following cash flows. Is either a meaningful IRR? The cash flows are −$50, $123, and −$75.

7.27 A coal deposit can be mined more rapidly if a third dragline and conveyor system is added. The initial cost would be $1.5M. Annual revenues for each of the next 12 years would increase from $800K annually to $1200K annually, and then for the next 6 years revenues would drop to $0. Thus, the mining is 50% faster, and the deposit is depleted 6 years earlier. The system will have no salvage value. What roots are there for the PW equation? Is there a meaningful IRR? Is the third dragline economically attractive if *i* is 20%? Graph the PW for interest rates of 0% to 100%?   (*Answer:* roots are 7.23% and 19.20%)

7.28 Construct the PB$_t$ graph for Problem 7.27.

7.29 A project's first stage will cost $250K, and its second stage at the end of 6 years will cost $200K. Net annual revenue will begin at $50K and increase by $2K per year. The time horizon is 10 years. Is there a meaningful IRR? Is the project attractive? Construct the PB$_t$ graph.

CHAPTER 8 BENEFIT/COST RATIOS AND OTHER MEASURES

The Situation and the Solution

There are many different economic measures that decision makers can and do use to evaluate projects.

The solution is to understand the strengths, weaknesses, and underlying rationale of each economic measure and to know which ones are more reliable because they are theoretically better.

Chapter Objectives

After you have read and studied the sections of this chapter, you should be able to:

Section 8.1
Recognize the variety of economic measures that are used to evaluate private- and public-sector projects.

Section 8.2
Define and use the benefit/cost ratio for public-sector projects.

Section 8.3
Define and use a present worth index for private-sector projects.

Section 8.4
Define and use the future worth measure.

Section 8.5
Define and use payback period, and recognize its weaknesses, as compared to discounted cash flow measures.

Section 8.6
Define and use discounted payback period.

Section 8.7
Define and use breakeven volume as an economic measure.

Key Words and Concepts

Benefits   Desired outcomes received by the public.

Disbenefits   Outcomes that would be avoided if possible.

Costs   Paid by the government.

Benefit/cost ratio   Benefits minus disbenefits, divided by the costs.

Present worth indexes   (1) The PW of periods 1 to N divided by first costs or (2) the PW of revenues divided by the PW of costs.

Payback period   The number of periods for a project's net revenues to equal or pay back its first cost.

Discounted payback period   The number of periods until the compounded sum of net revenues equals the compounded value of the first cost.

Breakeven volume   The number of units per period required for a PW of 0.

8.1 DEFINING OTHER MEASURES OF ECONOMIC ATTRACTIVENESS

Earlier chapters have focused on three measures of discounted cash flow (DCF) analysis: present worth, equivalent annual cost, and internal rate of return. This chapter discusses other discounted cash flow measures, such as benefit/cost ratios and present worth indexes. This chapter also discusses the weaknesses of measures, such as payback period, that ignore money's time value.

These measures will be defined more precisely in this chapter's other sections. For the moment, let the following definitions and decision rules suffice.

At a given interest rate, the benefits of a good public project should exceed its costs, so that the benefit/cost ratio exceeds 1 for good public projects. For example, a road improvement project might have a present worth of $400,000 for its benefits and a present *cost* of $300,000, for a ratio of 1.33. (The analogous measure for a private firm is a present worth index.)

A payback period is the time required for cumulative net revenues (that is, income minus cost) to equal a project's first cost. For example, if automating an assembly line has a first cost of $200,000 and saves $40,000 per year, then the payback period is 5 years—ignoring money's time value. This is the same as using an interest rate of 0%. Using a realistic interest rate leads to the discounted payback period.

Exhibit 8.1 summarizes how frequently some of these measures are used. This index of relative usage combines the results from several studies (see reference list), so the percentages shown in Exhibit 8.1 should not be considered as exact values. The studies were not conducted at the same time, and some asked about the most important technique; some about the primary techniques used; and some about all techniques used. These studies focus on constrained project selection within a capital budget, which is detailed in Chapter 9. Unfortunately, the research on techniques used for comparing mutually exclusive alternatives (see Chapter 10) is much scantier.

Exhibit 8.1   Frequency of use of economic measures (Total exceeds 100%, since more than one could be selected)

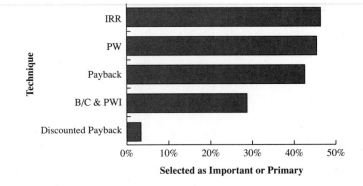

There are also trends that can be expected to continue. For example, over the last 30 years, the use of PW and IRR has been steadily increasing, while the use of payback period has been declining. The increasing use of spreadsheets is supporting this trend, for two reasons. First, as shown in Chapter 7, spreadsheets can easily do calculations of IRRs. Second, as shown in Chapters 4–6, spreadsheets encourage the use of more detailed cash flow projections that are not uniform, which makes payback more difficult to use.

8.2 BENEFIT/COST RATIO

Benefits are desired outcomes received by the public, and **disbenefits** are outcomes that would be avoided if possible. **Costs** are paid by the government, and the **benefit/cost ratio** is benefits minus disbenefits divided by the costs.

The first step in calculating a **benefit/cost ratio**, which is benefits minus disbenefits divided by the costs, is to define these benefits and costs for government projects. For public projects such as dams and roads, the **benefits** are the consequences to the public. Positive outcomes include recreation, electricity, shorter trips, and fewer accidents. Negative outcomes, or **disbenefits**, include lost white-water kayaking in the dammed river, traffic delays during construction, and neighborhoods divided by new highways. The **costs** are paid by the government for construction and for operation.

Not all government projects involve the public. For example, the post office might be evaluating a new mail sorting machine. The cost to buy and install the new machine would be compared with the savings in operating costs. In this case, the savings would be the benefits for any benefit/cost ratio.

The difference between the benefits and the costs (B − C) is the PW calculated in Chapter 5 or the EAW calculated in Chapter 6. However, governments want a measure of investment efficiency. It is intuitively appealing to find the amount of benefit a project produces per dollar of cost. This is the benefit/cost ratio, or B/C ratio.

The ratio can be calculated using PW (see Example 8.2) for both the numerator (benefits) and the denominator (costs), or by using equivalent annual measures for both the numerator and the denominator (see Example 8.1). As shown in Examples 8.1 and 8.2, the numerator includes public consequences and the denominator includes government costs. A desirable project has a ratio greater than or equal to 1. If the ratio is less than one, then the project is unattractive because the costs are greater than the benefits.

The benefit/cost ratio is calculated at a known interest rate, and when applied properly it will lead to the same recommendation as PW, EAW, or IRR measures. However, as will be shown in Chapters 9 and 10, applying it is not always easy. Chapter 14 describes the application of engineering economy to public projects in more detail.

The benefit/cost ratio is a discounted cash flow measure, and its popularity is based on the intuitive sense of return per dollar of cost. The higher the return per dollar of cost, the better the project.

Example 8.1   Shady Acres' Landfill

A new landfill site is being considered. It would save the citizens of Shady Acres $250,000 per year in the fees they pay a private firm for garbage pickup. The citizens who live near the road to the landfill site, and especially those who live closest to the site, are not supportive. A local university economist has estimated the disbenefits of truck traffic, noise, and odor to be $120,000 per year. The new landfill will cost $2.4 million; it will last 40 years; and Shady Acres uses an interest rate of 6%. Calculate the benefit/cost ratio using equivalent annual measures.

Solution
In the cash flow diagram both the disbenefits and the construction cost are negative cash flows. However, for calculation of the B/C ratio, they are assigned to the numerator and the denominator, respectively.

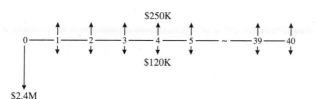

The construction cost's EAC, or the ratio's denominator, is:

$$C = 2.4M(A/P, 6\%, 40) = 2400K \cdot .0665 = \$159.6K$$

The benefits equal annual savings to citizens minus the disbenefits.

$$B = 250K - 120K = \$130K$$

The B/C ratio equals 130K/159.6K, or .815. Since this is less than 1, the landfill is not attractive.

Example 8.2   Shady Acres' Swimming Pool

A new municipal pool would cost $1.2M to build and $200K per year to operate. The pool would have no salvage value after 25 years. The benefit of having the pool is about $450K per year. Using PW, calculate the benefit/cost ratio with i equal to 6%.

Solution
The present value of the costs is

$$C = 1.2M + 200K(P/A, 6\%, 25)$$
$$= 1200K + 200K \cdot 12.783 = \$3757K$$

The PW of the benefits is

$$B = 450K(P/A, 6\%, 25) = 450K \cdot 12.783 = \$5752K$$

The B/C ratio equals 5752K/3757K, or 1.53. Since this exceeds 1, this project is attractive.

8.3 PRESENT WORTH INDEXES

The *dollar of return per dollar of cost* is the benefit/cost ratio in the public sector. In the private sector, similar measures are called **present worth indexes**. There are numerous variations of these indexes, but each is a ratio of income over outgo, or ongoing net revenue over initial expenses.

One **present worth index** is the PW of net revenues (in periods 1 to N) divided by a project's first cost. Another **present worth index** is the PW for revenues divided by the PW of the costs.

For example, suppose a machine costing $500 would save $200 per year for 4 years. The return is the PW of four $200 annual receipts, and the cost is $500. If i is 10%, $(P/A, .1, 4)$ is 3.170 and the return's PW is $634. The present worth index is 634/500 or 1.27. There is $1.27 in benefits or returns for each $1 of cost—over and above a 10% return on the investment.

The common assumption of present worth indexes is that investment projects are being evaluated. This implies that at least the time 0 cash flow is negative, and that the summation of net receipts exceeds the first cost. However, the division of costs and/or revenues between the numerator and the denominator is somewhat arbitrary. Three slightly different indexes are defined in Equations 8.1 to 8.3, and Example 8.5 illustrates another.

Mathematical Definition of PW Indexes.   The simplest PW index (PWI_1) divides the PW of all future cash flows by the cost at time 0. This cost at time 0 includes the cost to purchase, ship, and install, minus any rebates or other immediate savings. In Equation 8.1, the time 0 cash flow is shown as an absolute value—which converts it to a cost. This index works best if projects' start-up costs occur only in period 0.

$$PWI_1 = \frac{\sum\limits_{t \geq 1} PW(CF_t)}{|CF_0|} \tag{8.1}$$

Since some projects have a construction or start-up phase that goes beyond period 0, some firms use a different PW index (PWI_2). As shown in Equation 8.2, PWI_2 includes cash flows through the last period of construction, T, in the denominator.

$$PWI_2 = \frac{\sum\limits_{t > T} PW(CF_t)}{\left| \sum\limits_{t \leq T} PW(CF_t) \right|} \tag{8.2}$$

Some firms use yet another PW index (PWI_3). This index, defined in Equation 8.3, places all positive cash flows in the numerator, and all costs in the denominator.

$$PWI_3 = \frac{\sum\limits_{CF_t > 0} PW(CF_t)}{\left| \sum\limits_{CF_t < 0} PW(CF_t) \right|} \tag{8.3}$$

Each of these indexes is defined so that a value of 1 corresponds to a project that is economically attractive. Some firms create other variations on these indexes. For example, if each is multiplied by 100, then they correspond to what percentage of the "costs" are recovered by the "revenues."

Because Example 8.3 is a simple problem, the three indexes defined by Equations 8.1 to 8.3 all have the same value. However, Example 8.4 has different values for all three. These different values will always be either all greater than 1 or all less than 1, so at least a project is attractive under either none or all of the criteria.

These PW indexes are often used to select projects in the presence of budget constraints. The indexes themselves are theoretically correct, but sometimes their application is flawed. Example 8.5 illustrates one common kind of error.

Example 8.3   Present Worth Indexes for a Simple Project

Heavy Metal Stamping is evaluating a project with a first cost of $150,000, no salvage value, annual savings of $30,000, and a life of 10 years. Using an interest rate of 10%, calculate the three present worth indexes.

Solution

Since the only negative cash flow is at time 0, all three indexes can be found as follows:

$$PWI = 30K(P/A, .1, 10)/150K = 30K \cdot 6.145/150K = 1.23$$

Example 8.4　Present Worth Indexes for a Complex Project

Heavy Metal Stamping is evaluating a project with a first cost of $500,000, which is split evenly between period 0 and period 1. Annual revenues of $120,000 and O&M costs of $30,000 begin at the end of year 2 and continue for 10 years. There is no salvage value at the end of 11 years. Using an interest rate of 10%, calculate the three present worth indexes.

Solution
Since the first costs are spread over two periods, and since negative cash flows occur later as well, all three indexes have different values. For the first two indexes, the cash flows in years 2–11 are $90K. For the third index, the cash flows in years 2–11 are decomposed into a +$120K and a −$30K.

$$PWI_1 = \sum_{t \geq 1} PW(CF_t)/|CF_0|$$

$$= \frac{[-250K + 90K(P/A, .1, 10)](P/F, .1, 1)}{250K}$$

$$= \frac{[-250K + 90K \cdot 6.145].9091}{250K} = 1.102$$

For PWI_2 the last period of construction, T, is 1:

$$PWI_2 = \sum_{t > T} PW(CF_t) \bigg/ \left| \sum_{t \leq T} PW(CF_t) \right|$$

$$= \frac{90K(P/A, .1, 10)(P/F, .1, 1)}{[250K + 250K(P/F, .1, 1)]}$$

$$= \frac{90K \cdot 6.145 \cdot .9091}{[250K + 250K \cdot .9091]} = 1.053$$

PWI_3 places all positive cash flows in the numerator, and the absolute value of all negative cash flows in the denominator. The difference between this and PWI_2 is whether the *net* annual revenues of $90K, or the $120K in revenues and $30K in costs, is used.

$$PWI_3 = \frac{\sum_{CF_t > 0} PW(CF_t)}{\left| \sum_{CF_t < 0} PW(CF_t) \right|}$$

$$= \frac{120K(P/A, .1, 10)(P/F, .1, 1)}{250K + [250K + 30K(P/A, .1, 10)](P/F, .1, 1)}$$

$$= \frac{120K \cdot 6.145 \cdot .9091}{250K + [250K + 30K \cdot 6.145] \cdot .9091} = 1.040$$

Note that for all three indexes, the numerator minus the denominator equals the net PW of $25.5K.

Example 8.5   Brand X   PW Index

A certain firm specifies that a variation on PWI_3 be used. The change is that the index is multiplied by 100, so that the income's PW is stated as a percentage of the cost's PW. The firm uses an interest rate of 10%, but it specifies that attractive projects must have an index value that exceeds 150.

The first cost is $250K, annual O&M costs are $50K, annual revenues are $120K, salvage value is 0, and the time horizon is 15 years. Calculate the firm's Brand X index. Does it meet the 150 standard? What is this project's PW? What is its IRR?

Solution
At 10%, the PW of the costs is

$$PW_{costs} = -250K - 50K(P/A, .1, 15) = -\$630.3K$$

The PW of the revenues is

$$PW_{rev.} = 120K(P/A, .1, 15) = \$912.7K$$

Using the firm's PW index, the project does not meet the 150 standard.

$$PWI_X = 100 \cdot 912.7K / |-630.3K| = 144.8$$

However, the project has a PW of $282,400 at $i = 10\%$ and an IRR of 27.2%. If the firm's interest rate of 10% is correct, then this appears to be an outstanding project.

8.4 FUTURE WORTH

Maximizing future worth is clearly similar to maximizing present worth or equivalent annual worth, so long as the same interest rate is used. In fact, multiplying a PW by $(F/P, i, N)$ or an EAW by $(F/A, i, N)$ is the only mathematical step required.

Intuitively, maximizing future worth is closely linked to the theory of finance and to the rights and desires of the firm's owners. Emphasizing the "final" outcome deemphasizes the exact pattern by which wealth is accumulated. It is also linked to the point in time when a firm, having created wealth, distributes that wealth to its owners—be they partners or stockholders.

The problem with future worth is that engineers and managers do not have an implicit scale for judging the results. A $10K PW equates to a new car or a robot of a certain size and quality. A $40K EAW equates to an annual salary for an engineer with a certain level of experience. But how can a future worth of $200,000 in year 15 at $i = 12\%$ be judged?

Future worth is logically consistent with and very similar to PW and EAW, but psychologically it is different. Future worth can be accused of compounding small amounts of money until the result appears to be large in absolute terms, but the result would still be small in relative terms. The larger i or N, the more pronounced is the "exaggeration" effect of future worth.

To quote Lazarus Long from one of Robert Heinlein's novels, *Time Enough for Love:* "$100 placed at 7 percent interest compounded quarterly for 200 years will increase to more than $100,000,000—by which time it will be worth nothing." The exact amount is $106,545,305. Example 8.6 illustrates this difficulty.

Example 8.6   Future Worth of a Cost Savings

A chemical engineer is proposing a process change that will save $2500 per year at essentially no cost. The refinery has a remaining life of 20 years. If the firm's interest rate is 15%, what is this savings worth in the long run?

Solution
The future worth of $2500 per year is easy to calculate.

$$FW = 2500(F/A, .15, 20) = 2500 \cdot 102.44 = \$256,100$$

Saving $2500 per year is worthwhile, but viewing it as saving over a quarter of a million dollars seems to exaggerate its importance.

8.5 PAYBACK PERIOD

The **payback period** equals the number of periods for a project's net revenues to equal or pay back its first cost.

The **payback period** equals the time required for net revenues from a project to "pay back" its initial cost. Because it is simpler than discounted cash flow techniques, this was once the most common technique for evaluating projects. However, as was shown in Exhibit 8.1, this is no longer true.

If a project's net revenues are the same each period, then the payback period is easy to calculate. For example, a machine costs $8K to buy, and it saves $800 per month in costs. Then the payback period is $8K/$800 per month, or 10 months. In Example 8.7 the payback period is calculated for a more complex project.

Example 8.7   Payback for a Heat Pump

Rosenberg Engineering has offices in northern California, where a heat pump can be used for cooling in the summer and heating in the winter. Replacing their current system will cost $1500 in May and $500 in June. Starting in July, it will save them $200 per month for summer months (June–August), $100 for the fall and spring months, and $150 for the winter months (November–March). What is the payback period for the heat pump?

Solution
The month-by-month cumulative sum of the cash flows is shown in Exhibit 8.2. The system costs $1500 in May and an additional $500 in June, for a June-end total of $2000. During July and August, $200 is saved each month for month-end totals of $1800 and $1600, respectively. Each month the savings are subtracted from the cumulative total until payback is achieved.

Exhibit 8.2 Payback for a heat pump

Month	Year 1		Year 2	
	Cash Flow	Cumulative Total	Cash Flow	Cumulative Total
May	−$1500	−$1500	$100	−$450
June	−500	−2000	200	−250
July	200	−1800	200	−50
August	200	−1600	200	+150
September	100	−1500		
October	100	−1400		
November	150	−1250		
December	150	−1100		
January	150	−950		
February	150	−800		
March	150	−650		
April	100	−550		

The payback period is $15\frac{1}{4}$ months, from the project's beginning at the start of May until the $2000 has been recovered during August in the next year.

Difficulties with Payback Period. The main problems with payback period are that:

1. Payback ignores the time value of money.

2. Payback ignores receipts and costs that occur between payback and the project's time horizon. Some writers have described this as ignoring a project's life.

Payback is not a discounted cash flow technique. The summation that occured in Example 8.7 ignored the fact that the $2000 in costs occurred first and has a higher PW than the savings. In other words, Example 8.7 assumed that $i = 0\%$, which is not true.

Example 8.8 illustrates the difficulties caused when payback ignores cash flows that occur after the payback period. One project may have a short payback period, but its positive cash flows may end shortly thereafter. Other projects may have somewhat longer payback periods that are followed by years or decades of positive cash flows.

Ignoring cash flows that occur after payback can cause an evaluation to ignore overhaul costs and environmental reclamation expenses. In addition, working capital is difficult to account for. Similarly, evaluation of projects with staged development is unreliable using payback period.

Example 8.8   Comparing the Payback Periods of Three Projects

Calculate and compare the payback periods of the projects shown in Exhibit 8.3. For comparison purposes, calculate PWs at 10% and calculate IRRs.

Exhibit 8.3   Projects for payback evaluation

	Project		
	A	B	C
First cost	$10,000	$10,000	$10,000
Annual savings	$3000	$2000	$2000
Life (years)	4	10	5
Salvage value	0	0	$5000

Solution

The payback periods are very easy to calculate, since the only requirement is to divide the first cost by the annual savings. Notice that since salvage values generally occur after payback, those salvage values are ignored by the payback period calculations.

PWs are easy to calculate using the equations from Chapter 5, as are IRRs using the material from Chapter 7. The results are summarized in Exhibit 8.4.

Exhibit 8.4   Comparing payback, PW, and IRR

Measure	A	B	C
Payback period	3.33 years	5 years	5 years
PW at 10%	−$490	$2289	$686
IRR	7.7%	15.1%	12.2%

The payback period is obviously wrong in suggesting that project A is the most attractive. While its payback is short, its benefits only continue for two-thirds of a year after payback. It does not even earn a 10% return.

When Can Payback Be Used?   Payback should be limited to cases where the time horizon is measured in months, not in years. If payback occurs within 4 to 6 months and benefits continue for 2 or more years, then the project is generally well worth doing. This occurs most often in capital-short firms, such as start-up businesses, that cannot tie up funds in long-term investments.

However, payback is sometimes used when other measures are far better. One possible reason is quite negative and short-sighted. Suppose the payback period measures the immediate returns from an investment project that I am supporting or approving. Those returns will be received before I move to another job, so I will get some credit for them.

While engineering economists recommend against the use of payback period, most projects that appear attractive on one criterion will also be attractive on others. Most investments with short payback periods also have high PWs and high IRRs. Example 8.9 describes the largest decision that I know of that relied on payback period.

Example 8.9 Nihon Kokan Steel

In the late 1970s and very early 1980s, an island in Tokyo Harbor was greatly expanded to serve as the location for a new steel mill. Nihon Kokan described its then 13-year-old mill (the top-productivity mill in the world) as aging.

The cost of the island facility, which was financed by the firm, was over $5 billion. The firm's investment evaluation standard was a payback of 7 years. In this case, the mill was built even though the estimated payback was 10 years. As described by the mill's management, "We are sure we will find ways to make it pay back sooner."

8.6 DISCOUNTED PAYBACK

As discussed above, the payback period does not account for money's time value. However, the **discounted payback period** does. To use the language of Chapter 7, the interest rate is used to compute the project balances, or the PB_t, over time. When the project balance first equals 0, the discounted payback equals the number of periods.

> The **discounted payback period** is the number of periods until the compounded sum of net revenues equals the compounded value of the first cost.

Notice that when Example 8.10 applies this measure to the data from Example 8.7, the payback period is extended from 15 months to a discounted payback period of 16 to 17 months.

The discounted payback period is an improvement over the simple payback period, since the time value of money is considered. However, cash flows that occur after payback are still ignored, which is still wrong. Example 8.11 applies discounted payback to the three projects of Example 8.8.

Example 8.11 also applies Equation 8.4, which can be used to calculate discounted payback periods (N_{DPP}) when: (1) the only net negative cash flow is at time 0 and (2) the net revenue (A) in each period is constant.

$$(P/A, i, N_{DPP}) = |CF_0|/A \qquad (8.4)$$

Example 8.10 Discounted Payback for a Heat Pump (Example 8.7 Revisited)

The heat pump for Rosenberg Engineering will cost $1500 in May and $500 in June. Starting in July, it will save them $200 per month for summer months (June–August),

$100 for the fall and spring months, and $150 for the winter months (November–March). What is the discounted payback period for the heat pump if $i = 12.68\%$ per year, or 1% per month?

Solution

Exhibit 8.5   Discounted payback calculation for Examples 8.7 and 8.10

| | Year 1 | | Year 2 | |
Month	Cash Flow	PB$_t$	Cash Flow	PB$_t$
May	−$1500	−$1515	$100	−$634
June	−500	−2035	200	−440
July	200	−1855	200	−244
August	200	−1674	200	−46
September	100	−1591	100	54
October	100	−1507		
November	150	−1372		
December	150	−1236		
January	150	−1098		
February	150	−959		
March	150	−819		
April	100	−727		

The month-by-month PB$_t$ shown in Exhibit 8.5 is found using Equation 7.1 with $i = 1\%$:

$$PB_t = PB_{t-1} \cdot (1 + i) + CF_t$$

The end-of-period values for May and June are higher than $1500 and $2000, because those cash flows were assumed to occur at the beginning of the month. Thus, the cash flows were also multiplied by $(1 + i)$.

The discounted payback period is between 16 and 17 months, from the project's beginning at the start of May until September in the second year, when the $2000 has been recovered with 1% interest per month.

Example 8.11   Comparing the Discounted Payback Periods of Three Projects (Example 8.8 Revisited)

Calculate and compare the discounted payback periods of the projects shown in Exhibit 8.6 at $i = 10\%$. Compare the values with the payback periods, PWs, and IRRs from Example 8.8.

Exhibit 8.6 Data for payback period
(Example 8.8 repeated)

	Project		
	A	**B**	**C**
First cost	$10,000	$10,000	$10,000
Annual savings	$3000	$2000	$2000
Life (years)	4	10	5
Salvage value	0	0	$5000

Solution

The discounted payback periods are calculated by multiplying the PB_t by 10% and adding on the annual savings (see Exhibit 8.7). Notice that since salvage values generally occur after "payback," those salvage values are ignored by the discounted payback period calculations. Note that the final PB_t for project A, $-\$718$, is the future worth at $i = 10\%$ in year 4 of project A's PW of $-\$490$.

Exhibit 8.7 Calculations for discounted
payback period

Year	PB_t **for A**	PB_t **for B**	PB_t **for C**
0	-10,000	-10,000	-10,000
1	-8000	-9000	-9000
2	-5800	-7900	-7900
3	-3380	-6690	-6690
4	-718	-5359	-5359
5		-3895	1105
6		-2285	
7		-514	
8		1435	

Equation 8.4 can be used to check these results for Projects A and B, since both conditions are satisfied: (1) each project has a negative cash flow only in time 0, and (2) each project has a constant net revenue (the annual savings) for each period. Project C's discounted payback period cannot be checked using Equation 8.4, since its salvage value is needed to achieve payback, so it violates the second condition.

Project A:

$$(P/A,10\%,N_{DPP}) = 10K/3K = 3.3333$$

$$(P/A,10\%,4) = 3.170 \quad \text{and} \quad (P/A,10\%,5) = 3.791$$

so the discounted payback period is longer than the 4-year life.

Project B:

$$(P/A,10\%,N_{DPP}) = 10K/2K = 5.000$$

$$(P/A,10\%,7) = 4.868 \quad \text{and} \quad (P/A,10\%,8) = 5.335$$

$$N_{DPP} = 7 + (8 - 7)(5.000 - 4.868)/(5.335 - 4.868) = 7.28 \text{ years}$$

As shown in Exhibit 8.8, the discounted payback period is obviously wrong in suggesting that project C is the most attractive, but it at least eliminates project A, unlike the simple payback period.

Exhibit 8.8   Comparing discounted payback period with PW and IRR

Measure	A	B	C
Discounted payback period	Does not pay back	7.3 years	5 years
Payback period	3.33 years	5 years	5 years
PW at 10%	−$490	$2289	$686
IRR	7.7%	15.1%	12.2%

8.7 BREAKEVEN VOLUME

The **breakeven volume** is the number of units per period required for a PW of 0.

Sometimes the number of units to be sold or the number of operating hours per period are unknown. Rather than guessing a value, it is possible to calculate the volume required for a PW of 0. This **breakeven volume** is the simplest of the sensitivity analysis techniques that are detailed in Chapter 17.

Exhibit 8.9 illustrates the breakeven number of rainy lecture days required to justify buying a central campus parking permit. Each day it rains Joe drives his car and must either use the $2/entry lot or buy an annual permit for $60. Obviously, if more than 30 rainy days are expected, the annual permit is cheaper. The breakeven number of rainy days is 30.

Exhibit 8.9   Breakeven number of rainy days

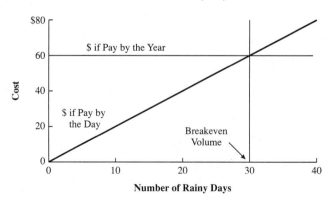

Example 8.12   **Annual Demand for a Project**

Find the breakeven volume for the annual demand (D) for a proposed project. The first cost is $30K, the salvage value is 0, N is 10 years, i is 10%, and the profit per item is $15.

Solution

The PW equation is:

$$PW = -30K + 15 \cdot D(P/A, .1, 10) = 0$$

$$D = 30K/[15 \cdot 6.145] = 325.47$$

or 326 items per year to break even.

8.8 SUMMARY

This chapter has introduced several other measures that are used in the economic evaluation of projects. Benefit/cost ratios are used for public projects, and when used correctly they are consistent with PW, EAW, and IRR measures. The ratios are used because they are based on an intuitively attractive question: "How much payoff per dollar of cost?"

Similar measures for the private sector are the various PW indexes. Because these measures are often applied incorrectly, the results of comparisons between projects are often misleading.

Future worth is mathematically equivalent to PW and EAW. However, judging the true value of a future worth of $100,000 in year 20 when $i = 12\%$ is very difficult.

The payback period is a very popular measure, but it is an incorrect approach. It ignores both the time value of money and any cash flows that occur after payback. The discounted payback period includes the time value of money, but it still ignores any cash flows that occur after payback. Neither measure should be used, unless the time to payback can be measured in months rather than years. The intuitive base for these measures is that the sooner a project covers its costs the better.

Sometimes the demand per year or per period is unknown or uncertain. In that case, the breakeven volume can be used to economically evaluate the project. This is the simplest sensitivity analysis technique.

Exhibit 8.10 summarizes some of the major points for each technique.

Exhibit 8.10   Summary of economic measures

Present Worth, Equivalent Annual Worth, and Internal Rate of Return
   Correct discounted cash flow techniques
   See Chapters 5–7, 9, and 10
Benefit/Cost Ratio
   Used for government projects
   Intuitive base is "bang for buck"
Present Worth Indexes
   Used for private sector
   Similar to benefit/cost ratio
   Easy to use incorrectly
Future Worth
   Mathematically equivalent to PW and EAW
   Intuitively misleading
Payback Period
   Ignores time value of money
   Ignores cash flows after payback
   Frequently this is one measure that is used
Discounted Payback Period
   Ignores cash flows after payback
Breakeven Volume
   Useful if number of units per period is uncertain or unknown

---

REFERENCES

Cook, Thomas J., and Ronald J. Rizzuto, "Capital Budgeting Practices for R&D: A Survey and Analysis of *Business Week*'s R&D Scoreboard," *The Engineering Economist,* Volume 34, Number 4, Summer 1989, pp. 291–304.

Farragher, Edward J., "Capital Budgeting Practices of Non-Industrial Firms," *The Engineering Economist,* Volume 31, Number 4, Summer 1986, pp. 293–302.

Kamath, Ravindra, and Eugene Oberst, "Capital Budgeting Practices of Large Hospitals," *The Engineering Economist,* Volume 37, Number 2, Spring 1992, pp. 203–232.

Khan, Aman, "Capital Budgeting Practices in Large U.S. Cities," *The Engineering Economist,* Volume 33, Number 1, Fall 1987, pp. 1–12.

Pike, Richard H., "Do Sophisticated Capital Budgeting Approaches Improve Investment Decision-Making Effectiveness?" *The Engineering Economist,* Volume 34, Number 2, Winter 1989, pp. 149–161.

PROBLEMS

8.1   Use PWs to calculate the benefit/cost ratio at $i = 5\%$ for a highway project. The first cost is $150K, and O&M costs are $5K per year.

There is no salvage value after 20 years. Time savings to users are worth $35K per year and neighborhood disruption has a disbenefit of $10K per year. Is this project attractive?   (*Answer:* 1.47)

8.2   Use EAWs to calculate the benefit/cost ratio at $i = 6\%$ for a new library. The first cost is $500K, and O&M costs are $100K per year. There is no salvage value after 30 years. Community benefits are estimated to be $130K per year. Is the library economically justified? (*Answer:* .954)

8.3   A new power plant will cost Newburg $2.1M to build and $230K per year to operate for its 25-year life, when its salvage value is 0. The higher efficiency of the new plant will lower utility bills by $21 per year for each of the 20,000 residents of Newburg. The plant's air pollution will cost residents $3 per year. If i is 5%, should the new power plant be built?

8.4   Using PW, determine the benefit/cost ratio of the proposal with the following cash flows. Use an interest rate of 10%.

Year	0	1	2	3	4	5	6
Cash flow	−$25K	4K	5K	6K	7K	8K	9K

8.5   Using EAW, determine the benefit/cost ratio for Problem 8.4.

8.6   A new water treatment plant proposed for Anytown, USA, has an initial cost of $6M. The new plant will service the 7500 residential customers for the next 30 years. It is expected to save each customer $125 per year. The plant will require a major overhaul every 5 years costing $1M. Determine the benefit/cost ratio at the city's interest rate of 6%.

8.7   ABC Block Co. may buy a new machine that has a first cost of $10K. It will produce benefits for each of the next 6 years of $3000, while requiring $1400 to operate. ABC uses a present worth index that is calculated by dividing the PW of benefits by the PW of costs. Determine the PWI at 8%.   (*Answer:* .842)

8.8   A new milling machine with an 8-year life will cost Brake-O-Matic $48,000. Net revenues for the 8 years are: $12K, $14K, $15K, $16K, $10K, $8K, $8K, $6K. The machine has an expected salvage value of $7500. Brake-O-Matic uses a present worth index that divides the PW of all future net revenues by the initial cost. If money costs Brake-O-Matic 12%, what is the present worth index?

8.9   NewTech uses a PW index that divides the PW of all future revenues by the cost in year 0 (PWI_1). What is the index value for the following project, if $i = 12\%$? The first cost is $180K. Annual revenues are $25K initially, and they increase by $2K per year. Annual costs are $5K. The salvage value is 0 after 10 years.   (*Answer:* .853)

8.10   Calculate PWI_3 for Problem 8.9.

8.11   Calculate all three PW indexes (at $i = 15\%$) for the projects described in the table. What reason(s) are there to prefer one index over another? Each project is to be evaluated over 10 years.

	A	B	C
First cost(s)	$30K	$30K in year 0, $20K in year 1	$30K
O&M/year	$13K	$10K	$46K
Revenue	$20K	$17K	$53K
Salvage	0	$20K	0

8.12   For the projects described in Problem 8.11, calculate the PWs and the IRRs. Are the rankings of the three projects consistent with the rankings using any of the PW indexes? (Note: Chapters 9 and 10 will discuss when PW and when IRR should be used.)

8.13   Calculate the future worth at 10% of a project that will save $25K per year for 20 years. The first cost is $120K, and the salvage value is $20K. Compare this with the PW and the EAW.   (*Answer:* FW = $645K)

8.14   You inherit $10K when you are 20. What is this worth when you are ready to retire at 65? Assume that the money can be invested at 7%, which is your rate for the time value of money. Compare this with an insurance policy that could be purchased with a lump-sum payment of $10K. That policy would pay you $100K at age 65 and your survivors $100K if you die sooner. How much value per year must you put on protecting your survivors for these to be equivalent?

8.15   A new plant can be constructed from the ground up, or an existing building can be purchased and retrofitted to suit the new plant requirements. Constructing a new plant will require a parcel of land costing $100K and initial design and construction costs of $175K. The remaining construction costs will be payable at $800K at the ends of years 1 and 2. Construction costs in year 3 decrease to $150K. If an existing building is retrofitted, the initial cost will be $1.25M. The remodeling will also cost $175K at the end of years 1 through 3. Determine the future value at the end of year 3 for each alternative at an interest rate of 15%.   (*Answer:* new $2546K, renovate $2509K)

8.16   Calculate the future worth at 12% for the following investment proposal.

First cost	$140,000
Annual income	$75,000
Annual costs	$35,000
Salvage value	$10,000
Useful life	5 years

8.17 A new soap press purchased by the Rub-a-Dub-Dub Soap Company costs $65K. Determine the payback period in months if the press can produce 120 gross of bars each month and each bar is sold for $.96 and cost $.42 to produce.   (*Answer:* 7.0 months)

8.18 The local hospital has just installed a totally automated switchboard that cost $205,000 to install. The switchboard replaced four operators who were paid $15K annually. Additional benefits cost the hospital 12% of each annual salary. Determine the payback period for the new switchboard.   (*Answer:* 3.05 years)

8.19 A new electric power generation plant is expected to cost $43,250,000 to complete. The revenues generated by the new plant are expected to be $3,875,000 per year, while operational expenses are estimated to be $2,000,000 per year. The plant will last 40 years, and the electric authority uses a 3% interest rate. Determine the benefit/cost ratio, where the benefits equal net annual revenues. Determine the payback period for the plant.

8.20 What is the payback for an automated voice mail system that will cost $25K to install and $300 per month to operate? It will save about 75% of the costs of a position that pays $12K per year.

8.21 The voice mail system in Problem 8.20 will incur training and inefficency costs of $2500 the first month, $1500 the second month, and $500 the third month. What is the payback period now?

8.22 Compute the payback periods for the three projects described in the table. Note that project B incurs its second first cost at the beginning of year 1, and its O&M expenses and revenues begin with the end of year 1.

	A	B	C
First cost(s)	$30K	$30K in year 0, $20K in year 1	$30K
O&M/year	$13K	$9K	$10K
Revenue	$20K	$20K	$7K in year 1; and increasing by $5K per year
Life	5 years	10 years	15 years
Salvage	0	$20K	0

8.23 Determine the discounted payback period for the soap press in Problem 8.17 if the nominal interest rate is 12% per year.   (*Answer:* 7.4 months)

8.24 Determine the discounted payback period for the switchboard in Problem 8.18 if the nominal interest rate is 18%.   (*Answer:* 53.7 months)

8.25 Let $i = 10\%$; calculate the discounted payback period for Problem 8.20. If using the tables in Appendix A, assume annual periods; but if using a spreadsheet or financial calculator, assume monthly periods.

8.26   Let $i = 10\%$; calculate the discounted payback period for Problem 8.21. Assume monthly periods.

  8.27   Let $i = 10\%$; calculate the discounted payback periods for Problem 8.22.

8.28   Carla is going on her first business trip. She will be renting a car to visit job sites in the Pacific Northwest over a 2-week period. The car can be rented for $198 per week with 100 miles free per day or for $218 per week with unlimited miles. If extra miles are charged at 25¢ per mile, what is the minimum mileage per day that she must spend visiting job sites to justify paying the higher rate? How many miles is this over the 2 weeks?   (*Answer:* 111 miles/day)

8.29   How many widgets must be needed by Acme Manufacturing per year to justify buying the required machinery for widget manufacturing for $10K? The manufacturing cost is $.26 each, and they can be purchased for $.35 each. The machinery must be overhauled at the end of every third year for $3000. Its salvage value after 12 years is 0. The firm uses an interest rate of 12%.   (*Answer:* 26,430)

8.30   What must the annual royalty for rights to use a new invention be to justify proceeding with development? The development period will be 2 years, with $650K to be spent at the beginning of each year. The firm expects the royalty period to be only 5 years due to the pace of technological development. Royalties would be received at the ends of years 3 to 7. The firm uses an interest rate of 20% to evaluate R&D projects.

8.31   Determine the number of souvenir coins that must be sold per year to justify the purchase of a $6000 stamping machine. Each coin will be sold for $5.00 and cost $1.50 to manufacture. The machine will have little or no salvage value at the end of its 4-year useful life. Use an interest rate of 8%.

8.32   The annual demand for widgets has been determined to be 10,000 per year. Each widget sells for $4.00 and costs $2.25 to produce. Using breakeven analysis and an interest rate of 12%, determine the maximum purchase price that could be paid for the required machinery. The machine is expected to last 8 years and have a salvage value of 5% of the purchase price.

PART THREE COMPARING PROJECTS AND ALTERNATIVES

There are two broad categories of engineering economy problems: constrained project selection and comparison of mutually exclusive alternatives.

In a constrained project selection problem, there are many good projects, but limited funds, so that only some of the projects may be undertaken. The problem is choosing the best projects from the list of acceptable projects. In a problem with mutually exclusive alternatives, only *one* can be chosen, even if there are 20 alternatives.

Constrained project selection problems are often considered by the firm's or the agency's top decision makers, while the engineer's economic choice between alternative engineering designs is a choice between mutually exclusive alternatives.

One large difference between the two problems is how the appropriate interest rate for the time value of money is chosen. In constrained project selection the limited budget and the available investment opportunities determine the interest rate. For problems with mutually exclusive alternatives, the interest rate is specified.

The interest rate that is used for mutually exclusive alternatives (Chapter 10) can be found through a constrained project selection problem (Chapter 9). Thus, we cover constrained project selection problems first.

Chapter 11 discusses replacement problems, a common special case of mutually exclusive alternatives. An aging existing asset is nearing the end of its economic life, and the replacement asset and its timing must be selected.

CHAPTER 9 CONSTRAINED PROJECT SELECTION

The Situation and the Solution

We usually have more good projects proposed than our funds allow. That is why projects must be compared, and we must choose some and postpone or abandon others.

The solution is to rank the projects from the highest internal rate of return (best) to the lowest IRR (worst). Then we do as many projects as we have funds to support. These limited funds form the capital budget, which is the limiting factor or constraint in constrained project selection.

The internal rates of return for the worst project that we select and the best project that we do not select define our minimum attractive rate of return and our reinvestment assumption.

Chapter Objectives

After you have read and studied the sections of this chapter, you should be able to:

Section 9.1
Define capital and operating budgets.

Section 9.2
Use IRR to rank projects from best to worst.

Section 9.3
Determine the minimum attractive rate of return (MARR) based on *opportunity cost* and use the MARR as the interest rate for money's time value.

Section 9.4
Describe the power and limitations of the perfect market model for the MARR.

Section 9.5
Impose a capital constraint on the perfect market model and link cost of capital and opportunity cost measures for the time value of money.

Section 9.6
Understand the assumptions that link the theory and application of capital budgeting.

Section 9.7
Recognize the limits of ranking on benefit/cost ratios or present worth indexes.

Section 9.8
Sort a list of projects in order of decreasing IRR using a spreadsheet.

Appendix 9A
Apply mathematical programming techniques to formulate simple capital budgeting problems and solve them using a spreadsheet.

Key Words and Concepts

Constrained project selection   Choosing the best projects from a larger set within the limits of a capital budget.

Capital budget   The maximum total spending permitted on the first costs of proposed projects.

Investment opportunity schedule   The ranked list of available projects.

Opportunity cost   The rate of return of the worst accepted project or the best one forgone.

Minimum attractive rate of return (MARR)   The minimum interest rate that a project must earn to be acceptable.

Perfect market model   A description of how firms could theoretically maximize their value.

Cost of capital   The rate paid for the use of invested funds.

Choosing the best projects within a limited budget is one of the most important problems solved by engineering economy. There is considerable research on various approaches, but there is not complete agreement among writers on engineering economy. This text describes the approach (ranking on IRR) that is most often practiced (see Exhibit 8.1)—including its theoretical justification. Other approaches are introduced, but their complete presentation is left to more advanced books [Bussey & Eschenbach and Park & Sharp-Bette]. The next chapter uses PW, EAW, and EAC at the minimum attractive rate of return (developed in this chapter) to choose between mutually exclusive alternatives.

To maintain clarity and simplicity, this chapter discusses only *investments*. As shown in Chapter 7, most engineering projects are investments, where cash is invested and then returns are generated. Engineering projects are rarely *loans* (the positive cash flows come first). Loans are omitted here to avoid the complication of wanting low interest rates for loans and high interest rates for investments.

9.1 BUDGETS AND PROJECT SELECTION

Constrained project selection is choosing the *best* projects from a larger set of acceptable projects within the limits of a capital budget.

Constrained project selection is choosing the *best* projects from the set of *acceptable* projects. Virtually all firms and governments have limited funds, and they cannot implement every attractive proposal. For brevity, the term *firms* is used for all organizations facing this problem.

Firms define two budgets each year. The first, an operating budget, covers expenses for current activities, such as wages, purchased materials, and energy costs. The second, a **capital budget,** supports major new initiatives and the cost of replacing expensive equipment. Both budgets are created out of the firm's income and borrowing, so funds can be and are transferred between the two budgets. However, most projects evaluated with engineering economy are funded by the capital budget.

The **capital budget** is made up of those funds allocated for purchasing equipment, constructing buildings, etc. This capital budget is the maximum total spending permitted on first costs.

This capital budget is first set for the entire firm. Then it is allocated to divisions, then to plants, then to departments, and then to groups (see Exhibit 9.1). Capital resources are limited at virtually every decision-making level of every firm. There are also limits on labor, managerial time, material, and equipment; and even more limits may be imposed by political or social issues. But engineering economy must first deal with a single limit—on capital.

Why is the capital budget limited? First, the capital budget may be externally limited. For example, the state legislature determines how much the state's highway department and its university have to spend. Similarly, the capital spending of a small high-tech start-up company is limited by the venture capital it acquires and by the cash flows it generates.

On the other hand, the spending limit may be determined by *internal* choices. Some divisions, plants, and products will have better growth prospects; some will be strategically crucial; and some may be in volatile, risky markets. A new pollution-control investment may be mandated, or the timing of a union renegotiation may influence plans for building a new plant or renovating an old one. Thus, each level in the hierarchy must consider noneconomic criteria as it allocates the capital budget to the level below it.

Whether externally or internally based, this limitation on capital spending is the norm, not the exception. The next several sections show that this spending

Exhibit 9.1   Allocating the capital budget

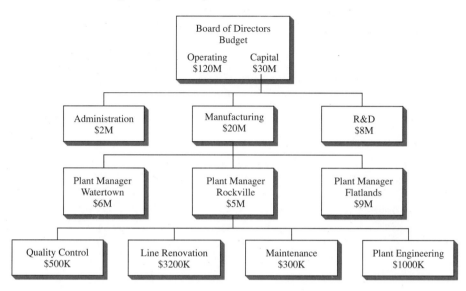

limit determines the <u>minimum attractive rate of return (MARR or i^*).</u> This MARR is the <u>interest rate for calculating present worth and equivalent annual cost</u> (see Section 9.3).

Problem Size.   The constrained project selection problem is very large. For example, if a refinery's engineering group can recommend 15 of 25 projects, there are 3,268,760 different project sets to evaluate. If 10 refineries each forward 15 proposals to the division's vice president, who immediately accepts or rejects two-thirds, or 100 projects, then there are 50 proposals still in consideration. If half of the 50 will be accepted, then there are over 126 trillion sets. Some sets would exceed the capital budget; others would leave money unspent; but only one set would include all of the best projects. The number of ways or combinations to choose <u>r projects out of n is $n!/[(n-r)! \cdot r!]$.</u>

Evaluating this many possible combinations is unrealistic, and the refinery managers and the division vice president need to use a ranking technique. Ranking on internal rate of return constructs the best combination for each manager. Example 9.1 illustrates this for the first refinery's engineering group.

Example 9.1   Selecting the Refinery's Best Projects

The engineering group has evaluated 25 projects with a positive PW at $i = 12\%$. Many projects serve multiple purposes, but they are categorized in Exhibit 9.2 by their main focus. Estimated lives range from 5 to 20 years. Which of the 25 projects are the best, since the refinery will not receive the $2M required to fund all of them?

Exhibit 9.2   25 engineering projects for a refinery

Focus	Project	First Cost	Life	Annual Benefit
Capacity expansion	V	$100K	10	$20.0K
	S	100K	10	22.2K
	A	100K	15	35.6K
	Y	105K	10	19.0K
	D	120K	15	36.7K
	J	120K	20	27.6K
	L	130K	15	29.5K
	O	175K	10	41.6K
Energy efficiency	Q	50K	20	10.0K
	B	50K	20	16.7K
	I	90K	20	21.5K
Lower costs on:				
Product 3	H	40K	5	15.0K
Product 1	E	40K	5	16.0K
Product 6	R	50K	5	16.2K
Product 2	F	50K	10	15.3K
Product 7	U	50K	10	10.3K
Product 4	M	60K	5	20.4K
Product 9	X	75K	5	21.6K
Product 8	W	80K	5	23.5K
Product 5	N	85K	10	20.5K
Pollution control	K	40K	10	10.3K
	G	60K	15	16.9K
	P	65K	15	13.7K
	C	75K	10	26.1K
	T	90K	15	16.4K

Solution

If IRRs are calculated, then the projects can be ranked on IRR. The ranking shown in Exhibit 9.3 clearly identifies which projects are better than others. The division's vice president could accept the top 10, with $745K in capital costs; reject the bottom 5 and their associated $410K in capital costs; and focus attention on projects K through T, with their total costs of $845K, in deciding how much of the capital budget should be allocated to this refinery. (Note: if projects A through O were funded at a cost of $1235K, another $65K to $165K, or 5% to 15%, might be allocated to cover possible cost overruns.)

Exhibit 9.3 also includes the PW and EAW of each project, evaluated at $i = 12\%$, to illustrate that ranking using these measures would lead to different projects being selected.

Exhibit 9.3   Refinery projects ranked by IRR

Project	IRR	First Cost	PW	EAW	Focus
A	35.2%	$100K	$142.5K	$20.9K	Capacity expansion
B	33.3	50K	74.7K	10.0K	Energy efficiency
C	32.8	75K	72.5K	12.8K	Pollution control
D	30.0	120K	130.0K	19.1K	Capacity expansion
E	28.6	40K	17.7K	4.9K	Lower costs on product 1
F	28.0	50K	36.4K	6.5K	Lower costs on product 2
G	27.4	60K	55.1K	8.1K	Pollution control
H	25.4	40K	14.1K	3.9K	Lower costs on product 3
I	23.5	90K	70.6K	9.5K	Energy efficiency
J	22.6	120K	86.2K	11.5K	Capacity expansion
K	22.3	40K	18.2K	3.2K	Pollution control
L	21.5	130K	70.9K	10.4K	Capacity expansion
M	20.8	60K	13.5K	3.8K	Lower costs on product 4
N	20.3	85K	30.8K	5.5K	Lower costs on product 5
O	19.9	175K	60.0K	10.6K	Capacity expansion
P	19.6	65K	28.3K	4.2K	Pollution control
Q	19.4	50K	24.7K	3.3K	Energy efficiency
R	18.6	50K	8.4K	2.3K	Lower costs on product 6
S	17.9	100K	25.4K	4.5K	Capacity expansion
T	16.3	90K	21.7K	3.2K	Pollution control
U	15.9	50K	8.2K	1.5K	Lower costs on product 7
V	15.1	100K	13.0K	2.3K	Capacity expansion
W	14.4	80K	4.7K	1.3K	Lower costs on product 8
X	13.5	75K	2.9K	.8K	Lower costs on product 9
Y	12.5	105K	2.4K	.4K	Capacity expansion

Budget Flexibility and Contingency Allowances.   Example 9.1 also illustrates the flexibility in decision making that forces firms to use a ranking technique (such as ranking on IRR), rather than more theoretically sophisticated approaches (see Appendix 9A). The division's vice president must decide how much of the capital budget to allocate to this refinery. The top 10 projects might be accepted with little examination, while the bottom 5 might be rejected as being a "wish list." The remaining 10 would be compared with competing projects from other refineries to set the final capital budget for this refinery. The 10 select/10 examine/5 drop or 5/15/5 may be chosen using a convenient 5 basis or by using the dollars required for each group.

As noted in Example 9.1, some firms identify part of the capital budget as an allowance for cost overruns on the selected projects. The contingency allowance might be 30% for a high-risk construction project, 100% for a high-risk new product development, and 0% for a more routine project. Sometimes the total contingency allowance is adjusted to match the available capital budget with the total first costs.

9.2 IRR IS THE BEST RANKING TOOL

In Chapter 3 the internal rate of return was defined as the interest rate at which a project's receipts and disbursements sum to a zero present worth. Chapter 7 showed how the relative size and timing of the project's cash flows determine its IRR.

The following discussion of IRR, PW, and equivalent annual measures as ranking techniques for potential projects concludes that:

1. The internal rate of return is the best ranking measure.
2. Other discounted cash flow measures may produce incorrect and different rankings.
3. These differences arise because IRR adjusts for project size, while equivalent annual and PW measures do not. PW indexes and benefit/cost ratios adjust for project size, but the wrong interest rate is often used (see Section 9.7).

Projects of Different Sizes.   IRRs provide a consistent measure of the rate of return earned for each dollar of investment, for each year the dollar is invested. Since the capital budget limits the total first costs, the goal is to maximize the return per dollar of first cost.

The easiest example for projects of different sizes is to take any project and consider doubling its cash flows. Since the PW $= 0$ at the IRR, multiplying all cash flows by 2 (or any other constant) will leave the IRR unchanged. On the other hand, multiplying all cash flows by 2 would double the project's PW or EAW. If PW or EAW were used to rank the projects, then the doubled project would appear twice as attractive. In reality, it is merely twice as large. Since the doubled benefits require twice as much investment, in reality the doubled project's attractiveness is unchanged.

In general, large projects that have large first costs will often have large PWs and EAWs. Project O in Exhibits 9.2 and 9.3 illustrates this. Ranked 15th by IRR, this project has the 8th largest PW and the 5th largest EAW at an interest rate of 12%.

The IRR considers every cash flow of the project, and it provides a single number to validly rank projects in order of decreasing attractiveness. A project's size does not change the validity of the IRR ranking.

Strengths and Weaknesses of the IRR.   The internal rate of return has a weakness as well. As we saw in Chapter 7, some projects with multiple sign changes in their cash flows do not have an IRR. However, most projects are either investments or loans and most have a unique IRR. For the moment, we will ignore loans and multiple sign changes, but we will revisit the issue at the end of the chapter.

The IRR's chief advantage over the present worth index is that the ranking can be made before the minimum attractive rate of return is chosen. In fact, ranking on IRR is used to select the interest rate, i. A second advantage is that the IRR is sensitive to the timing of the cash flows. Example 9.2 illustrates that projects with the same scale and duration may have similar EAWs and PWs, but still differ in their IRRs.

Example 9.2 Projects with Matching Lives and First Costs

Rank the following six projects, each of which has a life of 5 years. The returns of all six projects can be split into a level annual benefit and an arithmetic gradient. For example

$$PW_1(i) = -20{,}000 + 14{,}000(P/A, i, 5) - 4000(P/G, i, 5)$$

and

$$PW_6(i) = -20{,}000 - 6000(P/A, i, 5) + 7100(P/G, i, 5)$$

Project	First Cost	Annual Benefit	Gradient	EAW at 10%	IRR
1	$20,000	$14,000	$-4000	$1484	28.3%
2	20,000	9000	-1200	1552	22.6
3	20,000	6800	0	1524	20.8
4	20,000	4000	1550	1530	19.2
5	20,000	0	3750	1512	17.4
6	20,000	-6000	7100	1576	16.0

Solution

As summarized in Exhibit 9.4, these projects are correctly ranked on IRR, where the largest IRR is 75% greater than the smallest. Their EAWs (and PWs) vary by only 6%, and are ranked differently (6, 2, 4, 3, 5, 1). Project 1's IRR is the largest, because it has large positive cash flows in years 1 and 2, which is much earlier than the large cash flows for project 6.

Exhibit 9.4 Example 9.2 projects with matching first costs and lives

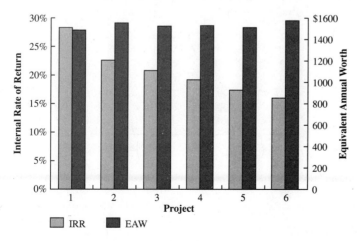

9.3 DETERMINING THE MINIMUM ATTRACTIVE RATE OF RETURN

Solving the constrained project selection problem is one way to define the minimum attractive rate of return. The first step was detailed in the last section:

1.   Ranking the projects in decreasing IRR order. Recall that for investments, higher rates of return are better, and that investment (not loan) cash flow patterns were assumed.

In this section, the steps also include:

2.   Computing the cumulative first cost as each project is added to the list.
3.   Comparing the cumulative first cost with the budget constraint.
4.   Finding the minimum attractive rate of return.

Example 9.3 illustrates the process for seven projects with different first costs, but with the same life. In Exhibit 9.5, the cumulative cost after five projects exactly matches the capital budget. (If the cumulative cost and the capital budget do not match, see Section 9.6.)

Example 9.3   Microtech

Microtech is considering seven projects within a capital budget of $90K. All projects have a 5-year life and all projects must be acceptable at a 10% interest rate. Determine which projects should be recommended for implementation.

Project	First Cost	Annual Benefit	PW at 10%	IRR
1	$20,000	$ 8000	$10,326	28.6%
2	30,000	11,000	11,699	24.3
3	10,000	3500	3268	22.1
4	5000	1650	1255	19.4
5	25,000	8100	5705	18.6
6	15,000	4650	2627	16.6
7	40,000	12,100	5869	15.6

Solution
For convenience, the data ranked the projects from the most to the least attractive IRR. The top project has a cost of $20,000; the top two have a cumulative cost of $50,000; the top three a cost of $60,000, etc. Since the budget is $90,000, the top five projects should be chosen.

Exhibit 9.5   Selecting the best projects for Microtech

Project	IRR	First Cost	Cumulative First Costs
1	28.6%	$20,000	$20,000
2	24.3	30,000	50,000
3	22.1	10,000	60,000
4	19.4	5000	65,000
5	18.6	25,000	90,000 budget
6	16.6	15,000	105,000
7	15.6	40,000	145,000

To confirm that ranking on IRR is best, note that the top 5 projects in Exhibit 9.5 have a cumulative annual benefit of $32,250. Projects 1, 2, and 7 have the largest PWs and the same total first costs of $90,000, but their total benefit is $1150 smaller per year at $31,100. This comparison is valid since all projects have the same 5-year life and uniform annual benefits. Thus, choosing the projects with the largest PWs would lower the total return.

Investment Opportunity Schedule.   The ranking of projects from the highest IRR to the lowest can also be shown graphically. If the x-axis is the cumulative first cost or investment, the step function that is generated is called the **investment opportunity schedule.** Since the projects are ranked from the highest IRR to the lowest, this forms the schedule or order of selection.

Exhibit 9.6 illustrates this for the refinery projects that were ranked in Exhibit 9.3 for Example 9.1. Exhibit 9.7 is the graphical version of Exhibit 9.5, the investment opportunity schedule for Microtech's projects (in Example 9.3).

> The **investment opportunity schedule** ranks the organization's proposed investment projects from the highest IRR (best) to the lowest IRR (worst).

Exhibit 9.6   Choosing the refinery's top projects

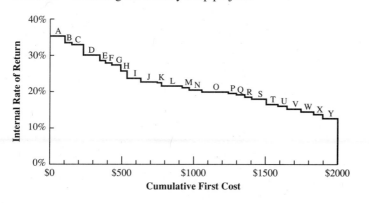

Exhibit 9.7   Choosing Microtech's top projects

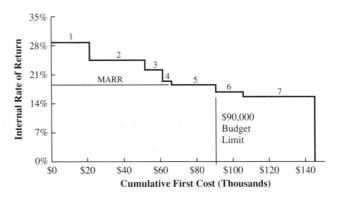

Comparing Exhibits 9.6 and 9.7 shows that the steps of this investment opportunity schedule become relatively smaller as the number of projects increases. For firms or agencies with more than 20 projects, this graph can be approximated by a smooth curve whose values are the marginal return of each additional dollar of investment.

Minimum Attractive Rate of Return (MARR).   Both Exhibits 9.5 and 9.7 show that Microtech's top five projects should be funded for a budget limit of $90,000. Specifically, project 5 is funded and project 6 is not. If the budget drops below $90,000 the opportunity is lost to invest in project 5, which returns its first cost with interest at a rate of 18.6%. On the other hand, if the budget increases by $15,000, project 6 can be added to gain returns at a 16.6% rate. This **opportunity cost** is the best basis for choosing the **minimum attractive rate of return (MARR).**

The **opportunity cost** of capital is the rate of return of the worst accepted project or the best forgone project.

Depending on whether projects are added or deleted, the opportunity cost for Microtech's investments is between 16.6% and 18.6%. Since most organizations consider more than seven projects, the IRRs of *adjacent* projects are closer together, and the MARR is usually known more exactly, as in Example 9.4.

The **minimum attractive rate of return (MARR)** is the basis for calculating present and equivalent annual worths, and it is the minimum interest rate that an investment must earn. For convenience, it is defined as the rate of return of the worst accepted project.

For convenience, the MARR is defined as the higher of the opportunity costs for adding or for deleting projects, which is the IRR of the worst project that is accepted. For Example 9.3, the IRR for project 5 and the MARR are both 18.6%.

This opportunity cost represents what the firm loses if it spends part of its capital budget on unlisted projects or on ones near the list's bottom. To calculate present and annual worths, this opportunity cost defines the MARR, which is the interest rate that should be used. Then the sign of the PW will give the proper signal for whether or not the project should be funded.

In equations, the MARR will be shown as i^*. This rate is also an upper limit for financing proposals, since it would certainly make no sense to borrow at 15% in order to invest at 10%.

This section has demonstrated how the MARR should be determined and how projects should be selected for funding. Once ranking on IRR has been used to find the MARR, then PW measures can be used to check the acceptability of other alternatives. But only ranking on IRR can be used to find the all-important MARR (i^*). Example 9.4 illustrates the normal procedure. Example 9.5 shows a shortcut that is possible if the cash flows for periods 1 to N are uniform and all projects have the same life, N. If the cash flows are uniform series and each project's life is the same, then only one IRR need be found—to determine the MARR. However, if spreadsheets are used, it is just as easy to find all the IRRs, to sort by IRR, and then to find the cumulative first cost.

Example 9.4 MARR for the Refinery Projects (Example 9.1 Revisited)

Assume that the division vice president has allocated $1.3M to the refinery. Since the 25 projects have total first costs of $2M, some must be forgone. Which projects should be done and what is the MARR?

Solution
The first costs of projects A through P sum to $1.3M, so those are the projects that should be done. Projects Q through Y should be forgone. Since the IRR of project P is 19.6%, that is the MARR. Note that project Q has an IRR of 19.4%, so the range of uncertainty in the MARR is only .2%.

Example 9.5 Highway Road Projects with Matching Lives and Uniform Benefits

Consider eight projects (see Exhibit 9.8) under consideration by a state highway department, which has a total budget of $1.3M. Each road project is assumed to have a life of 20 years with level annual benefits.

Exhibit 9.8 Highway projects

Project	First Cost	Annual Benefit
1	$300,000	$28,000
2	250,000	33,000
3	50,000	3750
4	500,000	45,000
5	250,000	29,000
6	150,000	15,500
7	400,000	24,000
8	350,000	38,000

Solution
Because all of the projects use a life of 20 years and uniform annual benefits, all use the same PW equation to find the IRR:

$$0 = -\text{First Cost}_j + \text{Annual Benefit}_j \cdot (P/A, i_j, 20) \quad \text{for project } j$$
$$(P/A, i_j, 20) = \text{First Cost}_j / \text{Annual Benefit}_j$$

Now, the $(P/A, i, 20)$ factors can be ranked, as shown in Exhibit 9.9. The largest factors are closest to the 0% factor for 20 years, $(P/A, 0\%, N) = N$ or 20. Only that factor closest to the \$1.3M cumulative budget must be solved to find the MARR of 6.9%. Exhibit 9.10 shows the (P/A) values ranked in ascending order, because higher IRRs have lower (P/A) values.

Exhibit 9.9   Ranked highway projects for Example 9.5

Project	First Cost	Cumulative First Cost	$(P/A, i, 20)$
2	\$250,000	\$ 250,000	7.576
5	250,000	500,000	8.621
8	350,000	850,000	9.211
6	150,000	1,000,000	9.677
1	300,000	1,300,000	10.714
4	500,000	1,800,000	11.111
3	50,000	1,850,000	13.333
7	400,000	2,250,000	16.667

Exhibit 9.10   IRR vs. (P/A) values for Example 9.5

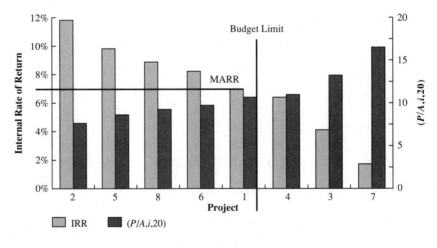

9.4 PERFECT MARKET MODEL FOR THE MARR

In Section 9.3, the minimum attractive rate of return was found by applying a capital limit to the investment opportunity schedule. In this section, rather than imposing a capital limit, the cost of acquiring funds for investment is used. Acquiring these funds (the firm's capital) is done through the firm's finance group. This basis for development of the MARR has the following steps:

1. Develop a financing opportunities curve.
2. Find the intersection of the financing opportunities curve and the invest-
    ment opportunity schedule. Under certain assumptions (those of a perfect
    market), the *x*-coordinate of this point is the optimal level of total invest-
    ment.
3. Conceptually define the cost of capital measures.

    Just as investments can be ranked, so can financing sources. However, unlike
potential investments, the best financing has the lowest interest rate, and the worst
has the highest rates. Thus, the financing curve increases as cumulative investment increases.

    Possible sources for these funds include: last year's profits; additional equity,
such as new stock or partners; additional long-term debt or bonds; or additional
short-term debt, such as unsecured loans or loans against accounts receivable.
However, the range of likely interest rates for financing sources is much narrower
than the range for investments. Thus, the financing curve slopes less steeply than
the investment opportunity schedule. The financing curve is combined with the
investment opportunity schedule in Exhibit 9.11.

Exhibit 9.11   Perfect market model

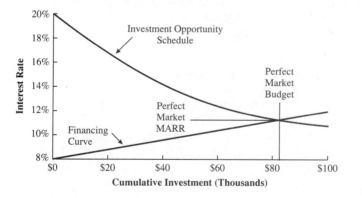

    There are two reasons for the financing curve's upward slope. First, the firm
seeks out and secures the lower-cost funds first. Second, the providers of all funds
charge somewhat higher rates as the firm increases its level of financing and debt.
This compensates for an increased risk of default.

    Exhibit 9.11 is called the **perfect market model** because of the four assump-
tions it is based on:

1. Investment opportunities and financing sources are available in small,
    divisible pieces. This assumption causes the two curves to be smooth,
    and it guarantees that they intersect.
2. The investment opportunities are independent of each other.

The **perfect market
model** describes how
a firm can theoretically
maximize its value by
undertaking every project
whose returns exceed the
marginal financing cost for
the entire set of projects.

3. The firm has some profitable investments; that is, at least the best projects have IRRs that exceed the firm's financing rate.

4. The firm can both invest and borrow at the rate where the two curves intersect, with no transaction costs.

Economic and financial theory (and simple logic) imply that a firm optimizes its performance by operating at the intersection of the two curves. If a firm operates to the right of the intersection, then it is borrowing money at a higher rate than it is investing. On the other hand, any firm that operates to the left of the intersection could increase its total profit by borrowing more money and investing it.

This optimization result is the justification for using the cost of acquiring funds to evaluate potential projects. Any project that will return more than the cost of additional financing should be undertaken, along with the borrowing required to do it. Because the financing curve is relatively flat, the firm can determine the rate at which funds are acquired and apply this to all potential investment opportunities. Theoretically, this separates the financing and investment decisions [see Bussey and Eschenbach]. Practically, this argument is the basis for using present worth at the cost of financing as the measure for evaluating individual projects.

Of the four perfect market assumptions, the last is perhaps the most questionable in reality. A firm must typically borrow money at a higher rate than it could loan money. Moreover, there are significant transaction costs involved in terms of fees, employee time, and managerial time.

9.5 CAPITAL LIMITS IN AN IMPERFECT BUT REAL WORLD

Why Capital Limits Are Imposed.   The perfect market model connects the cost of acquiring funds with the selection of an optimal investment level. While this is conceptually useful, the model must be modified to fit reality. This section discusses why capital budgets are imposed in the real world and the implications of the capital constraint.

It is well established that most firms do not operate at the perfect market intersection. Instead, capital budgets allocate limited available funds, and then engineering economy must be used to select the projects that should be undertaken. In fact, firms do not and should not operate at this intersection, for four good reasons:

1. To adjust for the firm's larger risk as compared to the bank's risk.

2. To allow a margin to protect against overly optimistic estimates.

3. To concentrate the firm's attention on the best projects.

4. To allow more flexibility to pursue new projects.

First, consider a firm that is issuing bonds or borrowing from banks. These lenders acquire rights to the firm's assets that come before the rights of common stockholders or of other owners. Thus, the risk to the firm's owners is larger than the risk to the lender. If the firm goes bankrupt, the lenders will receive partial or full repayment, while the owners come last. Suppose a firm borrows funds at 12% to invest in a project that earns 13%. The lender then earns 12% for supplying the funds and the firm earns about 1% (on someone else's money) for doing all the work and absorbing most of the risk. The firm will usually insist on a reasonable margin over the rate charged by the financing source.

Second, many decisions must be made at lower levels of the organization and the information on the perfect market rate exists at the top, if it exists at all. If the chief executive officer establishes a MARR, then subordinates are encouraged to be optimistic in their project evaluations. If their optimistic values show a positive PW at the MARR, then funding is warranted. Most people tend to be unconsciously optimistic about projects they want to undertake. Thus, a project that is predicted to barely meet the cost of financing is likely to result in a loss. The firm will usually insist on a reasonable margin over the financing rate to compensate for the risk that the project will not meet expectations.

In some organizations, overly pessimistic estimates are required. Suppose projects that cost $500,000 and require 10 people for 6 months are approved after cutting the budget to $400,000 and the resources to 7 people for 4 months. On the next project, it is likely that inflated cost, people, and time estimates will be submitted; then, after the expected cuts, there will be enough left over to get the job done. If pessimistic estimates are used for the first costs, then overly optimistic estimates may be used for the future benefits of the project to compensate.

Third, projects whose IRR is close to the perfect market intersection provide only minimal returns to the firm. If they are undertaken, then the firm's money and the attention, energy, and abilities of its management and workforce are to some extent diverted from the best projects, where a substantial payoff is possible.

Finally, a firm is poorly positioned to take advantage of new opportunities if it has borrowed all that it can or stretched its capital plant or its people to the limit. It has little flexibility to respond to the inevitable changes of the future.

Consequently, firms set a capital limit rather than borrowing up to the point of the perfect market model. The firm's capital budget may be allocated to each division or department, where each level can make the best investments available within the limits of its resources. A minimum MARR may be established for each division that considers the riskiness of projects within that division. For example, a petroleum company is likely to use a higher MARR for the evaluation of exploration projects than for distribution projects. However, after each level selects its projects within its capital limit, then the MARRs can be compared for levels with equivalent risks. If these rates differ substantially between divisions or departments with similar risks, then it might be desirable to reallocate funds.

Budget Limits and the Cost of Capital.   Exhibit 9.12 shows what happens when the budget constraint is added to the perfect market model shown in Exhibit 9.11. Three more interest rates—the *true* MARR, the marginal cost of capital, and the weighted average cost of capital—are defined in addition to the perfect market MARR.

The highest interest rate is the opportunity cost of forgone investments. As shown in Section 9.3, this should be the firm's MARR. If the firm's capital con-

Exhibit 9.12   Four possible MARRs

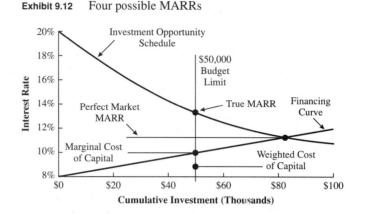

straint is even moderately severe, then the opportunity cost of forgone invest-ments can *greatly* exceed the *perfect market* or theoretical MARR (defined by the intersection of the two curves).

The **cost of capital** is the rate of return that is paid on funds.

The lower two interest rates are based on the firm's **cost of capital** for the bud-get that is used. The higher of these two is the firm's marginal cost of capital at the budget limit. This is the cost or interest rate for the firm's most expensive source of capital. The lower of the two cost of capital measures is the firm's weighted-average cost of capital for all of the capital used. This lower rate is a weighted average of the interest rates for bonds, loans, stock, and retained earnings.

The weighted-average cost of capital (WACC), which is the lowest rate, is often erroneously used as the minimum attractive rate of return (MARR). The WACC is closely linked to the marginal cost of capital, and it can be related to his-torical data about the cost of acquiring funds. At best, current estimates for stocks, bonds, and loans are used, but inexpensive and expensive sources are combined. At worst, debt financing that was secured at favorable rates nearly 10 years ago is included.

The cost of capital measures are useful because they provide a lower bound for the MARR. It certainly would not make sense to borrow money at 9% and then invest it at 6%. Managers can also subjectively increase this lower bound. For example, if a firm's cost of capital is 9%, then managers could require projects to earn at least 12%. The 3% margin compensates the firm for the risk of optimistic estimates for the projects and for the added risk it bears beyond that borne by the bank and other financing sources. The cost of capital measures also provide a measure of the firm's success in acquiring funds at reasonable interest rates.

However, only rarely is the cost of capital the MARR. In almost every case, there is a capital budget and an associated opportunity cost for forgone invest-ments. It is this opportunity cost that is the MARR for project evaluation and selection.

Example 9.6 illustrates how the cost of capital can be calculated. The inter-est rates for debt financing (bonds or loans) were calculated in Chapter 7. For common and preferred stock, the interest rate is found using models that include dividends and changes in the stock's price. The highest of the bond, loan, and stock rates is used for retained earnings.

Example 9.6   Marginal and Weighted Average Cost of Capital

What are the marginal and weighted average costs of capital for a manufacturing firm whose capital structure [see Kester] can be broken down as follows:

$75M from bonds at 9% on average

$125M from loans at 11% on average

$200M from stock estimated at 14%

$150M from retained earnings estimated at 14%

Solution

For this firm, the marginal cost of capital is 14% (since the marginal rate must be the highest rate of any source). The weighted average cost of capital can be found by calculating the weighted average of the four sources:

$$\text{WACC} = \frac{75M \cdot .09 + 125M \cdot .11 + 200M \cdot .14 + 150M \cdot .14}{(75M + 125M + 200M + 150M)}$$

$$= 12.6\%$$

9.6 MATCHING ASSUMPTIONS TO THE REAL WORLD

Solving the constrained project selection problem has required some assumptions—which may not precisely match the real world. These assumptions are:

1. Projects and financing come in divisible or *small* pieces.
2. Projects are independent.
3. Projects can be simultaneously evaluated.
4. The MARR is stable over time.
5. The MARR is unaffected by omitting loans and multiple-sign-change projects from the initial calculation.

This last assumption is the easiest to justify. In defining the MARR, individual projects were assumed to have only one sign change in the cash flow diagram and to be investments for which a high IRR is attractive. Once the MARR has been selected, PW or equivalent annual measures can be used to evaluate loan projects and projects with multiple sign changes. If the total cost of *good* projects in these two categories is high, then a slightly higher MARR might be used.

Assumption of Indivisible Projects and Increments of Financing.   A less realistic assumption was that the budget matched the cumulative first cost for the best projects. Instead, as suggested by Exhibit 9.13, projects are usually indivisible,

Exhibit 9.13   Increments of financing and indivisible projects

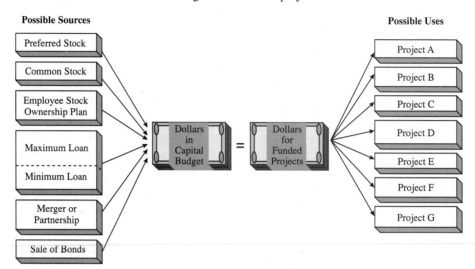

and financing may be available only in large increments. A new dam, robot, or assembly line is a "go or no-go" project: if the project is undertaken, the full cost is incurred; either the $1 million is spent or nothing is spent. The firm simply cannot do 38.3% of the project to make the total investment cost match the budget.

Recall that in Exhibits 9.5 and 9.7, a budget of $90,000 implied that Microtech should do projects 1 through 5. If the budget were $100,000 instead, then there would not be enough funds to add project 6, which costs $15,000. However, leaving $10,000 unspent may also be a problem, since it suggests that the firm, the agency, the department, etc., may not have enough good projects. This may even cause fewer funds to be allocated to this group next year.

The reality of indivisible projects and increments of financing can be faced in four different ways. Students and engineers can use the first two approaches. However, only in the real world are the latter two approaches available to everyone. They are:

1. "Juggle" the list.
2. Model the problem with a mathematical program and solve with a computer.
3. Adjust the scope or cost of projects and/or the budget available.
4. Add an adjustable contingency amount to match the projects and the budget (see Example 9.1).

First, the list can be "juggled" by the analyst. Suppose a $100,000 budget replaced the $90,000 limit shown in Exhibit 9.5. Then, substituting project 6 for project 4 increases the first costs by $10,000, which would fully use the budgeted funds. Since their IRRs, 19.4% and 16.6%, are relatively close, the new set of accepted projects is at least a good set. This set might be the best possible, but "juggling" cannot confirm that it is the best of all sets.

Second, mathematical models can produce an optimal selection. These models are computerized to deal with the enormous number of possible project combinations. These capital budgeting models, discussed in Appendix 9A, are a part of the field of operations research. However, these models are only optimal if all data is known and there is no uncertainty—which is rarely true.

Third, the budget or project costs and/or scope can be adjusted (up or down). This option is common in reality, but unavailable in textbooks. In textbooks, projects are treated as having clearly defined costs, and budgets as being fixed. In reality, these values can be and are adjusted through the decision-making process. Budgets may be increased or reallocated to fund an attractive project, which is on the "bubble" or even below it. (Note: the "bubble" suggests that support may be there or that the bubble may pop—leaving the project unsupported.) The common plea is to add $x to the budget so that this clearly desirable project can be accepted. On the other hand, projects are commonly approved with a stipulation 规定 that their costs be reduced, perhaps by 10%. Example 9.7 illustrates these and other possibilities.

Since data values are not exact in reality, IRRs and PWs are somewhat approximate. Thus, it is satisfactory for the MARR to be an approximate value. The MARR need not be precise to be useful. It is not fatal to omit a project from the list because its IRR is confounded by multiple sign changes. It is also not fatal to "juggle" the list so that the budget is fully utilized. Once the MARR is found, then all projects that have positive present worths at that value can be considered.

Example 9.7   Drew Metal Works

The casting division of Drew Metal Works is considering several projects, but the division has only been allotted a capital budget of $100,000. The best project has a first cost of $95,000 and the next best is expected to cost $10,000. Describe some of the options that the divisional vice president has.

Solution
If the organization rewards honesty, then the three options are:

1. Ask for an extra $5000 and approval of both projects. 意外
2. Ask for approval of the best project and a $5000 contingency fund.
3. Find a way to cut $5000 from the combined projects.

If the organization encourages game playing in its budgeting and capital project approval process, then variations on options 1 and 2 are likely:

4. Submit both projects for approval at $100,000 and "bootleg" (which means act without approval of higher management) $5000 from another part of your budget.
5. Submit only the best project for approval, but inflate its cost to $100,000. The extra $5000 may be either a hidden contingency or a source of funds for bootlegging to other projects.

Assumption of Project Independence.   Usually, proposed projects are assumed to be independent, but this is not always true. Investments in regional warehouses and in factory automation may work synergistically so that the benefits of doing both exceed the sum of the individual projects, or they may compete so that the total benefits are less than the sum of the parts. Similarly, a new hydroelectric dam and an electrical intertie to link different utility systems may have synergistic or competing effects. While synergism and competition complicate data gathering and model building, they can be included—and in some cases, they must be. Often, these interdependencies are subjectively considered; at other times, mathematical modeling techniques will be required.

If two projects are interdependent, then they can be converted into four mutually exclusive possibilities (the first, the second, both, and neither). This is illustrated in Example 9.8. The same approach can be used for three interdependent projects, but the number of mutually exclusive possibilities doubles to eight.

Example 9.8   Kelsey Manufacturing

Kelsey Manufacturing is considering two pairs of interdependent projects. The first pair consists of a plant modernization (PM) and a marketing campaign (MC). These are synergistic, since reducing manufacturing costs by 20% and increasing market share by 10% overlap positively. The second pair consists of a machine vision inspection station (MVIS) and a milling machine (MM) with tighter tolerances for MM performance. These are competitive, since having one reduces the need for the other, although both may still be justifiable. Create for each pair the four mutually exclusive possibilities.

Solution
First pair's possibilities: only PM, only MC, both, and neither; the second pair's: only MVIS, only MM, both, and neither. For each pair, the costs and benefits for the "both" case require special attention, since interdependence implies that the values for the individual projects cannot simply be added together.

For the first pair (plant modernization and marketing campaign), the net cash flows would be greater than the sum of the two individual projects. This pair is synergistic.

For the second pair (machine vision inspection station and milling machine), the net cash flows would be less than the sum of the two individual projects. This pair is competitive.

Assumption of Simultaneous Evaluation.   To construct the investment opportunity curve, projects must be available for simultaneous evaluation to find the MARR. This is suitable for larger projects and larger firms or agencies that emphasize an annual budget review cycle. However, many smaller projects are evaluated on a real-time basis as the opportunities arise. In these cases, perhaps the best solution is to assemble historical data and to calculate the MARR for future use, as is outlined in Example 9.9.

Using historical data to establish a MARR only works for ranking on IRR, and it cannot be done for the mathematical modeling approaches discussed in Appendix 9A. Mathematical modeling requires complete data on every project.

Example 9.9   SpaceCom's Workstations

Exhibit 9.14 lists 15 proposals that were considered over the last year at SpaceCom. Each proposal is for the purchase of a standard engineering workstation computer, so each project has the same $10,000 cost. Each proposal is linked to specific customer orders for space-based communications software, so its timing was determined by the specific customer orders rather than by SpaceCom's budget cycle. Also, the net annual returns of each project differ in amounts, pattern, and number of years. For example, project A has a return of $5000 per year for 7 years, while project F has a return of $6000 in year 1, which decreases by $1500 per year for years 2 through 4. The other gradients are defined similarly.

There is no salvage value for the workstations because the customer acquires title at the end of the project. These proposals were all evaluated at 10% (the interest rate suggested by corporate headquarters) and all but two seemed attractive, but only five were funded. Use this data to estimate the MARR for future project evaluations.

Exhibit 9.14   Fifteen workstation projects

Projects	IRR	PW at 10%	Year	Cash Flow	Year	Cash Flow
A	46.6%	$14,342	1–7	$5000		
B	36.3	4921	1–3	6000		
C	30.6	2397	1	10,000	2	$4000
D	30.0	1818	1	13,000		
E	27.3	3180	1–3	5300		
F	23.8	2452	1	6000	1–4	gradient = −$1500
G	19.4	7027	1–20	2000		
H	15.1	2289	1–10	2000		
I	15.0	455	1	11,500		
J	13.9	2770	1–20	1500		
K	12.0	653	1	1000	1–5	gradient = $1000
L	10.5	268	1–15	1350		
M	10.3	139	1–10	1650		
N	8.2	−368	1	2000	1–3	gradient = $2000
O	6.6	−1398	1–10	1400		

Solution

If five projects were funded, the budget is $50,000. Since these are all of the relevant projects for the same time span, the data can be used to simulate what we should do in future years.

Exhibit 9.15 shows the investment opportunity curve for the 15 projects in Exhibit 9.14. If the budget limit is $50,000, the MARR is 27.3%. This rate is substantially higher than the initial 10% value for the MARR. It is far more important to find that the correct MARR is over 25% than it is to find an exact value. Once this is established, then evaluating each project as it is proposed can be done by examining the PW at 25%. This would lead to the selection of a good set of projects.

Exhibit 9.15　MARR for Example 9.9

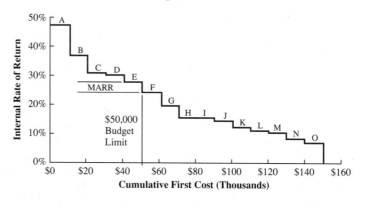

Stability and Reinvestment Assumptions.　Virtually every engineering economic analysis uses interest rates that are constant over time. This is appropriate because real rates of return and costs of capital over inflation are relatively constant, and even if they weren't, the data is too inaccurate for any other assumption.

This assumption is slightly more restrictive for constrained project selection. The reason is that a project's attractiveness may depend on whether the industry is on the peak or in the valley of cyclical demand. During booms, there are many high-potential projects, and the investment opportunity schedule will be higher. During plateaus, there may be few or no projects, and the investment opportunity curve would be lower.

Finally, whatever interest rate is used to calculate the PW, that interest rate represents an assumption about future reinvestment opportunities. If stability can be assumed, then the future MARR will be the same as today's MARR. Thus, constrained project selection and the opportunity cost of forgone investments are the proper bases for defining the reinvestment scenario of a firm's future. If a cost of capital measure (marginal or weighted average) is used to generate an artificially low MARR, then the implicit assumption is that the future will be much worse than the present. The reinvestment assumption of present worth was also discussed in Chapter 7.

9.7 PRESENT WORTH INDEXES AND BENEFIT/COST RATIOS

While it is the correct and most commonly used approach to constrained project selection, ranking on IRR is not the only technique that has been used. For example, until computers and calculators became commonplace, ranking on payback period was the most commonly used technique. Since the deficiencies of payback were enumerated in Chapter 8, this section focuses on more valuable approaches.

First, two techniques are applied in the public and private sectors, respectively: benefit/cost ratios and present worth indexes (both defined in Chapter 8). These are typically calculated for each project by dividing the present worth of the benefits or revenues by the present worth of the costs. It is the dollars of benefits per dollar of cost (both in present worth or equivalent annual terms) that is calculated. For example, SpaceCom's project A in Exhibit 9.14 has a present cost of $10,000, and the present worth of its $5000 per year benefits for 7 years is $24,342 at 10%. Thus, its present worth index is 2.43.

While benefit/cost ratios and present worth indexes are intuitively appealing and sometimes successfully used, they have a flaw. That flaw is that both are valid only after ranking on IRR has been used to find the MARR. Their chief advantage is that once the MARR has been found, they avoid any problems with multiple roots.

Exhibit 9.16 shows present worth indexes for Example 9.9, where those indexes are calculated at the a priori estimate of MARR, 10%, rather than at the true value of about 25%. Because the present worth index normalizes for the scale of the project, the correlation with the correct ranking is usually far better than if each project's net present worth were used. Benefit/cost ratios are virtually identical with present worth indexes (the ratios are the focus of Chapter 14, so they will not be discussed here).

Exhibit 9.16   SpaceCom's projects (Example 9.9): IRR vs. PW Index

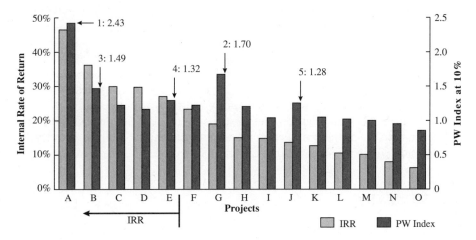

9.8 USING THE SORT SPREADSHEET TOOL

Once the data on the projects has been entered into a spreadsheet, the spreadsheet is useful for constrained project selection. First, it can be used to calculate the IRR or present worth index. Second, it can sort the projects and then calculate the cumulative first cost.

Using Spreadsheet Investment and Block Functions.   If the projects have a first cost, uniform annual cash flows, and a salvage value or cost, then an investment function can be used to calculate the IRR or present worth index. The IRR or present worth index for more complex cash flows, such as arithmetic or geometric gradients, can be calculated using the block functions @IRR and @NPV.

For example, project A in Example 9.9 has a 7-year life, a net annual benefit of $5K, a first cost of $10K, and no salvage value. The investment or annuity function is @IRATE(7,5000,−10000,0). Project F has an arithmetic gradient of −$1500, so the cash flows would be specified as −$10K, $6K, $4.5K, $3K, $1.5K. If these were in cells B1 to B5, then the IRR would be found as @IRR(B1..B5,guess) or =IRR(B1:B5,guess).

Using the SORT Tool.   Once the IRR, present worth index, or B/C ratio has been calculated for every project, then the SORT tool of the spreadsheet should be used to rank the projects before calculating the cumulative first cost. This tool works by identifying the block of data to be sorted, a column to be used as the sort key, and a sort order (ascending or descending). In Excel and Quattro Pro, the tool is found under the menu for DATA or DATABASE.

For Example 9.1, the data would be entered and a table similar to Exhibit 9.3 calculated. DOS-based spreadsheets select the SORT tool, and then identify the top-left and bottom-right corners of the data to be sorted. Windows- or Macintosh-based spreadsheets identify the top-left and bottom-right corners, and then select the SORT tool. In identifying the corners or selecting the data to be sorted, do *not* include headings, but do include all information in the row that goes with each project.

Once the data block and the SORT tool have been selected, the next step is to identify the column of IRRs as the first sort key and to specify a descending order for the sort. Then the SORT is started. Since Excel has the ability to sort by rows or columns, it is necessary to ensure that row sorting is selected.

Once the SORT has been completed, it is easy to add a column for the cumulative first cost. This column is compared with the capital limit to identify the MARR and which projects should be funded.

9.9 SUMMARY

This chapter established ranking on IRR as the method for solving the constrained project selection problem. Ranking on PW or indexes or on B/C ratios may produce similar, but not quite correct, decisions. Ranking on PW or on payback will usually produce rankings that differ more from the correct ranking. This chapter used simple projects, with first costs only at time 0. Ranking on IRR and other measures has been studied for more complex problems with simulation [see White and Smith], and ranking by IRR has been validated as the best method.

This chapter has ignored fluctuating investment climates, multiple criteria, risk, multiple roots, and other factors. However, it is easier to consider an IRR than a PW in a multiple criteria evaluation procedure. The IRR is easier to incorporate into calculations of risk/return tradeoffs, and ranking approaches are superior to the absolute "go/no-go" standards of PW analysis in situations where the projects and/or the funding can be adjusted.

Some real-world managers will support another approach. For example, managers eager to use payback period will emphasize the period of time until the original investment is returned. Shorter time periods suggest a smaller risk of losing that investment. Nevertheless, research has shown that payback does not minimize risk. Both PW and IRR are better at minimizing risk. A better analogy is that the difference between the IRR and the return on safe investments is the margin that may compensate for the risk associated with the project.

It should not be surprising that ranking on internal rates of return is the most commonly used technique for deciding which projects should be undertaken when operating under a constrained capital budget.

This chapter has also shown that the marginal and weighted-average costs of capital are less than the opportunity cost of forgone investments. The cost of capital (marginal or weighted average) should be relegated to a subordinate status, defining a minimum condition that is rarely relevant. It is the opportunity cost of forgone investments that should determine the choice of the minimum attractive rate of return (MARR).

This MARR, which is used for all engineering economic evaluations, should be found by ranking on IRR. Once the MARR has been found, it may be used in mathematical programming models (see Appendix 9A) to deal with interdependent projects.

REFERENCES

Bussey, Lynn E., and Ted G. Eschenbach, *The Economic Analysis of Industrial Projects*, 2nd ed., 1992, Prentice–Hall.

Kester, W. Carl, "Capital and Ownership Structure: A Comparison of United States and Japanese Manufacturing Corporations," *Financial Management,* Volume 15, Number 1, Spring 1986, pp. 5–16.

Mukherjee, T.K., and G.V. Henderson, "The Capital Budgeting Process," *Interfaces,* Volume 17, Number 2, March–April 1987, pp. 78–90.

Park, Chan S., and Gunter P. Sharp-Bette, *Advanced Engineering Economy,* 1990, Wiley.

White, Bob E., and Gerald W. Smith, "Comparing the Effectiveness of Ten Capital Investment Ranking Criteria," *The Engineering Economist,* Volume 31, Number 2, Winter 1986, pp. 151–163.

PROBLEMS

9.1 The ABC Block Company expects to fund 5 project proposals in the next budget. Managers have submitted the top 20 proposals for consideration. Determine the number of possible combinations.

9.2 Rank the following projects using IRR for the Acme Tool Company's allocation of its annual capital budget. While many worthy projects are available, the money available is limited. All projects have a 5-year life.

Project	First Cost	Net Annual Benefits
A	$10,000	$ 3200
B	68,000	22,000
C	20,000	6000
D	45,000	12,000
E	15,000	5000
F	5000	2000

9.3 Rank the projects of Acme Tool (see Problem 9.2) using PW at an interest rate of 10%. Does this change the ranking determined by IRR? If the ranking changes, explain why.

9.4 Develop the investment opportunity schedule for the proposed Acme Tool projects (see Problem 9.2).

9.5 If Acme Tool's capital budget is $98,000 (see Problem 9.2), which projects should be undertaken? What is the MARR?   (*Answer:* 18.03%)

9.6 Chips USA is considering seven projects to improve its production process for customized microprocessors. Because of the rapid pace of development, a 3-year horizon is used in the evaluation. Chips USA can afford to do several projects, but not all. Rank the following projects from best to worst.

Project	First Cost	Annual Benefit	PW at 15%
1	$20,000	$11,000	$5115
2	30,000	14,000	1965
3	10,000	6000	3699
4	5000	2400	480
5	25,000	13,000	4682
6	15,000	7000	983
7	40,000	21,000	7948

9.7 For Problem 9.6, what is the MARR if the budget is $70,000? Which projects should be done?

9.8 National Motors' Rock Creek plant is considering the following projects to improve its production process. Corporate headquarters requires a minimum 15% rate of return. Thus, all projects that come to the plant's capital screening committee meet that rate. The Rock Creek plant can afford to do several projects, but not all. Rank the seven projects from best to worst.

Project	First Cost	Annual Benefit	Life (years)	PW at 15%
1	$200,000	$ 50,000	15	$92,369
2	300,000	70,000	10	51,314
3	100,000	40,000	5	34,086
4	50,000	12,500	10	12,735
5	250,000	75,000	5	1412
6	150,000	32,000	20	50,299
7	400,000	125,000	5	19,019

9.9   For Problem 9.8, what is the MARR if the budget is $500,000? Which projects should be done?   (*Answer:* 20.85%)

9.10   For Problem 9.8, what is the approximate MARR if the budget is $650,000? Which projects should be done?

9.11   Estimate the internal rate of return for each project shown below within 1%. Which ones should be done if the capital budget is limited to $60,000? What is the minimum attractive rate of return? All projects have a life of 10 years.

Project	First Cost	Annual Benefit
A	$15,000	$2800
B	20,000	4200
C	10,000	2400
D	30,000	6200
E	40,000	7600

9.12   Estimate the internal rate of return for each project shown below within 1%. Which ones should be done if the capital budget is limited to $70,000? What is the minimum attractive rate of return? All projects have a life of 10 years.

Project	First Cost	Annual Benefit
A	$15,000	$4350
B	20,000	4770
C	10,000	2400
D	30,000	9200
E	40,000	8000
F	25,000	8100

(*Answer:* 26.16%)

9.13 If the budget is $6000, which of the projects shown below should be done? What is the minimum attractive rate of return?

Project	PW (12%) of Benefits	IRR	B/C Ratio	First Cost	Benefits				
					Year 1	Year 2	Year 3	Year 4	Year 5
A	$2605	25.0%	1.30	$-2000	$ 900	$ 800	$700	$600	$ 500
B	2803	25.8	1.40	-2000	600	700	800	900	1000
C	3405	24.6	1.70	-2000	0	0	0	0	6000
D	2704	25.4	1.35	-2000	750	750	750	750	750
E	3128	40.8	1.56	-2000	1400	1100	800	500	200

9.14 If the budget is $100,000, which of the projects shown below should be done? What is the minimum attractive rate of return?

Project	Life (years)	First Cost	PW at 11%	EAW at 11%	IRR	Annual Benefit	Gradient
1	10	$25,000	$ 5973	$1014	18.3%	$8000	$-750
2	20	25,000	6985	877	13.8	1200	450
3	30	25,000	10,402	1196	14.3	3300	100
4	15	25,000	5413	753	12.6	-5000	1800
5	5	25,000	2719	736	15.2	7500	0
6	10	25,000	7391	1255	17.7	5500	0

9.15 Which of the following projects should be done if the budget is $60,000? What is the MARR? All projects have lives of 20 years.

Project	First Cost	Annual Benefit
A	$15,000	$2500
B	22,000	3200
C	19,000	2800
D	24,000	3300
E	29,000	4100

(*Answer:* about 13.36%)

9.16 If all of the following projects have a life of 10 years and the budget is $100,000, which should be done with the limited budget? What is the minimum attractive rate of return? (Note: at most, one IRR must be calculated.)

Project	First Cost	Annual Benefit	PW at 10%
A	$ 15,000	$4350	$11,729
B	20,000	4770	9310
C	10,000	2400	4747
D	30,000	9200	26,530
E	40,000	8000	9157
F	25,000	8100	24,771
G	35,000	6200	3096
H	20,000	5250	12,259
I	30,000	6900	12,398
Total	$225,000		

9.17 The WhatZit Company has decided to fund the top six project proposals for the coming budget year. The final nine proposals are presented below. Determine the next capital budget for WhatZit. What is the MARR?

Project	First Cost	Annual Benefits	Life (years)
A	$15,000	$ 4429	4
B	20,000	6173	4
C	30,000	9878	4
D	25,000	6261	5
E	40,000	11,933	5
F	50,000	11,550	5
G	35,000	6794	8
H	60,000	12,692	8
I	75,000	14,058	8

9.18 The Eastern Division of HiTech Inc. has a $100,000 capital budget. Determine which project(s) should be funded.

Project	First Cost	Annual Benefits	Life (years)	Salvage Value
A	$50,000	$13,500	5	$5000
B	50,000	9000	10	0
C	50,000	13,250	5	1000
D	50,000	9575	8	6000

(*Answer:* A and B)

9.19   The Aluminum Division of Metal Stampings has a $200,000 capital budget. Which project(s) should be selected if corporate headquarters has suggested a MARR of 15%? The following projects have been proposed. What is the MARR for the division?

Project	First Cost	Annual Benefit	Life (years)	Salvage Value
A	$100,000	$24,000	10	$20,000
B	100,000	26,500	7	0
C	100,000	20,000	20	10,000

9.20   The National Bureau of Water Projects is considering the following projects. Congressional and executive policy mandates a discount rate of 10%. Thus, all projects have had their B/C ratios computed at that rate. However, the available funding is likely to be a third of the $450M requested. Rank the following projects from best to worst. Choose about $225M worth of projects to recommend for congressional consideration.

Project	First Cost ($M)	Annual Benefit ($1000)	Life (years)	B/C at 10%	PW at 10% ($1000)
1	$75	$20,000	30	2.51	$113,538
2	25	5000	25	1.82	20,385
3	50	10,000	20	1.70	35,136
4	25	4000	40	1.56	14,116
5	50	12,500	10	1.54	26,807
6	25	5000	15	1.52	13,030
7	75	12,500	20	1.42	31,420
8	50	10,000	10	1.23	11,446
9	75	9000	50	1.19	14,233

9.21   For Problem 9.20, what is the MARR if the budget is increased by $75M to $225M? Which projects should be done?   (*Answer:* 18.4%)

9.22   Northern Shelters manufactures homes, cabins, workshops, and garages. The prebuilt walls, floors, and roofs are assembled on-site. Each year five new designs are added (and five with low profit margins are dropped). Rank the designs from best to worst. Note that home designs (H) have a life of 5 years; cabins (C), 10 years; workshops (W), 15 years; and garages (G), 20 years. Northern Shelters requires at least a 15% rate of return on all projects.

Design	First Cost	Annual Benefit	Predicted Life (years)	PW at 15%
W1	$100,000	$25,000	15	$46,184
C2	100,000	30,000	10	50,563
H3	100,000	40,000	5	34,086
C4	100,000	24,000	10	20,450
H5	100,000	35,000	5	17,325
G6	100,000	22,500	20	40,835
W7	100,000	30,000	15	75,421
H8	100,000	50,000	5	67,608
G9	100,000	30,000	20	87,780

9.23   What is the minimum attractive rate of return in Problem 9.22 if the budget is $500,000? Which designs should be implemented?

9.24   Given the following information concerning the investment opportunity schedule, the financing curve, and the capital budget limit, determine the true MARR, the perfect market MARR, and the marginal cost of capital. The available capital budget is $35,000. The investment opportunity and financing curves pass through the following points (interpolate or extrapolate as necessary):

Cumulative investment	$10K	$20K	$30K	$40K	$50K	$60K	$70K
Project IRR	22.0%	19.5%	17.5%	16.0%	15.0%	14.5%	14.0%
Financing rate	9.0%		11.0%				

9.25   Determine the weighted-average cost of capital for the following capital structure of $12M. What is the minimum value of the marginal cost of capital?

$2,000,000 in bonds issued at 9%

$4,000,000 in bonds issued at 10%

$5,000,000 in stock estimated at 12%

$1,000,000 in retained earnings

9.26   The Ceramics Division of NewTech has a $200,000 capital budget. Which project(s) should be selected if corporate headquarters has suggested a MARR of 20%? The following projects have been proposed. What is the MARR for the division?

Project	First Cost	Annual Benefit	Life (years)	Salvage Value
A	$100,000	$40,000	5	$ 8000
B	100,000	35,000	8	0
C	100,000	29,000	10	11,000

9.27 If the budget is $100,000, which of the following projects should be done? What is the minimum attractive rate of return?

Project	Life (years)	First Cost	Annual Benefit	Salvage Value
1	20	$20,000	$4000	$ 0
2	20	20,000	3200	20,000
3	30	20,000	3300	10,000
4	15	20,000	4500	0
5	25	20,000	4500	−20,000
6	10	20,000	5800	0
7	15	20,000	4000	10,000

(*Answer:* 19.26%)

9.28 If the budget is $100,000, which of the following projects should be done? What is the minimum attractive rate of return? The annual benefit is for year 1, and the gradient is added on after that.

Project	Life (years)	First Cost	Annual Benefit	Gradient
1	10	$25,000	$4000	$ 0
2	20	25,000	1200	400
3	30	25,000	3300	100
4	15	25,000	−7000	2000
5	5	25,000	7500	−750
6	10	25,000	5800	0

(*Answer:* 10.55%)

9.29 If the budget is about $100,000, which of the following projects should be done? What is the minimum attractive rate of return?

Project	Life (years)	First Cost	Annual Benefit	Salvage Value
1	20	$10,000	$2200	$ 0
2	20	20,000	3200	20,000
3	30	30,000	4300	10,000
4	15	20,000	4500	0
5	25	25,000	4000	−20,000
6	10	15,000	2800	0
7	15	20,000	4000	10,000
8	20	10,000	2500	−7500
9	20	20,000	3200	20,000
10	30	25,000	3800	10,000

9.30 If the budget is $100,000, which of the following projects should be done? What is the minimum attractive rate of return? The annual benefit is for year 1, and the gradient is 0 for that year. Arithmetic gradients are stated as dollar amounts and geometric gradients as percentages.

Project	Life (years)	First Cost	Annual Benefit	Gradient
1	10	$25,000	$4000	+ 5%
2	20	25,000	1200	+ 10%
3	30	25,000	3300	+ 3%
4	15	25,000	−7000	+$2000
5	5	25,000	7500	− 5%
6	10	25,000	5800	− $100

APPENDIX 9A MATHEMATICAL PROGRAMMING AND SPREADSHEETS

Another approach to constrained project selection relies on mathematical programming. Mathematical models can deal with indivisible projects whose total cost does not equal the capital budget. The models can consider constraints on other resources such as engineering man-years, or on capital budgets in later years, or on allowable borrowings to adjust the capital budget. Project dependencies can be modeled for either synergistic or partially competitive projects. Mutually exclusive projects can be easily modeled. The only restriction is that these models typically must assume a MARR. The assumed MARR is usually the weighted cost of capital, which is valid only within the perfect capital market framework. If the perfect market assumptions are true, then mathematical programming is the only way to determine the optimal set of projects within all of the above constraints.

These mathematical models are typically made up of linear inequalities and equations. The models may be linear, integer, or mixed integer models, depending on whether the projects can be partially funded. If a potential investment is buying shares in another company, then fractional execution of the investment or project may make sense and the corresponding variable in the model is linear.

On the other hand, a potential project might be building a bridge across a river or developing a new product. Such projects must be executed in their entirety or not at all. Similarly, the design of a new highway will have 2, 3, 4, or more complete lanes (no half lanes). Steam turbines, vehicles, and other equipment are purchased as whole units. Thus, the corresponding variables in the model are restricted to integer values, such as 0, 1, or 2 pumps. If the project can be done only once, then the variable is binary (0 or 1) in value. If the model contains both linear and integer variables (projects), then it is a mixed integer model.

Example 9A.1 describes a simple mathematical program. In reality, these models are most useful when there are large numbers of variables (projects) and constraints, and when computerized solutions are required. If such a model is used, then extensive sensitivity analysis is recommended, because the solution can vary considerably in terms of projects selected if the data changes. This model assumes that present worths have been calculated, where more general models use the cash flows of each project over time to calculate net borrowings and capital budgets in later years.

Example 9A.1   Integer Programming Example

Four projects are being considered (A, B, C, and D). Write an integer program to find the best set of projects if the total capital budget is limited to $100 million. Besides the data in the following table, we must allow for the fact that Project D can be undertaken only if project C is (D depends on C), and that A and B are synergistic projects. Each project can be done once at most.

Project	First Cost	Present Worth
A	$30M	$2.5M
B	40M	3M
A and B	70M	6M
C	30M	3M
D	40M	3M

Solution
First, we define five zero-one variables (for convenience: A, B, AB (which is A and B), C, and D). Then we write the following objective function and inequalities:

$$\text{Max } 2.5A + 3B + 6AB + 3C + 3D$$

maximizes PW subject to:
   the budget constraint of spending at most $100M

$$30A + 40B + 70AB + 30C + 40D \le 100$$

can do at most one of A, B, or the combination

$$A + B + AB \le 1$$

if C = 0, then D = 0

$$C \ge D$$

the integer constraint

$$A, B, AB, C, D = 0 \text{ or } 1$$

Using Spreadsheets to Solve Linear Programs.   Exhibit 9A.1 illustrates the use of Quattro Pro to solve an approximation of Example 9A.1. This is only an approximation, since the model replaces the restriction of variable values to 0

Exhibit 9A.1   QuattroPro™ solution of linear program for Example 9A.1

	A	B	C	D	E	F
	A	B	C	D	E	F
1	Exhibit 9A.1 Linear Programming by QPRO					
2						
3		0 A				
4		0 B				
5		1 AB				
6		1 C				
7		0 D				
8		Limits	Values			
9	2.5*A+3*B+6*AB+3*C+3*D		9	objective		
10	30*A+40*B+70*AB+30*C+40D<	100	100	budget		
11	+A+B+AB<	1	1	one of A, B, & AB		
12	+C-D>	0	1	C>D		
13						
14	Answer Report					
15	Constraints:					Dual
16	Cell:	Value:	Constraint	Binding?	Slack:	Value
17	C10	100	C10 <= B10	Yes	0	0.08571
18	C11	1	C11 <= B11	Yes	0	0
19	C12	1	C12 >= B12	No	-1	0
20	A3	0	A3 <= 1	No	1	-0.0714
21	A4	0	A4 <= 1	No	1	-0.4286
22	A5	1	A5 <= 1	Yes	0	0
23	A6	1	A6 <= 1	Yes	0	0.42857
24	A7	0	A7 <= 1	No	1	-0.4286
25	A3	0	A3 >= 0	Yes	0	-0.0714
26	A4	0	A4 >= 0	Yes	0	-0.4286
27	A5	1	A5 >= 0	No	-1	0
28	A6	1	A6 >= 0	No	-1	0.42857
29	A7	0	A7 >= 0	Yes	0	-0.4286

File   Edit   Style   Graph   Print   Database ‖ Tools ‖ Options   Window   ? ↑↓

```
                                                                   ↑End
          Solution Cell              C9..C9  ►           ▲
          Variable Cell(s)           A3..A7              ►
          Constraints                                    ▼
          Options                              ►
          Answer  Repo  ┌ Constraints:          ..F42    ERS
          Detail  Repo  │ <Add New Constraints>          ───
          Go            │ C10   <=  B10                   CPY
          Restore       │ C11   <=  B11                   ───
          Model         │ C12   >=  B12           ►       MOV
          Reset         │ A3 ..A7  <= 1           ►       ───
          Quit          │ A3 ..A7  >= 0                   STY
                                                          ───
                                                          ALN
                                                          ───
                          Audit        ►                  FNT
                          Library      ►                  ───
                                                          INS
                                                          ───
                                                          BAR
```

or 1 values with limits of $\geq$ 0 and $\leq$ 1. This is a linear programming model, rather than an integer programming model. In this case, the solution to the linear programming model also satisfies the integer restrictions. Similar results can be obtained from other spreadsheet programs (some need add-in packages to do this).

The cells for each variable are identified, and in this case labeled A, B, AB, C, and D to simplify the formulas. The cells (A3 to A7) may initially be empty or they may contain guesses as to the correct values. Then the formulas for the objective function, the budget constraint, the constraint on A and B, and the constraint on C and D are entered into cells C9 to C12. The formulas are shown in cells A9 to A12. The limiting values for these constraints are then entered in the adjacent cells, B9 to B12.

At this point, the OPTIMIZATION menu of Quattro Pro is accessed (it is part of the TOOLS menu). Cell C9, which contains the formula for the objective function, is specified as the solution cell. The cells that contain the variables, A3 to A7, are specified as the variable cells. Then the constraint command is called and one by one the three formula constraints are specified, and the block constraints for 0 and 1 limits are identified. The screen display of the final CONSTRAINTS menu is included in Exhibit 9A.1. A cell for the answer report is specified to produce the detail on each constraint. Then the GO command starts the search. (For simplicity, part of the answer report has been deleted from Exhibit 9A.1.)

Disadvantages of Mathematical Programming Models. While these mathematical programming models can deal with complexity that cannot be handled well in any other way, they do have some disadvantages. First, the MARR must be determined externally and it cannot be found using the opportunity cost concept. Second, complete data is required for every project under consideration. Third, relatively small changes in the data can produce significantly different sets of recommended projects. Finally, because the model is often a "black box" to the organization, the results often appear more exact than warranted (false accuracy), and it is difficult for decision makers to develop an intuitive feel for the trade-offs in the problem. However, computerized models can support more extensive sensitivity analysis.

PROBLEMS

9A.1   Build a mathematical model for Problem 9.22 if the MARR is 15% and at least one design in each category must be developed.

9A.2   Solve the model in Problem 9A.1.

9A.3   Build a mathematical model for Problem 9.13 if the MARR is 12% and annual benefits must be at least $2000 for years 1–5.

9A.4   Solve the model in Problem 9A.3.

9A.5   Build and solve a mathematical model for Problem 9.27. The MARR is 12%. At least one of projects 2 and 3 must be done, at most one of projects 5 and 6 can be done, and at most two of projects 3, 4, and 5 can be done.

9A.6 Build and solve a mathematical model for Problem 9.28. The MARR is 5%. At most one of projects 2 and 4 can be done, and at least two of projects 1, 5, and 6 must be done. Project 3 depends on project 5.

9A.7 Build and solve a mathematical model for Problem 9.29. The MARR is 10%.

9A.8 Build and solve a mathematical model for Problem 9.30. The MARR is 6%. Project 6 depends on project 2, and project 4 depends on project 5. At most two projects of 1, 4, and 6 can be done.

9.A9 Five projects are under consideration by the ABC Block Company. Develop a linear program to model the capital budgeting decision if the budget is limited to $50,000.

Project	PW	First Cost
A	$13,400	$50,000
B	8000	30,000
C	−1300	14,000
D	900	15,000
E	9000	10,000

Either project A or B can be accepted, but not both. Project B can be partially funded with its PW reduced proportionately. Project D depends on the acceptance of project E. Either project C or D can be accepted, but not both.

CHAPTER 10

MUTUALLY EXCLUSIVE ALTERNATIVES

The Situation and the Solution

Often, two or more alternatives must be evaluated to decide which one is the best because only one can be implemented. For example, if a dam is built, it is one height even though several other heights may be considered.

The solution is to compare these mutually exclusive alternatives using present worth, equivalent annual worth, or equivalent annual cost. Internal rate of return and benefit/cost ratios can be applied, but an incremental analysis is required.

Chapter Objectives

After you have read and studied the sections of this chapter, you should be able to:

Section 10.1
Define and recognize problems with mutually exclusive alternatives.

Section 10.2
Identify the assumptions made in the comparison of mutually exclusive alternatives.

Section 10.3
Select the best alternative using PW, EAW, or EAC when the alternatives have lives of the same length.

Section 10.4
Use PW to explicitly match the lives of alternatives with lives of different lengths.

Section 10.5
Use EAW or EAC to implicitly match the lives of alternatives with lives of different lengths.

Section 10.6
Demonstrate the robustness of comparing the EAWs or EACs of alternatives with lives of different lengths.

Section 10.7
Show that IRR, PW, and EAW all use the same reinvestment assumption.

Section 10.8
Apply IRR and B/C ratios incrementally to correctly compare mutually exclusive alternatives.

Section 10.9
Use SOLVE FOR spreadsheet tools to find incremental IRRs for comparing mutually exclusive alternatives.

Key Words and Concepts

Mutually exclusive alternatives   Alternatives where one at most can be accepted from a set of choices.

Maximizing PW or EAW or **minimizing EAC**   Criteria for selecting the best alternative from mutually exclusive possibilities.

Doing nothing or **maintaining the status quo**   A mutually exclusive alternative that may or may not be specifically identified and may or may not be possible.

Problem horizon   The life or study period that is chosen as a common basis for comparing mutually exclusive alternatives.

Incremental analysis   Evaluates the difference between two or more mutually exclusive alternatives.

10.1 APPLYING ENGINEERING ECONOMY TO ENGINEERING DESIGN

Engineering design is the first application of engineering economy for beginning engineers. Design requires that at most one alternative can be selected. Because accepting any one alternative precludes accepting any other alternative, these are called **mutually exclusive alternatives**.

Mutual exclusivity usually occurs for a physical reason. For example, only one pump out of several sizes and manufacturers will be installed, or only one building can be built on a site, or only one roof design will be used for a building.

Accepting a **mutually exclusive alternative** from a set of possibilities precludes accepting any other.

Mutual exclusivity can arise for other reasons. For example, organizational or political factors may require at most one accepted project for each group, district, or division. Market demand may support at most one new product, or regulatory demands may only allow one new item through the permitting process. Similarly, scarce resources such as computer time, entrepreneurial talent, or R&D facilities may be the basis for mutual exclusivity.

Engineering design requires selecting one out of several mutually exclusive alternatives, and this is the most common context for mutually exclusive alternatives. The use of the engineered item defines a common study period or time horizon, which must be used for economically evaluating all alternatives. Consequently, all alternatives must be matched to a single study period. This is done automatically, or implicitly, when equivalent annual techniques are used. However, it must be done explicitly when present worth techniques are used.

The best mutually exclusive alternative is found by comparing the PWs, EAWs, or EACs. The alternative with the highest worth or the lowest cost is the best.

10.2 KEY ASSUMPTION IS THE INTEREST RATE OR MINIMUM ATTRACTIVE RATE OF RETURN

When mutually exclusive alternatives are analyzed, the correct interest rate cannot be found by comparing alternatives. Instead, the interest rate is determined by constrained project selection or by the cost of capital, and then it is used as the minimum attractive rate of return (MARR) or i^* (pronounced "i star").

Enough capital is available at this MARR to fund even the most expensive alternative—if justified by evaluation at the MARR. In other words, the choice between mutually exclusive alternatives is assumed not to influence the firm's other decisions nor to limit the capital available elsewhere.

Common Assumptions.   In Chapters 5 through 8, the assumptions of known cash flows, known lives, and a known interest rate were made. The same assumptions are made here.

There is only one interest rate, and it is used to evaluate all mutually exclusive choices. This known interest rate has been called the minimum attractive rate of return, but it is also the reinvestment rate that is assumed to apply in the future.

These assumptions are associated with evaluating mutually exclusive choices, and they are unaffected by the choice of PW, EAW, IRR, or B/C ratios. Thus, when correctly applied, each measure shows the same alternative as the best. Incremental analysis is required to properly apply IRR and B/C ratio approaches. As noted in Section 10.7, some authors incorrectly believe that IRR has a different reinvestment assumption.

10.3 COMPARING ALTERNATIVES WITH LIVES OF THE SAME LENGTH

When the lives of the mutually exclusive alternatives are the same, then the criteria to select the best alternative are simple and intuitively clear. Those criteria are to choose the alternative with the largest or **maximum PW** or **EAW** or the alternative with the smallest or **minimum EAC**.

Examples 10.1 and 10.2 illustrate these criteria, which can be applied by private firms and by public agencies. Whether EAW or PW is used does not matter. For Example 10.2, EAWs could be calculated by multiplying PWs by $(A/P, .12, 25)$, so the relative differences would remain constant. Similarly, for Example 10.1, PWs equal the EACs multiplied by $-(P/A, .12, 40)$.

Doing nothing or **maintaining the status quo** is a mutually exclusive alternative that may or may not be specifically identified, and it may or may not be possible. For example, in many cost-reduction efforts, it is possible to continue with the status quo, even though that option may not be specifically identified. On the other hand, when choosing the most economical roof or the most economical heating system, doing nothing is not allowed. The building must have a roof, and it must be heated.

Example 10.1 has two alternatives. Example 10.2 has three listed alternatives and an implicit fourth one—doing nothing with the land for now. Other problems may have 20 mutually exclusive alternatives. The decision-making principle remains constant: Choose the most attractive alternative. It will have the largest PW or EAW, or the smallest EAC.

Example 10.1   Insulating City Hall

The city of Northern Lights is building a new city hall. The civil engineer for the walls and ceiling and the mechanical engineer for the heating system are evaluating two plans for insulating and heating the building. They have identified standard construction (SC) and energy-efficient construction (EEC) as the two alternatives. Northern Lights uses an interest rate of 10%, and the new city hall is expected to have a life of 40 years. Using EAC, which alternative is better?

	Standard Construction	Energy-Efficient Construction
First cost: building	$2.45M	$2.50M
First cost: furnace	$100K	$85K
Annual heating cost	$10,000	$6000

Solution

$$EAC_{SC} = (2450K + 100K)(A/P, .1, 40) + 10K$$
$$= 2550K \cdot .1023 + 10K = \$270.9K$$
$$EAC_{EEC} = (2500K + 85K)(A/P, .1, 40) + 6K$$
$$= 2585K \cdot .1023 + 6K = \$270.4K$$

Since the energy-efficient construction cost is lower, that is the preferred alternative. Since EEC does not change the majority of the $2.45M in construction costs, the savings appear to be small relative to the project's cost.

The extra insulation, savings on the furnace, and change in heating costs can also be solved by analyzing the difference between the alternatives. Does the $50K for extra insulation (EI) save enough in furnace costs ($15K) and heating costs ($4K per year) to be justified?

$$EAC_{EI} = (50K - 15K)(A/P, .1, 40) - 4K$$
$$= 35K \cdot .1023 - 4K = -\$.42K$$

Since the EAC of the extra insulation is negative, it is certainly desirable.

In this case, doing nothing is not an option. The decision has been made to build a new city hall. It must have walls and a roof, and it must be heated.

Example 10.2   Developing an Industrial Property

HDP Development owns the last vacant parcel in a thriving industrial park. There are three proposals for developing the property (see table). In each case HDP builds and maintains the structure and receives income from it. HDP Development uses an interest rate of 12% and a horizon of 25 years for its analyses. Which proposal is preferred, using a PW criterion?

	R&D Lab	Instrument Maintenance (IM)	Snowboard Manufacturing (SM)
First cost	$3.0M	$3.3M	$4.6M
Annual O&M costs	$750K	$200K	$400K
Annual income	$1200K	$750K	$1100K
Salvage value	$1.0M	$.4M	$.5M

Solution

$$PW_{R\&D} = -3M + (1200K - 750K)(P/A, .12, 25) + 1M(P/F, .12, 25)$$
$$= -3000K + 450K \cdot 7.843 + 1000K \cdot .0588 = \$588.2K$$
$$PW_{IM} = -3.3M + (750K - 200K)(P/A, .12, 25) + 400K(P/F, .12, 25)$$
$$= -3300K + 550K \cdot 7.843 + 400K \cdot .0588 = \$1037.2K$$
$$PW_{SM} = -4.6M + (1100K - 400K)(P/A, .12, 25) + 500K(P/F, .12, 25)$$
$$= -4600K + 700K \cdot 7.843 + 500K \cdot .0588 = \$919.5K$$

Thus, the instrument maintenance facility is the most attractive mutually exclusive choice. It has the largest PW. If no alternative had a positive PW, then doing nothing would be permissible (at least for a while).

10.4 PWS AND EXPLICITLY COMPARING DIFFERENT-LENGTH LIVES

In many cases mutually exclusive alternatives do not have lives that are the same length. Sometimes the main difference between two alternatives may be their lives. Examples include items made with stainless steel rather than brass, aluminum, or steel. The higher-cost material allows use for a longer period before the item is replaced. Similarly, new equipment or vehicles should have a longer life than used equipment or vehicles.

Another example is purchasing memberships, subscriptions, or insurance. These can be bought 6 months at a time, or 1, 2, 3, 5, or even 10 years at a time. Buying a longer period of coverage costs less per year, and it postpones the next purchase until the longer period expires. The comparison is the EAC of a shorter period with the EAC of a longer period.

The approach is to define one **problem horizon** or time period that is used for comparing all alternatives. For example, when comparing steel, brass, and stainless-steel pumps, the starting point would be how long the pump is needed. Then the cost for each pump and any replacements would be computed over that period.

The **problem horizon** is the life or study period that is chosen as a common basis for comparing mutually exclusive alternatives.

Even though sewage must be pumped for the next couple of centuries, a shorter time horizon for comparing alternatives can be used. However, the same horizon must be used for each mutually exclusive alternative. One reason for using the same horizon for each alternative is that the benefits stream may not be specified. This is the norm for economic analysis during engineering design. The goal is to minimize the cost of building and operating a project, but the value of having the project may not be given. Example 10.1 illustrates this, where the goal is to minimize the cost of building and heating city hall, without specifying the value of a warm city hall.

Approaches for Defining a Problem Horizon.   When comparing alternatives with different-length lives, the problem horizon is defined by:

1. Assuming that each alternative repeats itself over time with the same costs and benefits, or
2. Terminating one or more alternatives early and using salvage values to account for the remaining life.

Assuming that alternatives repeat themselves with the same costs and benefits implies a time horizon that is either infinite or equal to the least common multiple of the alternatives' lives. Both are illustrated in Example 10.3, where a horizon of 12 years (the least common multiple of 3 and 4 years) is used. Example 10.3 also shows that the results from assuming an infinite horizon are proportional to the least common multiple results, so the same alternative is the best.

In some cases, the assumption of an infinite or least common multiple horizon can be interpreted as a long but indefinite life. For example, equipment with an estimated life of 7 years might be evaluated under the following horizon. "We're not really sure how long we'll need to do this. Maybe 10 years. Maybe 20."

The second assumption for a horizon picks a specific number of years to be the study period or horizon. Then salvage values are assigned for any alternative whose final replacement does not match the study period or horizon. Thus, salvage values are estimated for one or more mutually exclusive alternatives. For this approach, the calculated PW depends on the assumed salvage values. Possible study periods or horizons include:

Life of the "shorter" alternative

Life of the "longer" alternative

Expected service life

Arbitrary values, such as 10 years

In Example 10.4, where there are two alternatives with lives of 3 and 4 years, a horizon of 3, 4, or 5 years would require that a salvage value be estimated for one or more of the two alternatives. Often, an alternative will be most attractive in time horizons that are even multiples of its life. In Example 10.4, the brass pump is more attractive with a 3-year study period, and the stainless-steel pump is more attractive with a 4-year study period. The subsection after Examples 10.3 and 10.4 suggests guidelines for choosing the best horizon.

Example 10.3   Present Worth with Least Common Multiple Horizon

A brass pump and a stainless-steel pump are being compared for an application in which their operating costs will be the same. The firm uses an interest rate of 8%. Compare each alternative's PW, with a 12-year horizon and with an infinite horizon.

The brass pump lasts 3 years and costs $12,000. The stainless-steel pump lasts 4 years and costs $15,000. Neither will have a salvage value.

Solution
The brass pump's PW over a 3-year life is $-\$12,000$, and the stainless-steel pump's PW over a 4-year life is $-\$15,000$. But these are not comparable PWs. Comparing the $-\$12$K and the $-\$15$K would mean nothing.

A 12-year horizon is the least common multiple of a 3-year life and a 4-year life (it is also the least common multiple for 4 and 6 years). Brass pumps are purchased at time 0 and at the ends of years 3, 6, and 9. Stainless-steel pumps are purchased at time 0 and at the ends of years 4 and 8.

$$\begin{aligned} PW_{brass} &= -12K - 12K(P/F, i, 3) - 12K(P/F, i, 6) - 12K(P/F, i, 9) \\ &= -12K[1 + 1/(1+i)^3 + 1/(1+i)^6 + 1/(1+i)^9] \\ &= -12K(1 + .7938 + .6302 + .5003) = -12K(2.9243) = -\$35.09K \\ PW_{stainless} &= -15K[1 + (P/F, i, 4) + (P/F, i, 8)] \\ &= -15K[1 + 1/(1+i)^4 + 1/(1+i)^8] = -15K(1 + .73503 + .5403) \\ &= -15K(2.2753) = -\$34.13K \end{aligned}$$

The assumption of perpetual life has the same effect, since infinity is a multiple of 12 years. As shown in Section 6.6, the capitalized costs are computed by finding each EAW and dividing it by 8%.

$$\begin{aligned} PW_{brass,\infty} &= -12K(A/P, i, 3)/i = -12K \cdot .3880/.08 = -\$58,200 \\ PW_{stainless,\infty} &= -15K(A/P, i, 4)/i = -15K \cdot .3019/.08 = -\$56,610 \end{aligned}$$

Notice that the stainless-steel pump's PW is 97.3% of the brass pump's PW for both the 12-year comparison and the infinite-life comparison. Both comparisons assume that future brass pumps cost $12K and future stainless-steel pumps cost $15K.

Example 10.4 Present Worth with Assumed Salvage Values

For the brass and stainless-steel pumps of Example 10.3, use periods shorter than 12 years for the comparison. In particular, calculate comparable PWs for 3, 4, and 5 years.

Solution

For these calculations, the salvage value for the brass pump is assumed to be $6K after 1 year of use, $2K after 2 years, and $0 after 3 years. Similarly, the salvage values for the stainless pump are assumed to be $9K, $4.5K, $1.5K, and $0 after 1, 2, 3, and 4 years, respectively.

For the 3-year study period:

$$PW_{brass} = -\$12K$$
$$PW_{stainless} = -15K + 1.5K/(1.08)^3 = -\$13.81K$$

For the 4-year study period:

$$PW_{brass} = -12K - 12K/(1.08)^3 + 6K/(1.08)^4 = -\$17.12K$$
$$PW_{stainless} = -\$15K$$

For the 5-year study period:

$$PW_{brass} = -12K - 12K/(1.08)^3 + 2K/(1.08)^5 = -\$20.16K$$
$$PW_{stainless} = -15K - 15K/(1.08)^4 + 9K/(1.08)^5 = -\$19.90K$$

Thus, with these assumed salvage values, the brass pump is best at 3 years and the stainless-steel pump is best at 4 or 5 years. Unless the expected need for pumping is 3, 4, or 5 years, there is no basis for preferring one of these horizons. Changing the assumed salvage patterns could change the results.

Choosing the Best Horizon. Example 10.4 illustrates the difficulty in choosing the best horizon, since the best alternative depends on the assumed salvage values and the horizon. If a life for the pumping service is known to be t years, then t would be the best choice for a horizon. (See the calculations in Example 10.4 for 3, 4, or 5 years.)

However, a horizon of 7 years might be most cost effectively served by using a brass pump for 3 years and a stainless-steel pump for 4 years. The best strategy for a 7-year horizon is to choose the cheaper one now (stainless steel). Then, in about 4 years another choice is made with a new, more accurate estimate of the remaining service life.

If the service period is longer than the lives of the alternatives, it will be somewhat indefinite. For example, the 40-year life in Example 10.1 and the 25-year life in Example 10.2 are clearly estimated lives. Only through a legal contract can an exact period of 10 or 20 years be defined, and even then contracts can be extended or terminated early.

In most cases, using the assumption of an infinite horizon or of the least common multiple horizon is preferable to assuming somewhat arbitrary salvage values.

Mutually Exclusive Alternatives without a Common Horizon.   There are cases in which two or more mutually exclusive alternatives are compared without matching their lives to a study period. If the benefits and costs of each alternative are known, and the alternatives are mutually exclusive due to financial or political reasons, then the life of each alternative is used to calculate PWs. This does not apply to problems in engineering design, where the alternatives are mutually exclusive due to physical reasons.

For example, if a 10-year manufacturing modernization and a 15-year R&D project are mutually exclusive, then it is unrealistic to assume that each would be followed by an identical or even similar project. As long as the study period were at least 15 years, then the PWs for each project could be compared.

This is implicitly assuming that all returns from each alternative would be invested at the firm's minimum attractive rate of return (MARR). In this case, EAWs cannot properly be used because, as seen in Section 10.5, this would assume that projects repeat until a horizon that is equal to the least common multiple of the individual lives.

10.5 EAWS AND EACS AND IMPLICITLY COMPARING DIFFERENT-LENGTH LIVES

Rather than explicitly defining a study period or time horizon, one can be implicitly selected by using EACs or EAWs. This implicit study period is based on the results of Section 6.7. Specifically, the EAC for any alternative is the same for any integer multiple of its life, if identical repetitions are assumed.

This is the most commonly used approach to comparing mutually exclusive alternatives with different-length lives. It is the easiest approach, and it is often the best one. It is also consistent with the conclusions of Section 10.4 on explicitly choosing the horizon. The implicit study period is the least common multiple of the lives of the mutually exclusive alternatives, as illustrated in Exhibit 10.1.

Assuming identical cost repetitions for 10 or 12 years may be reasonable. However, assuming identical repetitions for 60 years or 2.5 centuries is unrealistic. Nevertheless, as shown in Section 10.6, the implicit assumption is quite robust. Assuming identical repetitions is often better, and always easier than assigning salvage values that may be arbitrary.

Example 10.5 applies this approach to the brass and stainless-steel pumps examined in Example 10.3. It also confirms that the assumptions in Example 10.3 (identical repetitions within a 12-year horizon) are the same assumptions made implicitly when an EAC comparison is made.

Example 10.6 shows a more complex analysis in which using EACs and making an implicit assumption about the problem horizon is much easier than using PWs.

Example 10.5   EAC Comparison with Implicit Least Common Multiple

In Example 10.3, the PWs for a brass and a stainless-steel pump were calculated to be $35.09K and $34.13K, respectively. Evaluate these pumps using EAC and the same 8% interest rate.

Solution

$$EAC_{brass} = 12K(A/P, .08, 3) = \$4656$$

$$EAC_{stainless} = 15K(A/P, .08, 4) = \$4529$$

The EAC for the brass pump is valid for lives of 3, 6, 9, etc. years, while the EAC for the stainless-steel pump is valid for lives of 4, 8, 12, etc. years. When they are compared and a difference is calculated, that difference is correct only for lives of 12, 24, 36, etc. years. Thus, calculating the difference of $127 per year implicitly assumes a study period of 12 years and identical repetitions.

Comparison with the PWs of Example 10.3 can confirm that claiming savings of $127.59 for the stainless pump is equivalent to a 12-year horizon with constant costs and repetition until the least common multiple of lives.

$$PW_{diff\ in\ EAC} = 127.59(P/A, .08, 12) = \$961.5$$

$$PW_{brass} - PW_{stainless} = -34.13K - (-35.09K) = \$960$$

Exhibit 10.1   Implicit horizons from least common multiples

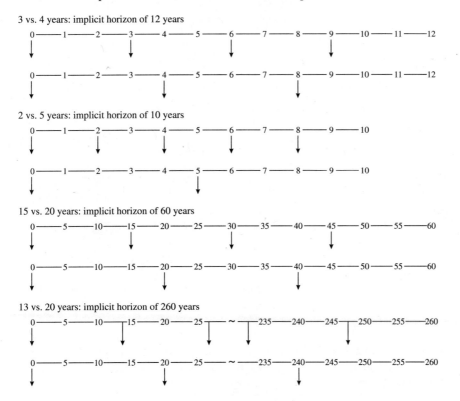

3 vs. 4 years: implicit horizon of 12 years

2 vs. 5 years: implicit horizon of 10 years

15 vs. 20 years: implicit horizon of 60 years

13 vs. 20 years: implicit horizon of 260 years

Example 10.6   Selecting the Cheapest Roof

Charley is selecting materials for a state-subsidized housing development. The buildings will be maintained for at least 75 years, and the agency uses an i of 6%.

Which of the following two grades of shingles and metal roofing is the most cost effective? All costs are per square (equals 100 square feet).

	Minimum-Quality Shingles	Maximum-Quality Shingles	Metal Roofing
Buy	$35	$70	$120
Install	$70	$70	$100
Annual maintenance	$ 5	$ 3	$ 1
Life (years)	15	25	50

Solution

Since there are three lives, theoretically the EACs are all comparable only if a 150-year horizon is used (the least common multiple for 15, 25, and 50). This 150-year horizon is also required for a PW analysis. However, comparing EACs is how this problem is solved when engineering economy is used to solve real-world problems.

$$EAC_{minQ} = 105(A/P, 6\%, 15) + 5 = \$15.81 \text{ per square}$$
$$EAC_{maxQ} = 140(A/P, 6\%, 25) + 3 = \$13.95 \text{ per square}$$
$$EAC_{metal} = 220(A/P, 6\%, 50) + 1 = \$14.96 \text{ per square}$$

Based on this comparison, Charley should choose the higher-quality shingles. Those shingles save $1 per year per square over the metal roofing and nearly $2 over the minimum quality shingles.

10.6 USING EAC FOR DIFFERENT-LENGTH LIVES IS A ROBUST APPROACH

Even though cash flows may not be identically repeated, EAC comparisons of alternatives with different-length lives are economically robust [Eschenbach and Smith]. This robustness or stability of conclusions is based on (1) discounting later cash flows more heavily due to money's time value, and (2) the uncertainty in lives and study periods.

Since estimated lives are often 5, 10, 15, 20, 25, or 50 years, being overly precise is not realistic. For example, a 50-year horizon with a 15-year alternative implies three complete lives and a fourth that is terminated after 5 years. This explicit pattern is less meaningful than simply using the EAC approach and picking the most cost-effective alternative.

Robustness Due to Discounting.   Cash flows may change over the 60 years implied by comparing a 15-year EAC with a 20-year EAC, but the largest changes occur near the horizon, and they are discounted heavily.

One of the largest changes that can occur is to "throw away" a repetition with no salvage value only partway through its life. Example 10.7 compares 15-year and 25-year alternatives over a 50-year study period rather than over the implicit 150-year study period of the EAC comparison. Even though the last repetition of the 15-year alternative is truncated after a third of its life, its EAC only increases by 1.7%. Larger annual O&M costs or truncating both alternatives would make the comparison even more robust.

This robustness holds true even for the comparison of 3-year and 4-year EACs, where the implicit study period is only 12 years. Example 10.8 returns to the data of Examples 10.3 through 10.5, where the use of a 10-year study period instead of the 12-year period only increases the EACs by $159, or 3.4%, for the brass pump and $247, or 5.5%, for the stainless-steel pump. The EAC difference changes by less than 2% of the roughly $4700 in EAC values.

Example 10.7   Fifty-Year Horizon for Selecting the Cheapest Roof

How much do the EACs from Example 10.6 change if a horizon of 50 years is used? The i is still 6%.

Solution
Since 50 is a multiple of 25 and 50, the EAC for the maximum-quality shingles is still $13.95 per square, and the EAC for the metal roofing is still $14.96 per square. To find the EAC for the minimum-quality shingles, replacement at years 15, 30, and 45 is assumed. Also, no salvage value at the horizon of 50 years is assumed for the 5-year-old shingles from the replacement in year 45.

The PW for buying and installing the minimum-quality shingles can be found by assuming that the $105 per square will be incurred in years 0, 15, 30, and 45. Then the EAC for 50 years can be calculated, including the $5 per year maintenance cost.

$$PW_{minQ} = -105[1 + (P/F, 6\%, 15) + (P/F, 6\%, 30) + (P/F, 6\%, 45)]$$
$$= -\$174.71$$
$$EAC_{minQ} = 174.71(A/P, 6\%, 50) + 5 = 11.08 + 5 = \$16.08 \text{ per square}$$

Thus, even though some 15-year shingles are discarded after only 5 years, the EAC is increased by only $.27, or 1.7%. This change is not significant.

Example 10.8   Present Worth with Assumed Salvage Values

For the brass and stainless-steel pumps in Examples 10.3 through 10.5, use a 10-year study period. Compare the EACs with those calculated for a 12-year study period, and those for 3-, 4-, and 5-year periods.

Solution
The salvage value for the brass pump was assumed to be $6K after 1 year of use; $2K after 2 years; and $0 after 3 years. Similarly, the salvage values for the stainless-steel pump are assumed to be $9K, $4.5K, $1.5K, and $0 after 1, 2, 3, and 4 years respectively.

The PW values calculated in Example 10.4 are converted to EACs (shown in the following table) by multiplying them by the (A/P) factors for 3, 4, and 5 years. The EACs for 10 years are calculated directly as follows:

$$\text{EAC}_{\text{brass}} = [12K + 12K/(1.08)^3 + 12K/(1.08)^6 + 12K/(1.08)^9 - 6K/(1.08)^{10}]$$
$$\cdot (A/P, 8\%, 10)$$
$$= (35.09 - 2.78)(.1490) = \$4815$$
$$\text{EAC}_{\text{stainless}} = [15K + 15K/(1.08)^4 + 15K/(1.08)^8 - 4.5K/(1.08)^{10}](A/P, 8\%, 10)$$
$$= (34.13 - 2.08)(.1490) = \$4776$$

Period	EAC Brass	EAC Stainless Steel	EAC Difference
3	$4656	$5359	$ - 703
4	5169	4529	640
5	5049	4984	65
10	4815	4776	39
12 or ∞	4656	4529	127

Thus, with these assumed salvage values the brass pump is best with 3 years (where the stainless-steel pump has a negligible salvage value). The stainless-steel pump is best with 4, 5, 10, 12, or a long but indefinite horizon.

Robustness Due to Estimated Lives.   Engineered products and projects rarely have precisely defined lives. A life of 4 years may really be between 3 and 7 years. The period of use is often defined even less precisely. The 10- or 20- or 50-year horizon is chosen as a limit to how far in the future costs and revenues will be estimated and considered, but the actual use period may be much longer.

The life of each alternative is not precise, and the study period is not precise. Thus, it is of little value to take the approach of Example 10.8, where 3- and 4-year alternatives are evaluated over a 10-year life, using salvage values for early terminations. However, this example was necessary to show that assuming a life of 3 years and a small salvage value for the stainless-steel pump can make a difference that should be avoided where possible. As stated earlier, the preferred approach is to assume an indefinite study period and use EACs or EAWs.

Moreover, lives of alternatives are extended to match changed situations. For example, you might buy a used car with a plan of keeping it for 4 years. However, if after 4 years you are planning to move to a new part of the country, or the kind of vehicle that you need will be changing due to changes in family size or interests, then you are likely to keep the used car for another year or so. Repair costs may be a little high, but buying a car and then changing cars after a year would be much more expensive.

In this case, think of the problem another way. The horizon for the use of the mutually exclusive alternatives (10 years) is well beyond the life of either alternative (3 or 4 years). Replacements will be purchased later, when better estimates will be available for the remaining use period and for the lives of new alternatives. Why not compare the EACs of each alternative over its "best life," and choose the cheapest?

This can be restated as: Compare EACs (with the given life for each alternative) whenever the study period is indefinitely longer than the lives. Or this could be stretched slightly to assume that $N = \infty$ is the "same as" N that is longer than the lives of the alternatives.

There are exceptions when this approach should not be used. For example, a contract to supply items may run for exactly 3 years, with a low probability of follow-on work; or the exclusive term for a pharmaceutical product may run for exactly 8 more years. Also, when comparing very short-lived alternatives with much longer-lived alternatives, it may be necessary to carefully define a study period and match alternatives to it. The replacement of aging assets with new ones is a common example of this short/long-life comparison (see Chapter 11 for details).

10.7 PW, EAW, AND IRR HAVE THE SAME REINVESTMENT ASSUMPTION

The interest rate, i^*, used for PW, EAW, EAC, and B/C ratio calculations is the assumed reinvestment rate. This rate is developed using the methods in Chapter 9 for finding the minimum attractive rate of return or MARR.

This MARR is also the reinvestment assumption when the IRR technique is used [Lohmann], but some writers on engineering economy and finance have incorrectly said that the IRR approach assumes that the IRR is the reinvestment rate [Bussey and Eschenbach].

The confusion arises because the IRR approach requires incremental analysis (see Section 10.8) when applied to mutually exclusive alternatives. The fact is that the IRR uses the same interest rate as the other approaches. In PW, EAW, EAC, and B/C ratio calculations, i^* appears in every factor. For the IRR approach, i^* is used as a standard of comparison.

For loans, a good IRR is below i^*; for investments, a good IRR is above i^*. The PW, EAW, EAC, B/C ratio, and IRR approaches make the same assumptions; when applied properly, they support the same recommendations.

The confusion about the reinvestment rate for the IRR arises because of examples like projects A and B whose cash flows are shown in Exhibit 10.2. Project B has a PW of $1630 at 10%, while project A's PW is $1301. Project B has a higher PW at the given MARR of 10%, and it is the preferred project at that MARR. Project A has a higher IRR, 20.5%, than project B, 16.1%. Some authors have incorrectly argued that this implies that project A is preferred and that it assumes reinvestment at the IRR for project A, i_A.

Section 10.8 details how to apply IRRs in an incremental analysis. This analysis shows that in Exhibit 10.2, project B is preferred for any interest rate less than the 12.3% rate, which is the rate where projects A and B have the same PW. This includes the 10% MARR. The extra effort of incremental analysis is required to properly use IRRs (and B/C ratios) for mutually exclusive alternatives. This extra effort is why PW, EAW, and EAC approaches are preferred for mutually exclusive alternatives.

Exhibit 10.2   PW, IRR, and the MARR

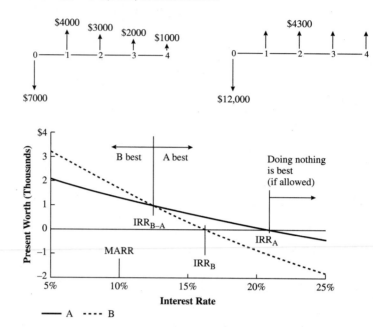

The only limitation on the use of IRRs to compare mutually exclusive alternatives is whether a unique root exists for the PW or EAW equation that is used to find the IRR. This is theoretically a problem for mutually exclusive comparisons, especially those involving different-length lives. However, the large number of problems solved by setting EAW$_{\text{Alt. 1}}$ equal to EAW$_{\text{Alt. 2}}$ suggests that the practical difficulties are minimal.

Example 10.9 illustrates how EAWs depend on the interest rate at which they are computed. If the MARR is known, as it must be for mutually exclusive alternatives, the preferred alternative has the highest PW or EAW or the lowest EAC—*at that MARR.*

Example 10.9   The Cheapest Roof Depends on the MARR

In Example 10.6, Charley was selecting materials for a state-subsidized housing development with a life of 75 years and an i of 6%. For what MARRs besides 6% are the maximum-quality shingles the best? For MARRs between 0% and 15%, which roof has the lowest EAC?

	Minimum-Quality Shingles	Maximum-Quality Shingles	Metal Roofing
Buy	$35	$70	$120
Install	$70	$70	$100
Annual maintenance	$ 5	$ 3	$ 1
Life (years)	15	25	50

Solution
Calculating the EACs is easy, particularly if a spreadsheet is used. Exhibit 10.3 summarizes the results of applying the following equations.

$$EAC_{minQ} = 105(A/P, i, 15) + 5$$
$$EAC_{maxQ} = 140(A/P, i, 25) + 3$$
$$EAC_{metal} = 220(A/P, i, 50) + 1$$

Exhibit 10.3   Cheapest EAC depends on the MARR

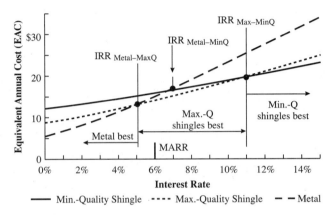

The best roof is the one with the lowest EAC. For MARRs between 0% and 4.86%, the metal roof is the best; for MARRs between 4.86% and 10.95%, the maximum-quality shingles are the best; and for MARRs above 10.95%, the minimum-quality shingle roof is the best. At a MARR of 6%, this graph confirms that the maximum-quality shingles are slightly cheaper than the metal roof.

10.8 INCREMENTAL ANALYSIS

Incremental analysis evaluates the difference, or the increment, between two or more mutually exclusive alternatives. Sometimes this approach is labeled as a challenger/defender analysis. This approach is required to correctly apply IRR or B/C ratio measures to mutually exclusive alternatives.

Incremental analysis evaluates the difference between two or more mutually exclusive alternatives.

The steps of incremental analysis are:

1. Order the alternatives in terms of increasing first costs.
2. Calculate the economic value of the do-nothing alternative or, if doing nothing is prohibited, the least-expensive alternative.
3. One by one, evaluate the incremental difference between each alternative (the "challenger") and the best alternative found so far (the "defender").

Incremental analysis begins by ordering the alternatives in terms of increasing first costs, which ensures that the increments have cash flow patterns corresponding to investments. Thus, the IRRs that are greater than or equal to the MARR

and the B/C ratios that are greater than or equal to 1 represent investments that
should be made. In that case, the challenger is better than the defender.

There are two other standards that are incorrect, but are sometimes suggested.
These are: (1) maximizing the IRR or the B/C ratio, and (2) choosing the largest
project with an acceptable IRR or B/C ratio. As shown in Example 10.10, these
approaches may not lead to a correct recommendation.

Example 10.10   Incrementally Comparing Mutually Exclusive Alternatives

Alternatives A, B, and C have lives of 5 years. Which is the best if the MARR is 10%?
Doing nothing is allowed, but the alternatives are mutually exclusive.

Alternative	First Cost	Annual Return	IRR	PW at 10%
A	$10K	$2913	14.0%	$1043
B	15K	4266	13.0%	1171
C	18K	5037	12.4%	1094

Solution

Without using incremental analysis, this problem could be solved by choosing the alternative with the largest PW—that is, project B.

The first step of incremental analysis, that of ranking in order of increasing first cost,
has already been done. The order is alternatives A, B, and C. The first defender is doing
nothing. Alternative A, the first challenger, has an incremental first cost of $10K and an
incremental annual return of $2913 over doing nothing. The incremental IRR for A is
14.0%, and A is preferred to doing nothing.

Next, B is the new challenger, and A is the new defender. The incremental first cost is
$5K, and the incremental annual return is $1353. Solving for the incremental IRR relies
on a PW equation:

$$PW = 0 = -5K + 1353(P/A, IRR_{B-A}, 5) \Rightarrow IRR_{B-A} = 11.0\%$$

Since 11.0% $\geq$ 10%, B is preferred to A. The incremental PW is $1171 - $1043 = $128,
which also confirms that B is preferred to A. Since A has the largest IRR, this shows that
that criterion is incorrect.

Next, C is the new challenger, and B is the new defender. The incremental first cost is
$3K, and the incremental annual return is $771. Solving for the incremental IRR relies on

$$PW = 0 = -3K + 771(P/A, IRR_{C-B}, 5) \Rightarrow IRR_{C-B} = 9.0\%$$

Since 9.0% < 10%, the increment is not justified. The defender, B, is preferred to the
challenger, C. Note the incremental PW is -$77.

C is the largest project, and it has an acceptable IRR (12.4%, compared with doing
nothing). Since B is better, this example shows that the criterion of choosing the largest
acceptable project is incorrect.

Exhibit 10.4 illustrates that these incremental rates correspond to where the PW_A =
PW_B for IRR_{B-A} and PW_B = PW_C for IRR_{C-B}. B is the preferred alternative, since it has
the highest PW at 10%, the MARR.

Exhibit 10.4 Comparing A, B, and C using PW and IRR

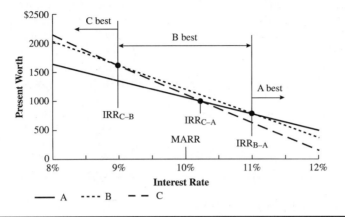

10.9 USING SPREADSHEET SOLVE FOR TOOLS TO CALCULATE INCREMENTAL IRRS

Several spreadsheets have equivalents to the SOLVE FOR tool in Quattro Pro. These tools identify a formula cell, a target value, and a variable cell. Then the variable cell is changed automatically by the computer so that the formula cell has the target value.

Exhibit 10.5 shows two ways to calculate an incremental IRR. The B−A cash flows (column D) are incremental cash flows that can be calculated if cash flows are given for each period and the periods are properly matched. With these incremental cash flows an @IRR function (see cell D8) can be used to find the incremental rate of return.

Or cell D11 can be identified as the formula cell and cell A11 as the variable cell. Quattro Pro assumes a target value of 0 unless otherwise specified. Since cell D11 equals the difference between the PWs, when it is 0, PW_A equals PW_B.

Exhibit 10.5 Using SOLVE FOR to find incremental IRR

	A	B	C	D	E
1	Using SOLVE FOR with PWs				
2		A	B	B-A	
3	0	-7000	-12000	-5000	
4	1	4000	4300	300	
5	2	3000	4300	1300	
6	3	2000	4300	2300	
7	4	1000	4300	3300	
8	IRR	20.53%	16.13%	12.28%	= @ IRR(0,D3..D7)
9					
10	i	PW	PW	D11=C11-B11	
11	12.28%	984.70	984.70	-0.000325	
12		B11=+ @ NPV($A11,B$4..B$7)+B$3			

In Quattro Pro this tool is accessed using /T(ools)/S(olve for). Note that the variable cell must somehow affect the formula cell, although the variable cell need not appear directly in the formula cell. In Exhibit 10.5, the interest rate (cell A11) appears in the PW formulas (cells B11 and C11), but not in the formula cell (cell D11).

Exhibit 10.6 shows a problem that would be extremely difficult to solve without the SOLVE FOR tool. If cash flows were identified so that incremental cash flows could be found, there would be 150 periods! On the other hand, financial functions make it easy to calculate EACs, and the formula cell can be the difference between two EACs.

Exhibit 10.6   Using SOLVE FOR to find incremental IRRs with EACs

	A	B	C	D	E	F
1	Using SOLVE FOR with EACs					
2		MinQ	MaxQ	Metal		
3	Buy	35	70	120		
4	Install	70	70	100		
5	O&M	5	3	1		
6	Life	15	25	50		
7					SOLVE	
8	i	EACminQ	EACmaxQ	EACmetal	VALUE	
9	10.95%	19.56	19.56		0.000281	=C9-B9
10	6.68%	16.29		16.30	5.39E-05	=D10-B10
11	4.86%		12.80	12.80	-0.000335	=D11-C11
12		= @ PMT (B$3+B$4, $A10, B$6)+B$5				

10.10 SUMMARY AND REVIEW OF CHAPTERS 9 AND 10

Choosing the most economical design is one of the most common problems solved by engineers. Such problems can only implement one alternative, so the choice is between mutually exclusive alternatives. The interest rate must be given for this class of problem.

This chapter has shown that using PWs, EAWs, or EACs is the easiest way to compare mutually exclusive alternatives. If the lives of the different alternatives are the same, then there is no advantage in choosing between EAWs/EACs and PWs.

If the alternatives have different-length lives, then comparing EAWs or EACs is much easier than comparing PWs. Usually the best implicit assumption is that costs repeat identically until the least common multiple of the lives being compared, even though theoretically it may not be realistic. As shown in Section 10.6, the implicit assumption is quite robust. Assuming identical repetitions is often better, and always easier, than assigning salvage values that may be arbitrary.

This chapter has shown that incremental analysis is required to apply IRR measures to mutually exclusive alternatives (see Chapter 14 for a similar example for

B/C ratios). It has also emphasized that the reinvestment asssumptions of PW, EAW, and IRR are the same; that is, reinvestment at the MARR or i^*.

This chapter has demonstrated that the SOLVE FOR tool available in some spreadsheets is very useful in calculating incremental IRRs.

Chapters 9 and 10 define two different classes of problems that are closely linked, and that require that different tools be used to solve them. Exhibit 10.7 summarizes these problem classes.

Exhibit 10.7   Selecting an evaluation technique

Choose at most one design from a mutually exclusive set (Chapter 10)

Best

    Maximize PW—if lives match

    Maximize EAW—whether or not lives match

    Minimize EAC—whether or not lives match

Acceptable to use incremental (challenger/defender) analysis

    Is the incremental IRR better than i^*?

    Is the incremental $B/C > 1$?

    Is the incremental PW index > 1?

Not acceptable

    Payback period (ignores money's time value and cash flow after payback)

Choose best projects within a constrained budget when too many otherwise acceptable projects are limited by available funds (Chapter 9)

Best

    Rank on IRR—identifies best projects and opportunity cost of forgone investments

Acceptable

    PW or EAW—after ranking on IRR identifies i^*

REFERENCES

Bussey, Lynn E., and Ted G. Eschenbach, *The Economic Analysis of Industrial Projects*, 2nd ed., 1992, Prentice–Hall, Chapter 8.

Eschenbach, Ted G., and Alice E. Smith, "Sensitivity Analysis of EAC's Robustness," *The Engineering Economist*, Volume 37, Number 3, Spring 1992, pp. 263–276.

Lohmann, Jack R., "The IRR, NPV and the Fallacy of the Reinvestment Rate Assumptions," *The Engineering Economist*, Volume 33, Number 4, Summer 1988, pp. 303–330.

PROBLEMS

10.1   Which is more economical, heating with natural gas or with electricity? Assume that both the equipment and the building have a 25-year life, that the salvage value is 0 for both, and that the interest rate is 8%.

Fuel	First Cost	Annual Heating Cost
Natural gas	$15,000	$1200
Electricity	8000	1900

10.2   Which is more economical, a tile floor or a wood floor? Assume that the public building has a life of 50 years, no salvage value, and either floor will last the 50 years with proper maintenance. Both floors cost $1500 per year for cleaning and waxing. The interest rate is 4%. The tile floor will cost $35,000 to install. The wood floor will cost $25,000 to install. The wood floor must be refinished every 5 years for $6000.   (*Answer:* The tile floor saves $602 per year.)

10.3   Flincham Fine Paints can buy new painting equipment for $10,000. Its useful life is 5 years, after which it can be sold for $2000. Estimated annual costs are $1000, and it should save Flincham $3000 per year. If Flincham's MARR is 6%, should the equipment be purchased? Use present worth analysis.

10.4   Hagen Hardwood Floors must purchase new finishing equipment. The cash flows for each alternative are presented below. If Hagen's MARR is 10%, which setup should be chosen?

	First Cost	Annual Cost	Useful Life
Smoothie	$3000	$1660	4 years
Blaster	5000	1000	4 years

(*Answer:* Blaster saves $29/year)

10.5   A manufacturer, whose MARR is 20%, may replace a manual material handling system with one of two automated sytems. What course of action is appropriate?

	Cost	Savings	Useful Life
System A	$100,000	$10,290	15 years
System B	115,000	16,880	15 years

10.6   The Do-Drop-Inn must choose between two alternatives for a new swimming pool. If Do-Drop's cost of capital is 10%, which alternative should be chosen?

	A	B
First cost	$30,000	$38,000
Annual operating cost	$ 3000	$ 1750
Rebuilding cost in year 6	$ 5000	$ 3500
Useful life (years)	12	12

(*Answer:* B's PW is $1360 better)

10.7 Green Acre Farms must replace the piping system in its milking barn. Based on EAW and a cost of capital of 12%, which of the following two alternatives should be chosen?

	Pamper Pipes	Fluid Flush
First cost	$75,000	$90,000
Operating costs per year	$ 5000	$ 4000
Savings per year	$ 1000	$ 2000
Salvage value	$ 0	$ 2000
Useful life (years)	10	10

10.8 Ousley Sporting Goods, Inc., is planning to purchase a new computer system for its warehouse. MBI and TTA have submitted the two most attractive bids. Use incremental analysis to decide which one should be selected if $i = 13\%$.

Year	Year	MBI	TTA
Cost	0	$10,000	$15,000
Annual savings	1–5	2985	4160

(*Answer:* MBI saves $246.5/year)

10.9 Better Burger must replace its current grill setup. The two best alternatives are the Burna-Burger and the Chara-Burger. The Burna-Burger costs $12,000 to buy and $1000 in annual operating costs. At the end of the 8 years the setup can be sold as scrap for $500. The Chara-Burger costs $15,000 to buy and $500 per year for operating costs. The setup can be sold for $1000 at the end of 8 years. Using a MARR of 8%, which setup should be chosen?

10.10 The state highway department may purchase new lawn-mowing equipment. The best alternative requires an initial investment of $90,000. Each year the new equipment is expected to save the state $19,500. The equipment will be used for 6 years and has little or no expected salvage value. Using benefit/cost analysis and an interest rate of 10%, should the equipment be purchased?   (*Answer:* B/C = .944, no)

10.11 The Green Thumb Seed Company may purchase a new security system for its lab. Each alternative has a 10-year useful life. Which system should be purchased based on payback analysis? Is the answer different using EAC at an i of 6%?

Alternative	Cost	Annual Savings	Salvage Value
I Spy	$120,000	$20,000	$10,000
Gotcha	140,000	22,000	25,000

10.12   The Tom Thumb Railroad Company is considering new automated ticketing equipment, which has a life of 20 years. The Kwik-Print ticket machine costs $50,000 and is expected to produce net savings of $14,000 the first year. Each year after the first, net savings will drop by $2000 until they reach $4000. These $4000 savings will continue for the machine's remaining life. The Tick-Stamp costs $54,000 and is expected to save $8000 per year. Based on discounted payback analysis and a MARR of 8%, which machine should be chosen? Does the EAC measure lead to the same decision?

10.13   Two proposals are presented to your company. Both proposals have a first cost of $20K. Proposal A has an annual cash flow of $5K for 6 years. Proposal B has cash flows that begin at $3K, and that increase by $1K per year for a 6-year life. If your company uses 12% as its cost of money, which would you recommend?   (*Answer:* EAW$_B$ is $172 higher)

10.14   The Oliver Olive Stuffing Company is considering the purchase of new, more efficient stuffing equipment. Which one should be purchased ($i = 11\%$), if $N = 20$?

	First Cost	Annual Operating Costs	Annual Savings
Stuff-It	$100K	$100K	$122K
Stuffer-Full	135K	100K	128K

10.15   Which project should be selected if the following projects are mutually exclusive and the minimum attractive rate of return is 12%? Each project has a life of 10 years.

Project	First Cost	Annual Benefit
A	$15,000	$2800
B	20,000	4200
C	10,000	2400
D	30,000	6200
E	40,000	7600

10.16   The following projects are mutually exclusive. Which one should be done if the minimum attractive rate of return is 12%? All of the following projects have a life of 10 years.

Project	First Cost	Annual Benefit
A	$15,000	$4350
B	20,000	4770
C	10,000	2400
D	30,000	9200
E	40,000	8000
F	25,000	8100

10.17 If the following projects are mutually exclusive and the MARR is 12%, which one(s) should be done? Why?

Project	PW (12%)	IRR	B/C Ratio	First Cost	Benefits Each Year				
					1	2	3	4	5
A	$2605	25.0%	1.30	−$2000	$900	$800	$700	$600	$500
B	2803	25.8	1.40	−2000	600	700	800	900	1000
C	3405	24.6	1.70	−2000	0	0	0	0	6000
D	2704	25.4	1.35	−2000	750	750	750	750	750
E	3128	40.8	1.56	−2000	1400	1100	800	500	200

(*Answer:* Do project C; it has the largest PW at 12% MARR)

10.18 A city engineer must choose between two different routes for a new sewer line. Route 1 is a 20,000-foot-long gravity line, which includes 8000 feet of tunneling at $250 per foot. The remaining 12,000 feet will cost $80 per foot. The annual maintenance cost is expected to be $2000. Route 2 is a 23,000-foot line requiring a pumping station, which will cost $840K to build and $15K annually to operate and maintain. The sewer line costs $80 per foot and will cost $2100 annually to maintain. The expected life of the station and both lines is 35 years. Using an interest rate of 8%, which route should be chosen?

10.19 The P&J Brewery must select one of the four alternatives presented below. If money costs P&J 8%, what decision should be made, based on EAW? Each investment is to be evaluated for 6 years.

Alternative	Cost	Annual Benefits
A	$10,000	$2100
B	15,000	3250
C	17,500	3800
D	20,000	4350

10.20 The local electric company may renovate a number of substations within its system. Various costs and savings can be anticipated, depending upon the type of renovation performed. Using benefit/cost analysis and an interest rate of 12%, determine which alternative should be chosen.

Alternative	First Cost	Annual Cost	Annual Savings	Life
A	$1,100,000	$24,000	$180,000	20 years
B	1,500,000	36,000	245,000	20 years
C	1,750,000	41,000	295,000	20 years

10.21 Which pump should be installed if the MARR is 12%?

Type	First Cost	Annual O&M Costs	Life
Brass	$13,000	$2200	8 years
Stainless steel	17,000	1950	12 years

(*Answer:* Stainless saves $123 per year)

10.22 Compare the PWs of the following two pumps at an interest rate of 15%. Which one should be installed? What horizon did you use? If necessary, assume that the salvage value for early termination is a linear interpolation between the given first cost and the salvage value.

Pump	First Cost	Annual O&M Costs	Salvage Value	Life (years)
Brass	$5500	$450	$1000	6
Titanium	7500	300	2000	9

10.23 Which of the following two processing pumps should be purchased? The interest rate is 8%. How much better is the EAC for the recommended pump? What assumptions have you made?

Type	First Cost	Annual Operations	Life
Brass	$4875	$1925	10 years
Stainless steel	6450	1730	15 years

(*Answer:* Stainless steel saves $168 per year)

10.24 For the pumps in Problem 10.23, graph the EACs as a function of i. Use rates between 0% and 25%.

10.25 Two alternative bridges have the same supports and approaches. However, one is made of a new galvanized steel and the other is made of more conventional painted metal. The galvanized bridge has a life of 45 years and the painted bridge has a life of 30 years. For the following costs, compare the present worths of the two choices. What life did you use and what assumptions did you make? ($i^* = 8\%$)

Type	First Cost	Annual Operations
Galvanized steel	$85 million	$1.5 million
Painted metal	60 million	2.1 million

10.26   If you drive 10,000 miles per year, which tire should you buy? Tire A costs $65 and should last 40,000 miles. Tire B costs $85 and should last 60,000 miles. You have an outstanding balance on your credit card, and you are paying 20% as an effective annual rate.
  a.   What are comparable EACs for the two alternatives?
  b.   What are comparable PWs for the two alternatives?
  (*Answer to part a:* Save $.45 per year or 1.8% with Tire A)

10.27   Which pickup should be purchased by Northwest Construction for use by its job superintendent? How much does this choice save (using either PW or EAC)? Northwest uses a 12% rate to evaluate its capital investments. What assumptions have you made and why?

Pickup	First Cost	Maintenance Cost/Year	Salvage	Life
New	$18,000	$300	$5000	5 years
Old	10,000	750	3000	3 years

10.28   State University must choose between two new snow-removal machines. The Sno-Blower has a $60,000 first cost, a 20-year life, and a $5000 salvage value. At the end of 9 years, the machine needs a major overhaul costing $14,000. Annual maintenance and operating costs are $8000. The Sno-Mover will cost $40,000, has an expected life of 10 years, and has no salvage value. The annual maintenance and operating costs are expected to be $10,000. Using a 12% interest rate, which machine should be chosen?

10.29   The Thunderbird Winery must replace its present grape-pressing equipment. Two alternatives are under consideration, the Quik-Skwish and the Stomp-Master. The annual operating costs increase by 15% each year as the machines age. If the MARR is 8%, which press should be chosen?

	Quik-Skwish	Stomp-Master
Cost	$250,000	$400,000
Annual operating costs	$ 18,000	$ 12,500
Salvage value	$ 25,000	$ 40,000
Useful life (years)	5	10

10.30   Two mutually exclusive investment opportunities have been presented to "Money Bags" Battle. Which project (if any) should Money Bags invest in, if his MARR is 10%? Assume that neither project can be repeated.

	A	B
Cost	$10,000	$8000
Annual benefits	$ 1500	$1300
End-of-investment receipt	$10,000	$8000
Life of investment (years)	10	5

10.31   Given the following mutually exclusive alternatives, which should be chosen? Assume that neither alternative can be repeated. Use an interest rate of 6%. Which measure should be used and why?

	A	B
Cost	$15,000	$10,000
Annual savings	$ 6000	$ 4000
Annual costs	$ 2000	$ 1000
Salvage value	$ 0	$ 1000
Useful life (years)	6	4

10.32   Westnorth Airlines may purchase a New York to London route from Divided Airlines for $1.5M. The new route should generate income of $1.2M, with associated costs of $850K each year for the next 8 years. (Government regulations require all international routes to be rebid in 8 years.) If Westnorth requires a 14% return on all investments, should the route be purchased? Analyze, using appropriate IRRs.

10.33   For the data in Problem 10.26, find the incremental IRR for buying the more expensive tire.

10.34   For the Binder Book Company, choose from the following mutually exclusive investment opportunities, using an incremental IRR analysis. Each has an 8-year life, and the MARR is 6%.

	A	B	C	D
First cost	$1000	$1100	$2000	$1500
Benefits	180.67	90.03	345.21	275.09

10.35   For each set of investments, determine which alternative is preferred (if either) at a MARR of 8%. Why is it preferred?

a.

	A	B
Investment	$500	$700
IRR	10%	10%
Incremental IRR	10%	

b.

	A	B
Investment	$300	$600
IRR	12%	9%
Incremental IRR	7%	

10.36   Five mutually exclusive alternatives with 5-year lives are under consideration by the ABC Company. Using incremental IRR analysis, which should be chosen if the MARR is 12%?

	A	B	C	D	E
Cost	$20,000	$18,000	$22,000	$15,000	$12,000
Annual Benefits	5558.0	5756.4	6563.6	3957.0	4012.8

10.37 Which one of the following mutually exclusive projects should be done if the MARR is 12% (all choices have lives of 20 years)? Solve the problem using incremental IRRs.

Project	First Cost	Annual Benefit
A	$15,000	$2500
B	22,000	3200
C	19,000	2800
D	24,000	3300
E	29,000	4100

10.38 For the data in Problem 10.37, define the ranges of MARRs within which each project is best. Let i^* range from 0% to 30%.

10.39 The lease of a computer system will cost $150 per month for a 2-year commitment. If the lease is renewed, it should cost less on each contract. The estimated price if renewed 2 years hence is $120 per month and $90 per month if renewed again in year 4. Buying the computer system will cost $7000 and it will be worth $1500 in year 6, which is the horizon for this problem.

   In deciding between leasing and buying the computer, what is the incremental rate of return? If the MARR is 12%, should the computer be leased or purchased? What are the noneconomic factors, and are they likely to favor leasing or buying?

10.40 *The Engineering Economist* is a quarterly journal that costs $20 per year, $38 for 2 years, or $56 for 3 years. Problem 7.19 calculated incremental rates of return for 2 years vs. 1 year at a time and for 3 years vs. 1 year at a time. Calculate the incremental rate of return for buying 3 years vs. 2 years at a time.

10.41 Each of the following projects has a first cost of $25K. They are mutually exclusive and the MARR is 11%. Which one(s) should be done? Why? Does your answer depend on an assumption about whether or not the projects can be repeated? If so, answer the question again for the other assumption.

Project	Life (years)	PW at 11%	EAW at 11%	IRR	Annual Benefit	Gradient
1	10	$ 5973	$1014	18.3%	$8000	$ -750
2	20	6985	877	13.8	1200	450
3	30	10,402	1196	14.3	3300	100
4	15	5413	753	12.6	-5000	1800
5	5	2719	736	15.2	7500	0
6	10	7391	1255	17.7	5500	0

CHAPTER 11 REPLACEMENT ANALYSIS

The Situation and the Solution

Often, an aging existing asset must be evaluated for the timing of its replacement. For example, a 10-year old diesel generator costs more to run and to maintain than a new generator, but a new generator has much higher capital costs.

The solution is to find the best life and its equivalent annual cost for the most cost-effective challenger and then to decide whether the existing asset's life can be extended at a lower equivalent annual cost.

Chapter Objectives

After you have read and studied the sections of this chapter, you should be able to:

Section 11.1
Explain why equipment is replaced, retired, or augmented.

Section 11.2
Recognize that replacement analysis involves mutually exclusive alternatives whose lives usually differ.

Section 11.3
Recognize the sunk costs of existing equipment, the risks of new equipment, and the difference between profit making and cost saving.

Section 11.4
Calculate the best economic life for new equipment and for existing equipment.

Section 11.5
Apply the techniques developed for optimal challengers to the problem of selecting the optimal design alternative.

Section 11.6
Include the estimated performance of future challengers in replacement analysis.

Section 11.7
Apply replacement and repair models, such as block replacement.

Key Words and Concepts

Replacement   When new equipment is purchased and is used instead of existing equipment, which is disposed of or shifted to another use.

Augment   When new equipment is purchased to increase the capacity or capabilities of existing equipment, which is retained.

Retirement   When an existing asset is disposed of or shifted to a backup status without purchasing a new asset.

Challenger   The potential new equipment, when considered for its economic life.

Defender   The existing equipment, when considered for its economic life.

Economic life   Has the lowest costs (the minimum EAC for the challenger).

Physical life   The time from creation to disposal.

Accounting life   Based on depreciation.

Ownership life   The time from purchase to sale.

Service period   The time that equipment must be available for use.

Sunk costs   Expenses that have been incurred or dollars that have been spent.

Inferiority gradient   The annual improvement in the costs of new challengers, plus the annual increase in O&M costs for existing defenders.

Block replacement   A policy of replacing all units—failed and functioning—at the same time.

11.1 WHY IS EQUIPMENT REPLACED, RETIRED, OR AUGMENTED?

Replacement analysis focuses on situations in which aging equipment will be **replaced** by new equipment. The aging equipment may be sold, donated to charity, dismantled for parts, or shifted to another use, such as backup status for emergencies.

Replacement analysis also includes **retirements**, in which aging equipment is not replaced. This is common when fleet sizes are being reduced. Finally, replacement analysis includes **augmenting** aging equipment with new equipment

Existing equipment (1) may be **replaced** by new equipment, (2) may have its capacity **augmented** by new equipment, or (3) may be **retired** by disposal or by shifting to a backup status.

when additional capacity or capabilities are required. The choice may be between replacing current equipment with new, larger-capacity equipment or augmenting the current equipment with new, lower-capacity equipment.

The reasons that equipment may be replaced include:

Reduced performance of the aging equipment.

Altered requirements for use of the equipment.

Obsolescence due to a new, improved model.

The risk of catastrophic failure or unplanned replacement.

Shifts between renting/leasing and owning.

Reduced Performance.   Older equipment often has lower efficiency or productivity. It must be repaired or adjusted more often, and there is more time when it is not working or available for use. As shown in Exhibit 11.1, this causes the cost of ownership to increase over time. For short lives, the cost of ownership may be dominated by the recovery of the first cost for purchase and installation. However, for long lives, the costs of reduced performance often dominate.

Exhibit 11.1   Equipment costs over time

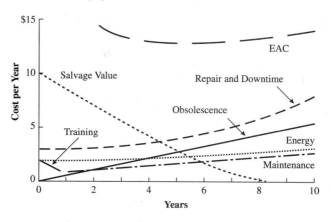

Examples of reduced performance include the repair needs of a 10-year-old car with 100,000 miles, of a 7-year-old machine tool, of a 30-year-old bridge, and of a 10-year-old power generator. In fact, almost every piece of mechanical equipment shows the effects of wear in reduced efficiency and productivity.

Altered Requirements.   Sometimes equipment is considered for replacement because the conditions of use have changed. If demand has increased, then a larger capacity may be needed. A machine tool may be required to handle a new material or larger stock. At the personal level, an engineer who owns a sports car may marry and need more of a family car. An engineering professor might move from Alaska to Missouri—selling the four-wheel-drive cars and buying cars with air conditioning.

A manufacturer may need eight stamping lines for new products and one for spare parts. Thus, seven sets of dies may be retired when a model is revised. The remaining set is shifted from producing new parts to spare parts. All that has changed is the demand for the output of the dies.

A variation on this is when the requirements have not changed, but the current equipment was incorrectly selected. It does not quite meet the original or the current requirements, and the question is whether to continue to use the "not quite right" equipment or to replace it.

Obsolescence. Substantial advances occur frequently in the capabilities of electronic equipment, and they are significant for mechanical equipment as well. New computers, calculators, testing equipment, and software are often purchased because they offer new capabilities, not because the old equipment has become unreliable. In fact, the requirements may not have changed at all. It just may be that the new equipment can get the same job done faster (and thus save on labor costs).

Technological and market advances can revolutionize industries. Obvious examples include the shift from slide rules to calculators, from mechanical cash registers to electronic point-of-sale terminals, from tubes to transistors to integrated circuits—even the shift from bias-ply tires to radial tires. Tire manufacturers had to discard and replace equipment before they could shift from producing mostly bias-ply tires to producing mostly radial tires. Because tire firms with leading market shares and competitive advantages were reluctant to replace their equipment, 4 years later a new set of firms were the market leaders [Foster].

Exhibit 11.2 illustrates the S-curves of technological performance that can make current equipment obsolete. When initially introduced, the new equipment does not have an advantage over the existing equipment, but it rapidly achieves superior performance as its technology matures. This superior technical performance translates into better economic performance.

Exhibit 11.2   Cost, performance, and technological innovation

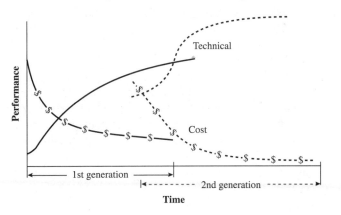

Each firm in the industry must decide how rapidly to shift from the old to the new technology. The reluctance of firms that have large market shares to replace

existing equipment has been termed the "attacker's advantage" [Foster]. Chapter 16 explains how growth or S-curves can be calculated and used in cost estimating.

Risk of Catastrophic Failure or Unplanned Replacement.   A scheduled replacement can be planned so that it is rapid, inexpensive, and minimally disruptive. On the other hand, an unplanned replacement often disrupts or stops operations and requires expedited, expensive transportation costs. Overnight delivery may be cheap compared to lost production time, but a normal shipment for a planned replacement is cheaper yet.

The operation of some equipment is so critical that it is replaced before it fails simply to reduce the possibility of failure during operation. Auto engines may be run until they throw a rod or blow a gasket, but engines in airplanes are replaced or overhauled before this occurs.

Lease or Rental vs. Ownership.   It is common to replace leased equipment with purchased equipment. The original equipment may have been leased, so that it might be tested in a particular application. Once it has demonstrated satisfactory performance, and the long-term nature of the need has been demonstrated, then identical (or nearly so) equipment might be purchased. Sometimes owned equipment will be replaced by leased equipment, but this is less common than replacing leased with purchased equipment.

Summary of Reasons for Replacement.   Equipment is replaced because the new equipment will be more economical. This may occur because the old equipment has deteriorated or was incorrectly chosen. It may occur because the new equipment has new features, or the circumstances of use have changed. The old equipment may be disposed of, saved for backup use, or kept to be used in conjunction with new equipment that augments its capabilities.

11.2 OLD AND NEW ARE MUTUALLY EXCLUSIVE

Replacement analysis is a choice between mutually exclusive alternatives (discussed in Chapter 10). Either the existing equipment will continue to be used, or new equipment will be purchased to replace it.

Different-Length Lives.   In these comparisons, the remaining life of the existing asset is almost always shorter than the life of the potential new equipment. The existing asset is usually nearing the end of its life, since its performance is decreasing, it is becoming obsolete, or requirements are changing. The new equipment is at the beginning of its life. Since the lives are different, the normal method of comparison is with EAWs or EACs.

Sometimes the use of whichever equipment is chosen will cease in a short time, such as 3 years. Then the lives will match and comparison with present worth or IRR is possible. EAC or EAW will work in this case as well.

The language used in the comparison of the existing and the new equipment comes from incremental analysis (see Chapter 10). The existing equipment, because it is in place, is referred to as the **defender**. The potential new equipment that may replace the defender is the **challenger**. This language originated in the context of replacement analysis [Terborgh].

*The **challenger** is the potential new equipment; the **defender** is the existing equipment. Both are considered for their economic life.*

Economic Life.   An asset's life could be defined in many ways. For our purpose, the most important life is an economic one. For the challenger, the **economic life** results in the minimum EAC for the piece of equipment. For the defender, the **economic life** is how long the marginal cost to extend service is less than the minimum EAC for the challenger. The calculations for this are detailed in Section 11.4.

*An **economic life** has the lowest costs (minimum EAC for the challenger); a **physical life** is the time from creation to disposal; the **accounting life** is based on depreciation; an **ownership life** is the time from purchase to sale; and a **service period** is the time that equipment must be available for use.*

Other possible definitions of an asset's life include the following: The **physical life** is the time until the asset is dismantled or scrapped. The life for accounting purposes (**accounting life**) is how long until it is fully depreciated (see Chapter 12). The **ownership life** is the time from purchase to sale or disposal. The **service period** may include multiple economic lives. For example, an engineering computer lab might have a service period of 30 years, but the computers within it might be replaced every 5 years. Example 11.1 illustrates these lives.

Example 11.1   The Lives of a Car

Why would the economic, physical, accounting, and ownership lives and the service period of a car or other vehicle differ?

Solution
Cars in the U.S., unless they are destroyed in an accident, usually have physical lives of 10 to 15 years. Most are owned by more than one person during that physical life, and these ownership lives are different. If they are owned and operated for business purposes, they will be depreciated over 5 years, which is the accounting life.

The economic life may be short or long, depending on how the vehicles are used. For example, many car rental firms, such as Hertz, Avis, or National, sell their vehicles after 2 years, while taxis may be owned and operated much longer—even though they operate for more miles each year.

11.3 SUNK COSTS, RISKS, AND COST SAVING VS. PROFIT MAKING

Sunk Costs.   **Sunk costs** were introduced in Chapter 1 because they occur in many engineering economy problems. Sunk costs are mentioned again here because every existing asset has sunk costs that are irrelevant.

***Sunk costs** are expenses that have been incurred or dollars that have been spent.*

As always, the key question is: What costs are influenced by this decision? What matters is the equipment's or asset's value now, not what you paid for it. If there will be a loss due to sunk costs, then income taxes will be reduced. The larger the loss, the greater the income tax benefit, and the greater the incentive to replace the existing asset.

When considering an asset's value, there are several possibilities. There is the original first cost, but this is a sunk cost. The book value, which accounts for depreciation from this first cost, is only an entry in the firm's accounting records. The present cost to buy and install a similar new asset or a similar used asset would be relevant if the existing asset were not in place. The trade-in value of the existing asset on a new replacement is often distorted by the price of the new replacement. For example, a salesman might offer a high value on your used car trade-in rather than discounting the new car's list price.

The correct value of equipment is its market value, after accounting for removal costs. This is what would be received if it were sold and the equipment were replaced. This is sometimes called the outsider's or consultant's viewpoint because it emphasizes that the sunk costs are irrelevant. As noted in Example 11.2, the costs of change must be included.

Example 11.2 Value of a CNC Machine Tool

A computerized, numerically controlled (CNC) machine tool cost $65,000 to buy and $4000 to install 10 years ago. It is fully depreciated, and its book value is $0. New tools are easier to program; they have larger memories for machining more parts; and they now cost $50,000. The installation costs are still $4000, and the removal cost is $1000. Ten-year-old used machines can be bought or sold for $7000. Ignoring taxes, what is the relevant cost?

Solution
The first cost of $69,000 to buy and install 10 years ago is a sunk cost. The book value is relevant only for accounting and tax purposes. The $54,000 to buy and install a new tool is the first cost for a potential new replacement, but it is not the existing tool's value. The $7000 purchase price and $4000 installation cost (total: $11,000) is the cost to buy and install a used tool—not the existing tool's value.

The relevant cost of the machine is $6000, which is the market value minus the removal cost. This is the income that would be received if the asset is sold. It is the income that is forgone if the asset is kept.

Ignoring sunk costs is an extremely difficult economic principle for some to accept. This may be caused by psychological, personal, and organizational "sunk costs" (prestige, reputation, and the cost of admitting a mistake). A common example in many organizations focuses on upgrading computer systems.

Engineers and organizations need computers to compete. Usually a new computer will offer new features, faster processing, more memory, more disk space, etc.—at a lower price than the previous generation. Even though an engineer's desktop computer is only 2 or 3 years old, it may be nearing the end of its economic life for use by that engineer. That engineer's productivity may be increased enough by more computer power that it may be time to replace the current computer. (Note: the existing computer may then be used to increase the productivity of someone else, such as a technician or secretary, who has an even older computer.)

The decision should be based on productivity increases, the value of the machine if sold or used by someone else, and the cost of the new computer. The decision should ignore the sunk cost of the old machine, even though someone argues, "only 2 years ago, it was worth $2500."

Risks of the New often Far Exceed Those of Extending the Old.   Sunk costs are often inappropriately included, which tends to favor keeping the existing old asset. On the other hand, there is a valid reason for tending to favor the existing old asset. Two different adages provide some guidance:

"If it ain't broke, don't fix it."

"Better the problem you know than the solution you don't."

To be worthwhile, any new asset must offer the expectation of better economic performance. However, because it has not been implemented in exactly this situation, more is unknown about the new asset. This often corresponds to more risk.

Of course, the actual performance of the new challenger may be better than expected. However, sad experience has shown that more often the new machine's technical and cost performance is worse than expected. Construction costs, allowances for lost or lower productivity during changeover, etc., are more likely to be over budget than under budget. In addition, when they are under budget, the savings are usually small. When they are over budget, the overruns can be very large.

There are exceptions when the risks of extending the old equipment's life are larger than the risks of buying new equipment. These are cases when the replacement is being done to avoid the costs of catastrophic failure (such as airplane engines) or unscheduled replacement (such as oil refineries). More often, the replacement is being considered because the old equipment has reduced performance, the requirements have changed, or the new challengers are making existing equipment obsolete. Then the risks of the new tend to be greater.

Cost Saving vs. Profit Making.   When evaluating aging equipment, cost saving must be distinguished from profit making. Repairing broken equipment may avoid the cost of buying new equipment, but it does not increase the firm's profit. Decisions that increase the firm's profits are easier to make in isolation. Is the PW or EAW for this alternative positive?

However, decisions that save costs by repairing old equipment absolutely require a projection of future repair costs. As shown in Example 11.3, failure to do so may result in "saving" the same cost so often that is not economical.

Example 11.3   Saving the Old Clunker

An engineering professor owns a 17-year-old pickup. During the last year, he has purchased new tires ($280), had the brakes redone ($450), replaced the muffler ($95), and had the transmission fixed ($550). The professor has just been told that the tie rods and other steering parts must be replaced ($800).

What should the professor do? In the past, the professor has compared each repair cost with the cost of buying a new car and concluded that the repair was economically justified. The cost of replacement has been saved four times already and a fifth time is now being considered.

Solution

First, the professor must recognize that the tires, brakes, muffler, and transmission represent sunk costs. They may increase the price the clunker could be sold for, or the tires might be sold separately if the clunker is bound for the junk yard. They do reduce the chances of further problems with the tires, brakes, muffler, and transmission.

However, the correct analysis in the past would have estimated these potential expenses as a group, rather than considering each individually. The correct analysis may not have been done in the past, but that is not an excuse for making another mistake by failing to do the correct analysis now.

In deciding whether or not to fix the steering mechanism, the professor must estimate other future costs. How soon will the engine need repair? Is any body part that must be replaced about to rust off? What will these or other costs be? An EAC must be calculated for the marginal cost of extending the life. This marginal cost, or EAC, must be compared with the EAC for a replacement vehicle—perhaps a used pickup that is only 10 years old.

11.4 OPTIMAL CHALLENGERS AND OPTIMAL DEFENDERS

The economic life of the potential new equipment was defined as the life that minimized the EAC. The economic life for the existing equipment is calculated by asking, For how long is it cheaper to extend the service of the existing asset?

Often, these economic lives are not obvious. For the existing asset, the minimum EAC is not used because it makes unrealistic assumptions, and it often indicates a different answer than the cost to extend service for another year. In some cases, it is necessary to calculate the best EAC for the defender *and* the costs to extend the life for several years.

Challenger's Optimal or Economic Life.   Each potential challenger's economic life must be calculated. If a physical life may be up to 20 years, then the possible economic life might be anywhere from 1 to 20 years, depending on the equipment's O&M costs and salvage values. This economic life has the lowest EAC or the highest EAW. It is the optimal life in an economic sense.

Example 11.4 illustrates this calculation for potential new equipment. Perhaps this is best visualized as selecting the best challenger from five different alternatives. The choice of a 1-year, 2-year, or other life of the challenger is a choice between mutually exclusive alternatives. Which of these alternatives or which life has the lowest EAC?

Example 11.4   Economic Life of a Challenger

Assume that a machine has no installation or removal costs. It costs $10,000 to buy, and the firm's interest rate is 8%. Its salvage value in each year is tabulated below, as are its maintenance costs. What is its economic life?

This data set is typical of problems where the benefits of having the machine, the labor costs to operate it, and its required power costs do not depend on the machine's age.

Year	Salvage Value	Maintenance Costs	
1	$7000	$ 0	
2	5000	750	
3	3500	1500	$750 \, (P/F, i, 2)(A/P, i, 3) + 1500 (A/F, i, 3)$
4	2500	2250	
5	2000	3000	

Solution

The calculation of the EAC for lives of 1 to 5 years is simplest if the first cost, salvage value, and O&M costs are calculated separately. For the first cost and the maintenance gradient, only the number of years changes. For the salvage value column, both the salvage value and the number of years change.

$$EAC_{P_t} = 10,000(A/P, 8\%, t)$$

$$EAC_{O\&M_t} = 750(A/G, 8\%, t)$$

$$EAC_{S_1} = -7000(A/F, 8\%, 1)$$

$$EAC_{S_2} = -5000(A/F, 8\%, 2)$$

$$EAC_{S_3} = -3500(A/F, 8\%, 3)$$

and so on.

Year	EAC of P	EAC of S	EAC O&M	Total
1	$10,800	$-7000	$ 0	$3800
2	5608	-2404	361	3565
3	3880	-1078	712	3514
4	3019	-555	1053	3517
5	2505	-341	1385	3549

The costs for a 3-year and a 4-year life are virtually identical, so either life could be chosen.

The Cost Curve for the Challenger's Economic Life.   The data of Example 11.4 illustrates a standard pattern of (1) increasing operating and maintenance costs and (2) salvage values that decrease by a smaller amount each year. This is sufficient to imply a concave cost curve, as illustrated in Exhibit 11.3 for the data of Example 11.4.

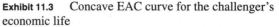

Exhibit 11.3   Concave EAC curve for the challenger's economic life

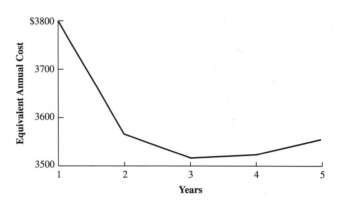

With a concave cost curve, two values can confirm whether the lowest EAC occurs at either the shortest or the longest possible life. Three values (EAC_{t-1}, EAC_t, and EAC_{t+1}) can confirm that t is the best life. Thus, if the problem is being solved by hand rather than by using a spreadsheet, it may be worthwhile to guess whether the best life is short, long, or in between. Then begin calculating EACs, at the shortest life, the longest life, or $N/2$ as a first guess.

For economic lives that are in between the shortest and the longest possible, it is usually true that at least two and often three EACs are fairly close together. This implies that there is some flexibility as to the best timing for replacing equipment.

Irregular decreases and increases in O&M costs, salvage values, major overhauls, or other costs imply that every year must be individually checked. If spreadsheets are used, this is very simple. The best life for a challenger is usually just before an overhaul. For example, consider a machine that has a physical life of 20 years, with overhauls every 5 years. The best life for this machine is likely to be 5, 10, 15, or 20 years, with no overhaul in the last year. Rarely would it make sense to overhaul the machine in year 5, and then retire it in year 6 or 7.

Defender's Economic Life.   The best economic life for the defender (the existing equipment) is calculated differently. This calculation relies on the marginal cost to extend service for another year, MC_t, including salvage values and O&M costs.

The salvage value at the year's beginning is the "first cost" for extending service, and the salvage value at the year's end is the salvage value for extended service. The maintenance term is considered on an annual basis. Equation 11.1 is calculated as an end-of-year cost, so that it can be compared with the EAC of the challengers.

$$MC_t = S_{t-1}(1 + i) - S_t + O\&M_t \tag{11.1}$$

As long as the MC_t is less than the best EAC for all challengers, the existing equipment should be kept. This is illustrated in Example 11.5. The same data is used as in Example 11.4, but it is applied to 1-year-old existing equipment.

The next section and Example 11.6 will explain why it is often wrong to minimize the EAC over the remaining possible lives of the existing equipment. The section after that and Example 11.7 explain why it is sometimes necessary to calculate the minimum EAC for the defender.

Example 11.5   Economic Life of a Defender

Assume that the machine analyzed in Example 11.4 has been purchased and that it is now 1 year old. Assume that the data has not changed. What is the economic life of the 1-year-old machine? Is this 1 year less than the economic life when new?

Solution
The methodology for finding the defender's economic life is different than for finding the challenger's. The marginal cost to extend service for another year is calculated rather than a series of EACs.

For each year, the machine's first cost is its salvage value at the year's beginning, which is the value at the end of the previous year. Thus, the first cost for year 1 is $7000. The $10,000 purchase price is a sunk cost. The machine's salvage value in each year is tabulated below, as are its maintenance costs.

Year	Salvage Value	O&M Costs
1	$5000	$ 750
2	3500	1500
3	2500	2250
4	2000	3000

Thus, the marginal cost to extend for the first year is

$$MC_1 = S_0(1 + i) - S_1 + O\&M_1$$
$$= 7000 \cdot 1.08 - 5000 + 750 = 3310$$

The values for other years are included in the following summary table:

Asset's Age	MC of S_{t-1}	MC of S_t	MC O&M	Total
1	$7560	$-5000	$ 750	$3310
2	5400	-3500	1500	3400
3	3780	-2500	2250	3530
4	2700	-2000	3000	3700

The EAC of the new challenger is $3514 for 3 years. The marginal cost to extend the old asset's service for another year is less than $3514 for a 1-year-old asset and for a 2-year-old asset. Thus, the 1-year-old defender's life should be extended for 2 more years for a total life of 3 years (as in Example 11.4).

. The EAC for 3 years of $3514 is equivalent at the 8% interest rate to the marginal costs to extend service from 0 to 1 year (same as the 1-year EAC, calculated to be $3800 in Example 11.4), from 1 to 2 years ($3310), and from 2 to 3 years ($3400). Stated using engineering economy factors, this is:

$$EAC = [3800(P/F, 8\%, 1) + 3310(P/F, 8\%, 2) + 3400(P/F, 8\%, 3)](A/P, 8\%, 3)$$
$$= [3800 \cdot .9259 + 3310 \cdot .8573 + 3400 \cdot .7938] \cdot .3880$$
$$= \$3513.3$$

The cost to extend service for a 3-year-old asset is only slightly higher at $3530. This slight increase is why the EAC for the new asset's 4-year life is only $3 higher than the EAC for a 3-year life.

Incorrect Assumptions for Minimizing the Defender's EAC.   The EAC for each possible life of the existing asset should not be calculated, as two incorrect assumptions are required. These incorrect assumptions are (1) that used assets are available for purchase, and (2) that the price of the used assets matches the salvage values assumed for disposal of the existing assets. The correct assumptions arc as follows:

1. Used equipment is often unavailable.

2. When available, installation and/or removal costs ensure that used equipment to be purchased has a different value than used equipment to be sold or disposed of.

Example 11.6 illustrates the difficulties in trying to calculate an EAC for each possible life of the existing asset. Using the data correctly analyzed in Example 11.5 to incorrectly calculate EACs creates the new, unrealistic alternative of buying a 1-year-old asset for $7000, installing it for nothing, using it for 1 year, and selling it for $5000.

Nearly new used equipment is rarely available with exactly the right options, if it is available at all. A firm rarely buys a machine tool or other equipment and sells it a year or two later. If it does, it is likely the firm is closing down a line of business, going bankrupt, or disposing of unreliable equipment. Exceptions, such as 1-year-old used cars, often come from leasing firms, where very heavy usage and wear are common.

When used equipment is available, it is better to consider it as an explicit challenger to replace the existing equipment. This calculation will automatically include the correct purchase price, installation cost, O&M costs, and salvage values.

Example 11.6   Incorrectly Minimizing EAC to Find the Economic Life of a Defender

Assume that the machine analyzed in Example 11.4 has been purchased and that it is now 1 year old. Assume that the data has not changed and that the salvage values represent the cost to buy or to sell the equipment. What is the minimum EAC? Why does the life for this example not match the economic life of the 1-year-old machine calculated by the marginal cost to extend service shown in Example 11.5?

Solution

The machine's first cost is its salvage value at the end of year 1, or $7000. (The $10,000 purchase price is a sunk cost.) The firm's interest rate is 8%. Its salvage value and maintenance costs in each year were tabulated in Example 11.5.

The calculation of the EAC for the first cost and the salvage value is nearly identical to Example 11.4, and the maintenance term need only add an annual component. For the first cost and maintenance term only the number of years changes. For the salvage value column, both the salvage value and the number of years change.

$$EAC_{P_t} = 7000(A/P, 8\%, t) \qquad [\text{Note: } \$7000, \text{ not } \$10,000]$$

$$EAC_{O\&M_t} = 750(A/G, 8\%, t) + 750$$

$$EAC_{S_1} = -5000(A/F, 8\%, 1)$$

$$EAC_{S_2} = -3500(A/F, 8\%, 2)$$

$$EAC_{S_3} = -2500(A/F, 8\%, 3)$$

and so on.

Year	EAC of P	EAC of S	EAC O&M	Total
1	$7560	$-5000	$ 750	$3310
2	3925	-1683	1111	3353
3	2716	-770	1462	3408
4	2113	-444	1803	3472

According to this table, the minimum EAC corresponds to 1 additional year for a 1-year-old asset. This uses the same data as in Example 11.4, where a 3-year life was slightly less expensive than a 4-year life. Yet here, the erroneous conclusion is to keep the equipment for a total of 2 years.

Using this table requires two assumptions. The first assumption is that there is no installation cost. The second assumption is that a used asset can be purchased for $7000. Often this assumption is false.

If it is possible to buy a 1-year-old machine for $7000, to install it for nothing, and to sell it for $5000, then this is a new challenger. It has an EAC of $3310, which makes it the least-cost alternative. However, this is unlikely to be true, and even it were, it is not the cost for different lives of the current defender. Those costs, as shown in Example 11.5, are calculated using the marginal cost to extend its life by another year.

When It Is Necessary to Calculate the Defender's Best EAC. In most cases, the defender is nearing the end of its economic life, and the marginal cost to extend service for another year increases over time. However, sometimes the existing asset is new enough that the marginal cost to extend service for another year declines (at least for the first few years). In this case, replacement may be considered because of (1) altered requirements, (2) a new product that is a new challenger, or (3) a poor selection of the exisiting equipment.

The three possibilities and approaches are:

1. The cost to extend the defender's life for next year is less than the chal-
    lenger's EAC, so keep the defender for now, and use the MC_t compu-
    tations to determine when to replace. *This corresponds to challenger's
    EACs in region 1 of Exhibit 11.4.*

Exhibit 11.4   When to calculate the defender's best EAC

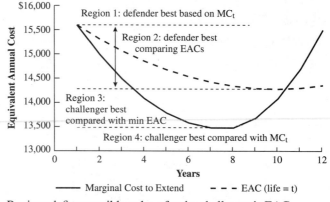

Regions define possible values for the challenger's EAC.

2. The cost to extend the defender's life for next year is more than the chal-
    lenger's EAC; however, the marginal cost of extending service for the
    defender falls below the challenger's EAC over time. *Then compute the
    minimum EAC for the defender.* If the defender's minimum EAC is less
    than the challenger's EAC, then replace when the marginal cost of ex-
    tending service goes from being less than the challenger's EAC to be-
    ing more. *This corresponds to challenger's EACs in region 2 of Exhibit
    11.4.* However, if the defender's minimum EAC exceeds the challenger's
    EAC, then replace immediately with the challenger. *This corresponds to
    challenger's EACs in region 3 of Exhibit 11.4.*

3. The minimum marginal cost to extend service for the defender is greater
    than the challenger's EAC, so replace the defender immediately with the
    challenger. *This corresponds to challenger's EACs in region 4 of Exhibit
    11.4.*

Cases 1 and 3 require the same computations as before—the marginal cost to
extend service. Only for case 2 is it necessary to calculate the defender's minimum
EAC. This case is illustrated in Example 11.7. There are even more complicated
cases not included here, such as equipment with a remaining economic life of 15
years with overhauls every 5 years. Or situations where replacement can be done
only every 3 years, during a scheduled plant shutdown. These cases (like Example
11.7) require computing the marginal cost to extend service and the EACs over
the potential life of the equipment.

Example 11.7   How Soon Should the Communications System Be Replaced?

The communications system for Conglomerate Engineering cost $50,000 four years ago. Its current salvage value is $26,000, which will decline as follows: $20,000, $16,250, $14,000, and $12,500. The O&M costs will be $6000 this year, and these costs will increase by $2000 per year. At Conglomerate's interest rate of 10%, the best challenger has an EAC of $14,200. When should the communications system be replaced?

Solution

$$MC_t = S_{t-1}(1.1) - S_t + O\&M_t$$

$$EAC_t = 26K(A/P, 10\%, t) - S_t(A/F, 10\%, t) + 6K + 2K(A/G, 10\%, t)$$

First cost $26,000		O&M $-6000	Gradient $-2000	i 10%	i 10%
Year	S_{t-1}	S_t	**O&M**	MC_t	**EAC**
0	$26,000		$0		
1	26,000	$20,000	$-6,000$	$14,600	$14,600
2	20,000	16,250	$-8,000$	13,750	14,195
3	16,250	14,000	$-10,000$	13,875	14,098
4	14,000	12,500	$-12,000$	14,900	14,271

   Since the defender's lowest EAC is less then $14,200, which is the challenger's EAC, the communications system should *not* be replaced yet. With current data the best life for the defender is 3 more years, since it is year 4 before the marginal cost to extend service goes from less than the challenger's EAC to more. This strategy saves $102 per year over the strategy of immediate replacement.
   Note that if the $EAC_{challenger}$ were more than $14,600, which is MC_1, no EAC calculations would be required to know that keeping the defender for at least a year would then be the best choice. Similarly, if the $EAC_{challenger}$ were less than $13,750, which is the minimum MC_t, no $EAC_{defender}$ calculations would be required to know that replacing the defender immediately would then be the best choice.

11.5 ANALOGOUS OPTIMAL CAPACITY PROBLEMS

The optimal or economic life of a challenger was chosen by calculating the EAC for each possible life. Exhibit 11.3 illustrated a common curve shape, where increasing O&M costs and decreasing salvage values implied that the curve would be concave.
   A similar method and result often apply to the design problem of selecting the best capacity from among mutually exclusive alternatives. The height of a dam, the number of freeway lanes, the size of a building, and insulation thickness are

examples in which increasing investment often has decreasing returns to scale. For example, each additional inch of added insulation saves less energy. As with the new challengers in a replacement analysis, minimizing the EAC is the correct approach.

Example 11.8 details this for selecting the insulation amount that minimizes first costs and heating and cooling expenses. More insulation costs more money, but it allows the downsizing of heating, ventilating, and air-conditioning (HVAC) equipment and reduces annual energy expenses. Often, filling structurally available space becomes almost automatic, but the question of whether 2×4, 2×6, or 2×8 framing should be used requires detailed analysis [Ruegg and Marshall].

Example 11.8   Optimal Insulation Thickness

The roof and ceiling of a commercial building will be insulated by blown-in insulation, which can be specified in 1-inch-depth increments. The minimum amount of insulation that will meet code is 6 inches, and the maximum amount that will fit is 12 inches. What is the optimal depth of this insulation?

The firm uses an interest rate of 8% and a 30-year building life. The installed cost of the insulation for the 10,000-square-foot ceiling is $1.75 per cubic foot. Thicker insulation will reduce the capacity of the HVAC equipment and the annual heating and cooling costs. With 6 inches of insulation, the annual bill for heat gains and losses through the roof will be $1200.

Insulation thickness	6″	7″	8″	9″	10″	11″	12″
HVAC first cost	$10,000	$9643	$9375	$9167	$9000	$8864	$8750

Solution

The easiest solution is to calculate the costs per inch of insulation for the 10,000-square-foot ceiling/roof. In the following equation the cost per cubic foot is stated as a cost per 12 in-ft^2 so the units balance.

$$\text{EAC}_{\text{Insulation}} = (\$1.75/12 \text{ inch-ft}^2) \cdot (A/P, 8\%, 30) \cdot 10,000 \text{ ft}^2$$

$$= \$129.54 \text{ per inch of thickness}$$

The EAC_{HVAC} are found by multiplying each first cost by $(A/P, 8\%, 30)$.

The annual heating and cooling costs are inversely proportional to the insulation thickness. Doubling the insulation thickness doubles the resistance to heat gain and loss, and halves the amount of energy lost due to conduction. Thus, for example, the annual heating and cooling cost for 7 inches of insulation is 6/7 of the heating and cooling bill for 6 inches of insulation, and for 8 inches of insulation it is 6/8 of the bill for 6 inches.

Thickness	Annual Insulation	Annual HVAC	Annual Energy Cost	EAC
6"	$ 777.24	$888.27	$1200.00	$2866
7"	906.78	856.55	1028.572792	
8"	1036.32	832.76	900.00	2769
9"	1165.86	814.25	800.00	2780
10"	1295.40	799.45	720.00	2815
11"	1424.94	787.33	654.55	2867
12"	1554.48	777.24	600.00	2932

The best insulation thickness is 8 inches, although 7 inches to 10 inches is a nearly constant cost.

11.6 ESTIMATING FUTURE CHALLENGERS

One advantage of keeping the current defender a little longer is the opportunity to obtain improved challengers. This is formalized in several replacement models, but the best example is perhaps the models of the Machinery and Allied Products Institute (MAPI). These models are outlined in this section.

Exhibit 11.5 contrasts the EAC of future challengers under two scenarios of technological progress: normal or incremental progress vs. rapid revolutionary progress. Rapid revolutionary progress implies rapid obsolescence and short life spans, as well as substantial uncertainty.

Exhibit 11.5   Incremental vs. revolutionary progress

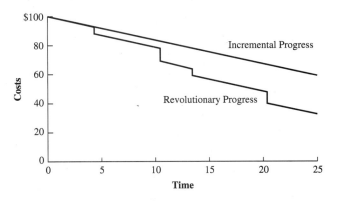

A Simple Rule of Thumb.   It is possible to arbitrarily shorten the challenger's assumed life to account for better future challengers. This recognizes that economic obsolescence shortens the economic life of otherwise functional equipment. For some examples, shortening the challenger's life by one-third seems to be about right.

For instance, a computer might have an economic life of 6 years—ignoring obsolescence. That cost could be approximated by shortening the economic life by one-third, to 4 years.

This approach should be used with caution because this rule of thumb will depend heavily on how significant obsolescence is and how rapidly it occurs.

MAPI.   The MAPI approach to replacement analysis was developed to serve many different practical situations [Terborgh]. It consists of sets of assumptions, which are linked with recommendations summarized in formulas, tables, and graphs. The different sets of assumptions were developed to correspond with the different practical situations that occur.

Thus, the approach is to choose the set of assumptions that most closely corresponds to the replacement problem under analysis. Then values for the key parameters or variables are identified. Finally, these values are substituted into the formulas, tables, and/or graphs.

The advantage of the MAPI approach is that a team of individuals has analyzed myriad situations and summarized the results. The disadvantage is that the detailed formulas can become a "black box" that inhibit understanding of the problem. The details are too voluminous to include here.

However, there is at least one concept that is quite useful in describing the advantage of waiting for better challengers. That concept is the inferiority gradient. The **inferiority gradient** equals the sum of the improvement in annual costs of next year's challenger over this year's challenger, plus the amount by which the annual costs for the defender will increase.

As shown in Exhibit 11.6, the annual cost for any machine (defender or challenger) is assumed to increase over time. Similarly, the best challenger available each year is assumed to be better over time. When the inferiority gradient is large, then economic lives will tend to be shorter. When the inferiority gradient is small, then economic lives will tend to be longer.

The **inferiority gradient** is the annual improvement in annual costs of new challengers, plus the annual increase in O&M costs for existing defenders.

Exhibit 11.6   MAPI inferiority gradient for challenger and defender series

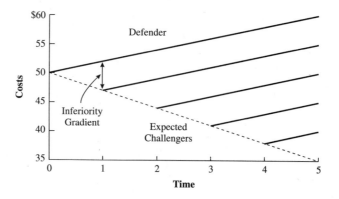

One of the important MAPI assumptions was characterizing the inferiority gradient. Was it a constant amount, a constant rate, or some other pattern?

11.7 REPLACEMENT AND REPAIR MODELS

Government agencies or private firms may operate hundreds or even thousands of vehicles; utility firms own thousands of poles; every large building has banks of fluorescent lighting; and every chemical processing plant has many pumps. These situations of fleets or pools of similar items emphasize replacement and, often, repair. Retirement (where the equipment is not replaced) is not considered, since the need for the service has a much longer life than the equipment—cities need garbage, phone, and electrical service for much longer than trucks or poles last.

Fleet operators and building and plant owners have policies that guide item replacement and repair decisions. For example, some city transit systems replace a bus at the first of (1) 500,000 miles, (2) 7 years, (3) when cumulative maintenance costs exceed the purchase price, or (4) when a major repair or overhaul is required on an older bus. Choosing the policies that will minimize EACs requires mathematical models of the replacement problem.

Classifying Replacement Models. These models are so numerous that it is useful to classify them. Luxhoj and Jones suggest that structure, realism, simplifying assumptions, and descriptive characteristics be used. A model's structure would include whether standby systems are available. Realism might focus on budget constraints on replacements and repair facility capacity constraints. A simplifying assumption would include ignoring taxes for a private-sector firm.

Key descriptive characteristics focus on problem details. For example, the MAPI model shown in Exhibit 11.6 focuses on including technological improvement. Other models have included the purchase of used items, preventive maintenance, and inflation. In some cases the intangible costs of employee morale must be included—everyone prefers to use newer vehicles and equipment and everyone is frustrated by the breakdown or poor performance of old machines. Similarly, there may be very large penalty costs for unplanned replacements. For example, the planned replacement of a motor on an assembly line might take 1 hour that could be scheduled when the line is not operating, but failure during operation will stop the line. It may still take only 1 hour to repair (but everyone is standing around for that hour), or it may take 48 hours until a part arrives.

Block Replacement. Light bulbs and fluorescent tubes in factories and buildings are often replaced on a block replacement schedule. In **block replacement,** all items are replaced, whether or not they are still functioning. Items that fail before the scheduled block replacement are sometimes individually replaced, and sometimes they are not. For example, lighting that can be replaced by an individual on a ladder might be replaced, while lighting that requires the erection of scaffolding might not be.

Block replacement is a policy of replacing all units—failed and functioning—at the same time.

The increasing rate of failures for older items that is shown in Exhibit 11.7 is the reason for block replacement. The curve of failure rates shown in Exhibit 11.7 is so common for equipment that it is the norm, and the pattern of high initial rates and high final rates is called the "bathtub curve."

Exhibit 11.7   The "bathtub curve" of reliability

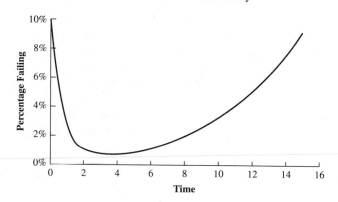

At some point it is more economical to replace an item than to wait for it to fail. Example 11.9 illustrates this.

Example 11.9   Block Replacement of Light Bulbs

Assume that a light bulb has an average life of 22,000 hours, where a fifth of the bulbs fail at 20, 21, 22, 23, and 24 thousand hours. The facility operates every hour of the year, and the bulbs are always on. Each bulb costs $3.50 to buy, $1.50 to install individually, and $.25 to install in bulk replacement. At least 60% of the bulbs must be operating for adequate light. The firm's interest rate is 10%. Compare bulk replacement at 21,000 hours with individual replacement.

Solution
If bulbs are replaced individually, then their average life is 22,000 hours. There are 8760 hours in a year, so the average life is 2.51 years.

$$(A/P, 10\%, 2.51) = (A/P, 10\%, 2) - .51[(A/P, 10\%, 2) - (A/P, 10\%, 3)]$$

$$= .5762 - .51(.5762 - .4021) = .4874$$

The first cost is $5 for purchase and installation:

$$\text{EAC}_{\text{indiv.}} = 5(A/P, 10\%, 2.51)$$

$$= 5 \cdot .4874 = \$2.44 \text{ per bulb}$$

For bulk replacement, the life is 21,000 hours after 40% of the bulbs have failed, or 21,000/8760 = 2.40 years.

$$(A/P, 10\%, 2.40) = (A/P, 10\%, 2) - .40[(A/P, 10\%, 2) - (A/P, 10\%, 3)]$$

$$= .5762 - .40 \cdot (.5762 - .4021) = .5066$$

The first cost is $3.75 for purchase and installation:

$$EAC_{bulk} = 3.75(A/P, 10\%, 2.40)$$

$$= 3.75 \cdot .5066 = \$1.90 \text{ per bulb}$$

The (A/P) factor is higher for bulk replacement, since the life is shorter. Even so, the lower cost for installation ensures that it is much cheaper to replace all of the bulbs every 21,000 hours.

The initial years in Exhibit 11.7 illustrate failures that may or may not be observed by the equipment owners. When there are significant initial failures, as with much electronic equipment, it is common for manufacturers to "burn in" the equipment by operating it for an appropriate time before it is shipped to the user. That way the initial failures occur in the factory, and the defective equipment is repaired before shipment. Users may also face a slightly higher initial failure rate because the conditions of use may not match the design assumptions.

The phenomenon of bulk replacement is obvious for light bulbs, but street repaving or utility distribution systems provide other examples. Potholes are patched continuously, but periodically the entire street is repaved. Similarly, downed power lines are repaired immediately, but periodically the entire group of poles and lines is replaced.

11.8 SUMMARY AND CONCLUSIONS

Replacement decisions are often the most common ones facing engineers in existing manufacturing plants, utilities, etc. The goal is to keep the facility operating as economically as possible.

The existing equipment can usually be used for at least one more year, and it is the defender in the economic comparison. The possible new equipment is the set of challengers. The defender and the challengers are mutually exclusive alternatives. Moreover, the defender and the challengers are intrinsically different; the defender is nearing the end of its economic life, while each challenger is at the beginning of its economic life.

Occasionally, an approaching termination of need will mean that the lives of the defender and the challenger(s) match, but the norm is a substantial difference. One consequence is that EAC or EAW is the measure of choice.

For the challenger(s), the economic life is found by minimizing the EAC over the possible lives. For the defender, the economic life is found by calculating the marginal cost to extend service for each year. As long as this is below the EAC for the best challenger, the defender should be kept.

The economic life for the defender is not found by minimizing the EAC over its possible remaining lives; this would require that used equipment be readily available and that there be no cost for installation or removal. If used equipment is available, then it should be analyzed as another potential challenger.

The techniques for calculating the optimal life of a challenger can be applied to calculating the optimal insulation thickness and other design choices. The key is a first cost that increases, but that ensures some level of annual savings in operating costs. What is the best tradeoff between first costs and annual costs?

At times, it is more economical to do a block replacement of functioning equipment than it is to individually replace items as they fail. For example, it is faster to replace all fluorescent tubes at once than to simply replace each as it fails. Early in the life, the most cost-effective pattern will be to replace individual failed items, but late in the life all bulbs will be replaced.

The pattern of improvement in the challenger is the key part of the MAPI models (described briefly in Section 11.6). The inferiority gradient combines improvement in the challenger with annual increases in the cost of the defender.

REFERENCES

Foster, Richard N., *Innovation: The Attacker's Advantage,* 1986, Summit Books.

Luxhoj, James T., and Marilyn S. Jones, "A Framework for Replacement Modeling Assumptions," *The Engineering Economist,* Volume 32, Number 1, Fall 1986, pp. 39–49.

Ruegg, Rosalie T., and Harold E. Marshall, *Building Economics: Theory and Practice,* 1990, Van Nostrand Reinhold.

Terborgh, George, *Business Investment Management,* 1967, Machinery and Allied Products Institute.

PROBLEMS

11.1   Describe a problem in replacement analysis where the replacement was being considered due to reduced performance of the existing equipment.

11.2   Describe a problem in replacement analysis where the replacement was being considered due to altered requirements.

11.3   Describe a problem in replacement analysis where the replacement was being considered due to obsolescence of the existing equipment.

11.4   Describe a problem in replacement analysis where the replacement was being considered due to the risk of catastrophic failure or unplanned replacement of the existing equipment.

11.5   For the situations you described in Problems 11.1–11.4, define the economic lives, physical lives, ownership lives, and service periods for the alternatives that you identified.

11.6 For the situations you described in Problems 11.1–11.4, describe any economic sunk costs that had to be ignored. How did any psychological, personal, or organizational sunk costs affect the decision-making process?

11.7 For the situations you described in Problems 11.1–11.4, describe the risks associated with the existing equipment and with the prospective new equipment.

11.8 Describe a replacement problem from your personal experience where the question of cost saving vs. profit making was a key issue.

11.9 A pulpwood-forming machine was purchased and installed 8 years ago for $45,000. The declared salvage value was $5000, with a useful life of 10 years. The machine can be replaced with a more efficient model that costs $75,000, including installation. The present machine can be sold on the open market for $14,000. The cost to remove the old machine is $2000. Which are the relevant costs for the old machine?

11.10 E&J Fine Wines recently purchased a new grape press for $120,000. The annual operating and maintenance costs for the press are estimated to be $5000 the first year. These costs are expected to increase by $2000 each year after the first. The salvage value is expected to decrease by $20,000 each year to a value of zero. Using an interest rate of 8%, determine the economic life of the press. (*Answer:* 1 year)

11.11 A particular chemical process deposits scale on the inside of pipes. The scale cannot be removed, but increasing the pumping pressure can maintain the flow through the narrower diameter. The pipe costs $20 per foot to install, and it has no salvage value when it is removed. The pumping costs are $8 per foot of pipe initially, and they increase annually by $5 per year starting in year 2. What is the economic life of the pipe if the interest rate is 12%?

11.12 An electric oil pump's first cost is $45,000, and the interest rate is 10%. The pump's end-of-year salvage values over the next 5 years are $42K, $40K, $38K, $32K, and $26K. Determine the pump's economic life. (*Answer:* 3 years)

11.13 A $20,000 machine will be purchased by a company whose interest rate is 10%. It will cost $5000 to install, but its removal costs are insignificant. What is its economic life if its salvage values and O&M costs are as follows?

Year	1	2	3	4	5
S	$16K	$13K	$11K	$10K	$9.5K
O&M	$5K	$8K	$11K	$14K	$17K

(*Answer:* 3 years)

11.14   A $40,000 machine will be purchased by a company whose interest rate is 12%. The installation cost is $5K, and removal costs are insignificant. What is its economic life if its salvage values and O&M costs are as follows?

Year	1	2	3	4	5
S	$35K	$30K	$25K	$20K	$15K
O&M	$8K	$14K	$20K	$26K	$32K

11.15   A machine that has been used for 1 year has a salvage value of $10,000 now, which will drop by $2000 per year. The maintenance costs for the next 4 years are $1250, $1450, $1750, and $2250. Determine the marginal cost to extend service for each of the next 4 years if the MARR is 8%.   (*Answer:* $MC_1 = \$4050$)

11.16   A drill press was purchased 2 years ago for $40,000. The press can be sold for $15,000 today, or for $12,000, $10,000, $8000, $6000, $4000, or $2000 at the ends of each of the next 6 years. The annual operating and maintenance cost for the next 6 years will be $2700, $2900, $3300, $3700, $4200, and $4700. Determine the marginal cost to extend service for each of the next 6 years if the MARR is 12%. If a new drill press has an EAC of $7000, when should the drill press be replaced?

11.17   Eight years ago, the ABC Block Company installed an automated conveyor system for $38,000. When the conveyor is replaced, the net cost of removal will be $2500. The minimum EAC of a new conveyor is $5500. When should the conveyor be replaced if ABC's MARR is 12%? The O&M costs for the next 5 years are $5K, $6K, $7K, $8K, and $9K.

11.18   The machine in Problem 11.13 has been purchased and it is now 1 year old. Calculate the marginal cost to extend service for this year and each of the next 3 years. The best challenger has an EAC that is 10% higher than the machine in Problem 11.13 (due to a change in the value of the yen). What is the economic life of the existing machine?

11.19   The machine in Problem 11.14 has been purchased and it is now 1 year old. Calculate the marginal cost to extend service for this year and each of the next 3 years. New challengers are not significant improvements over the machine in Problem 11.14, and they cost the same. What is the economic life of the existing machine?

11.20   For the machine in Problem 11.19, calculate the EACs for remaining lives of 1, 2, 3, and 4 years. Calculate the EACs for a used machine that is purchased for $35K and installed at a cost of $5K. Which of these is theoretically wrong? Is either theoretically correct? Why?

11.21   Green Cab Taxi Company owns several taxis that were purchased for $25,000 each 4 years ago. The cabs' current market value is $20,000 each, and if they are kept for another 6 years they can be sold for $2000 per cab. The annual maintenance costs for the cabs are $1000 per year. Green Cab has been approached about a leasing plan that would replace the cabs. The leasing plan calls for payments of $4500 per year. The annual maintenance costs for the leased cabs are $750 per year. Should the cabs be replaced if the interest rate is 10%?

11.22   How much does the answer for Example 11.8 change if energy costs double?   (*Answer:* Optimal thickness is 11 inches)

11.23   What is the optimal height for a flood-control dam? The government agency uses an interest rate of 5% and a horizon of 50 years. Assume that the dam has no salvage value after 50 years.

Height	30 m	35 m	40 m	45 m	50 m
First Cost	$10M	$12M	$14M	$16M	$18M
Annual Benefit	$1M	$1.1M	$1.15M	$1.17M	$1.18M

11.24   For Example 11.9, what is the EAC of bulk replacement if 60% of the bulbs can fail before the replacement?   (*Answer:* $1.83/bulb)

11.25   The lights used at the local baseball stadium cost $125 per bulb. Of the 120 bulbs at the stadium, 20 will fail after 2400 hours, 80 will fail after 2800 hours, and the remaining bulbs will fail after 3600 hours. The labor cost to replace individual bulbs is $150 per bulb. If the bulbs are replaced in bulk, the labor cost is $85 per bulb. The stadium hosts 120 games and events each year. The average time the lights are in use for each event is 5 hours. Compare the cost of individual replacement with bulk replacement at 2400 hours using $i = 12\%$.   (*Answer:* $EAC_{individual} = \$79.39$ and $EAC_{bulk} = \$69.13$)

11.26   Anytown's Street Department repaves a street when the annual cost to patch potholes exceeds $60,000 per mile (every 8 years). Potholes cost $10,000 per mile beginning at the end of year 3 after construction or repaving. The cost generally increases by $10,000 each year. Repaving costs are $120,000 per mile. Anytown uses an interest rate of 6%. What is the EAC for Anytown's policy? What is the EAC for the optimal policy? What is the optimal policy?

PART FOUR ENHANCEMENTS FOR THE REAL WORLD

Real-world engineering economy problems must account for taxes in private-sector firms, for several complications in public-sector agencies, and for inflation in both the private and public sectors. Two steps are required to compute income taxes. First, the capital asset must be depreciated or "written off" (Chapter 12), and second, the taxable income and taxes must be calculated (Chapter 13).

Public-sector engineering economy problems are often more difficult to solve than private-sector problems. As shown in Chapter 14, many benefits are difficult to quantify in dollar terms, and costs or user fees may have to be allocated among competing stakeholders.

Geometric gradients were introduced in Chapter 4, but one important application—inflation—has been reserved for Chapter 15. Virtually every engineering economy problem is theoretically affected by inflation, but for many problems all cash flows are inflating at about the same rate and inflation can be ignored.

CHAPTER 12 DEPRECIATION

The Situation and the Solution

An asset's first cost or value does not disappear in the first year; rather, it is used up over the asset's life. This "using up" is a cost to the firm that is recognized through depreciation, which reduces the firm's taxable income.

The solution is to apply the depreciation techniques that are permitted by the Internal Revenue Service (IRS).

Chapter Objectives

After you have read and studied the sections of this chapter, you should be able to:

Section 12.1
Define the key variables in depreciation.

Section 12.2
Calculate straight-line, declining balance, and sum-of-the-years'-digits depreciation.

Section 12.3
Calculate depreciation under the current tax law, the Modified Accelerated Cost Recovery System (MACRS).

Section 12.4
Define and calculate loss on sale and recaptured depreciation.

Section 12.5
Choose an optimal depreciation strategy.

Section 12.6
Calculate the PW of a depreciation schedule.

Section 12.7
Calculate depletion deductions for natural resource deposits.

Section 12.8
Calculate the Section 179 deduction.

Section 12.9
Calculate depreciation using spreadsheet functions.

Key Words and Concepts

Depreciation   Accounts for the using up of a productive asset.

Depletion   Accounts for the using up of a natural resource.

Basis   For depreciation, the **first cost (FC)**, which is the purchase price plus installation expenses.

Recovery period (N_D)   The life for computing depreciation. It is the item's useful life for straight-line, declining balance, and sum-of-the-years' digits depreciation. The IRS defines it for MACRS.

Salvage value (S)   Part of the formulas for straight-line and sum-of-the-years' digits depreciation, and it is considered for declining balance depreciation. Salvage value is assumed to equal 0 for MACRS.

Book value (BV_t)   An asset's value for accounting and tax purposes. The undepreciated book value at the end of year t equals the initial basis minus the accumulated depreciation.

Depreciation (D_t)   The deduction or charge for year t. This reduces taxable income, but it is not a cash flow.

Straight-line depreciation   Has equal annual depreciation.

Declining balance depreciation   Computes depreciation as a fraction of current book value—which declines over an asset's life.

Sum-of-the-years'-digits (SOYD) depreciation   An accelerated depreciation technique that includes an arithmetic gradient.

Modified Accelerated Cost Recovery System (MACRS) and the earlier **ACRS** Tax-code-specified combinations of declining balance and straight-line depreciation with recovery periods shorter than expected economic lives.

Recaptured depreciation   An adjustment to income when an asset is disposed of for more than its *current* book value.

Capital gain   An adjustment to income when an asset is disposed of for more than its *initial* book value.

Loss on sale   An adjustment to income when an asset is disposed of for less than its *current* book value.

12.1 INTRODUCTION

Depreciation accounts for the using up of a productive asset and depletion provides for using up a natural resource.

Depreciation has many meanings that are linked to: physical life, amortization, market value, value to owner, appraisals, impaired usefulness, and legal limits for taxes. These are usually related, but there are significant differences. This chapter emphasizes the last meaning, so that income taxes can be computed in Chapter 13. These depreciation values are also used to compute book values, which are the basis for property taxes.

This presentation focuses on the current U.S. tax code, but it also includes other methods since the tax codes of all nations are periodically revised. In addition to depreciation of physical assets, the **depletion** of natural resource assets is also covered.

The basic depreciation methods that the Internal Revenue Code has permitted are straight-line, declining balance, and sum-of-the-years' digits. Straight-line and declining balance are combined in the cost recovery systems of recent tax code revisions. When given a choice, firms depreciate assets as rapidly as possible (the income tax deductions occur sooner). All methods allow the same total depreciation—with different timings.

Because of income taxes, depreciation and depletion must be considered in engineering economy—even though they are *not* cash flows. The tax savings generated through depreciation in the U.S. exceeds twice the retained earnings for many U.S. firms [Bussey and Eschenbach].

It must be clearly understood that depreciation only matters for after-tax analyses. The accounting visualization of an asset's cost is a fictitious picture of reality. When equipment is purchased from a supplier, the supplier is paid at time 0. The depreciation charges do not correspond to when the supplier is paid; rather, they are only useful in computing income and property taxes.

Depreciation applies to what were defined as capital costs in Chapter 1. The distinctions between capital and operating costs are summarized in Exhibit 12.1. The large capital cost for a machine or a building is not an expense for income taxes. Instead, it is depreciated over its life, and those depreciation charges are deducted from taxable income. In contrast, the expenses to run the machine and

Exhibit 12.1   Capital vs. operating costs

Capital Cost	**Operating Costs and Income**
First cost Salvage	Operations and maintenance Revenue Advertising, labor, etc.
Depreciation	Expensing
Buy an asset ≡ exchanging money for another asset ⇒ no change in income & profit	Buy an item ≡ spending money ⇒ changes income and profit
Life ≥ 3 years & large costs	Short life or small costs
Depreciation charge is the annual expense that links these.	

to heat the building are operating costs, which are deducted from taxable income in the year they occur.

Definitions.   Depreciation is based on first costs and salvage value—that is, on the capital recovery for an asset. The detailed IRS definitions must account for many special cases, but the following simplified definitions are adequate here. Another simplification is to focus on annual depreciation, rather than including the details for quarterly computations. Exhibit 12.2 depicts these quantities for straight-line depreciation. These definitions are clarified in the examples of Section 12.2.

Exhibit 12.2   Depreciation definitions for straight-line depreciation

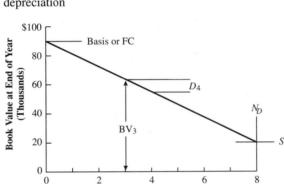

Basis. The basis for depreciation equals first cost (FC) or the purchase price plus installation expenses. (Note that casualty losses or capital improvements at a later date change the basis.)

Recovery Period (N_D). The life for computing depreciation is the recovery period. The recovery period (1) equals an item's useful life for some depreciation methods, and (2) is specified by the IRS for other methods.

Salvage Value (S). Salvage value is part of the formulas for the straight-line and sum-of-the-years' digits methods, and it is considered for the declining balance method. Salvage value is assumed to equal 0 for MACRS and ACRS.

Book Value (BV_t). The book value at the end of year t equals the initial basis minus the accumulated depreciation. This is an asset's value for accounting and tax purposes.

Depreciation (D_t). The depreciation deduction or charge for year t. This reduces taxable income, but it is not a cash flow.

12.2 BASIC DEPRECIATION METHODS

The allowable depreciation methods change with each revision of the tax code. For example, until 1981, straight-line, declining balance, and sum-of-the-years'-digits methods were used for calculating depreciation—by themselves and with switching between them. In 1981, the Accelerated Cost Recovery System (ACRS) was adopted. The Tax Reform Act of 1986 adopted the Modified Accelerated Cost Recovery System (MACRS). The MACRS schedules combine declining balance and straight-line methods.

Unfortunately, knowing the current tax code is not enough, for these reasons:

1. Straight-line and declining balance methods must be understood to provide a solid foundation for the MACRS schedules.

2. Retirement or replacement of an aging asset has tax consequences. And this depreciation schedule is determined by the allowable depreciation strategy chosen at the time the asset was purchased.

3. The tax code will be revised again.

4. Many firms operate under other national jurisdictions, whose tax codes will not match that of the U.S., but whose allowable depreciation methods will be some variation of these methods [Fleischer & Leung and Remer & Song].

Three depreciation techniques underlie most tax codes or are useful in conducting engineering economy studies. These are straight-line, sum-of-the-years'-digits (SOYD), and declining balance. Since $N_D = N$ for all of these techniques, the subscript is dropped.

Straight-line depreciation has annual depreciation amounts that are equal for each year.

Straight-Line.   This is the simplest and least attractive (from a tax minimizing perspective) depreciation strategy. Before 1986, it could be used on any depreciable property and was the only method that could be used on intangible property. The rate of depreciation is constant at $1/N$ per year. The amount of depreciation is constant as well.

As shown in Exhibit 12.3, the book value over time is a straight line. This straight line is the basis for defining accelerated depreciation, which has a book value that falls at a faster rate, at least initially. Straight-line depreciation has historically had the broadest applicability, and it is the standard against which the more attractive accelerated methods are measured.

The deduction amount is also constant over the life, and it is shown in Equation 12.1:

$$D_t = (FC - S)/N \tag{12.1}$$

Applying this formula to Example 12.1 leads to Exhibit 12.3.

Example 12.1   Straight-Line Method for Computer Workstation Network

Smith, Jones, and Rodriguez is an engineering design firm considering the purchase and installation of a network of computer workstations for $90,000. The expected life of the system is 8 years, when it will have a salvage value of $18,000. Using straight-line depreciation, calculate the annual depreciation and graph the book value over the 8 years.

Solution
Applying Equation 12.1 leads to:

$$D_t = (FC - S)/N = (90K - 18K)/8 = \$9000 \text{ per year}$$

As shown in Exhibit 12.3, the book value begins at $90,000 at time 0. It then falls by $9000 per year until the book value of the computer system is $18,000 at the end of year 8. This is graphed as a straight line, but the book values are only measured at time 0 and at the end of each year. They are $90K, $81K, $72K, ... , $18K.

Exhibit 12.3   Book value for straight-line depreciation

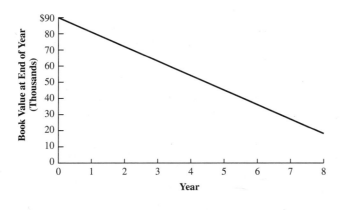

Declining Balance.   Before 1981, declining balance was the most commonly used depreciation technique, and it is the foundation for the current accelerated cost recovery system (MACRS). For the **declining balance** method, each year's depreciation is computed by multiplying the beginning-of-year book value by the depreciation rate. The maximum rate was 200%/N for new property, with lower rates of 150%/N or 175%/N for used property or special cases.

Declining balance depreciation, D_t, and the corresponding book values, B_t, are calculated using Equation 12.2, where α is the percentage used for finding the rate.

Declining balance depreciation computes the annual depreciation amount as a constant fraction of the remaining book value. For example, double-declining balance uses 200%/N.

$$D_1 = FC \cdot \alpha/N \qquad\qquad BV_1 = FC - D_1$$
$$D_2 = BV_1 \cdot \alpha/N \qquad\qquad BV_2 = BV_1 - D_2$$
$$\vdots \qquad\qquad\qquad\qquad \vdots$$
$$D_t = BV_{t-1} \cdot \alpha/N \qquad\qquad BV_t = BV_{t-1} - D_t \qquad\qquad (12.2)$$

or

$$D_t = BV_{t-1} \cdot \alpha/N \qquad\qquad BV_t = FC(1 - \alpha/N)^t \qquad\qquad (12.2')$$

In Example 12.2, Exhibit 12.4 shows the results of applying this formula to Example 12.1's data. In this case, the salvage value provides a minimum value below which the book value cannot drop. This is why the depreciation in year 6 is reduced and why it is $0 for years 7 and 8 in Example 12.2.

In other cases, the salvage value is low enough that a pure declining balance approach would lead to a final book value greater than the salvage value. Double-declining balance over 8 years uses a rate of 200%/8, or 25% per year. $BV_8 = .75^8 \cdot FC$, or 10.0% of the initial basis. In Example 12.2, this means that any salvage value below $9000 would be below the final book value. The solution is to shift to straight-line depreciation. This shift is detailed in the MACRS section.

Exhibit 12.5 graphs the book values for double-declining balance, sum-of-the-years'-digits, and straight-line depreciation applied to the data of Example 12.1. This exhibit shows why double-declining balance was usually the preferred technique—it depreciated assets the fastest and saved on taxes the soonest.

Example 12.2   Declining Balance Method for Computer Workstation Network

As stated in Example 12.1, the network of computer workstations has a first cost of $90,000. The expected life of the system is 8 years, when it will have a salvage value of $18,000. Calculate the annual depreciation and tabulate the book value over the 8 years, assuming double-declining balance depreciation.

Solution
Since double-declining balance and a life of 8 years were specified, the depreciation rate is 200%/8, or 25% per year. Applying Equation 12.2 leads to:

$$D_1 = FC \cdot \alpha/N = 90K \cdot .25 = \$22,500$$
$$BV_1 = FC - D_1 = 90,000 - 22,500 = \$67,500$$
$$D_2 = BV_1 \cdot \alpha/N = 67,500 \cdot .25 = \$16,875$$
$$BV_2 = BV_1 - D_2 = 67,500 - 16,875 = \$50,625$$

and so on.

These calculations are summarized in Exhibit 12.4, for the 8-year life of the network. In year 6, the depreciation is limited because the book value cannot drop below the salvage value. The depreciation in years 7 and 8 is $0 for the same reason.

Exhibit 12.4 Calculations for declining balance depreciation

Year	Depreciation (D_t)	Book Value (BV_t)
0		$90,000
1	$22,500	67,500
2	16,875	50,625
3	12,656	37,969
4	9492	28,477
5	7119	21,358
6	3358	18,000
7	0	18,000
8	0	18,000

Exhibit 12.5 Book value—straight-line, declining balance, SOYD

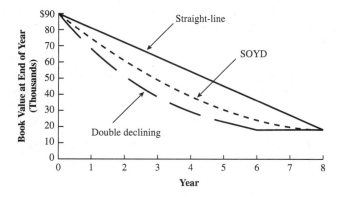

Sum-of-the-years'-digits (SOYD).   This depreciation technique uses fractions whose denominator equals the sum of the digits from 1 to N or **the sum-of-the-years' digits (SOYD)**. For $N = 5$, then, the denominator equals $1+2+3+4+5 = 15$. The numerators for the fractions begin at N for the first year and decline by 1 each year. With $N = 5$, the fractions become 5/15, 4/15, 3/15, 2/15, and 1/15. These fractions are multiplied by the total amount to be depreciated, which is the initial basis or first cost minus the salvage value (FC − S).

Sum-of-the-years'-digits (SOYD) is an accelerated depreciation technique that includes an arithmetic gradient.

Each year the fraction and the depreciation charge drop by a constant amount. For the 5-year example here, the fraction drops by 1/15 each year. The resulting gradient is 1/SOYD times the total to be depreciated. Thus, SOYD is the easiest accelerated depreciation technique to incorporate into economic analysis. An arithmetic gradient factor can be used, rather than requiring the development of a spreadsheet-like table. SOYD can even be used to approximate MACRS or double-declining balance (when the salvage value is lower, the approximation tends to be better).

In general, with N years' useful life, the denominator equals $N(N + 1)/2$, or SOYD, and the numerator equals $N - t + 1$. Equation 12.3 describes two forms of the formula for calculating depreciation.

$$D_t = (FC - S) \cdot 2(N - t + 1)/[N(N + 1)]$$
$$D_t = (FC - S)(N - t + 1)/SOYD \tag{12.3}$$

The book value and depreciation at year t can easily be found by building a table for years 1 to t, or the book values can be calculated using Equation 12.4 (see Example 12.3). The book value can also be found by adding the salvage value and the remaining depreciation, which uses the numerators of the remaining fractions or $j = 1$ to $N - t$, which is $(N - t)(N - t + 1)/2$. The result is Equation 12.4′.

$$BV_t = BV_{t-1} - (FC - S)2(N - t + 1)/[N(N + 1)] \tag{12.4}$$

or

$$BV_t = S + (FC - S)(N - t)(N - t + 1)/[2 \cdot SOYD] \tag{12.4′}$$

Example 12.3   SOYD Method for Computer Workstation Network

As stated in Example 12.1, the network of computer workstations has a first cost of $90,000. The expected life of the system is 8 years, when it will have a salvage value of $18,000. Calculate the annual SOYD depreciation and book value for the 8 years.

Solution

Since SOYD depreciation and a life of 8 years were specified, the first step is to calculate the SOYD, which equals $8 \cdot 9/2$, or 36. Applying Equations 12.3 and 12.4 leads to:

$$D_1 = (FC - S) \cdot 8/36 = (90K - 18K) \cdot 8/36 = \$16,000$$
$$BV_1 = FC - D_1 = 90,000 - 16,000 = \$74,000$$
$$D_2 = (FC - S) \cdot 7/36 = (90K - 18K) \cdot 7/36 = \$14,000$$
$$BV_2 = BV_1 - D_2 = 74,000 - 14,000 = \$60,000$$

and so on.

These calculations are summarized in Exhibit 12.6 for the network's 8-year life. (The book values were graphed in Exhibit 12.5.)

Exhibit 12.6   Calculations for SOYD depreciation

Year	Fraction	Depreciation (D_t)	Book Value (BV_t)
0			$90,000
1	8/36	$16,000	74,000
2	7/36	14,000	60,000
3	6/36	12,000	48,000
4	5/36	10,000	38,000
5	4/36	8000	30,000
6	3/36	6000	24,000
7	2/36	4000	20,000
8	1/36	2000	18,000

Units of Production.   The units-of-production technique is designed for equipment whose life may best be measured by cumulative volume or operating hours, and whose use in a given year is unpredictable. Then the depreciation charge for each year is automatically adjusted for the actual level of use.

For example, this approach could be used for heavy construction equipment whose life may best be measured by operating hours. Equipment with an expected life of 10,000 hours and usage of 1100 hours this year would generate an 11% depreciation charge on its depreciable cost of its first cost minus its salvage value (FC − S). Once the book value equals the salvage value, no more depreciation would be charged.

12.3 ACCELERATED COST RECOVERY

Accelerated Cost Recovery System (ACRS).   The Economic Recovery Tax Act of 1981 reduced the earlier asset depreciation ranges (ADRs) used to define depreciable lives to only four classes for personal property. For example, 5-year property (ADR life between 5 and 18 years) was the default for personal, most production, and most public-utility property. Since ACRS was superseded by MACRS in 1986, the four ACRS classes and their depreciation percentages are not detailed here.

One important change was that ACRS eliminated estimated salvage values from calculations of depreciation. Salvage values are assumed to be $0, and any salvage value that is received is treated as a gain on sale or recaptured depreciation (see Section 12.4).

The ACRS depreciation percentages[1] are analogous to those of the MACRS approach (detailed in the next subsection). If you are analyzing the early replacement of an asset purchased before 1986, the exact ACRS percentages can be obtained from the IRS or the firm's accounting department.

The differences between ACRS and MACRS can be summarized as follows. MACRS added 7- and 20-year recovery periods. The write-off period was lengthened for some assets—for example, most vehicles were shifted from the 3- to the 5-year class and office furniture was shifted from the 5-year to the new 7-year class. The declining balance rate was increased from 150% to 200% for recovery periods of 10 years or less. Another half-year (a year at half rate) was added to the end of each recovery period for matching assumptions of half-year depreciation in the first and last years. Like ACRS, MACRS ignores estimated salvage values.

In examining the accelerated cost recovery systems, it is worth noting that they were established under the same restraints that govern all tax codes. That is, the percentages were political compromises between a desire for revenue generation and for increasing incentives for capital investment.

---

[1] ACRS used 150% declining balance with a switch to straight-line and a half-year rate for the first year. The optimal switching point was chosen for 3-, 10-, and 15-year property.

For 5-year property, the declining balance rate was 150%/5 years or 30% per year. For the first year, the rate was halved to 15%. Then the switch to straight-line occurred, with the remaining 85% split into 4 years. All rates were rounded to integer percentages, so they became 15%, 22%, 21%, 21%, 21%. For easier use of (P/A) factors, the last 4 years might be rounded to 21.25%.

Modified Accelerated Cost Recovery System (MACRS).   The current depreciation system, MACRS (pronounced "makers"), was established by the 1986 Tax Reform Act. The derivation of the tables of percentages and the descriptions of the underlying depreciation assumptions may seem a little complicated. However, the system is very easy to use, since there are only three steps to finding D_t. In fact, you will probably want to tag, paper clip, or otherwise mark Exhibits 12.7 and 12.9, since they are needed so often.

1.   Determine the recovery period from Exhibit 12.7.
2.   Find the percentage for that recovery period for year t in Exhibit 12.9.
3.   Multiply the first cost or basis by the percentage.

Exhibit 12.7 is based on *IRS Publication 534: Depreciation,* which provides far more detail. If identified by name, the asset's life is clear. For example, computers and R&D equipment have a 5-year recovery period, while office furniture and most manufacturing equipment are in the 7-year category. If the name is not listed but the life is well established, the ADR ranges can be used. Finally, the 7-year recovery period is the default for items not otherwise classified.

Exhibit 12.7   Recovery periods for MACRS

Recovery Period	Description of Assets Included
3-Year	Tractors for over-the-road tractor/trailer use and special tools such as dies and jigs; ADR < 4 years
5-Year	Cars, buses, trucks, computers, office machinery, construction equipment, and R&D equipment; 4 years ≤ ADR < 10 years
7-Year	Office furniture, most manufacturing equipment, mining equipment, and items not otherwise classified; 10 years ≤ ADR < 16 years
10-Year	Marine vessels, petroleum refining equipment, single-purpose agricultural structures, trees and vines that bear nuts or fruits; 16 years ≤ ADR < 20 years
15-Year	Roads, shrubbery, wharves, steam and electric generation and distribution systems, and municipal wastewater treatment facilities; 20 years ≤ ADR < 25 years
20-Year	Farm buildings and municipal sewers; ADR ≥ 25 years
27.5-Year	Residential rental property
31.5-Year	Nonresidential real property purchased on or before 5/12/93
39-Year	Nonresidential real property purchased on or after 5/13/93

Note: This table is used to find the recovery period for *all* items under MACRS.

There are many detailed exceptions listed in *IRS Publication 534.* For example, assets for drilling oil and gas wells are in the 5-year class, assets for oil and gas exploration are in the 7-year class, and petroleum refineries are in the 10-year class (chemical plants are in the 7-year class). For the purpose of learning engineering economy, Exhibit 12.7 is detailed enough. However, the current IRS regulations and the rules for different assets are best established by the firm's accounting department.

MACRS assumes a salvage value of 0, and it is based on declining balance combined with straight-line, which was the fastest technique that was permitted before ACRS. The major difference between MACRS and declining balance is that the switch to straight-line depreciation is built into the table and is not optional to the taxpayer. That previous flexibility was needed to deal with nonzero salvage values, which are now assumed to be 0.

Modified Accelerated Cost Recovery System (MACRS) assumes that $S = 0$. It combines declining balance and straight-line depreciation with recovery periods that are shorter than expected economic lives.

Exhibit 12.8 derives the percentage each year's depreciation is of the initial book value or basis for the 5-year class. Since the 5-year recovery period MACRS percentages are based on the double-declining balance method, the depreciation rate is 200%/5 or 40% of the book value. Since the purchase is assumed to happen halfway through the year (the half-year convention), this rate is halved to 20% for the first year. Each year, the declining balance rate is calculated as 40% of the book value at the beginning of the year. Then each tabulated percentage is stated as a percentage of the initial first cost. For example, in year 2 the depreciation is 40% of the remaining book value, which is 80% of the initial book value. So the depreciation is 40% of the 80% which equals 32% of the initial book value.

Exhibit 12.8 Example derivation of MACRS percentages for 5-year asset

Year	Beginning Book Value	Declining Balance Rate	Straight-line Computation	Straight-line Rate
1	100 %	20 %	100%/5 yrs/2	10 %
2	80	32	80%/4.5 yrs	17.8
3	48	19.2	48%/3.5 yrs	13.7
4	28.8	11.52	28.8%/2.5 yrs	11.52
5	17.28			11.52
6	5.76			5.76

As shown in Exhibits 12.8 and 12.9, there is an extra "half-year" beyond the 5-, 7-, 10-, 15-, or 20-year recovery period. This extra half-year is the "other half" of the year 1 deduction (reflecting the half-year convention). When calculating the switching point, the annual depreciation for the straight-line method is the remaining book value divided by the number of years remaining, plus a half-year. Thus, in year 2 of a 5-year asset, there are 4 years left plus a half-year for straight-line depreciation, and in year 4 there are 2.5 years left (year 4, year 5, and half of year 6). In Exhibit 12.8, the underlined values are the ones tabulated in Exhibit 12.9. These values represent the most rapid depreciation possible.

Exhibit 12.9 summarizes the annual MACRS percentages for the six classes of property with accelerated depreciation. These percentages are based on declining balance depreciation with a shift to the straight-line method (see Exhibit 12.8), so that the final book value equals $0. For 3, 5, 7, and 10 years, the declining balance rate is 200%; for 15 and 20 years, the rate is 150%. For real property (27.5 and 31.5 years), a straight-line method is used.

Exhibit 12.9   MACRS percentages

	Recovery Period					
Year	3-year	5-year	7-year	10-year	15-year	20-year
1	33.33%	20.00%	14.29%	10.00%	5.00%	3.750%
2	44.45	32.00	24.49	18.00	9.50	7.219
3	14.81	19.20	17.49	14.40	8.55	6.677
4	7.41	11.52	12.49	11.52	7.70	6.177
5		11.52	8.93	9.22	6.93	5.713
6		5.76	8.92	7.37	6.23	5.285
7			8.93	6.55	5.90	4.888
8			4.46	6.55	5.90	4.522
9				6.56	5.91	4.462
10				6.55	5.90	4.461
11				3.28	5.91	4.462
12					5.90	4.461
13					5.91	4.462
14					5.90	4.461
15					5.91	4.462
16					2.95	4.461
17						4.462
18						4.461
19						4.462
20						4.461
21						2.231

Exhibits 12.7 and 12.9 are the basis for the calculations shown in Example 12.4, as they will be for the majority of the problems and examples in Chapter 13. Exhibit 12.10 compares the book value over time for Examples 12.1–12.4. Notice that MACRS has the lowest book value after the first year or, equivalently, that its accumulated depreciation is the highest. Not only does MACRS use a shorter recovery period, but it also assumes a salvage value of $0. When the cash flow for the salvage value occurs in year 8, there will be recaptured depreciation for MACRS, so that all depreciation systems depreciate a net total of $72,000 for this example.

Exhibit 12.10 Book value vs. time for network of workstations

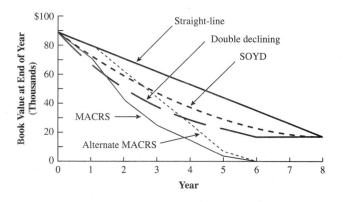

Example 12.4 MACRS for Computer Workstation Network

As stated in Example 12.1, the network of computer workstations has a first cost of $90,000. The expected life of the system is 8 years, when it will have a salvage value of $18,000. Calculate the annual depreciation and book value over the 8 years, assuming MACRS depreciation.

Solution
Since MACRS depreciation was specified, the first step is to identify the recovery period. From Exhibit 12.7, the recovery period for computers is 5 years. The percentages for years 1 to 6 can be found in either the example calculation (Exhibit 12.8) or in the MACRS table (Exhibit 12.9).

The calculations summarized in Exhibit 12.11 are all based on a percentage of the $90,000 first cost, which is also the initial book value or basis. Because the book value is depreciated to $0, when the computer network is sold in year 8 for $18,000, there will be $18,000 in recaptured depreciation counted as taxable income (see Section 12.4).

Exhibit 12.11 MACRS for workstation network

Year	MACRS (%)	Depreciation (D_t)	Book Value (BV_t)
0			$90,000
1	20 %	$18,000	72,000
2	32	28,800	43,200
3	19.2	17,280	25,920
4	11.52	10,368	15,552
5	11.52	10,368	5,184
6	5.76	5184	0
7	0	0	0
8	0	0	0

Alternate MACRS.   This approach is not detailed here; it is rarely used because it is economically unattractive. This approach is essentially the straight-line method over the same period as with MACRS (see Exhibit 12.10). The exception is that items without an identified category have a recovery period of 12 years. In addition, the first and last years are subject to the same half-year convention used in MACRS.

12.4 GAINS AND LOSSES ON SALES AND RECAPTURED DEPRECIATION

When depreciable assets are sold, given away, or disposed of, the current book value is compared with the net amount received. If the receipts are less than the book value, then there is a **loss on the sale**. On the other hand, if the book value is less than the amount received, then too much depreciation has been taken. The tax law requires that this **excess prior depreciation be recaptured**. To use the language of an earlier tax code, there is a gain on the sale. The resulting tax essentially equals the taxes saved in prior years by way of the excess depreciation.

When an asset is sold for more than its book value, the difference is **recaptured depreciation**. A **loss on sale** results when the asset is disposed of for less than its book value.

This recaptured depreciation or loss on sale allows an economic analysis to ignore any depreciation during the year of sale. The difference between the book value at the beginning of the year and the sale price is either added to taxable income for recaptured depreciation or subtracted for a loss on sale. If additional depreciation were taken during the year of sale, then there would be that much more recaptured depreciation or a matching reduction in the loss on sale. The taxable income is the same, so it is easier to treat a sale during a year as a single transaction, rather than as a sale and a partial depreciation deduction with a different book value for the computation of recaptured depreciation or loss on sale.

Exhibit 12.12 illustrates the four possibilities with the computer network used in Example 12.4. Three of the possibilities can be illustrated by assuming that the sales price equals the estimated $18,000 salvage value:

1. *Loss on sale.* If the network is sold for $18,000 at the beginning of year 3, when the book value is $43,200, then there is a $25,200 loss on the sale.

2. *No gain or loss.* If the network is sold for $18,000 during year 4, when the book value at the beginning of the year is $25,920, then there is $7920 available in a depreciation deduction. Or, if $10,368 is claimed as D_4, then $2448 will be recaptured, for the same net of $7920.

3. *Recaptured depreciation.* If the network is sold as originally estimated for $18,000 any time after the end of year 6, then there is $18,000 in recaptured depreciation that must be added to the taxable income for the year of the sale.

The fourth possibility, a capital gain, requires a different assumed sales price.

4. *Capital gain.* Suppose that the firm selling the computer network made a major pricing error, or suppose that inflation skyrockets, so that the computer network could be sold during year 2 for $105,000. The book value at the end of year 1 (or the beginning of year 2) is $72,000. The $33,000 gain on sale is divided for tax purposes. The recaptured depreciation is

$18,000, and the capital gain is $15,000, which equals the sales price minus the original first cost.

Exhibit 12.12   Recaptured depreciation and loss on sale

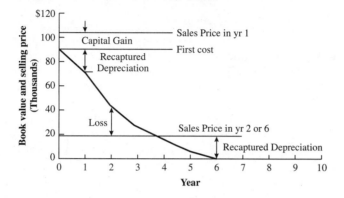

Obviously, this last situation of a capital gain is relatively rare for manufactured or constructed items. It occurs principally with assets that are not depreciated, such as land, bonds, and stock. Capital gains may be taxed at a lower rate than the rate for regular income and recaptured depreciation.

The importance of recaptured depreciation increased with the passage of ACRS in 1981, and that importance was maintained with MACRS in 1986. ACRS, MACRS, and alternate MACRS assume salvage values of $0 when depreciating. Thus, there is more recaptured depreciation than with the earlier straight-line, declining balance, and SOYD techniques.

12.5 OPTIMAL DEPRECIATION STRATEGIES

An optimal depreciation strategy maximizes the PW of the resulting tax savings. Since the tax rate is usually assumed to be constant, this is equivalent to maximizing the PW of the depreciation deductions.

This strategy equates to choosing the shortest legal recovery period with the most accelerated depreciation method. Section 179 expensing, which is detailed in Section 12.8, should be used whenever possible; if it can only be applied to some assets, it should be applied to those assets with the longest lives. The strategy can be summarized as follows:

1. Take the Section 179 depreciation deduction and the investment tax credit wherever possible.
2. Always use as small a value of N as possible.
3. When using the declining balance method, use as large a value of α as possible.

To a very large extent, the optimal depreciation strategy is achieved by using MACRS. This system is legal, it has relatively short recovery periods, it depreciates to a book value of $0, and it combines the 200% or 150% declining balance method with an optimal switch to the straight-line method.

The advantages of MACRS are exemplified by Exhibit 12.10 and again in Example 12.5. While not as attractive as MACRS, both the declining balance and SOYD methods are far more attractive than the straight-line technique.

Example 12.5   Midwest Manufacturing

Some manufacturing equipment has an installed first cost of $90K, and a salvage value after 15 years of $10,520. Midwest Manufacturing uses an interest rate of 10%. Find the PW of the depreciation deductions for straight-line, SOYD, double-declining balance, MACRS, and alternate MACRS methods. Rank the strategies in order of their attractiveness. Which is the best?

Solution
The salvage value of $10,520 matches that implied by double-declining balance depreciation, so no switch to the straight-line method is required. Because MACRS and alternate MACRS assume that the salvage value is $0, there is $10,520 of recaptured depreciation in year 15 for these methods.

Exhibit 12.13 summarizes the annual depreciation charges. The top portion of this table is the data block for the depreciation calculations. The next row is the basis for each

Exhibit 12.13   Example 12.5, optimal depreciation strategy

FC	S	N	i		
90,000	10,520	15	10%		
Basis	$79,480	$79,480	$90,000	$90,000	$90,000
"helps"	5298.667	120	0.133333		7
Year	**Straight-Line**	**SOYD**	**DDB**	**MACRS**	**Alternate MACRS**
1	$ 5299	$ 9935	$12,000	$12,861	$ 6429
2	5299	9273	10,400	22,041	12,857
3	5299	8610	9013	15,741	12,857
4	5299	7948	7812	11,241	12,857
5	5299	7286	6770	8037	12,857
6	5299	6623	5867	8028	12,857
7	5299	5961	5085	8037	12,857
8	5299	5299	4407	4014	6249
9	5299	4636	3819		
10	5299	3974	3310		
11	5299	3312	2869		
12	5299	2649	2486		
13	5299	1987	2155		
14	5299	1325	1867		
15	5299	662	1618	−10,520	−10,520
Total depreciation	$79,480	$79,480	$79,480	$79,480	$79,480
Present worth (depreciation)	$40,302.08	$48,972.4	$49,989.46	$62,412	$57,230.37

depreciation technique. The values for the straight-line and SOYD methods differ from the others, since the salvage value, S, is subtracted before they are computed. The "help" row contains intermediate values useful for computing the depreciation in that column.

Notice that the next-to-last row shows that each technique claims the same total depreciation of $79,840—but the timing differs. Thus, the PW is highest for the most accelerated technique. MACRS is the most cost effective, then alternate MACRS, double-declining balance, SOYD, and finally straight-line depreciation. Since a higher PW means the tax deductions are more valuable, this is the order of attractiveness. The PWs for SOYD and double-declining balance methods are very similar.

Much of the large difference between the MACRS values and the other approaches is due to assuming a salvage value of $0 and recapturing that excess depreciation 7 years later.

12.6 PW OF A DEPRECIATION SCHEDULE

To support the calculation of income taxes in Chapter 13, it is possible to generalize the calculations in Exhibit 12.13. Since the tax rate is usually assumed to be a known constant, minimizing the PW of taxes is the same as maximizing the PW of the depreciation deductions. These calculations for the different depreciation methods also confirm the strategy of using MACRS if possible, then double-declining balance, then SOYD, and straight-line only if required.

Straight-Line. D_t equals $(FC - S)/N$ for N years. Thus a simple (P/A) factor can be used to calculate a PW using Equation 12.5.

$$PW_{SL} = (P/A, i, N)(FC - S)/N \qquad (12.5)$$

Declining Balance. An equation can be derived for declining balance with a switch to straight-line—once the rate (200%, 175%, or 150%), life (N), and salvage value (S) are defined. Because MACRS is the current system and MACRS is an example of a specific declining balance application, this derivation is not presented here.

Sum-of-the-Years'-Digits (SOYD). Equation 12.6 indicates that the annual D_t values can be expressed as a uniform periodic payment and a gradient. The annual periodic amount is

$$A = (FC - S) \cdot N/SOYD$$

and the annual arithmetic gradient is

$$G = -(FC - S)/SOYD.$$

Assuming that the tax savings occur at the ends of periods $t = 1, 2, \ldots, N$, the present value is

$$PW_{SOYD} = (FC - S)[N(P/A, i, N) - (P/G, i, N)]/SOYD \qquad (12.6)$$

MACRS.   It is easiest to apply single-period (P/F) factors to the MACRS percentages in Exhibit 12.9 (especially with a spreadsheet). The results are summarized in Exhibit 12.14.

As detailed in Section 12.4, the salvage value will be taxed as recaptured depreciation, since the book value is depreciated to $0 under MACRS. Since the timing of that recapture depends on when the salvage value is received, the recapture is omitted from Exhibit 12.14. That salvage value in year N_{last} should be added using a separate (P/F) factor, as in Equation 12.7. Exhibit 12.14 sums over t years the products of the MACRS depreciation percentage for year t ($\%_t$) and the (P/F) factors, which is the summation in Equation 12.7.

$$\text{PW}_{\text{MACRS}} = \text{FC} \cdot \sum_t \%_t \cdot (P/F, i, t) - S(P/F, i, N_{\text{last}}) \qquad (12.7)$$

Exhibit 12.14   Table of PW factors for MACRS depreciation

i	3-Year	5-Year	7-Year	10-Year	15-Year	20-Year
1%	.9807	.9726	.9648	.9533	.9259	.9054
2	.9620	.9465	.9316	.9101	.8596	.8234
3	.9439	.9215	.9002	.8698	.8001	.7520
4	.9264	.8975	.8704	.8324	.7466	.6896
5	.9095	.8746	.8422	.7975	.6985	.6349
6	.8931	.8526	.8155	.7649	.6549	.5867
7	.8772	.8315	.7902	.7344	.6155	.5441
8	.8617	.8113	.7661	.7059	.5797	.5062
9	.8468	.7919	.7432	.6792	.5471	.4725
10	.8322	.7733	.7214	.6541	.5173	.4424
11	.8181	.7553	.7007	.6306	.4902	.4154
12	.8044	.7381	.6810	.6084	.4652	.3911
13	.7912	.7215	.6621	.5875	.4424	.3691
14	.7782	.7055	.6441	.5678	.4213	.3492
15	.7657	.6902	.6270	.5492	.4019	.3311
16	.7535	.6753	.6106	.5317	.3839	.3146
17	.7416	.6611	.5949	.5150	.3673	.2995
18	.7300	.6473	.5798	.4993	.3519	.2856
19	.7188	.6340	.5654	.4844	.3376	.2729
20	.7079	.6211	.5517	.4702	.3243	.2612
21	.6972	.6087	.5384	.4567	.3118	.2503
22	.6868	.5968	.5257	.4439	.3002	.2403
23	.6767	.5852	.5136	.4317	.2893	.2310
24	.6669	.5740	.5019	.4201	.2791	.2223
25	.6573	.5631	.4906	.4090	.2696	.2142

For Example 12.4,

$$\begin{aligned}
PW_{MACRS} = &\, 90,000[.2(P/F,.1,1) + .32(P/F,.1,2) + .192(P/F,.1,3) \\
&+ .1152(P/F,.1,4) + .1152(P/F,.1,5) + .0576(P/F,.1,6)] \\
&- 18,000(P/F,.1,8) \\
= &\, 90K[.18182 + .26446 + .14425 + .07868 + .07153 + .03251] \\
&- 18K \cdot .46651 = 90K \cdot .77326 - 18K \cdot .46651 = \$61,196
\end{aligned}$$

Notice that looking up $i = 10\%$ under the 5-year column in Exhibit 12.14 shows a value of .7733, which is the rounded value of the calculated value of .77326. Thus, the tabulated values in Exhibit 12.14 can be used instead of the detailed (P/F) factors in Equation 12.17.

Example 12.6   PW of Depreciation Deductions for Midwest Manufacturing (Example 12.5 Revisited)

Using equations, calculate the PW of straight-line, SOYD, and MACRS depreciation schedules for Example 12.5. The first cost is $90K, the salvage value after 15 years is $10,520, and i equals 10%.

Solution
The values for straight-line, SOYD, and MACRS depreciation found using Equations 12.5–12.7 should check with those shown in Exhibit 12.13: $40,302, $48,972, and $62,412, respectively.
   Using Equation 12.5 for the PW of straight-line results in:

$$PW_{SL} = (P/A, 10\%, 15) \cdot 79,480/15 = 7.606 \cdot 79.48K/15 = \$40.30K$$

which matches!
   Using Equation 12.6 for the PW of SOYD results in:

$$\begin{aligned}
PW_{SOYD} &= 79.48K[15(P/A,.1,15) - (P/G,.1,15)]/SOYD \\
&= 79.48K[15 \cdot 7.606 - 40.152]/120 = \$48.97K
\end{aligned}$$

which matches!
   Finally, using Exhibit 12.14 ($i = 10\%$ and 7-year recovery period) for the summation portion of Equation 12.7 leads to:

$$\begin{aligned}
PW_{MACRS} &= 90K \cdot .7214 - 10,520(P/F,.1,15) = 90K \cdot .7214 - 10.52K \cdot .2394 \\
&= \$62.40K
\end{aligned}$$

which matches!

12.7 DEPLETION OF RESOURCES

Depletion is analogous to depreciation. It reduces taxable income by accounting for the using up of natural resources, such as mines, wells, and timberlands. There are two methods for calculating depletion: cost and percentage. Cost depletion applies to all depletable assets. Percentage depletion applies to mineral deposits and in some cases to domestic oil and gas wells. In many cases, both depletion bases are calculated and the larger one is used for each year.

Cost Depletion.   Cost depletion divides a natural resource deposit's acquisition cost by estimated recoverable reserves and multiplies that by the number of units sold during the taxable year.

In other words, the number of recoverable tons, barrels, board feet, etc. is estimated. This recoverable number is divided into the property's cost to find a cost depletion per unit. This unit cost and the year's volume of production determine the year's depletion on a cost basis. Sometimes the property's cost must be allocated to the resource (depletable) and to land (not depletable)—for example, with timber, where only the trees' value is depleted.

Percentage Depletion.   Exhibit 12.15 summarizes some of the categories where percentage depletion is applied. The tabulated percentages are multiplied by the deposit's gross income. In any year, percentage depletion cannot exceed half of the property's net income (calculated without depletion).

Exhibit 12.15   Percentage depletion rates

Rate	Assets Included
22%	Asbestos, bauxite, cobalt, lead, manganese, mercury, molybdenum, nickel, platinum, sulfur, tin, and uranium
22	Regulated natural gas
15	Small producers' oil and natural gas deposits
15	Copper, gold, iron ore, oil shale, and silver
14	Rock asphalt, vermiculite, bentonite, and refractory clay
14	Nonmetallic minerals for building stone
10	Coal, lignite, sodium chloride
7½	Clay for sewer pipe and tile
5	Gravel, sand, peat, clay for roofing tile and flower pots

Cost depletion stops once the deposit's book value reaches $0; in other words, when the accumulated depletion equals the acquisition cost. However, percentage depletion is still allowed, which may result in negative book values; these are typically credited to retained earnings.

Example 12.7   Northern Gold Mining

Northern Gold Mining bought a mine site with 500,000 tons of recoverable ore for $800,000. Of this total, $50,000 is the land's estimated value. If 40,000 tons were processed this year, what is the cost depletion?

The revenue from 5500 ounces of gold is $1.8M, and the expenses (other than depletion) were $1.6M. What is the percentage depletion, and should cost or percentage depletion be claimed?

Solution

The cost of the deposit is $750,000, and the recoverable reserves are 500,000 tons. So the cost per unit is $1.50. At a volume of 40,000 tons, the depletion on a cost basis is $60,000 this year.

The 15% depletion rate that applies to gold is multiplied by $1.8M, for a maximum percentage depletion of $270K. The 50% income limit leads to a limit of $(1.8M - 1.6M)$ .5 = $100K. Thus, the percentage depletion is limited to $100K. Because that amount exceeds the cost depletion of $60K, percentage depletion should be claimed.

12.8 SECTION 179 DEDUCTION

The Section 179 deduction was included in the tax code to benefit small businesses. It permits firms to expense up to $17,500 in equipment in the first year, rather than depreciating that amount over time with MACRS. When this deduction was established in 1981, it was limited to $5000. It was increased to $10,000 in 1986 and to $17,500 in 1993.

While Section 179 may be important to a firm, it may not be appropriate to include it in the analysis for any particular project, except for the smallest firm. As detailed below, Section 179 is limited to $17,500 out of potentially as much as $200,000 in capital investment. If the firm has only one project, there is no problem. However, if a firm has 20 projects and each costs $17,500, which one receives credit for the Section 179 deduction? The choice would be arbitrary, so it is better to analyze all potential projects without considering Section 179 (unless total expenditures are less than $17,500 or only one project will be undertaken). However, the firm, if eligible, should claim the Section 179 deduction.

Items qualifying for Section 179 have MACRS recovery periods of 15 years or less, lives of 3 years or more, and are tangible (not patents or copyrights). The dollar amount for each new or used item is its acquisition cost after subtracting any trade-in allowance. The amount for all qualified items is then totaled, so that the Section 179 deduction can be calculated.

The maximum Section 179 deduction must be less than:

1. Total spent on qualifying property that is placed in service in this year
2. $17,500
3. $|$217,500 - Total spent on qualifying property$|$
4. Taxable income

The third limit is a dollar-for-dollar reduction in the deduction, and it applies to larger firms that spend over $200,000 per year on qualifying property. Thus, any firm placing more than $217,500 worth of qualifying property in service cannot use Section 179.

The taxable income works as follows. A small business places $19,000 of qualifying property in service, but its taxable income is only $6000. The business could deduct $6000 this year and could carry forward an $11,500 deduction for next year. Only that amount disallowed by the income limit may be carried forward, and not the $1500 disallowed by the $17,500 limit.

The Section 179 deduction reduces the basis for computing depreciation, and taxable income is reduced by any depreciation. Thus, whether the small business depreciated $13,000 or something else, its taxable income would be reduced and the Section 179 limit would drop from $6000. Recapture provisions also apply for equipment, such as a computer, which is purchased for business use, expensed using Section 179, and later shifted partially or totally from business to personal use.

This deduction is very attractive when it is available. First, it provides the maximum immediate reduction in income taxes. Second, for small businesses with limited annual expenditures on equipment, the Section 179 deduction can eliminate (after the first year) some or all of the record-keeping and forms associated with depreciation.

Recapture for Section 179 Assets.   If an item for which Section 179 has been claimed is sold or disposed of before its MACRS recovery period ends, then recaptured depreciation must be reported as income.

Example 12.8   Recaptured Copier Write-off

Smith, Jones, and Rodriguez purchased a sophisticated copier for $9000 and expensed it under Section 179. During the fourth year of ownership, the copier is sold for $3000. What is the recaptured depreciation?

Solution
Copiers have 5-year recovery periods under MACRS, so the total depreciation that could have been claimed through year 3 is 71.2% = 20% + 32% + 19.2%. Therefore, the implied book value is 28.8% of the $9000 initial cost, or $2592. Since the sales price exceeds this by $408, that amount is taxed as recaptured depreciation.

12.9 SPREADSHEET FUNCTIONS FOR DEPRECIATION

Most spreadsheet packages do include functions for straight-line, declining balance, and SOYD depreciation. Exhibit 12.16 lists these functions for two packages. Hopefully, at some point, functions for MACRS will become common, because specifying a first cost, a recovery period, and a year is enough to determine the MACRS tax deduction.

Exhibit 12.16   Spreadsheet functions for depreciation

Depreciation Technique	Excel	Quattro Pro
Declining Balance	$DDB(cost, S, N, year, factor)$	$@DDB(cost, S, N, year)$
Straight-line	$SLN(cost, S, N)$	$@SLN(cost, S, N)$
Sum-of-the-years'-digits	$SYD(cost, S, N, year)$	$@SYD(cost, S, N, year)$

Spreadsheets can easily construct tables for depreciation without using the specialized functions. Such tables, when printed, could look like Exhibits 12.4 and 12.6. However, as shown in Example 12.9, the task is even easier with the depreciation functions.

Example 12.9 Spreadsheet Functions for Computer Workstation (Examples 12.2 and 12.3 Revisited)

The workstations have a first cost of $90K and a salvage value after 8 years of $18K. What are the depreciation charges for the next 8 years for double-declining balance depreciation?

Solution
From Exhibit 12.16, the Excel function for double-declining balance depreciation has terms for the cost, salvage value, life, year, and factor—that is, DDB(cost,S,N,year,2). The final term is optional. If it is omitted, then double or 200% declining balance is assumed. Another useful value is 1.5 for 150% declining balance depreciation.

Exhibit 12.17 is a simple spreadsheet with a data block, a list of years, and a column for the depreciation function. Another column with the formulas of the depreciation function is included solely to help explain Exhibit 12.17.

Exhibit 12.17 Spreadsheet for declining balance depreciation

	A	B	C
1	Example 12.9 Computer Workstation Depreciation		
2			
3	$90,000	First Cost	
4	$18,000	Salvage Value	
5	8	Life	
6			
7		Declining Balance	
8	Year	Depreciation	Formula
9	1	$22,500.00	=DDB($A$3,$A$4,$A$5,A9)
10	2	$16,875.00	=DDB($A$3,$A$4,$A$5,A10)
11	3	$12,656.25	=DDB($A$3,$A$4,$A$5,A11)
12	4	$9,492.19	=DDB($A$3,$A$4,$A$5,A12)
13	5	$7,119.14	=DDB($A$3,$A$4,$A$5,A13)
14	6	$3,357.42	=DDB($A$3,$A$4,$A$5,A14)
15	7	$0.00	=DDB($A$3,$A$4,$A$5,A15)
16	8	$0.00	=DDB($A$3,$A$4,$A$5,A16)

12.10 SUMMARY

This chapter has defined the terms for depreciation—basis, recovery period, salvage value, book value, and depreciation. It has emphasized that depreciation is

NOT a cash flow, and it should only be included in after-tax analyses in which the taxes depend on the depreciation deducted.

The chapter included several basic depreciation methods. These methods are summarized in Exhibit 12.18. To compute straight-line depreciation, the first cost minus the salvage value is divided by the asset's life. Declining balance depreciation in year t equals the book value at the year's beginning multiplied by α/N. Sum-of-the-years'-digits (SOYD) depreciation has a schedule that is very similar to double-declining balance, and its PW can be calculated using tabulated (P/G) factors.

Exhibit 12.18   Summary of depreciation methods

	MACRS	Straight-Line	$\alpha\%$ Declining Balance	Sum-of-the-Years'-Digits (SOYD)
Recovery period	$N_D(< N)$	N	N	N
Basis	FC	$(FC - S)$	FC	$(FC - S)$
Assumed salvage	\$0	S	S	S
Depreciation in year t	$\%_{ND_t} \cdot FC$	$\dfrac{(FC - S)}{N}$	$BV_{t-1} \cdot \alpha/N$	$\dfrac{(FC - S)(N + 1 - t)}{SOYD}$

The current U.S. tax code uses the Modified Accelerated Cost Recovery System (MACRS). This is a combination of double-declining balance for recovery periods of 3, 5, 7, or 10 years (150% for 15 or 20 years) and straight-line methods. A salvage value of \$0 is assumed, and the recovery periods are shorter than the estimated lives that were used for straight-line, SOYD, and declining balance depreciation.

Because the MACRS salvage value is assumed to be \$0, recaptured depreciation is common when assets are disposed of. On the other hand, capital gains are rare for engineered projects. Capital gains are important for land, stocks, fine art, etc.—not machinery that is used for 10 years and then disposed of.

The optimal depreciation strategy maximizes the PW of the available income tax deductions. This is achieved by using the shortest legal recovery period, and by using MACRS wherever possible. The PW of available income tax deductions can be calculated using the annual depreciation deductions, or Equations 12.5–12.7 can be used instead, or for MACRS Exhibit 12.14 can be used.

Resource depletion is used for mineral deposits. The computations are analogous to those for depreciation, but are somewhat more complicated since both a cost depletion basis and a percentage depletion basis may be available (the larger may be chosen).

Finally, Section 179 is an attractive depreciation method for small businesses. However, since it is applied to the firm, not to the item, it is usually inappropriate to consider Section 179 in the project selection phase. Which capital asset should receive the credit? These depreciation methods will be applied in Chapter 13 to income taxes. At that point, cash flows will again be the focus.

REFERENCES

Bussey, Lynn E., and Ted G. Eschenbach, *The Economic Analysis of Industrial Projects,* 2nd ed., 1992, Prentice-Hall.

Fleischer, Gerald A., and Lawrence C. Leung, "Depreciation and Tax Policies in China and the Four Little Dragons," *IIE Integrated Systems Conference Proceedings,* 1988, pp. 314–320.

Fleischer, G.A., A.K. Mason, and L.C. Leung, "Optimal Depreciation Policy Under the Tax Reform Act of 1986," *IIE Transactions,* Volume 22, Number 4, December 1990, pp. 330–339.

Remer, Donald S., and Yong Ho Song, "Depreciation and Tax Policies in the Seven Countries with the Highest Direct Investment from the U.S.," *The Engineering Economist*, Volume 38, Number 3, Spring 1993, pp. 193–208.

Smith, Gerald W., *Engineering Economy: Analysis of Capital Expenditures,* 1987, Iowa State University Press, pp. 161–163.

PROBLEMS

12.1   Metal Stampings, Inc., can purchase a new forging machine for $400,000. After 20 years of use, the forge should have a salvage value of $25,000.
   a.   Under MACRS, what depreciation is claimed in year 3?
   b.   Under the straight-line (pre-1981) method, what depreciation is claimed in year 3?
   (*Answers:* a. $69,960; b. $18,750)

12.2   Muddy Fields Earthmoving can purchase a bulldozer for $30,000. After 7 years of use, the bulldozer should have a salvage value of $5000.
   a.   Under MACRS, what depreciation is claimed in year 3?
   b.   Under the straight-line (pre-1981) method, what depreciation is claimed in year 3?
   c.   Under the SOYD (pre-1981) method, what depreciation is claimed in year 3?

12.3   An asset costs $15,000 and has a salvage value of $500 after 10 years. What is the depreciation charge for the fifth year, and what is the book value at the end of the fifth year:
   a.   Using the 150% declining balance method
   b.   Using the 200% declining balance method
   (*Answers:* a. $D_5 = \$1175$, $BV_5 = \$6656$; b. $D_5 = \$1229$, $BV_5 = \$4915$)

12.4   A milling machine costs $8000, and it will be scrapped after 10 years. Compute the book value and depreciation for the first two years using:

a.   MACRS
b.   Straight-line depreciation
c.   Double-declining balance depreciation
d.   Sum-of-the-years'-digits depreciation

12.5   The Jafar Jewel Mining Company has just purchased a new crystal extraction machine that cost $5000 and has an estimated salvage value of $1000 at the end of its 8-year useful life. Compute the depreciation schedule using the following methods:
a.   Straight-line
b.   Double-declining balance
c.   Sum-of-the-years'-digits
d.   MACRS (ADR = 8 years)

12.6   To meet increased delivery demands, the Moo-Cow Dairy has just purchased 10 new delivery trucks. Each truck cost $18,000 and has an expected life of 4 years. The trucks can each be sold for $2000 after 4 years. Using MACRS: (a) determine the depreciation schedule for each truck, and (b) determine the book value at the end of each year.   (*Answer:* D_2 = $5760, BV_2 = $8640)

12.7   A small microprocessor costing $12,000 is to be depreciated using 150% declining balance depreciation over the next 4 years. The microprocessor will have no value after 4 years. Determine the depreciation schedule. What is the book value at the end of the 4 years?   (*Answer:* $1831)

12.8   The Delta Cruise Company purchased a new tender (a small motor boat) for $15,000. Its salvage value is $500 after its useful life of 5 years. Calculate the depreciation schedule using (a) double-declining balance, and (b) SOYD methods.

12.9   Blank Mind, Inc., just purchased a new psycho-graph machine for $10,000. The expected resale value after 4 years is $500. Determine the book value after 2 years using the following depreciation methods:
a.   Straight-line
b.   Sum-of-the-years'-digits
c.   Double-declining balance

12.10   A used drill press costs $12,000, and delivery and installation charges add $1500. The salvage value after 8 years is $1000. Compute the accumulated depreciation through year 4 using:
a.   7-year MACRS
b.   Straight-line
c.   150% declining balance
d.   Sum-of-the-years'-digits

12.11   A machine purchased by the Bee Buzzsaw Company originally cost $245,000. Delivery and installation charges amounted to $5000. The declared salvage value was $25,000. Early in year 4, the company changed its product mix and found that it no longer needed the machine. One of its competitors agreed to buy the machine for $90,000. Determine the loss, gain, or recapture of MACRS depreciation on the sale. The ADR is 12 years for this machine.   (*Answer:* $19,325 loss on sale)

12.12   A numerically controlled milling machine was purchased for $95,000 four years ago. The estimated salvage value was $10,000 after 15 years. What is the machine's book value now, after 4 years of depreciation? If the machine is sold for $20,000 early in year 5, how much of a gain on sale or recaptured depreciation is there?   Assume:

   a.   7-year MACRS
   b.   Straight-line
   c.   Double-declining balance
   d.   Sum-of-the-years'-digits

12.13   A computer costs $9500 and its salvage value in 8 years is negligible. What is the book value after 4 years? If the machine is sold for $3500 in year 5, how much of a gain or recaptured depreciation is there?   Assume:

   a.   MACRS
   b.   Straight-line
   c.   Double-declining balance
   d.   Sum-of-the-years'-digits

12.14   A conveyor was purchased for $40,000. Shipping and installation costs were $15,000. It was expected to last 6 years, when it would be sold for $5000 after paying $3000 for dismantling. Instead, it lasted 9 years, and several workers were permitted to take it apart on their own time for reassembly at a private technical school. How much of a gain, loss, or recaptured depreciation is there?   Assume:

   a.   7-year MACRS
   b.   Straight-line
   c.   Double-declining balance with switch to straight line
   d.   Sum-of-the-years'-digits

12.15   An automated assembly line is purchased for $250,000. The company has decided to use units-of-production depreciation. At the end of 8 years, the line will be scrapped for an estimated $20,000. Using the following information, determine the depreciation schedule for the assembly line.

Year	Production Level
1	5000 units
2	10,000 units
3	15,000 units
4	15,000 units
5	20,000 units
6	20,000 units
7	10,000 units
8	5000 units

(*Answer:* $D_1 = \$11,500$)

12.16   During the construction of a highway bypass, earthmoving equipment costing $40,000 was purchased for use in transporting fill from the borrow pit. At the end of the 4-year project, the equipment will be sold for $20,000. The schedule for moving fill calls for a total of 100,000 cubic feet during the project. In the first year, 40% of the total fill is required; in the second year, 30%; in the third year, 25%; and in the final year, the remaining 5%. Determine the units-of-production depreciation schedule for the equipment.

12.17   A small delivery van can be purchased for $14,000. At the end of its useful life (8 years), the van can be sold for $2000. Using the methods listed below, determine the PW of the depreciation schedule based on 12% interest.
  a.   Straight-line
  b.   Sum-of-the-years'-digits
  c.   MACRS
(*Answer:* a. $7451, b. $8424, c. $9526)

12.18   The ABC Block Company purchased a new office computer costing $4500. During the third year, the computer is declared obsolete and is donated to the local tech training school. Using an interest rate of 10%, calculate the PW of the depreciation and donation deductions. Assume that no salvage value was initially declared and that the machine was expected to last 5 years.
  a.   Straight-line
  b.   Sum-of-the-years'-digits
  c.   MACRS

12.19   A pump in an ethylene plant costs $15,000. After 9 years, the salvage value is nil.
  a.   Determine D_t and BV_t using straight-line, sum-of-the-years'-digits, and 7-year MACRS.
  b.   Find the PW of each depreciation schedule if the interest rate is 8%.

12.20   Some R&D equipment costs $250,000; the interest rate is 10%; and the salvage value is $25,000. The expected life is 10 years (with periodic upgrades). Compute the PW of the depreciation deductions assuming:

a. MACRS
b. Straight-line depreciation
c. Sum-of-the years'-digits depreciation

12.21 Western Coal expects to produce 75,000 tons of coal annually for 15 years. The deposit cost $2M to acquire; the annual gross revenues are expected to be $8 per ton; and the net revenues are expected to be $3 per ton.
a. Compute the annual depletion on a cost basis.
b. Compute the annual depletion on a percentage basis.
(*Answer:* cost = $133.3K; percentage = $60K)

12.22 Eastern Gravel expects to produce 50,000 tons of gravel annually for 5 years. The deposit cost $120K to acquire; the annual gross revenues are expected to be $8 per ton, and the net revenues are expected to be $3 per ton.
a. Compute the annual depletion on a cost basis.
b. Compute the annual depletion on a percentage basis.

12.23 A 2000-acre tract of timber is purchased by the Newhouser Paper Company for $800,000. The acquisitions department at Newhouser estimates that the land will be worth $225 per acre once the timber is cleared. The materials department estimates that a total of 5 million board feet of timber is available from the tract. It is expected to take 5 years for all the timber to be harvested. The harvest schedule calls for equal amounts of the timber to be harvested each year. Determine the depletion allowance for each of the 5 years.   (*Answer:* $70,000)

12.24 The Piney Copper Company purchased an ore-bearing tract of land for $7,500,000. The geologist for Piney estimated the recoverable copper reserves to be 450,000 tons. During the first year, 50,000 tons were mined and 40,000 tons were sold for $4,000,000. Expenses (not including depletion allowances) were $2,500,000. What are the percentage depletion and the cost depletion allowances?

12.25 The Red Dog oil field will become less productive each year. Martinson Brothers is a small company that owns Red Dog, which is eligible for percentage depletion. Red Dog cost $2M to acquire, and it will be produced over 15 years. Initial production costs are $4 per barrel, and the wellhead value is $9 per barrel. The first year's production is 80,000 barrels, which will decrease by 5000 barrels per year.
a. Compute the annual depletion (each year may be cost-based or percentage-based).
b. What is the PW at $i = 12\%$ of the depletion schedule?

12.26 A small design firm's only capital purchase this year is a graphics workstation, which will cost $28,000. The firm's taxable income currently averages $60,000 per year. What is the MACRS recovery period? Contrast the depreciation schedules with and without Section 179.   (*Answer:* year 1: $19,600 vs. $5600)

12.27   Redo Problem 12.26 with a taxable income before depreciation for this year of only $7000. Assume the Section 179 deductions for later years will be used for later equipment purchases.

12.28   A small, but very profitable, research firm purchased $175,000 worth of research equipment in the past year. Using MACRS and Section 179 expensing, determine the depreciation schedule for the equipment.

12.29   A small, profitable construction contractor purchases equipment costing $205,000. Using MACRS (5-year class) and Section 179 expensing, determine the depreciation schedule for the equipment.

CHAPTER 13 INCOME TAXES

The Situation and the Solution

Income taxes are cash flows that depend on the profits generated by projects and alternatives. The computation of taxes depends on ever-changing tax codes.

The solution is to understand the principles underlying the tax code. In particular, this includes depreciation (see Chapter 12), marginal income tax rates, tax credits, and deductible interest. After-tax cash flows can be used to compute PWs, EACs, and IRRs.

Chapter Objectives

After you have read and studied the sections of this chapter, you should be able to:

Section 13.1
Select a point of view and describe the steps of calculating income taxes.

Section 13.2
Compute income taxes using a progressive marginal tax rate schedule.

Section 13.3
Find taxable income, including depreciation.

Section 13.4
Select an after-tax minimum attractive rate of return and calculate after-tax cash flows using year-by-year cash flows.

Section 13.5
Calculate after-tax cash flows using formulas for straight-line, MACRS, and sum-of-the-years'-digits depreciation.

Section 13.6
Include investment tax credits and capital gains as necessary.

Section 13.7
Use spreadsheets to compute after-tax IRRs, including tax-deductible interest on financing.

Appendix 13A
Calculate personal income taxes.

Key Words and Concepts

Differing perspectives on taxes Government's: to raise revenue and promote general welfare; firm's: to pay as little as is legal, and that as late as possible; analyst's: that taxes always matter to the firm, but they may have little impact on the choice between alternatives.

Marginal tax rate One that applies to a block of taxable income, not to all income.

Capital asset When bought is not an expense, and it is not tax deductible.

Investment tax credit A 3% to 10% portion of an investment's first cost that is credited against income taxes owed.

Capital gain The amount by which an asset's selling price exceeds its first cost.

Leverage The increase in an attractive project's PW or IRR by financing part or all of it at an interest rate below the MARR. The variability of the PW and IRR also increases if revenues go down or costs go up.

13.1 PRINCIPLES OF INCOME TAXES

Point of View. Before discussing the detailed principles behind income tax calculation, there are three perspectives that may be helpful: the government's, the firm's, and the analyst's.

The government's perspective on income taxes focuses on *generating revenue for government operations and promoting the general welfare through tax policy.* For example, governments have used investment tax credits to encourage capital investment, limited the deductibility of corporate "perks" to discourage wasteful practices, and eliminated deductions for interest paid on consumer debt to encourage savings.

For both firms and individuals, the private view of taxes is the same. From an economic perspective, the goal is to *pay as little as legally permissible, and that as late as possible.* This underlies the selection of MACRS as the optimal depreciation strategy in Section 12.5. Because of this perspective, firms that operate in multiple states or countries may set transfer or internal prices so that subsidiaries in low-tax-rate jurisdictions "earn" the profits for the corporation.

The analyst's perspective on income taxes is that *taxes always matter to the firm, but sometimes taxes can be ignored when making a decision.* For example, the choice between a stainless-steel pump that lasts 8 years and a brass pump that lasts 6 years is unlikely to be changed by including income taxes. *To save time and to focus attention on more important items, the analyst may ignore income taxes when they will not have a significant impact.* Taxes can rarely be ignored (1) when doing something is being compared with doing nothing, (2) when capital- and labor-intensive alternatives are compared; or (3) when a budget is being set.

This chapter will focus on the firm's perspective, not on the government's. After mastering the computation of income taxes, you should be able to identify cases in which the omission of taxes will not affect the recommendations.

Principles of Calculation.   Income taxes depend on profits, not on an item's value. Thus, accounting for an item's impact on taxes requires considering depreciation's impact on profits. Income taxes are assessed by the U.S. government, most states, and many municipalities.

Other taxes, such as property, sales, payroll, and excise taxes, are also part of the cost of doing business. These taxes are part of an alternative's after-tax cash flows. For example, a firm might pay 1.2% of the book value of its property to the local government every year. The details of these taxes can easily be added to a problem once income taxes are understood, so this chapter focuses on income taxes.

The basic steps of calculating income taxes from a corporate perspective are as follows:

1. Find the firm's taxable income, considering revenues, expenses, and depreciation deductions.
2. Compute the tax owed using the appropriate schedule of marginal tax rates.
3. Reduce the tax owed by any credits, such as the investment tax credit.

In some cases, only the costs of alternatives are analyzed. For example, an engineer might evaluate automating a production line with robots without analyzing the assembly line's benefits or revenues. In this case, an after-tax analysis requires the following assumption:

Firms subject to income tax may lose money periodically, but in the long run they are profitable. Thus, it is safe to assume that these firms have adequate revenues so that an alternative's expenses and depreciation are completely deductible from taxable income, even if that income is unknown.

One justification for this assumption is the existence of carryforward and carryback provisions that permit the losses of one year to be considered when computing future taxes or recomputing prior taxes.

Firms pay income taxes on their total operations, not on individual projects. And, as the next section details, the income tax rate is linked to the firm's income level and not to an alternative's level of income generation.

13.2 PROGRESSIVE MARGINAL TAX RATES

Income taxes for firms and individuals are calculated similarly. Each uses a progressive schedule of income tax rates. Corporate taxes are included here; personal income taxes for individuals and for small businesses that are taxed as individuals are included in Appendix 13A.

Exhibit 13.1 summarizes the federal corporate tax rate structure that was established in 1986. The last three brackets were added by the Revenue Reconciliation Act of 1993. Each rate is a marginal tax rate. That is, each **marginal tax rate** applies to a block of taxable income. For example, a firm whose profit is $60,000 is taxed at 15% on the first $50,000 and at 25% on the next $10,000. See Example 13.1 as well.

A marginal tax rate applies to a block of taxable income, not to all income.

Exhibit 13.1 U.S. corporate income tax schedule

Tax Rate	Income Block
15%	$x \le \$50,000$
25%	$\$50,000 < x \le \$75,000$
34%	$\$75,000 < x \le \$100,000$
39%	$\$100,000 < x \le \$335,000$
34%	$\$335,000 < x \le \$10,000,000$
35%	$\$10,000,000 < x \le \$15,000,000$
38%	$\$15,000,000 < x \le \$18,333,333$
35%	$\$18,333,333 < x$

The corporate tax rate schedule includes a 5% surtax applied to the $235K worth of income between $100K and $335K and a 3% surtax applied to $3.33M of income. The first surtax erases the tax "cut" from 34% to lower tax rates for the first $75K of taxable income. Mathematically, lowering the rate from 34% to 15% for the first $50K and to 25% for the next $25K saves firms $50K · 19% + $25K · 9% = $11,750 in taxes. The surtax is $235K · 5%, or $11,750. Thus, for taxable incomes between $335K and $10M, the tax rate is essentially a flat 34%. The second surtax erases the tax "cut" from 35% to lower tax rates for the first $10M of taxable income. Thus, for taxable incomes above $18.33M, the tax rate is essentially a flat 35%.

Example 13.1   ElectroSwitch

ElectroSwitch manufactures electric switches for industrial applications. Last year its taxable income was $450K. Calculate the income tax owed by ElectroSwitch.

Solution

Since $450K is between $335K and $10M, a flat rate of 34% can be used. Tax = .34 · 450K = $153K

This tax can also be calculated using the marginal tax schedule, as follows:

$$
\begin{aligned}
\$\quad 7.50\text{K} &= .15 \cdot 50\text{K} \\
6.25\text{K} &= .25 \cdot 25\text{K} \\
8.50\text{K} &= .34 \cdot 25\text{K} \\
91.65\text{K} &= .39 \cdot 235\text{K} \\
\underline{39.10\text{K}} &= .34 \cdot 115\text{K} \\
\$153.00\text{K} \ &\text{Total tax}
\end{aligned}
$$

Effective Tax Rate for State, Local, and Federal Taxes.   Most states and many cities and local governments levy income taxes. These state (s) and city (c) income tax rates can be combined with the federal rates (f) using Equation 13.1. The rates are not added together, because state income taxes can be deducted from federal taxable income. Equation 13.1 assumes that city taxes can also be deducted from federal taxable income, but that city taxes do not reduce state taxable income and state taxes do not reduce city taxable income.

$$\text{After-tax income} = \text{Before-tax income} \cdot (1 - f) \cdot (1 - s - c) \qquad (13.1)$$

This can be restated as an effective income tax rate (t_e) using Equation 13.2.

$$t_e = f + s + c - f(s + c) \qquad (13.2)$$

As noted in Grant, Ireson, and Leavenworth (p. 579), the top state income tax rate exceeds 5.5% in three-fourths of the states. It exceeds 7.1% in half of the states, and 8.9% in one-fourth of the states. The highest rate was 12%, and five states did not have conventional taxes on corporate income.

As shown in Example 13.2, if a 9.1% state income tax rate is assumed, then an effective income tax rate of 40% results. This text sometimes uses this rate, rather than requiring a lookup of the federal rate.

Example 13.2   Effective Tax Rate

Suppose that ElectroSwitch operates in a state with a 9.1% income tax rate. At an income level of $450K, compute the effective tax rate if state income taxes are deductible from federal taxable income.

Solution
Compute state income taxes first. They equal .091 · $450K, or $40,950. Taxable federal income is now $409,050, which at the 34% rate is taxed at $139,080. The total tax of $180K is 40% of $450K. Thus, the effective tax rate, t_e, is 40%.

Using Equation 13.2, where $c = 0$:

$$
\begin{aligned}
t_e &= f + s - f \cdot s \\
&= .34 + .091 - .34 \cdot .091 = .431 - .031 = .4
\end{aligned}
$$

13.3 FINDING TAXABLE INCOME WHEN INCLUDING DEPRECIATION

Taxable income begins with the revenue earned each year by the firm. Yearly expenses are subtracted from the annual revenue to find the taxable income. These expenses include hourly labor, salaries, fringe benefits, energy and other utilities, purchased materials and parts, freight, interest on borrowed money, etc.

All of these expenses are included in computing the firm's taxable income, but only some of them can be identified with individual projects or alternatives. For example, the salaries and fringe benefits of many individuals cannot be linked to individual projects. These include salaries for top management, the personnel, accounting, and marketing departments, etc. Interest costs on bonds and loans taken out by the firm are another example of costs that cannot normally be linked to individual projects.

Then there are the capital expenditures that exchange cash assets for machinery assets. These capital expenditures do not change taxable income directly, but the annual depreciation of the capital costs is tax deductible.

Categorizing Before-Tax Cash Flows.   Before-tax cash flows can be divided into those that are revenues or expenses and those that represent capital purchases—which are depreciated or expensed over a recovery period.

As shown in Examples 13.3 and 13.4, the first cost is not included in the calculation of taxable income. Examples 13.11 and 13.12 will show that the principal payments on a loan are not included in taxable income. Buying a **capital asset**, whether with a lump-sum payment or with loan payments over time, is not an expense; it is the exchange of a cash asset for a physical asset. Because buying a capital asset is not an expense, it is not tax deductible.

> Buying a **capital asset** is not an expense, and it is not tax deductible. The tax deduction is spread out over time as depreciation.

Salvage values are somewhat more complicated. As shown in Example 13.3, if the salvage value matches the book value (using straight-line, declining balance, or SOYD methods), then sale of the asset generates no income. Instead, it is the exchange of a physical asset for a cash asset. As shown in Example 13.4, if the salvage value does not match the book value (MACRS assumes that the salvage value is $0), then the recaptured depreciation is taxable income.

Whenever an asset is sold, the book value is compared with the net receipts—cash received minus removal expenses. If the difference is nonzero, then the loss or the recaptured depreciation is taxable income, as discussed in Section 12.4.

In Examples 13.3 and 13.4, the same sign convention of previous chapters applies. Thus, the asset's cost and the annual O&M costs are negative in sign. A negative taxable income will result in a negative tax, which is a savings in taxes. Remember the assumption of Section 13.1, that the firm does have profits and taxable income to offset any negative taxable income from a particular project.

Taxable income equals the taxable portion of any before-tax cash flows (BTCFs) minus the depreciation for that year. Notice that if the taxable incomes for Examples 13.3 and 13.4 are totaled, they are both $-\$152K$. The depreciation method does not change the total taxable income—just its timing.

Example 13.3   Straight-Line Taxable Income for Computer Workstation Network

Example 12.1 described a computer workstation network with a first cost of $90K, a life of 8 years, and a salvage value of $18K. Here, let us add an annual expense of $10K for operations and maintenance, training, software updates, etc. Assuming straight-line depreciation, calculate the annual taxable income for the workstation network.

Solution
From Example 12.1, D_t = $9000 for straight-line depreciation. That is: $(FC - S)/N$, or $(92K - 18K)/9$ years. The $90K in first cost does not reduce taxable income, nor does the $18K in salvage value increase the taxable income. Exhibit 13.2 summarizes this.

Exhibit 13.2   Straight-line depreciation and taxable income

Years	BTCF	D_t	Taxable Income
0	−$90K	$ 0	$ 0
1–7	−$10K	9K	−19K
8	−$10K +$18K	9K	−19K

(handwritten: −10 −9   Salvage → compare w/ BV → diff =0)

Example 13.4   Taxable Income with MACRS for Workstation Network

Example 13.3 described a computer workstation network with a first cost of $90K, a life of 8 years, a salvage value of $18K, and an annual expense of $10K. Calculate the annual taxable income, assuming MACRS depreciation.

Solution
The D_t values in Exhibit 13.3 are from Example 12.4. The $90K in first cost does not reduce taxable income. However, the $18K in salvage value is taxed, because MACRS reduces the book value to $0 over the first 6 years and the $18K is recaptured depreciation.

Exhibit 13.3   MACRS and taxable income

Year	BTCF	D_t	Taxable Income
0	−$90K	$ 0	$ 0
1	−10K	18,000	−28,000
2	−10K	28,000	−38,800
3	−10K	17,280	−27,280
4	−10K	10,368	−20,368
5	−10K	10,368	−20,368
6	−10K	5184	−15,184
7	−10K	0	−10,000
8	−10K + 18K	0	8000

13.4 CALCULATING AFTER-TAX CASH FLOWS AND EACS USING TABLES OR SPREADSHEETS

Calculating an after-tax cash flow (ATCF) requires that the calculated tax be *subtracted* from the before-tax cash flow (BTCF) for that year. Then the principles of engineering economy can be applied to the resulting cash flows. Usually i, the discount rate, is significantly lower after taxes than before taxes. The after-tax i often equals the before-tax i multiplied by $(1 - t_e)$.

Unless loan payments or first costs are involved, the starting point for taxable income is the before-tax cash flows. Subtracting the depreciation then leads to the taxable income, as was shown in Examples 13.3 (straight-line) and 13.4 (MACRS). As was shown in those examples, the first costs do not affect the taxable income. Similarly, the principal portion of any loan payment also is not part of taxable income calculations.

The next step is to multiply the firm's marginal tax rate by the taxable income and to subtract the resulting tax from the BTCF. This is shown in Examples 13.5 (straight-line) and 13.6 (MACRS). If the taxable income is negative, then the negative tax stands for a tax savings. Whatever entry is in the tax column, it is subtracted from the BTCF column.

Once the ATCFs have been calculated, then it is a standard problem in engineering economy to calculate a PW or an EAC. In Example 13.6 this is done using the @NPV(i,range) function for the ATCF block.

Example 13.5 ATCF for Straight-Line Depreciation for Workstation Network

Example 13.3 calculated the taxable income for a computer workstation network with a first cost of $90K, a life of 8 years, a salvage value of $18K, and an annual expense of $10K. Assume that the 40% effective tax rate calculated in Example 13.2 for ElectroSwitch can also be used for the Smith, Jones, and Rodriguez engineering firm. Calculate the ATCFs and the EAC for the workstation network if $i = 8\%$.

Solution
From Example 13.3, the first four columns are repeated below. The fifth column in Exhibit 13.4, taxes, simply multiplies the taxable income by the 40% effective income tax rate. The sixth column, ATCF, is simply BTCF minus taxes.

Exhibit 13.4 After-tax cash flows with straight-line depreciation

BTCF − Taxes

Years	BTCF	D_t	Taxable Income	Taxes	ATCF
0	$ −90K	$ 0	$ 0	$ 0	$−90K
1–7	−10K	9K	−19K	−7.6K	−2.4K
8	−10K + 18K	9K	−19K	−7.6K	15.6K

At $i = 8\%$,

$$\begin{aligned}
PW &= -90K - 2.4K(P/A, 8\%, 7) + 15.6K(P/F, 8\%, 8) \\
&= -90K - 2.4K(P/A, 8\%, 8) + 18K(P/F, 8\%, 8) \\
&= -90K - 2.4K \cdot 5.747 + 18K \cdot .5403 = -\$94.07K \\
EAC &= 90K(A/P, 8\%, 8) + 2.4K - 18K(A/F, 8\%, 8) \\
&= 90K \cdot .1740 + 2.4K - 18K \cdot .0940 = \$16.37K
\end{aligned}$$

Example 13.6   ATCF with MACRS for Workstation Network

Use the data of Example 13.5 (a first cost of \$90K, a life of 8 years, a salvage value of \$18K, and an annual expense of \$10K), and use MACRS depreciation. Calculate the ATCF and the EAC assuming $i = 8\%$.

Solution

The first four columns of Exhibit 13.5 are from Example 13.4. As in Example 13.5, the fifth column for taxes is calculated by simply multiplying the fourth column by 40%, the effective income tax rate. The sixth column for ATCFs is the BTCF minus the tax.

Exhibit 13.5   After-tax cash flows with MACRS

Years	BTCF	D_t	Taxable Income	Taxes	ATCF
0	$ -90,000	$ 0	$ 0	$ 0	$-90,000
1	-10,000	18,000	-28,000	-11,200	1200
2	-10,000	28,800	-38,800	-15,520	5520
3	-10,000	17,280	-27,280	-10,912	912
4	-10,000	10,368	-20,368	-8147	-1853
5	-10,000	10,368	-20,368	-8147	-1853
6	-10,000	5184	-15,184	-6074	-3926
7	-10,000	0	-10,000	-4000	-6000
8	-10,000 + 18,000	0	8000	3200	4800

$$PW = @NPV(8\%, ATCF \text{ block}) = -\$89,437$$
$$EAC = 89,437(A/P, 8\%, 8) = 89,437 \cdot .1740 = \$15,562$$

The PW could also be calculated using the PW factors for MACRS (see Exhibit 12.14). Since $i = 8\%$ and the recovery period is 7 years, the factor is .7661. However, this will be detailed in Example 13.8.

Note that using MACRS rather than straight-line depreciation decreases the EAC of the computer workstation network to \$15,562 from \$16,370. This 5% decrease in equivalent annual after-tax cost is due solely to the faster write-off that is allowed under MACRS. It is also worth noting that a 5% decrease in costs may *double* profits, since profits are often between 1% and 10% of costs.

Selecting an After-Tax Interest Rate.   Just as with before-tax cash flows, there are two after-tax situations. Constrained project selection can be solved by ranking on IRRs—if each project's after-tax cash flows are used. Similarly, mutually exclusive alternatives can be compared after taxes.

To compare mutually exclusive alternatives after taxes, i must be specified as an after-tax rate. That rate can be derived from constrained project selection for that firm, or it may be assigned a value by upper management.

The relationship between before-tax and after-tax discount rates can be intuitively explained as follows. If a 10% rate of return is earned and the effective tax rate is 34%, then the firm will earn 6.6% and pay 3.4% to the government in taxes.

The relationship between the before- and after-tax internal rates of return and the effective tax rate may not apply to individual projects. Depreciation and interest on loan payments are not evenly distributed over a project's duration, and so the lower after-tax IRR must be computed from the project's cash flows.

13.5 CALCULATING ATCFs AND EACs USING FORMULAS

Straight-line and SOYD depreciation have regular patterns that match standard engineering economy factors for A and G. For problems that do not include loans, it is possible to replace the tables used in Section 13.4 with fairly simple formulas—for straight-line and SOYD depreciation. Using Exhibit 12.14 for the PW of the MACRS depreciation schedule, a formula can be developed for MACRS depreciation as well.

In these formulas, the ATCF will still be calculated by subtracting the calculated tax from the BTCF for each year. As shown in Examples 13.5 and 13.6, each year's tax equals the firm's marginal tax rate multiplied by its annual taxable income.

The following formulas can be simplified by partitioning each BTCF into a taxed (shown with a T subscript) and an untaxed (U subscript) portion. If there are loans with annual payments, then this separation is too complicated to be useful.

For straight-line and SOYD depreciation, the untaxed portions are simply the first cost (FC) and salvage value (S). For MACRS, only the FC is untaxed. Thus, the taxed portion equals the annual revenues minus expenses for the straight-line and SOYD methods, while the salvage value is also included for MACRS.

Before deriving the formulas for each depreciation technique, Equation 13.3 describes the results in a general way for the cash flows in year t:

$$\begin{aligned}
\text{ATCF}_t &= \text{BTCF}_t - \text{Tax}_t \\
&= \text{BTCF}_t - \text{Taxable Income}_t \cdot \text{Tax Rate} \\
&= (\text{BTCF}_{tT} + \text{BTCF}_{tU}) - (\text{BTCF}_{tT} - D_t) \cdot \text{Tax Rate} \\
&= \text{BTCF}_{tU} + \text{BTCF}_{tT}(1 - t_e) + D_t \cdot t_e
\end{aligned} \tag{13.3}$$

Equation 13.3 is expressed in terms of cash flows, but this can easily be converted for PWs or EACs. For example, for PWs, present worth factors are added on each cash flow term and on the depreciation term.

Straight-Line.   This depreciation technique is the simplest, and so is its corresponding formula for calculating an after-tax EAC. Assuming that the annual expenses form an annual series in which each annual term equals O&M, then Equation 13.4 can be used to calculate the PW or Equation 13.5 the EAC.

$$PW_{ATCF} = -FC + S(P/F, i, N)$$
$$+ [(\text{Revenues} - \text{O\&M}) \cdot (1 - t_e) - D_t \cdot t_e](P/A, i, N) \quad (13.4)$$
$$EAC_{ATCF} = FC(A/P, i, N) - S(A/F, i, N)$$
$$- (\text{Revenues} - \text{O\&M}) \cdot (1 - t_e) - D_t \cdot t_e \quad (13.5)$$

Example 13.7 applies Equation 13.5 to Example 13.5. Examples 13.5 and 13.7 calculate the same answer.

Example 13.7   EAC for Straight-Line Depreciation for Workstation Network

Example 13.5 calculated the ATCFs, PW, and EAC for a computer workstation network with a first cost of $90K, a life of 8 years, a salvage value of $18K, an annual expense of $10K, and an effective 40% tax rate. Calculate the EAC for the workstation network if $i = 8\%$.

Solution

$$EAC_{ATCF} = FC(A/P, i, N) - S(A/F, i, N) - (\text{Revenues} - \text{O\&M}) \cdot (1 - t_e) - D_t \cdot t_e$$
$$= 90K \cdot .1740 - 18K \cdot .0940 - (0 - 10K)(1 - .4) - 9K \cdot .4$$
$$= 90K \cdot .1740 - 18K \cdot .0940 + 10K \cdot .6 - 9K \cdot .4 = \$16.37K$$

This is the same answer as in Example 13.5, without constructing the full table.

MACRS Depreciation.   Equation 13.5 for the after-tax PW with straight-line depreciation can easily be adapted for MACRS, which is Equation 13.6. First, the depreciation term, $D_t \cdot t_e \cdot (P/A, i, N)$, is replaced by the first cost times a PW factor for MACRS (from Exhibit 12.14), or $FC \cdot MACRS_{PWF}$. Second, since MACRS assumes the salvage value is $0, the salvage value in year N is taxed as recaptured depreciation, so the salvage value is multiplied by $(1 - t_e)$.

In the EAC formula, Equation 13.7, the depreciation term, $D_t \cdot t_e$, is replaced with $MACRS_{PWF} \cdot t_e \cdot FC(A/P, i, N)$. Also, the salvage term is again taxed, so that a $(1 - t_e)$ term is added.

$$PW_{ATCF} = -FC + S(P/F, i, N)(1 - t_e) + MACRS_{PWF} \cdot t_e \cdot FC$$
$$+ [(\text{Revenues} - \text{O\&M}) \cdot (1 - t_e)](P/A, i, N) \quad (13.6)$$
$$EAC_{ATCF} = FC(A/P, i, N) - S(A/F, i, N)(1 - t_e)$$
$$- MACRS_{PWF} \cdot t_e \cdot FC(A/P, i, N)$$
$$- (\text{Revenues} - \text{O\&M}) \cdot (1 - t_e) \quad (13.7)$$

Example 13.8 applies these formulas to the data in Example 13.6.

Example 13.8 EAC with MACRS for Workstation Network

Example 13.6 calculated the ATCFs, PW, and EAC for a workstation network with a first cost of $90K, a life of 8 years, a salvage value of $18K, an annual expense of $10K, and MACRS depreciation. (The PW was $-\$89,437$ and the EAC was $15,562.) Calculate the EAC using the MACRS formulas assuming $i = 8\%$.

Solution

$$\begin{aligned} PW_{ATCF} &= -FC + S(P/F, i, N)(1 - t_e) + MACRS_{PWF} \cdot t_e \cdot FC \\ &\quad + [(Revenues - O\&M) \cdot (1 - t_e)](P/A, i, N) \\ &= -90K + 18K \cdot .5403(1 - .4) + .8113 \cdot .4 \cdot 90K + [0 - 10K] \cdot .6 \cdot 5.747 \\ &= -\$89,440 \end{aligned}$$

$$\begin{aligned} EAC_{ATCF} &= FC(A/P, i, N) - S(A/F, i, N)(1 - t_e) \\ &\quad - MACRS_{PWF} \cdot t_e \cdot FC(A/P, i, N) - (Revenues - O\&M) \cdot (1 - t_e) \\ &= 90K \cdot .1740 - 18K \cdot .0940 \cdot .6 - .8113 \cdot .4 \cdot 90K \cdot .1740 \\ &\quad - (0 - 10K) \cdot (1 - .4) \\ &= \$15,563 \end{aligned}$$

Sum-of-the-Years'-Digits (SOYD). The only difference between these equations and the ones for straight-line is that the depreciation term is no longer a uniform annual series. Rather, there is an annual series that matches the depreciation in year 1, and then there is a gradient series for later years. The first year's depreciation, D_1, equals $(FC - S)N/SOYD$. Each year the numerator of the fraction decreases by 1, so the gradient is $1/N$ of D_1.

$$\begin{aligned} PW_{ATCF} &= -FC + S(P/F, i, N) + (Revenues - O\&M) \cdot (1 - t_e)(P/A, i, N) \\ &\quad + D_1 \cdot t_e \cdot (P/A, i, N) - (D_1/N) \cdot t_e \cdot (P/G, i, N) \end{aligned} \tag{13.8}$$

$$\begin{aligned} EAC_{ATCF} &= FC(A/P, i, N) - S(A/F, i, N) - (Revenues - O\&M) \cdot (1 - t_e) \\ &\quad - D_1 \cdot t_e + (D_1/N) \cdot t_e \cdot (A/G, i, N) \end{aligned} \tag{13.9}$$

Example 13.9 applies Equation 13.9 to the computer workstation example. The first three terms of this equation are the same as for straight-line depreciation; it is only the last two terms for the SOYD depreciation that are different.

Example 13.9 EAC for SOYD Depreciation for Workstation Network

Using the same data as in Example 13.7, solve the problem again using SOYD. The computer workstation network has a first cost of $90K, a life of 8 years, a salvage value of $18K, an annual expense of $10K, and an effective 40% tax rate. Calculate the EAC for the workstation network if $i = 8\%$.

Solution

For this data, D_1 is \$16,000 and the arithmetic gradient is $-\$2000$. As shown in Example 12.3, $D_1 = (FC - S)N/\text{SOYD}$, or $(90K - 18K) \cdot 8/120$. Equation 13.9 states the gradient in terms of D_1/N, or \$16K/8, since it has already accounted for the minus sign.

$$\begin{aligned}
\text{EAC}_{\text{ATCF}} &= FC(A/P, i, N) - S(A/F, i, N) - (\text{Revenues} - \text{O\&M}) \cdot (1 - t_e) \\
&\quad - D_1 \cdot t_e + (D_1/N) \cdot t_e \cdot (A/G, i, N) \\
&= 90K \cdot .1740 - 18K \cdot .0940 - (0 - 10K)(1 - .4) - 16K \cdot .4 \\
&\quad + 2K \cdot .4 \cdot 3.099 \\
&= \$16.047K
\end{aligned}$$

13.6 INVESTMENT TAX CREDITS (ITC) AND CAPITAL GAINS

History of ITC.   The Tax Reform Act of 1986 repealed the investment tax credit (ITC) except for special categories (like reforestation and solar power). However, as summarized in Exhibit 13.6, this tax credit has often been repealed and re-instated. ITCs are supported because they stimulate investment (rather than the consumption that is stimulated by most tax reductions). On the other hand, ITCs are sometimes accused of stimulating inflation instead.

Exhibit 13.6   History of investment tax credit

1962	Instituted at 7%	1975	Increased to 10%
1966	Suspended for 15 months	1976	Reduced to 7%
1967	Restored after 5 months	1981	Increased to 10%
1969	Repealed	1986	Repealed
1971	Restored	199?	????

Exhibit 13.6 may be an extreme example of instability in a single provision of the tax code, but it is only a small part of the continuous changes in the tax code. It is not enough for an engineer to know the past and current tax code; sometimes the future tax code must also be estimated. Most changes in the tax code are motivated by ideals of fairness or economic efficiency; however, some cynics claim that another motivation is to generate campaign contributions from lobbyists.

An **investment tax credit** is 3% to 10% of an investment's first cost that is credited against income taxes owed.

Investment tax credits are instituted by governments to encourage investment in new plants and equipment. A certain fraction (usually 7% to 10%) of the asset's first cost is credited against or reduces the income taxes owed. Credits are subtracted from *taxes*, rather than deducted from taxable *income*, so tax credits are far more valuable than deductions. The lower the tax rate, the more valuable the credits become relative to a deduction.

Computing and Using an ITC.   The investment tax credit has at times supplemented the allowable depreciation, and at other times it has reduced the asset's depreciable basis. Example 13.10 assumes that the ITC reduces the asset's basis.

Other common restrictions on the ITC include:

1. A $100,000 limit on the amount of used property
2. A limit of $25,000 + .5 · (Tax Liability − $25,000)
3. A reduced ITC rate for shorter lives

For example from 1981–86, the rate was 0% for assets with lives less than 3 years, one-third of the maximum 10% for 3 or 4 years, two-thirds of the maximum 10% for 5 to 7 years, and 10% for 8 or more years.

Example 13.10 Computing an ITC

Assume that the ITC is reinstituted in 199X with the same provisions that existed from 1981 to 1986, including the restrictions listed earlier. Compute the PW of the workstation network in Examples 13.3–13.9. Assume that it is used in conjunction with MACRS depreciation.

Solution
The workstation network has a life of 8 years, even though its recovery period is 5 years. Assume that the life and not the recovery period determines the percentage rate, so that a 10% rate is used. The equipment is new, and the ITC is less than $25,000, so the other two restrictions do not apply.

The ITC cash flow is at year 1's end. (Note: if quarterly tax payments are assumed, then the ITC is at time 0, and the timing of depreciation is also affected.) Thus, there is a $9000 decrease in the taxes owed on the first tax return. This $9000 also reduces the basis for computing depreciation from $90K to $81K.

The first two columns in Exhibit 13.7 are from Example 13.4, but column 3 is changed to match the $81K basis, and the other columns change correspondingly. As in Example 13.5, the fifth column, taxes, is calculated by simply multiplying the fourth column by 40%, the effective income tax rate. The taxes in year 1 have two terms: one for 40% of the taxable income and the other for the 10% ITC.

Exhibit 13.7 Investment tax credit cash flow computations

Year	BTCF	D_t	Taxable Income	Tax	ATCF
0	$ −90,000	$ 0	$ 0	$ 0	$−90,000
1	−10,000	16,200	−26,200	−10,480−9,000	9480
2	−10,000	25,920	−35,920	−14,368	4368
3	−10,000	15,552	−25,552	−10,221	221
4	−10,000	9332	−19,332	−7733	−2267
5	−10,000	9332	−19,332	−7733	−2267
6	−10,000	4664	−14,664	−5866	−4134
7	−10,000	0	−10,000	−4000	−6000
8	−10,000 + 18,000	0	8000	3200	4800

$$PW = @NPV(8\%, \text{ATCF block}) = -\$84,024$$
$$EAC = 84,024 \cdot (A/P, 8\%, 8) = 84,024 \cdot .1740 = \$14,621$$

The PW is a lower cost than the −$89,437 without the ITC, and the EAC is down from the $15,562 without the ITC. Thus, this ITC would reduce both measures of the network's discounted cost by over 6%.

A capital gain is the amount by which an asset's selling price exceeds its first cost or adjusted basis.

Capital Gains.   Another often-changed provision of the tax code covers the taxation of capital gains. From Section 12.4 and Exhibit 12.12, a **capital gain** occurs when an asset is sold for more than its purchase and installation cost. The capital gain is the amount over the first cost.

When a capital gains provision is in force, assets with a holding period of more than 1 year are taxed at a lower rate. Sometimes the rate is specified. At one time only one-third of the gain was taxed (at whatever rate applied to the firm or individual).

This provision is important to investors in stocks or real estate, but it is generally unimportant for engineering projects. Machines go down in value as they are used, and as technical advances make them obsolescent. They do not go up in value. Thus, capital gains are not detailed in this text. In addition, since 1986, capital gains have been taxed at or close to the firm's tax rate for income or recaptured depreciation. A 5% change in the tax rate may increase profits, but it will not change the comparison between two alternatives significantly.

13.7 INTEREST DEDUCTIONS AND AN AFTER-TAX IRR

Interest paid by firms on loans and on bonds is tax deductible. Usually this interest is not linked to specific projects, so its tax deductibility is not part of the project analysis. In some cases, a project is linked to specific financing, and the interest deductions can be included in calculating an after-tax IRR.

When financing is included, three changes are made in the cash flows:

1. Some or all of the first cost is paid with a promise to pay later.
2. The payments are included as cash flows, but the principal portion of the payments is not included in computing taxable income.
3. The interest portion of the payments is tax deductible, so taxes are reduced.

Example 13.11 includes tax deductible financing, but not depreciation. Example 13.12 includes both. Notice that the calculation of an after-tax IRR is very simple with the spreadsheet function @IRR. Without that function, separate (*P/F*) factors would be needed, since the interest portion of the loan payment is different every period.

Example 13.11   Financing of a Wetlands Purchase

Metal Stampings uses a coal-fired cogeneration facility for steam heating and electric power. A scrubber has been installed to meet new standards on sulfur emissions, but it has created a water-pollution problem. The regional treatment district will treat the waste stream if Metal Stampings reclaims a large area of cooling ponds and fields as wetlands for a corporate nature refuge.

The wetlands reclamation would cost $6M, 80% of which would be financed by a 10-year loan at 9% (with uniform annual payments). Using the corporate guidelines of a 10-year horizon and an after-tax i of 8%, what are the EAC and PW of the project? The effective tax rate is 40%.

Solution

First, the annual loan payment is calculated for the 80% of the $6M that will be borrowed.

$$\text{Payment} = -.8 \cdot 6M(A/P, 9\%, 10) = -\$747,936$$

The year 0 cash flow is the down payment (20% of the $6M, or $-$1200K). The taxable income is just the interest to be deducted, which is 9% of the balance due after the previous year. The principal portion of each $747,936 payment simply reduces the previous balance due. The taxes are the effective tax rate of 40% times the taxable income. Finally, as shown in Exhibit 13.8, the ATCF is the BTCF minus the tax (or plus the tax savings).

Exhibit 13.8 After-tax cash flows with payments on a loan

Year	BTCF	Tax. Income = −Interest	Balance Due	Taxes	ATCF
0	$-1,200,000		$4,800,000	$ 0	$-1,200,000
1	-747,936	$-432,000	4,484,064	-172,800	-575,136
2	-747,936	-403,566	4,139,694	-161,426	-586,510
3	-747,936	-372,572	3,764,330	-149,029	-598,907
4	-747,936	-338,790	3,355,184	-135,516	-612,420
5	-747,936	-301,967	2,909,215	-120,787	-627,149
6	-747,936	-261,829	2,243,108	-104,732	-643,204
7	-747,936	-218,080	1,893,252	-87,232	-660,704
8	-747,936	-170,393	1,315,709	-68,157	-679,779
9	-747,936	-118,414	686,187	-47,366	-700,570
10	-747,936	-61,757	8	-24,703	-723,233

Handwritten annotations: 4800 K − (747936 − 432000); 4800K × 0.09; 4,484,064 × 0.09; 432 K × 0.4; BTCF − Taxes

Notice that the after-tax cash flows approach the before-tax values in later years, when the amount of income tax is small. Using a spreadsheet or a financial calculator, the PW is $-$5.431M, or an EAC of $809.4K. The present cost is less than $6M, because the loan's after-tax cost is less than the 8% time value of money (even though its before-tax cost is 9%).

Example 13.12 Financing of a Project

A manufacturing project has a first cost of $900K. After 10 years, the equipment will have a salvage value of $100K. Annual revenues will be $300K, and annual O&M costs will be $125K. The firm uses an after-tax i of 9%. The effective tax rate is 40%.

Financing is available through a 5-year loan at 10%. The uniform payments are made annually. Assume that 60% of the project will be financed through a loan and 40% internally. Find the PW and EAW for the project after taxes, and find the project's IRR.

Solution

Exhibit 13.9 is the spreadsheet that is used to solve this problem (stating all cash flows in $1000s). It has been simplified by finding the interest paid each year with the @IPAYMT function, so that no computation of balance due is required. Since this problem includes both depreciation and loan payments, it is easier to treat the columns for before-tax cash flows (BTCF) and for taxable income (TaxInc.) as two separate computations.

The before-tax cash flows equal revenue minus O&M expense minus loan payments, except for years 0 and 10, which include the first cost paid by the firm (−$360K) and the salvage value ($100K). The taxable income equals revenue minus O&M expense minus interest on the loan. This formula does not subtract from taxable income the $540K of principal payments that are spread out over the 10 years; only the interest payments reduce taxable income.

Again, the taxable income and the BTCF columns differ in that taxable income considers depreciation and the interest portions of the payments, while the BTCF column includes both the principal and interest portions of the payments.

Note that it does not matter that the before-tax interest rate for the loan exceeds the firm's discount rate. The after-tax interest rate for the loan is about $10\%(1 - t_e)$, or 6%. It is not this value exactly, because the after-tax cash flows for the loan are not a uniform series. The rate is clearly less than the firm's 9% discount rate, so the loan is economically attractive.

Exhibit 13.9 Including loan payments and depreciation

	A	B	C	D	E	F	G	H
1	900	FC	125	O&M	10%	loan rate		
2	100	S	9%	i	5	loan N		
3	300	Revenue	40%	tax rate	60%	loan frac		
4			540	loan amt	142.451	loan payment		
5								
6	year	BTCF	MACRS	Deprec	Interest	TaxInc.	Tax	ATCF
7	0	-360					0.0	-360.0
8	1	32.55	0.1429	128.61	54	-7.6	-3.0	35.6 =C8*$A$1
9	2	32.55	0.2449	220.41	45.15	-90.6	-36.2	68.8
10	3	32.55	0.1749	157.41	35.43	-17.8	-7.1	39.7 =F10*$C$3
11	4	32.55	0.1249	112.41	24.72	37.9	15.1	17.4
12	5	32.55	0.0893	80.37	12.95	81.7	32.7	-0.1 =B12-G12
13	6	175.00	0.0892	80.28		94.7	37.9	137.1
14	7	175.00	0.0893	80.37		94.6	37.9	137.1
15	8	175.00	0.0446	40.14		134.9	53.9	121.1
16	9	175.00		0		175.0	70.0	105.0
17	10	275.00		0		275.0	110.0	165.0
18						PW at 9%		109.01
19		@ IPAYMT ($E$1, $A12, $E$2, -$C$4)				IRR		13.87%
20				+$A$3-$C$1-D17-E17+A2				

Leverage.   A common financial tactic is to borrow money to undertake projects so that the firm can do more projects. The expected rates of return from the projects exceed the interest rates that are paid on the loans, so the firm has higher PWs and IRRs.

As shown in Example 13.13 and Exhibit 13.10, this financial **leverage** increases the rate of return that is earned on the portion of the investment supplied by the firm. Although not detailed here, this leverage also increases the sensitivity of the PW or IRR to the costs and revenues. If the revenues go down or the costs go up for a highly leveraged project, then the PWs or IRRs go down disproportionately.

At 0% financed, the IRR is the return from the project's cash flows without considering loan payments or tax-deductible interest. Unless the financing is inextricably linked to the project, it is suggested that this rate be used for decision making. If financing and leverage are included, then the same fraction financed should be used for all projects.

Leverage is the increase in an attractive project's PW or IRR by financing part or all of it at an interest rate below the MARR. The variability of the PW and IRR also increases if revenues go down or costs go up.

Example 13.13   Financial Leverage

For the project in Example 13.12, calculate the IRR for financing percentages ranging from 0% to 90%. At 0% financed, the IRR of 10.77% is the return from this project's cash flows. Again, unless the financing is inextricably linked to the project, it is suggested that this rate be used for decision making.

Solution
Either a DATA table function or a series of calculations using the spreadsheet in Exhibit 13.9 can be used to construct Exhibit 13.10. It is possible to enter the fraction financed (0% to 90%) into spreadsheet cell E3; the PW and IRR then appear in cells H18 and H19, respectively. These results can be copied as values into a range for the graph. The process is: (1) enter the fraction financed; (2) the computer calculates the PW and IRR; (3) the value is copied as a value and not as a formula; and (4) go to (1).

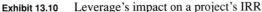

Exhibit 13.10   Leverage's impact on a project's IRR

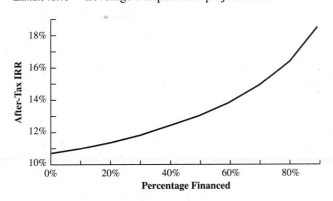

13.8 SUMMARY

This chapter has explained the calculation of federal corporate income taxes. The viewpoint taken is that of the firm, whose goal is to pay the legal minimum amount of income tax and that as late as possible. The government's goal of promoting the public good has been used to explain why the tax code includes investment tax credits. The chapter has emphasized that taxes always matter to the firm, but that comparisons of some alternatives can be made either before or after taxes— without changing the recommendations.

Federal corporate income taxes are computed using tax rates in the progressive marginal tax rate schedule (the rates range from 15% to 35%). Within certain income ranges, a 5% or a 3% surtax is applied, so that corporations with taxable incomes between $335,000 and $10M pay tax at a flat rate of 34%, and corporations with taxable income above $18.33M pay tax at a flat rate of 35%.

Taxable income and before-tax cash flows are related but different. Depreciation reduces taxable income, but it is not a cash flow. Similarly, purchasing capital assets outright or through loans is a cash flow, but it does not change the taxable income. Salvage values are not taxed for straight-line, declining balance, and SOYD depreciation, but they are taxed as recaptured depreciation for MACRS.

Taxes and PW, EAC, and IRR measures are often most easily calculated using tables of before-tax cash flows, depreciation, taxable income, taxes, and after-tax cash flows. For straight-line, SOYD, and MACRS depreciation, shortcut formulas can be used if revenues and costs are uniform series.

Investment tax credits are important incentives that make engineering projects more attractive by reducing the taxes that will be paid. On the other hand, capital gains tax breaks are much less important for engineering projects, since physical equipment usually declines in value over time.

Finally, spreadsheets have been used to calculate after-tax cash flows, PWs, and IRRs in the presence of loans and tax-deductible interest. Tax-deductible interest must be considered for projects in which the investment and the financing are inextricably linked together, but the norm for engineering economic decisions (especially between mutually exclusive alternatives) is to consider projects without considering the financing. One reason is to avoid the bias of allowing different projects or alternatives to be evaluated with different amounts of leverage.

This chapter has only touched upon the intricacies of our tax code. Many corporate experts, tax accountants, and lawyers work full-time keeping up with the ever-changing tax code and using it to the corporation's advantage. Our focus is much more limited. This chapter has presented basic material suitable for project analysis, but matching the complexity of the tax code is beyond this author and many engineers.

REFERENCES

Grant, Eugene L., W. Grant Ireson, and Richard S. Leavenworth, *Principles of Engineering Economy*, 8th ed., 1990, Wiley.

Smith, Gerald W., *Engineering Economy: Analysis of Capital Expenditures*, 4th ed., 1987, Iowa State University Press.

PROBLEMS

Note on computer use: Many or most of these problems would be solved more easily using the tabular mathematical functions in your spreadsheet or word processing program. Computer support is recommended for the problems with computer icons.

13.1   Find an article in the financial press (*The Wall Street Journal, Fortune, Business Week*, etc.) on a proposed change in the federal tax code. Summarize the government's, firm's, and analyst's perspectives on the proposed change.

13.2   SmallTech Manufacturing had overhead expenses of $80,000, depreciation of $55,000, and interest expenses of $25,000. The manufacturing and distribution costs were $50,000. What is the taxable income, tax, and after-tax cash flow if sales were
   a. $235,000?
   b. $635,000?
   (*Answer:* a. tax = $3750; b. tax = $144,500)

13.3   TinyTech's taxable income is $72,000. A new product would increase this by $15,000 (to $87,000) per year for 10 years.
   a.   What tax is paid without the new product?
   b.   What tax is paid with the new product?
   c.   What are the product's marginal and average income tax rates?

13.4   LowTech Inc. had revenues of $400,000 in the year just completed. The company incurred $133,000 in cash expenses and $24,575 in depreciation expenses. Determine the federal taxes owed by LowTech and the after-tax cash revenues.   (*Answer:* ATCF = $189,200)

13.5   ABC Block's taxable income was $85,000 in the year just completed. The state where ABC was incorporated taxes corporate profits at a rate of 8% on the first $50,000 and 10% on any remaining income above $50,000. Calculate the state and federal taxes owed by ABC. Also calculate the effective marginal tax rate on ABC's income.   (*Answer:* t_e = 26%)

13.6   The Red Ranger Company recorded revenues of $45,000 and recaptured depreciation of $1200 for the year just ended. During the year it incurred cash expenses of $23,500 and depreciation expenses of $11,575. It also sold real estate for a profit (income not capital gain) of $40,000. Determine the federal taxes owed by Red Ranger.

13.7   LargeTech Manufacturing has the following estimates for a new semiconductor product: The machinery costs $800K, and it will have no salvage value after 8 years. LargeTech is in the highest tax bracket. Income starts at $200K per year and increases by $100K per year, except that it falls by $300K per year in years 7 and 8. Expenses start at $225K per year and increase by $50K per year. What are the before-tax cash flows and taxable income under
   a.   Straight-line depreciation?
   b.   SOYD depreciation?

    c. Double-declining balance depreciation?
    d. MACRS?
    e. Each method, if the machine's salvage value is $40,000?

13.8 A milling machine costing $18,000 will be depreciated using straight-line depreciation over 10 years. The firm is in the top tax bracket. What are the after-tax cash flows if the salvage value is estimated to be
    a. $4000? (*Answer:* ATCF$_3$ = $490)
    b. $0? (*Answer:* ATCF$_3$ = $630).

13.9 An engineering and construction management firm with a $1M taxable income is opening a branch office. Nearly $250,000 worth of furniture will be depreciated under MACRS.
    a. What are the after-tax cash flows for the first 5 years?
    b. If the office is closed early in year 6 and the furniture is sold to a second-hand office furniture store for 10% of its initial cost, what tax is paid? (*Answer:* −$10,463.5)

13.10 Gillespie Gold Products Inc. is considering the purchase of new smelting equipment. The new equipment is expected to increase production and decrease costs, with a resulting increase in profits. Determine the after-tax cash flow using a tax rate of 42% and (a) straight line and (b) sum-of-the-years'-digits depreciation.

    Cost $40,000
    Savings per year $10,000
    Life   6 years
    Salvage value $4000

13.11 Five years ago, the XYZ Company purchased a $30,000 automated production line. The line has just been sold for $5000. The machinery was depreciated using the straight-line method to a salvage value of $0. The actual savings due to the purchase of this line are presented below. Was the purchase justified if XYZ's MARR is 8%? Taxes are paid at the 34% marginal rate.

Year	1	2	3	4	5
Savings	$5K	$6K	$7K	$7K	$4K

(*Answer:* No: PW = −$4298)

13.12 The R&D lab of BigTech Manufacturing will purchase a $1.8 million process simulator. It will be replaced at the end of year 5 by a newer model. Use MACRS and a tax rate of 40%. The simulator's salvage value is $.5M.
    a. What is the after-tax cash flow due to the simulator in year 5?
    b. What is the equipment's after-tax EAC over the 5 years if the interest rate is 10%?

13.13 Find the EAC after taxes for a computer that costs $15,000 to purchase and $3500 per year to run, and that has a salvage value of

$1000 at the end of year 7. The company's profit exceeds $20 million per year, and it is purchasing $850,000 worth of capital assets this year. The company uses an interest rate of 12%.   (*Answer:* $4648)

13.14   A computer costing $32,000 will be written off, using MACRS, by a small firm whose tax rate is 15%. Annual operating costs are $8000 for 9 years—after which there will be no salvage value.
   a.   What are the after-tax cash flows without Section 179?
   b.   What are the after-tax cash flows with Section 179?

13.15   A machine was purchased for $60,000 and depreciated, using SOYD depreciation, over 15 years with a $5000 salvage value. If the machine is sold for $2000 in year 10, what is the after-tax cash flow due to depreciation and sale in year 10? Assume that the firm is in the top tax bracket.

13.16   NewTech has decided to redesign its open office areas. It has decided to install workstation cubicles for each engineer. These cubicles are classed as built-in furniture and cost $2000 each. The 11 cubicles will last 15 years, with maintenance costs of $75 per year. Assume that the cubicles have $0 salvage values. Assume MACRS and the current tax law. The corporation's taxable income places it in the top federal income tax bracket. NewTech's interest rate is 9%.
   a.   Find the EAC for each cubicle.
   b.   Find the after-tax cash flows for years 0 through 15.

13.17   The purchase of a large-volume copier is being considered.

First Cost	Maintenance Cost (per year)	Salvage	Life
$18,000	$1400	$5000	10 years

   a.   Find the after-tax cash flow for year 3 for the copier, using MACRS and the current tax law. The corporation's taxable income is about $5 million.
   b.   Find the EAC after taxes for the copier, assuming a MARR of 8% and traditional straight-line depreciation over 10 years.
   c.   Find the EAC after taxes, assuming MACRS depreciation and an *i* of 8%.

13.18   Which pump should be installed and how much cheaper is it after taxes? Assume straight-line depreciation and a tax rate of 40%. The after-tax MARR is 10%.

Pump Type	First Cost	Annual O&M Costs	Life (years)	Salvage Value
Brass	$ 9500	$150	10	$ 875
Stainless Steel	12,000	90	20	3000

   (*Answer:* EACs virtually identical)

13.19   Assume that the pump in Problem 13.18 is used as manufacturing equipment. Using MACRS, which pump should be installed, and how much cheaper is it?   (*Answer:* stainless saves $129 per year)

13.20   If the pump in Problem 13.18 were evaluated before taxes, would the conclusion be any different? What interest rate did you assume, and why?

13.21   If the pump in Problem 13.19 were evaluated before taxes, would the conclusion be any different? What interest rate did you assume, and why? Is there any reason why MACRS may matter more or less than straight-line depreciation for problems similar to this?

13.22   Which water-treatment system should be installed, and how much cheaper is it after taxes? Assume SOYD depreciation and a tax rate of 40%. The interest rate is 8%.

System Type	First Cost	Annual O&M Costs	Life (years)	Salvage Value
Brass	$20,000	$1500	10	$2000
Stainless Steel	40,000	1200	30	3000

13.23   Assume that the treatment system in Problem 13.22 is used as manufacturing equipment. Using MACRS, which treatment system should be installed, and how much cheaper is it?

13.24   If the treatment system in Problem 13.22 were evaluated before taxes, would the conclusion be any different? What interest rate did you assume, and why?

13.25   If the treatment system in Problem 13.23 were evaluated before taxes, would the conclusion be any different? What interest rate did you assume, and why? Is there any reason why MACRS may matter more or less than straight-line or SOYD depreciation for problems similar to this?

13.26   Jaradet Jewel Inc. is considering the purchase of new drilling equipment. The new equipment (two choices) is expected to increase production with a resulting increase in profits. If Jaradet pays taxes at the 34% marginal tax rate and has a MARR of 8%, which alternative should it choose?

	A	B
Cost	$50,000	$40,000
Savings/year	$13,000	$10,000
Depreciation method	Straight-line	Straight-line
Useful life	5 years	5 years
Declared salvage value	$5000	$4000
Expected sale price	$6000	$6000

(*Answer:* PW$_A$ is $583 higher)

13.27   Old McDonald Farm Inc., a large, profitable crop farm, must replace its planting equipment. The farm expects all investments to return at least 8%.

	Qwik Plant	Robo Plant
Cost	$160,000	$95,000
Operating cost/year	$12,000	$9000
Depreciable salvage value	$20,000	$10,000
Depreciation method	SOYD	SOYD
Actual salvage value	$24,000	$10,000
Life	8 years	4 years

13.28   An investment opportunity in securities has been presented by MT Pockets Brokerage to the investment firm of Johnson, Todd, and Xanders Inc. (JTX). The first cost of the securities will be $150,000. Each year, it is expected that the securities will yield $15,000 in income. At the end of 5 years, the securities will be sold for $250,000. If JTX pays taxes at the 20% marginal rate and requires a 15% return on all investments, should the securities be purchased? Assume capital gains are taxed at the 20% rate.

13.29   The Anhouser-Shrub Brewing Company (AB), makers of Budsmarter Beer, must purchase a new bottle-labeling machine. The machine is expected to save production costs and to increase efficiency. The three best alternatives are presented below. AB pays taxes at the 34% marginal rate and has an interest rate of 12%. MACRS depreciation is used for all capital purchases.

	Kwik-Stik	Rite-Stik	Straight-Stik
Cost	$100,000	$75,000	$110,000
Annual savings	$25,000	$15,000	$30,000
Life	12 years	12 years	12 years
Salvage value	$10,000	$6000	$17,500

13.30   We-Clean-U Inc. expects to receive $42,000 each year for 15 years from the sale of its newest soap, OnGuard. There will be an initial investment in new equipment of $150,000. The expenses of manufacturing and selling the soap will be $17,500 per year. Using MACRS depreciation, a marginal tax rate of 42%, and an interest rate of 12%, determine the EAW of the product.   (*Answer:* $225)

13.31   The Sandwich Company, whose earnings put it in the 46% combined state and federal tax bracket, is considering purchasing a piece of food-handling equipment for $25,000. The equipment has a useful life of 4 years and a salvage value of $5000. The resale value of the equipment will be $6550. The new equipment is expected to increase the company's earnings by $8000 in each of the 4 years. Determine the EAW using (a) straight line depreciation and (b) MACRS depreciation. The MARR for Sandwich is 10%.

13.32 A large company must build a bridge to have access to land for expansion of its manufacturing plant. The bridge can be fabricated of normal steel for an initial cost of $30,000 and should last 15 years. Maintenance (cleaning and painting) will cost $1000 per year. If a more corrosion-resistant steel were used, the annual maintenance cost would be only $100 per year, although the life would be the same. In 15 years, there will be no salvage value for either version of the bridge. The company pays taxes at the 48% marginal rate and uses straight-line depreciation. If the minimum acceptable after-tax rate of return is 12%, what is the maximum amount that should be spent on the corrosion resistant bridge?

13.33 If the simulator in Problem 13.12 is purchased on a 3-year loan, what is the EAC? The loan is for 80% of the cost; the interest rate is 12%; and the payments are uniform.   (*Answer:* $259.4K)

13.34 The simulator in Problem 13.12 will save $600,000 per year in experimental costs.
   a. What is the after-tax IRR?
   b. What is the after-tax IRR if the loan in Problem 13.33 is used to buy the equipment?
   c. Graph the IRR for financing percentages ranging from 0% to 90%.

13.35 A small research firm purchases new equipment valued at $80,000. The equipment is expected to produce a net income of $15,000 for each year of its 6-year useful life. Using MACRS, determine the IRR for the project if the company is taxed at the 25% marginal rate. Do not use Section 179.   (*Answer:* 2.75%)

13.36 A heat exchanger purchased by Hot Spot Manufacturing cost $24,000. The exchanger will produce savings of $6500 in each of the 10 years it is in service. Using a tax rate of 34% and straight-line depreciation over the 10-year life, determine the IRR. The exchanger will have a salvage value of $500 at the end of its life.

13.37 The Boomer Golfclub Corporation is considering the installation of new, automated club-assembly equipment. Boomer pays federal taxes at the 34% marginal rate. Determine, using EAW, whether alternative A or B, or neither, should be chosen, based on an interest rate of 6%.

	A	B
Cost	$30,000	$40,000
Savings/year	$6000	$6375
Salvage value	$2000	$3000
Life	6 years	8 years
Depreciation method	straight-line	straight-line

(*Answer:* EAW$_B$ is $302 and EAW$_A$ < $0)

13.38 NewTech is in need of new jigs for some assembly equipment. The jigs favored by the manufacturing engineer cost $30,000 and are expected to provide service for 6 years. The annual operating costs are estimated to be $2000. The industrial engineer favors a choice that costs $35,000 and that will also provide service for 6 years at an estimated annual cost of $1500. Using MACRS depreciation, a tax rate of 34%, and a MARR of 10%, which jigs should be chosen?

13.39 LargeTech is choosing between countries A and B for a new manufacturing plant. To analyze tax policies, LargeTech will briefly assume matching costs of $120M for construction, $25M for annual operations, and $0 for salvage. The plant's life will be 20 years. The interest rate is 15%, and both countries have tax rates of 40%. Country A requires straight-line depreciation over 5 years. Country B permits SOYD but requires a 20-year depreciation period. Where is the plant's PW higher? By how much?

13.40 Smith (see References; pp. 161–163) showed that after taxes, a low salvage value can be better than a high one. To confirm this, calculate the EAC after taxes for a machine costing $500K. Compare salvage values of $400K and $0 after a life of 30 years. Use straight-line depreciation, an interest rate of 10%, and a tax rate of 40%.

13.41 Senator Paula Stevens wants to revise the investment tax credit. If most firms use an interest rate of 15% and are taxed at 34%, which plan is more attractive to them? Plan A allows a 10% credit with a 10-year MACRS schedule. Plan B allows a 7% credit with a 5-year MACRS schedule. For both plans, the credit is subtracted from the basis for depreciation.

APPENDIX 13A INDIVIDUAL INCOME TAXES

This appendix briefly introduces individual income taxes. It includes schedules and examples for single and married taxpayers—but it does not cover all cases. For example, there is no coverage of married taxpayers filing separately, only coverage for joint returns.

Exhibit 13A.1 details the progressive marginal tax rate schedules that apply to single persons and to married couples filing joint returns.

Unlike businesses, most expenses of individuals are not tax deductible. Instead, the tax code incorporates a personal exemption, which protects $2350 of income from taxation for each family member. Some expenses, such as medical expenses, charitable contributions, moving expenses, and home mortgage interest, are deductible within some limits. Unless these expenses are high, however, it is often more attractive to claim the standard deduction ($3700 for single taxpayers and $6200 for joint returns).

The first step in computing taxable income is to find the adjusted gross income. This is computed by adding together wages, interest received or credited

Exhibit 13A.1   U.S. individual income tax schedule

Tax Rate	Single	Married (Joint)
15%	≤ $22, 100	≤ $36, 900
28%	$22,101 to $53,500	$36,901 to $89,150
31%	$53,501 to $115,000	$89,151 to $140,000
36%	$115,201 to $250,000	$140,001 to $250,000
39.6%	over $250,000	over $250,000

Note 1: These figures are for taxes due in April 1994. The dollar amounts that define the tax brackets are indexed to inflation, so they change yearly.

Note 2: For incomes above $108,450 per single filer and $162,700 for joint filers, personal exemptions of $2350 each are phased out.

to accounts, alimony, etc., for one or both taxpayers. Then either the standard deduction or the sum of an itemized list is subtracted. A typical limit on deductible expenses is that only those medical expenses in excess of 7.5% of adjusted gross income are deductible. Finally, the $2350 exemption per family member is subtracted (unless the total income received exceeds $108,450 for single filers or $162,700 for joint filers).

The resulting taxable income is taxed at the rates shown in Exhibit 13A.1. Example 13A.1 illustrates this for a young, single engineer. Example 13A.2 illustrates this for an older, married pair of engineers with two kids.

This appendix is too brief to provide details, but young engineering students often need to analyze the following situations:

1. Is it better to get married in December and file a joint return, or to get married in January and each file a separate return?

2. Is it better to claim a standard deduction or to itemize so that moving expenses can be claimed in connection with starting a job after graduation?

3. Is it economically more attractive to lease or to buy? Home mortgage interest is tax deductible, but there are large costs in buying or selling a house.

Example 13A.1   Marco Is Single

Marco earns $38,000 and is single. He does not itemize. What is his taxable income and tax?

Solution

$$\text{Taxable Income} = 38{,}000 - \text{Standard Deduction} - 1 \text{ Exemption}$$
$$= 38{,}000 - 3700 - 2350 = \$31{,}950$$
$$\text{Tax} = .15 \cdot 22{,}100 + .28(31{,}950 - 22{,}100) = \$6073$$

Marco's marginal tax rate is 28%. His average tax rate is $6073/38{,}000 = 16.0\%$.

Example 13A.2   Christine and Mike Are Married

The Knight-Smiths are both chemical engineers. Their salaries total $95,000 per year. Interest on their savings totals another $3000 per year. They have two school-age kids. Their itemized deductible expenses total $15,000 this year (mostly mortgage interest, property tax, and state income taxes). What is their taxable income and how much tax do they pay?

Solution
First, there is no reduction on their exemptions because their joint income is less than $162,700 per year. Second, they are better off itemizing than claiming the standard deduction.

$$\text{Taxable Income} = 98,000 - \text{Itemized Deductions} - 4\,\text{Exemptions}$$
$$= 98,000 - 15,000 - 4 \cdot 2350 = \$73,600$$
$$\text{Tax} = .15 \cdot 36,900 + .28(73,600 - 36,900) = \$15,811$$

   Their marginal tax rate is 28%, and their average tax rate is $15,811/98,000 = 16.1\%$. Note that the child care credit would reduce their tax by 20% of the $4800 limit, or $960. Their actual expenses were nearly double the $4800 limit.

PROBLEMS

13A.1   Ringo is in his third year with a rapidly growing high-tech company. He earns $43,000 per year, he is single, and he rents a condo. What is his taxable income, his tax, his marginal tax rate, and his average tax rate?   (*Answer:* tax = $6582)

13A.2   Laura is in her 10th year with a rapidly growing high-tech company. She earns $63,000 per year, she is single, and she rents a condo. What is her taxable income, her tax, her marginal tax rate, and her average tax rate?

13A.3   If Laura and Ringo were married and they still rented the condo, what would be their taxable income, their tax, their marginal tax rate, and their average tax rate?

13A.4   If Laura and Ringo had three kids and $26,000 of itemized expenses, what would be their taxable income, their tax, their marginal tax rate, and their average tax rate?

13A.5   If Laura and Ringo are planning to marry should they do it in December or January? Assume that a December marriage implies a joint return for this year, while waiting until January allows them to file separately. How much of their honeymoon expenses can they fund with the difference?

CHAPTER 14 PUBLIC-SECTOR ENGINEERING ECONOMY

The Situation and the Solution

Government agencies must decide which projects should be supported with available funds—in spite of politics, competing objectives, and benefits that are difficult to quantify.

The solution is to accept that the answers may be less accurate than in the private sector, and to use benefit/cost methodology correctly.

Chapter Objectives

After you have read and studied the sections of this chapter, you should be able to:

Section 14.1
Define benefits, disbenefits, and costs.

Section 14.2
Explain why evaluating and choosing projects in the public sector is more difficult than in the private sector.

Section 14.3
Select the correct method and interest rate for (1) mutually exclusive projects, and (2) sets of public projects constrained by a budget, and apply the correct method to problems of deferred maintenance.

Section 14.4
Select the correct point of view, with corresponding internal and external costs.

Section 14.5
Allocate costs among different project purposes and different benefit recipients.

Section 14.6
Use the value of a human life for project evaluation.

Section 14.7
Apply cost-effectiveness techniques for benefits that can be quantified, but not in terms of money.

Appendix 14A
Analyze the higher value of incremental IRRs, as compared to incremental benefit/cost ratios, for mutually exclusive alternatives.

Key Words and Concepts

Benefits   Positive consequences that accrue to the public.

Disbenefits   Negative consequences that accrue to the public.

Costs   Positive and negative consequences that accrue to the government.

Spillovers   Consequences that are not accounted for in the economic analysis.

Benefit/cost ratio   A common measure that equals *net* benefits divided by costs (both in present worth or equivalent annual terms).

Consumers' surplus   The total benefit received by the users of a public facility, minus the total user fees or costs.

Cost effectiveness   A methodology applied when the benefits cannot be equated to money, but they can be quantitatively measured.

14.1 DEFINING BENEFITS, DISBENEFITS, AND COSTS

In common usage, benefits are positive outcomes and costs are negative ones. In public-sector applications of engineering economy, a more specialized definition is useful. **Benefits** and **disbenefits** are consequences to the public at large, while **costs** are consequences to the government.

Reading newspaper and magazine articles about public projects requires insight about **spillovers** [Sassone and Schaffer]. These are not specifically accounted for in the economic analysis, and in fact, they may not even be identified. Often, these are disbenefits. For example, nearby residents may be inconvenienced by trucks traveling to and from a new landfill site. Spillovers can also be positive. For example, R&D projects that supported the space program have led to unanticipated new products in the consumer market.

Benefits are positive public outcomes, **disbenefits** are negative public outcomes, and **costs** are outcomes to the government.

Spillovers are consequences that are not accounted for in the economic analysis.

Example 14.1   A New Dam

A new hydroelectric dam has been proposed. What are some of the benefits, disbenefits, and costs likely to be?

Solution

The dam's benefits are likely to include electric power, lake-based recreation, flood control, water storage for irrigation, and allocation to different water users. Potential disbenefits include the loss of river-based recreation; businesses, homes, and farms inundated by the reservoir; and possible flooding due to catastrophic dam failure. Costs are likely to include construction costs, compensation to the owners of land that will be inundated, and operation costs for the dam.

However, each dollar should be counted only once. For example, suppose a fifth-generation family farm and home is valued by the family at $350K. If the government compensation for inundating it with the reservoir is $200K, based on the market value, then the cost is $200K and the uncompensated disbenefit is $150K.

Agencies that Do Not Serve the Public Directly.   The definitions of benefits and costs are designed for projects done by the government for the public. These definitions work well for the National Park Service, the Public Health Service, and the Bureau of Reclamation. But not all government units serve the public directly. For example, the Defense Logistics Agency and the General Services Administration approve or complete projects that serve other agencies.

Such units usually define benefits as accruing to other agencies and costs as accruing internally to the unit. For example, a computer services unit might consider the hardware and software consequences as costs, and the time savings to clerks in other governmental agencies as benefits. That way, computer projects would be evaluated by the same mechanisms that were used for other government projects.

Some of the most difficult problems in public-sector economic evaluations occur in the defense industries. How do you place a value on a faster jet, a more capable tank, or a nuclear submarine? Since we do not have a market value for national defense, it is sometimes necessary to rely on cost-effectiveness measures (see Section 14.7).

"Benefits to Whomsoever They Accrue."   The benefit and cost terms used in public-sector engineering economy can be traced to J. Dupuit's "On the Measurement of Utility of Public Works" (published in 1844), but one of the clearest statements is in the U.S. Flood Control Act of 1936. That is: "if the benefits to whomsoever they may accrue are in excess of the estimated costs, and if the lives and social security of people are otherwise adversely affected."

One theoretical difficulty with the language "to whomsoever they may accrue" is that each outcome causes another layer of outcomes. For example, in Exhibit 14.1, the primary benefits, disbenefits, and costs cause a secondary set, which in turn cause another set.

Exhibit 14.1   A new traffic light at a dangerous intersection

	Primary Level	
Work for:	Less property damage	Fewer lost lives
Manufacturer		
Contractor		
City employee		
⇓	⇓	⇓
	Secondary Level	
Income spent:	Fewer insurance claims	Less inheritance
By manufacturer	Less work for:	Fewer promotions
By employees	Doctors	
Other projects not done,	Hospital	
delayed, or more	Lawyers	
expensive		
Higher taxes		
⇓	⇓	⇓
	Tertiary Level	

More spending on goods and services by doctors, auto body shop employees, etc.; less income for car manufacturers, etc.

Another difficulty is correctly accounting for transactions between the government and the people. For example, if a property owner is compensated by the government after it takes the land by *eminent domain* for a dam or a road, is the payment a valuation of the disbenefit to the landowner or a cost to the government? Similarly, if I pay $.08 per kWh for electricity, does that measure the value of the electricity to me, or should it be classified as revenue (a negative cost) to the government?

While the answer in these two examples may be arbitrary, there is a clear principle: each dollar should be included once, and only once, in our calculations. In other words, there is no double-counting of benefits and user fees.

14.2 WHY ARE PUBLIC-SECTOR PROBLEMS DIFFICULT?

There are several reasons why public-sector engineering economy problems are often more difficult than private-sector problems. These include the following:

Benefits that are difficult to quantify and to value in money terms

Uncertain probabilities for rare events

Projects with multiple objectives

Interest groups with conflicting perspectives

Difficulties in selecting the interest rate

These differences are introduced here, detailed in later sections, and summarized in Exhibit 14.2.

Quantifying and Valuing Benefits.   The chief difficulty in public-sector engineering economy is poor, uncertain data. Many benefits are difficult to quantify at all, let alone in terms of money. For example, what is the value of improving access to education with a new library, reducing the probability of a fatal accident, reducing the level of carbon monoxide, reducing visual clutter through a new sign ordinance, adding a fishing day on a new lake, or avoiding a war through deterrence?

In some cases, engineering economy in the public sector is used to calculate the impact of regulations. The question is: Do the benefits of an Environmental Protection Agency (EPA) or a Consumer Products Safety Commission regulation exceed its costs? Most benefits and compliance costs will be incurred by firms and individuals (the public). In these cases, data may only be available from regulated industries, whose self-interest may inflate the estimated costs of compliance.

For example, the Occupational Safety and Health Administration (OSHA) was considering reducing the standard for industrial workers' exposure to polyvinyl chloride (PVC) from 14 ppm to 1 ppm. The industry predicted economic losses of $69–90 billion and 2 million lost jobs. The standard was imposed anyway, and the actual consequences have been estimated as $325 million in economic losses and 290 lost jobs [Tafler]. The health benefits required to justify the regulatory change under the predicted data are 250 to 7000 times larger than those required under the actual costs.

Probabilities of Rare Events.   Some difficulties in estimating benefit values occur because the project protects the public from *rare* events such as earthquakes, traffic accidents, major floods, or occupational cancers. The benefit must be valued, and its probability estimated. Estimating probabilities is easier and far more accurate if the event is common rather than rare. It is easier to estimate the size and consequences of a 10-year flood, than of a 100-year flood.

Some projects create hazards. A nuclear power plant has a very small possibility of an uncontrolled release of radioactivity, and a predictable stream of radioactive waste. The alternatives have their own hazards. Coal-fired power plants contribute to acid rain, and the reservoirs behind dams can cause immediate, catastrophic floods if the dam fails (see Example 14.2).

Example 14.2   Flooding and Dam Failures

One common justification for dams is providing flood control. By collecting surges from rainstorms or snowmelt and releasing them in a controlled manner, dams can reduce or eliminate flooding downstream.

However, the dam's reservoir creates a new hazard—the chance of a catastrophic flood due to dam failure. In the U.S., this chance is about .0001 per year for each dam [Mark and Stuart-Alexander]. In fact, there is some evidence that as our understanding and technology improve, we are building larger dams with about the same failure rate. Larger dams and deeper reservoirs increase the probabilities of earthquakes—which can cause dam failure. The catastrophic flood can occur without dam failure, as was demonstrated at the Vaiont Dam in Italy in 1963, when a landslide into the reservoir overtopped the dam and 2000 people were killed.

How significant is the new hazard compared to the basic flood-control benefit?

Solution
Consider the 1976 failure of the Teton Dam in Idaho, which caused 11 deaths and between $400M and $1B in property damage when it released 288,000 acre-feet of water. This project's economic analysis (which omitted the catastrophic-flood hazard) calculated PWs at an interest rate of $3\frac{1}{4}\%$, and the benefits exceeded the costs by 59%.

The impact of the catastrophic hazard to property can be estimated using the average estimate for property damage ($700M) and the Bureau of Reclamation's failure probability for earth-filled dams with reservoirs larger than 1000 acre-feet (.00015 per year), then the benefits exceed the costs by only 51%.

If the site is labeled "high risk" and the probability is increased by a factor of 10, then there are *no* flood-control benefits and total benefits exceed costs by only 15% [Mark and Stuart-Alexander].

Multiple Objectives that May Conflict. Many public projects have multiple purposes. The dam in Example 14.1 was intended to generate power, control flooding, provide recreational opportunities, and support irrigation. If a public project has multiple purposes, allocating costs to each purpose is needed to charge the correct budgets and establish equitable user fees.

The multiple purposes may conflict. To continue with the dam example, maintaining low water levels provides the maximum capacity for flood control, while higher levels are better for recreation and power generation. Similarly, water releases for power and for irrigation may be needed at different rates over the year.

Interest Groups with Differing Perspectives. In addition, the benefits of public projects are received by different groups [Lang]. For example, should *your* home be torn down so that *my* commuting trip will be faster? Or which one of us is closer to the new elementary school, or farther from the city dump or the regional hazardous waste site? Should an upstream community improve its sewage treatment for the benefit of a downstream community's water supply? Because different groups of people are affected differently, equity is a key concern.

These differences in viewpoint also occur between governmental units, such as between the civilian Secretary of Defense, the army, and the navy; the federal government and the state highway and city transit departments; and state departments of education and local school districts. The local agency is charged with filling the needs at a local level, while the state agency strives to do the best projects on a statewide basis and to maintain some regional equity. The federal government takes the broadest view of all.

Selecting an Interest Rate. Finally, selecting an interest rate is harder for public agencies than for private firms. Public taxation and borrowing both remove money from the private sector for public purposes—consequently, the cost of public borrowing and taxation is difficult to calculate. Also, the opportunity cost for forgone projects is difficult to calculate because the projects are proposed by many different agencies and the benefits are often difficult to calculate in dollar terms.

The differences between private and public sector engineering economy are summarized in Exhibit 14.2.

Exhibit 14.2   Engineering economy in the public sector vs. the private sector

Factor	Public Sector	Private Sector
Data	Benefits must be: 1. Quantified and 2. Equated to money	Most benefits are monetary
Probability	Rare events often crucial (1 chance in 100 to 1 in a billion)	1 chance in 10 often the limit
Objectives	Multiple	Maximize PW or IRR
Stakeholders' perspectives	Often conflicting	All want firm to be successful
Interest rate	Complicated by nonmonetary benefits	Derived as an opportunity cost or from cost of borrowing

14.3 CORRECT METHODS AND INTEREST RATES

Evaluating a Single Project.   For a single project, deciding whether the benefits of a project will exceed its estimated costs is a go/no go decision. The interest rate must be supplied, since this is a mutually exclusive decision between yes and no.

The interest rate may be the city's interest rate on prospective bonds or it may be the 10% value that has been mandated by the Office of Management and Budget within the Executive Office of the President. (OMB's 10% rate is mandated to apply to constant-value dollars, so relative price level changes may be taken into account if necessary, but general inflation is not.)

The decision is whether the benefits are at least as large as the costs, or $B \geq C$. As long as C is positive, this can be converted to Equation 14.1.

The **benefit/cost ratio** equals net benefits divided by costs (both stated in present worth or equivalent annual terms). The B/C ratio must be greater than or equal to 1 for acceptable projects.

$$B/C \geq 1 \qquad\qquad (14.1)$$

The benefits and the costs must both be stated either as present values or in equivalent annual terms. The **benefit/cost (B/C) ratio** value is unaffected by the decision between present values and equivalent annual terms, as shown in Example 14.3.

Example 14.3   Highway Improvement

Calculate the B/C ratio for a highway project with the following benefits and costs. The project life is 40 years; the interest rate is 10%; and the project's right-of-way is worth $5M in 40 years.

$ 20M	Construction cost (includes acquiring right-of-way)
350K	Annual maintenance cost
3M	Repaving every 8 years
1M	Annual value of lives saved (1 per year)
1.25M	Time savings for commercial traffic $\left[\dfrac{\$20}{hour} \cdot \dfrac{.25\ hours}{trip} \cdot \dfrac{250,000\ trips}{year}\right]$
1M	Time savings for commuter and recreational traffic $\left[\dfrac{\$5}{hour} \cdot \dfrac{.25\ hours}{trip} \cdot \dfrac{800,000\ trips}{year}\right]$

Solution

Most of the costs are stated in annual terms; thus, it will be slightly easier to compute the ratio in annual terms. The benefits include lives saved and time savings for two groups. The cost equation shown below includes the construction cost, the annual maintenance, the periodic repaving, and in the final term the "correction" for the repaving that does not occur in year 40 and the right-of-way value.

$$B = 1M + 1.25M + 1M = \$3.25M$$

$$C = 20M(A/P, 10\%, 40) + .35M + 3M(A/F, 10\%, 8)$$

$$- (3M + 5M)(A/F, 10\%, 40)$$

$$= 2.046M + .35M + .262M - .018M = \$2.64M$$

$$B/C = 3.25M/2.64M = 1.23$$

Since 1.23 is greater than 1, build it!

To convert the benefits and costs to present values, it is easiest to simply multiply B and C by $(P/A, 10\%, 40)$. Since both the numerator and the denominator are multiplied by the same number, the ratio is unchanged.

Criteria for Mutually Exclusive Alternatives.   If there is more than one mutually exclusive alternative, then, as discussed in Chapter 10, it is necessary to calculate B/C ratios for the *incremental* investments. As shown in Example 14.4, it is wrong to choose an alternative because it has the largest ratio or because it is the largest project with a ratio greater than or equal to 1.

Instead, each incremental investment is evaluated by asking if its B/C ratio is greater than or equal to 1. As shown in Chapter 10, calculating each alternative's present or equivalent annual worth is an easier way to find the best mutually exclusive alternative. The difference between an alternative's benefits and costs, B − C, is not changed if another alternative is added, or if the data for any other alternative is modified.

Example 14.4   Incremental Comparisons with B/C Ratios

Show that if B/C ratios are used, only incremental analysis chooses the best mutually exclusive alternative for the following data.

	Present Worth		
Project	B (M)	C (M)	B − C (M)
A	$ 3	$ 1	$2
B	10	5	5
C	13	9	4
D	18	14	4
E	25	19	6*
F	30	27	3

*Best, since max PW.

Solution

Two wrong criteria are sometimes suggested for analyzing B/C ratios for mutually exclusive alternatives. These use the ratios for each project (which are implicitly compared to doing nothing). The two criteria are (1) choosing the project with the largest ratio, and (2) choosing the largest project whose B/C ratio is greater than or equal to 1. Both lead to wrong answers in this example, as shown below.

	Present Worth			
Project	B (M)	C (M)	B/C Ratio	
A	$ 3	$ 1	3.0	(highest B/C ratio)
B	10	5	2.0	
C	13	9	1.44	
D	18	14	1.29	
E	25	19	1.32	(best, from max PW)
F	30	27	1.11	(largest project with B/C ≥ 1)

Project A, with a B/C ratio of 3.0, has the largest B/C ratio. But project A's PW is only $2M, which is the smallest of all the alternatives.

Project F is the largest one with an acceptable B/C ratio that is greater than or equal to 1 (project F's ratio is 1.11). But project F's PW is only $3M, which is the second-smallest of all the alternatives.

The correct criteria, (1) maximizing the PW and (2) evaluating the incremental B/C ratios, select project E. Its PW, at $6M, is the highest, and, as shown in the next table, an incremental B/C analysis makes the same recommendation. In this table, the incremental benefits and costs are calculated one line at a time; in each case, the increment is between the current challenger and the best defender found so far. For example, the incremental benefit to do project B rather than project A is $10M − $3M, or $7M.

Since project A's incremental ratio exceeds 1, A is better than doing nothing. Since project B's incremental ratio exceeds 1, B is better than A. However, neither C nor D is better than B. Project E is better than B, and, finally, project F is not better than E. So do project E! (Note: if IRRs were used for this mutually exclusive problem, then incremental IRRs would have to be calculated.)

| | Incremental PW | | |
Project	B (M)	C (M)	Incremental B/C
A vs. nothing	$3	$1	3.0
B vs. A	7	4	7/4 = 1.75
C vs. B	3	4	3/4 = .75
D vs. B	8	9	8/9 = .89
E vs. B	15	14	15/14 = 1.07
F vs. E	5	8	5/8 = .63

Criteria for Constrained Project Selection.   Government agencies, like private-sector firms, have budgets that prevent them from doing all acceptable projects. Thus, the principle of using opportunity costs to establish an interest rate applies. Ranking on IRR is needed to find the opportunity cost of forgone investments.

Ranking on IRR is correct when a set of projects must be selected within a budget limit. Ranking on B/C ratios is wrong, just as ranking on PW indexes was shown to be wrong in Chapter 9. The difference in recommended sets of accepted alternatives for ranking on IRR vs. ranking on B/C ratios is illustrated by Example 14.5.

Ranking on B/C ratios is incorrect because it assumes that the opportunity cost of forgone investments is the externally defined interest rate rather than the higher value implied by the budget limit.

Example 14.5   DOTPUF Project Selection

The Department of Transportation, Public Utilities, and Facilities (DOTPUF, pronounced "dot puff") is responsible for a wide variety of public projects. Historically, the minimum acceptable interest rate for computing life-cycle costs has been 15%. However, in many years, only one-third or one-half of the projects that meet that standard have been funded.

This year the state legislature has asked that projects be prioritized by ranking on the B/C ratio computed at 15%. If DOTPUF expects funding for about $1M in projects, is there a difference between the top projects ranked on IRR and those ranked on B/C ratios?

To simplify calculations, the data below includes the IRR for each project and also the B/C ratio computed at 15%.

Project	First Cost	Annual Benefit	Gradient	Life (years)	IRR	B/C at 15%
A	$400	−$100	$70	10	22.7%	1.72
B	300	100	−5	25	27.3	1.52
C	250	80	−5	15	23.9	1.34
D	500	70	10	40	22.8	1.80
E	350	0	30	10	21.9	1.46
F	200	50	0	20	24.7	1.56
G	125	−90	20	30	17.1	1.79
H	250	100	−7	50	31.0	1.43

Solution

When ranked by the B/C ratio, the projects are ranked as follows:

D, G, A, F, B, E, H, and C.

The first three of these projects have a combined first cost of $1.025M.

As detailed in Chapter 9, when organizations are solving the constrained project selection problem, ranking on IRR is the correct approach. In this case, when ranked on IRR, the projects are ranked as follows:

H, B, F, C, D, A, E, and G.

The first four of these projects have a combined first cost of $1M. Since this matches the available budget, the recommended set of projects is H, B, F, and C.

When the capital budget is limited to about $1M, the minimum attractive rate of return is 23.9%—not 15%. Thus, if the B/C ratios were computed at 24%, the B/C ranking could be used. But ranking on IRR must be used first to find the interest rate. The ratios computed at 15% are misleading.

Deferred Maintenance.   Many government agencies find their capital and operating budgets to be limited when compared to the tasks they are asked to perform. Government income comes from taxes, which are set politically, rather than through an economic analysis.

As a result, engineers who work in public agencies may not be able to do what is obviously best. Capital projects are often divided into two capital budgets—one for major new projects and one for smaller projects, such as maintenance of existing facilities. New buildings, new roads, and new bridges have clear benefits that can earn political support from users, taxpayers, and political figures.

Maintenance is much less visible, even though it may be very important economically. Because it is less visible, it often receives less political support. It is common for agencies to postpone or defer maintenance when budgets are tight. As shown in Example 14.6, this can be a very costly strategy. Effective maintenance is very important in minimizing the life-cycle cost of the public infrastructure.

This example illustrates a poor choice between mutually exclusive alternatives (repair now or replace later) that is caused by a poor solution to the constrained project selection problem. The maintenance project, at a cost of $200K, should receive very high priority for capital funding, even though it is not a high-visibility project with high political appeal.

Example 14.6   Bridge Maintenance

Many bridges are built of steel-reinforced concrete. As long as the steel is protected from corrosion, the bridge lives are very long. The cost difference between proper maintenance and early replacement is also very large. For example, a particular bridge would cost $200,000 now to repair or it can be replaced in 10 years for $1.8M. The state highway department claims it uses an interest rate of 8%, but it has decided to replace the bridge later, due to a current shortage of funds. The state agency also anticipates some federal funding for the replacement bridge. What is the implied interest rate for choosing to replace later rather than repair now?

Solution

The present worth equation for saving $200,000 now and spending $1.8M after 10 years is:

$$0 = +.2M - 1.8M/(1 + i)^{10}$$

Solving this equation leads to the interest rate at which the agency is borrowing money from the future. This rate, $i = 24.6\%$, is very high.

14.4 WHOSE POINT OF VIEW?

What Is Internal? What is External? Deciding which consequences are internal to the problem and which are external depends on your perspective. This is the principal difference in engineering economy between the public and private sectors. A private-sector firm can define as internal the short- or long-run financial consequences to its bottom line. A government agency must usually take a broader view.

The difference can be illustrated by a firm that chooses to meet air-pollution standards with a taller smokestack that disperses the pollutants over a broader area so that concentrations are lower. The firm is concerned with the question of the most cost-effective way to meet the regulatory standard. The local air-pollution district must balance the needs of local industry with the needs of local residents for jobs and for clean air. It must consider all pollution sources, not just one. Moreover, the health and safety consequences may be the district's principal focus. A federal agency may be concerned with the cumulative effect on acid rain over a large region of many such smokestacks.

Federal Subsidies. Transportation problems, as in Example 14.7, illustrate the difficulty of selecting the proper perspective. The federal government may pay most of the construction costs for highways or transit systems that will be operated by states or cities. Similarly, state governments often subsidize the construction of schools that will be operated by local governments. There are many examples where local users pay only the unsubsidized portion of the cost. The most cost-effective solution to them may not be the best solution from a broader perspective.

This problem is exacerbated by the political nature of public-sector decision making. There are many competing interests with parochial viewpoints. Since the data are poor, there is a lot of room to interpret the situation in conflicting ways.

Example 14.7   Subsidized Transportation Improvements

MidCity has severe rush-hour traffic jams in a congested valley that connects to its main industrial area. The transportation planning department's engineers have identified two plans. The first involves a light rail system (LR), and the second is a new freeway (NF).

Both alternatives have expected lives of 50 years, when their rights-of-way will have comparable values. MidCity uses an interest rate of 10%.

The federal government subsidizes freeway construction by paying 90% of the costs. It also provides 70% of the costs for light rail construction. It provides no funds for the operation of either system. The benefits of each system are being debated, but the air-pollution benefits of the rail system about equal the extra user benefits for the larger number of freeway users. The total benefits for each system are about $15 million.

Which alternative is preferred by MidCity's taxpayers? Which is better from a systems or overall perspective?

Factor	Light Rail	New Freeway
First cost	$40M	$70M
Annual operating cost	8M	6M

Solution

From a systems viewpoint or overall perspective, either system can be justified since the $15M in annual benefits exceeds the annual cost for both (see equations below). Since the benefits of both alternatives are the same, only the costs need be compared. As shown below, the light rail's EAC is $1.1M per year less than the new freeway's. The light rail option is better from an overall perspective.

$$C_{LR} = 40M(A/P, .1, 50) + 8M = 40M \cdot .10086 + 8M = \$12.0M$$

$$C_{NF} = 70M(A/P, .1, 50) + 6M = 70M \cdot .10086 + 6M = \$13.1M$$

However, the viewpoints of MidCity's taxpayers and "average" U.S. citizens differ. To MidCity's taxpayers, the relevant capital cost is the fraction they pay that is *not* subsidized by the federal government. This is the 30% of the $40M in light rail costs or the 10% of the $70M in new freeway costs. Thus, the higher subsidy for freeways implies that MidCity taxpayers would prefer the lower costs *to them* of the new freeway alternative.

To MidCity:

$$C_{LR} = .3 \cdot 40M(A/P, .1, 50) + 8M = 12M \cdot .10086 + 8M = \$9.2M$$

$$C_{NF} = .1 \cdot 70M(A/P, .1, 50) + 6M = 7M \cdot .10086 + 6M = \$6.7M$$

Consumers' Surplus.   Because governments must consider "benefits to whomsoever they may accrue," they must consider the value to the users of government services. Consider a motorist paying a bridge toll. The opportunity to cross the bridge must be worth more than the toll, or the driver will take another route or stay home.

If the toll authority were private, it would consider only total tolls. Assume instead that the transportation authority is public and that it is deciding whether to build the toll bridge. Then all benefits received by the motorist, including those in excess of the toll, must be considered.

Intuitively, the **consumers' surplus** is the total extra benefit received by everyone, over and above the user fees paid for the service. Mathematically, this equals total benefits received by all users minus the costs that are paid to use the government service.

The excess in value received over the user fee is the individual's consumer surplus. When summed over all individuals, the excess in value received is the **consumers' surplus.**

Setting user fees has social and political consequences. Thus, the user fee for a government service may be set much lower than the cost to provide the service. For example, many municipal libraries charge a nominal $5 fee for a library card, and then no other charges as long as books, tapes, etc. are returned on time.

For public projects, all benefits must be considered. This total equals the consumers' surplus, plus user fees. However, different users place different values on government services. For example, I might be willing to pay $5 to cross a bridge, while you might be willing to pay only $3. If the toll is $1, then my surplus is $4 and yours is $2. For the two of us, the total toll is $2, the total surplus is $6, and the total benefit is $8.

Estimating the number of motorists willing to pay $10, $9, $8, ..., $2, $1, etc., leads to a curve similar to the one shown in Exhibit 14.3. Economists call this a demand curve because it identifies how many individuals would pay for (or demand) a service at different prices. In this case, no one is willing to pay more than $10; 10% of the current traffic level would pay $8; 30% would pay $6; 60% would pay $4; and the current traffic level occurs at a toll of $2.

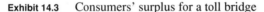

Exhibit 14.3　　Consumers' surplus for a toll bridge

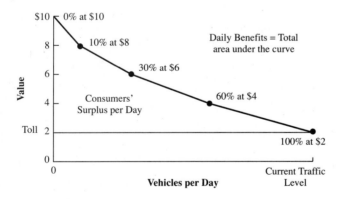

Now two public perspectives are possible.

First, a bridge authority that must recover the cost of the bridge and pay for its maintenance would simply consider the total revenue from tolls. This revenue equals the toll times the number of motorists.

Second, a regional transportation authority would consider the total benefits (including the consumers' surplus) in deciding on the relative priority for building this bridge vs. other potential projects. This second problem is much more difficult because it implies that the entire demand curve must be estimated—not just a point or a segment.

14.5 ALLOCATING COSTS TO BENEFIT RECIPIENTS

Public projects often involve multiple objectives with different classes of benefit recipients. If user fees are collected, then the costs of the project must be allocated to the different users.

For example, a dam might provide power, irrigation, recreation, and flood control. User fees could be established for the first three uses. Property taxes for land in the floodplain might be possible for the fourth use. However, questions like the following must be answered: Do recreational users pay for the extra costs of boat ramps, picnic areas, and public facilities, or do they pay a pro rata share of the project's total costs?

Example 14.8 illustrates the allocation of costs to benefit recipients.

Example 14.8   Paying for the Speed-Skating Oval

Northern Heights wants to sponsor the winter Olympics. A speed-skating oval must be added to its existing facilities. Which alternative, the Ice Palace or the temporary inflatable, is better? Since political support is crucial, allocate the costs to the various uses. The two alternatives are a permanent and a temporary facility.

1.   Construct a permanent Ice Palace for later use by residents, tourists, and local colleges. A permanent ice rink in the oval's middle will be a practice rink for the figure skaters. Use of the practice rink will save $4M in operating and security costs for alternate practice facilities. Moveable bleachers can be placed on the rink during the speed-skating competition, or, after the Olympics, on the oval during rink events.

2.   Construct a temporary speed-skating oval in an inflatable building. After the Olympics, convert the oval to a cinder track for spring, summer, and fall use by the local schools and colleges. No practice rink is included.

Northern Heights' interest rate is 10%. The temporary facility would be up for one year (there is also an intercollegiate competition scheduled then). The Ice Palace and the cinder track have no salvage value after the 40-year study period (assume 39 years of use and 1 year of competition). Note: both alternatives have the same $3M cost of operations for the year of competition, so it can be ignored.

	Ice Palace	Temporary Inflatable and Cinder Track
Facility cost	$20M	$ 5M
Salvage value after competition	0	1M
Extra practice rink costs	0	4M
Annual operating cost	1M	50,000

Solution
Supporters of the Ice Palace will provide at least two justifications. First, the fancier facility will be more attractive to the International Olympic Committee. Northern Heights will then have a higher probability of being selected as the host city. Second, the Ice Palace as a unique facility will add more to the community than would the cinder track. Detractors

will emphasize that the Ice Palace costs far more. The ability of boosters to "sell" the Ice Palace may hinge on the allocation of Ice Palace costs between the Olympics (Oly.) and Northern Heights (NH).

The costs of the temporary facility and the cinder track are easy to allocate. The Olympics are charged the cost of the facility, the extra practice rink costs, and the salvage value for the inflatable building. Northern Heights incurs a cost of $50,000 per year for 39 years.

$$PW_{Oly.} = -5M + 4M + 1M/1.1 = -\$8.04M$$
$$PW_{NH} = -50K(P/A, .1, 39)(P/F, .1, 1) = -\$.44M$$

The Ice Palace's first cost must be allocated between the Olympics and Northern Heights. At one extreme, Northern Heights might claim that the Olympics should account for all capital costs. The resulting allocation is:

$$PW_{Oly.} = -\$20M$$
$$PW_{NH} = 1M(P/A, .1, 39)/1.1 = -\$8.87M$$

This implies that the Olympics budget should pay for 100% of the capital costs for the Ice Palace.

At the other extreme, the Olympics might claim that Northern Heights should pay all costs over and above those of the temporary facility and the practice rink savings, since the permanence is for Northern Heights' 39 years of future use. The resulting allocation is:

$$PW_{Oly.} = -\$8.04M$$
$$PW_{NH} = -20M + 8.04M - 8.87M = -\$20.83M$$

This would allocate 40.2% of the Ice Palace's capital costs to the Olympics, and $11.96M, or 59.8%, to Northern Heights. In both cases, the total PW for the costs is $-\$28.87M$.

The cinder track's PW is $20.39M larger. However, if the Ice Palace increases the probability of becoming the Olympics host city or if its benefits are large enough, then the Ice Palace may be better. From the local perspective, the two allocations are nearly $12M different. This difference represents benefits that would have to be generated for the Ice Palace to be worthwhile.

14.6 VALUING THE BENEFITS OF PUBLIC PROJECTS

A Life's Present Value.   Many public benefits are difficult to quantify, such as reducing pollution, adding airbags to cars, and preserving pristine areas as wilderness. Often, the chief difficulty is in valuing a human life. In an individual wrongful-death case, the value of a particular human life may be based on the present worth of that individual's expected earnings, as in Example 14.9. Or, as in Example 14.10, the breakeven value of a life (cost per fatality) can be calculated.

Example 14.9   Early Death of an Engineer

A married civil engineer with two young children was killed on a job site when a crane's load shifted and fell. Punitive damages will be assessed if and when negligence by a contractor's employee is established. Compensation to the family for loss of income will be made using an interest rate of 6%, even if negligence is not proven. The engineer was 35, and retirement at age 65 is to be assumed. The engineer's salary at death was $45,000. If the salary was expected to increase $1000 per year in constant-value dollars, what is the present worth of the family's lost income? Is this a realistic estimate of the value of a human life?

Solution

$$P = 45,000(P/A, 6\%, 30) + 1000(P/G, 6\%, 30)$$
$$= 45,000 \cdot 13.765 + 1000 \cdot 142.36 = \$761,800$$

This approach does not compensate for noneconomic losses, and it implies that children, other nonemployed individuals, and a happy retirement have little value. However, there are no living expenses for the deceased individual.

Example 14.10   Reducing Highway Accidents

Data over the last 5 years indicates that for each fatality, there are 40 nonfatal injury accidents ($15,000 present cost each) and 300 property damage accidents ($2000 present cost each). What is the breakeven cost per fatality needed to justify a highway project if *i* is 8%?

The death rate on a particular three-lane road is 8 per 100 million vehicle miles. Adding a lane would reduce this to 5 per 100 million, and other accidents would be reduced proportionately. The lane would cost $1.5M per mile to build and annual maintenance would be 3% of the first cost. Assume that the lane would last 40 years. The road carries 10,000 vehicles per day.

Solution
The annual cost per mile for construction and maintenance equals

$$\text{EAC}_{\text{mile}} = 1.5M(A/P, 8\%, 40) + .03 \cdot 1.5M = 125.8K + 45K$$
$$= \$170.8K$$

Since 3.65 million vehicles travel each mile each year, the fatality rate is originally $3.65M \cdot 8/100M$, or .292 fatalities per mile each year. This is the basis for all three types of accidents, since there are 40 injury and 300 property damage accidents per fatality. The reduction factor is 3/8 for all accidents. Let CF equal the cost per fatality.

Setting the annual cost per mile equal to the annual benefit per mile will lead to the breakeven cost per fatality:

$$170.8K = (3/8) \cdot .292(\text{CF} + 40 \cdot 15K + 300 \cdot 2K)$$
$$\text{CF} = \$359.8K$$

Standards of Federal Agencies.   Exhibit 14.4 shows explicit standards for the value of a life for National Highway and Transportation Safety Administration (NHTSA) and the Air Force [Greer and U.S. Congress]. Note that the Air Force valued civilian lives more highly than military lives, and officers more highly than enlisted personnel. The value for the Nuclear Regulatory Commission (NRC) can be converted through risk analysis to a value of a human life.

Exhibit 14.4 Standards for value of a human life

Agency	Standard	Value per Life
National Highway & Transportation Safety Administration		$359K
Nuclear Regulatory Commission	$1000/body-rem within 50 miles	$5–10M
U.S. Air Force*		$47K–330K

*Note: permanent total disability higher

Exhibit 14.5 lists the cost per life saved for the regulations of several agencies. At an average performance of $64K per life, NHTSA clearly has projects and regulations that are much more cost-effective than the minimum level represented by the $359K-per-life standard.

The high cost per life for the Occupational Safety and Health Administration (OSHA) is misleading, because many of its standards are intended to prevent injury rather than death. Similarly, the effective performance of the Consumer Product Safety Commission may be due to a low funding level that ensured a focus on high-priority targets.

Exhibit 14.5 also illustrates the cost effectiveness of several approaches to reducing highway deaths. Mandatory seat belts cost far less per life saved than airbags. However, they also save fewer lives because they are less effective.

Exhibit 14.5 Effectiveness of regulation for several agencies

Agency/Regulation	Cost per Life Saved
Occupational Safety & Health Administration	$12.1M
Environmental Protection Agency	2.6M
Health & Human Services	102K
National Highway Transportation & Safety Administration	64K
Consumer Product Safety Commission	50K
Airbags	130K
Passive restraint	30K
Mandatory seatbelts	500

Source: Tafler.

Risk and Valuing Public Benefits. Dealing explicitly with risk requires the tools presented in Chapter 18. However, Exhibit 14.6 introduces some of the risks that engineering economy must value to solve public-sector problems.

Note that the engineering economic analysis of regulations must consider tradeoffs among these risks. For example, a safety change for aircraft travel might increase air fares. This would increase the number of families that travel by automobile—which is about seven times more dangerous per mile. To help put these risks in perspective, Exhibit 14.6 includes several examples that are personally controllable, while others are regulated by government.

Exhibit 14.6   Risks estimated to increase annual chance of death by 1 in a million

Activity	Cause of Death
Smoking 1.4 cigarettes	Cancer, heart disease
Drinking .5 liter of wine	Cirrhosis of the liver
Spending 1 hour in a coal mine	Black lung disease
Spending 3 hours in a coal mine	Accident
Living 2 days in New York or Boston	Air pollution
Traveling 6 minutes by canoe	Accident
Traveling 10 miles by bike	Accident
Traveling 150 miles by car	Accident
Flying 1000 miles by jet	Accident
Flying 6000 miles by jet	Cancer from cosmic radiation
Living 2 months with a cigarette smoker	Cancer, heart disease
Eating 40 tablespoons of peanut butter	Liver cancer from aflatoxin B
Eating 100 charcoal-broiled steaks	Cancer from benzopyrene
One chest X ray taken in a good hospital	Cancer from radiation
Living 20 years near a PVC plant	Cancer from vinyl chloride (1976 standard)
Living 150 years within 20 miles of a nuclear power plant	Cancer from radiation

Source: R. Wilson, "Analyzing the Daily Risks of Life," *Technology Review*, Volume 81, Number 4, 1979, pp. 40–46; in *Acceptable Risk* by Baruch Fischhoff, Sarah Lichtenstein, Paul Slovic, Stephen L. Derby, and Ralph L. Keeney, 1981, Cambridge University Press, p. 81.

14.7 COST EFFECTIVENESS

A **cost-effectiveness** measure is a numerical evaluation measure, such as cost per life saved or student served per dollar.

In some cases, a ranking method for public projects is used instead of stating quantifiable benefits, such as the number of lives saved, in dollars. Example **cost-effectiveness** measures include the cost per life saved, the cost per fishing day, the cost per additional high-school graduate, and the average time savings per road-improvement dollar.

If cost is in the measure's numerator, then smaller ratios are better. This is the case in the cost per life saved measure in Exhibit 14.5 and the cost per resident measure used in Example 14.11. If cost is in the denominator, such as lives saved per $1M, then larger ratios are better.

Comparing mutually exclusive choices with a cost-effectiveness measure requires incremental analysis. For constrained project selection, the measure of cost effectiveness used to rank projects implies a dollar value for the unit of effectiveness (at the agency's interest rate). In Example 14.11, project A is the worst one funded, and it is the basis for calculating the marginal value for each resident served. This can be a capital cost per life saved, or an EAC over a specified life and at a given interest rate.

This approach is similar to the risk/return relationships discussed in Chapter 18 and to more complex multiple-objective techniques, which are discussed in Chapter 19.

Example 14.11 Alaskan Village Airports

Many Alaskan communities are accessible only by air or water. The state is considering a program to build airports for year-round air access to communities with only several months a year of river or ocean access. The costs to build each airport vary due to differences in soils, local terrain, locally available construction equipment, and distance from the state's urban centers. Each airport serves a different number of residents.

Construct a cost-effectiveness measure to rank the following projects for funding. Assume that the state's interest rate is 10%, and that no additional state funds will be provided over an assumed life of 20 years. If a budget of about $12 million is established, what is the implicit annual value of each citizen served by these projects? Which communities should be served?

Community	Airport Cost (M)	Number of Residents
A	$3.4	300
B	2.3	250
C	4.5	275
D	1.9	400
E	3.2	175
F	3.1	190
G	2.7	150
H	3.9	575

Solution

Cost per resident, which should be minimized, is a good measure. In the following table, (1) the cost per resident is calculated, (2) the projects are listed in order of decreasing effectiveness, and (3) the cumulative first cost is calculated. If the budget is about $12M, then $11.5 million for communities D, H, B, and A should be spent. (Save $.5M for use as a contingency fund.) The marginal cost per resident is $11,333, the value for project A.

Community	Airport Cost (M)	Number of Residents	Cost per Resident	Cumulative Cost (M)
D	$1.9	400	$ 4750	$ 1.9
H	3.9	575	6783	5.8
B	2.3	250	9200	8.1
A	3.4	300	11,333	11.5
F	3.1	190	16,316	14.6
C	4.5	275	16,364	19.1
G	2.7	150	18,000	21.8
E	3.2	175	18,286	25.0

Based on the last project funded (for community A), the implied value can be calculated by setting the costs and the benefits equal.

$$0 = -3.4M(A/P, .1, 20) + \text{Benefit}_A \cdot 300$$

$$\text{Benefit}_A = 3.4M \cdot .1175/300 = \$1331 \text{ per year per resident}$$

14.8 SUMMARY

This chapter discussed how to apply the material of other chapters to public-sector problems. The principles are the same, but the applications are more difficult, primarily because of the difficulties in quantifying benefits and equating them to money.

Within this application area, the benefit/cost ratio is often applied. The numerator of this ratio is the sum of the benefits and disbenefits (positive and negative consequences) that accrue to the public. The denominator of this ratio is the costs and the revenues that accrue to the government. The ratio can be calculated using present values or using equivalent annual amounts. But the key question is: Do the benefits exceed the costs?

Engineering economy in the public sector is difficult for several reasons:

Benefits that are difficult to quantify and to equate to money.

Rare-event probabilities that must be estimated.

Multiple objectives that often conflict.

Multiple stakeholders whose interests often conflict and whose viewpoints are often parochial.

Interest rates whose selection is complicated by the other four reasons.

Two different approaches to the selection of an interest rate and of recommended alternatives apply.

First, for problems with one go/no go option or several mutually exclusive alternatives, the easiest approach is to maximize the present worth or equivalent annual worth. If B/C ratios are used, then incremental analysis is necessary.

Second, for problems where several alternatives must be selected within a budget constraint, ranking on IRR is used to calculate the opportunity cost of forgone projects.

Public projects must take a broad systems view, which includes the effects of subsidies and of consumers' surplus. Finally, because of the multiple objectives of public projects, it is often necessary to allocate costs to objectives and to benefit recipients.

The wide variety of values for a human life is simply one of the many examples of the futility of precisely defining the benefits of many public projects. However, when the benefits can be quantified, and only one nonmonetary benefit exists, then a cost-effectiveness measure can be used to select the best project(s) from a set of projects.

REFERENCES

Fischoff, Baruch, Sarah Lichtenstein, Paul Slovic, Stephen L. Derby, and Ralph L. Keeney, *Acceptable Risk*, 1981, Cambridge University Press.

Greer, William R., "Value of One Life? From $8.37 to $10 Million," *New York Times*, June 26, 1985.

Lang, Hans J., *Cost Analysis for Capital Investment Decisions*, 1989, Marcel Dekker (Chapter 16 on Objectivity).

Mark, R.K., and D.E. Stuart-Alexander, "Disasters as a Necessary Part of Benefit-Cost Analyses," *Science*, Volume 197, September 16, 1977, pp. 1160–1162.

Sassone, Peter G., and William A. Schaffer, *Cost-Benefit Analysis: A Handbook*, 1978, Academic Press.

Tafler, Susan, "Cost-Benefit Analysis Proves a Tough Task," *High Technology*, Volume 2, Number 4, July/August 1982, pp. 76–77.

U.S. Congress, Joint Hearings of Subcommittees of U.S. House of Representatives, *Use of Cost-Benefit Analysis by Regulatory Agencies*, Serial 96-157, 1980.

PROBLEMS

14.1 A toll road between Kansas City and Chicago is being considered. Assume that it will be government financed. Name at least three benefits, three disbenefits, and three costs. How could each value be estimated?

14.2 A domed municipal stadium is being touted to attract professional sports franchises. Name at least four each of benefits, disbenefits, and costs. How could these values be estimated? What are some likely spillovers?

14.3 Name at least five benefits of the national initiative for Space Station Freedom. Name three disbenefits. How could these values be estimated?

14.4 The right-of-way for a new highway will cost $1.5 million. The roadbed and earthworks will cost $3 million. The pavement will cost $2.5 million, and its life is 20 years. The discount rate is 9% and the project has an assumed life of 20 years. The new highway will save travelers $750K worth of travel time each year. What is the B/C ratio if the right-of-way cost is the disbenefit to property owners? What is the B/C ratio if the right-of-way cost is the price paid by government in eminent domain proceedings? (*Answer: .973, .978*)

14.5 In Problem 14.4, assume that the right-of-way and roadbed have perpetual lives. What are the B/C ratios?

14.6 A busy intersection in MetroCenter needs renovation. Plans call for adding a turn lane, a new computer-controlled signal, and new sidewalks. The estimated cost is $1M. The intersection's annual maintenance cost will be $5K. Users will save $45K in time each year.

Fewer accidents are expected to save $55K in property losses annually and $90K for a reduced loss of life for motorists and pedestrians. The renovation is expected to handle traffic for the next 8 years. Using an interest rate of 5%, calculate the project's B/C ratio.   (*Answer:* 1.19)

14.7   Lorra, Missouri, must either contract out its landfill operations or build a new municipal landfill. The city uses a discount rate of 5%, and the city engineer is required to use benefit/cost ratios. Which plan should she choose, and on which ratio(s) should she base the decision? Both sites have lives of twenty years.

	Contract	New Landfill Site
First cost	$250,000	$5,500,000
Annual:		
Operating cost	750,000	350,000
Citizen's travel	80,000	65,000
Citizen's time	110,000	85,000

14.8   A new airport expansion will cost $85 million. The largest amount, $55 million, is for land acquisition and major earthworks, which will last as long as the airport is used. The next $20 million is for terminal buildings, which will last 50 years; these buildings will cost $2.5 million per year in O&M. The last $10 million will be spent on runways. These will also last forever, with a major repaving every 20 years at a cost of $4 million. What annual benefit is required for B/C $\geq$ 1? Assume an interest rate of 8% and a perpetual life.   (*Answer:* $9.42M)

14.9   Assume a study period of 100 years with $0 salvage values for Problem 14.8. If the initial annual benefit is $6 million, what must the annual gradient be for B/C $\geq$ 1?

14.10 A road costs $450,000 to build and $12,000 per year for street cleaning. The road is restriped every 2 years for $28,000 and repaved every 10 years for $290,000. Assume that the road will last 60 years and that the interest rate is 8%. How many users per year are needed to justify the road if each trip is valued at $2? Assume restriping cost not included in repaving cost.   (*Answer:* $40,790)

14.11 Which of the following mutually exclusive alternatives should be selected using B/C analysis?

Alternative	PW of Cost	PW of Benefits
1	$100K	$120K
2	200K	225K
3	250K	270K
4	325K	345K
5	400K	420K

14.12 The state Department of Interior (DOI) must ensure the continued quality of the environment. The state Budget Department has informed the DOI that its budget for the next fiscal year will be between $2.0M and $2.25M for environmental proposals. Which of the proposals listed below should be funded?
  a. Rank using IRR.
  b. Rank using benefit/cost ratios calculated at an interest rate of 10%.

Proposal	First Cost	Annual Benefits	Life
A	$ 125,000	$ 72,500	2
B	400,000	90,000	6
C	600,000	112,750	8
D	200,000	32,000	12
E	800,000	110,000	15
F	775,000	125,000	10
G	225,000	90,500	4
H	1,000,000	123,000	18

14.13 Under pressure to "complete" a road project, the bike path that goes with it is being postponed. Doing it now would cost $35,800. Instead it will be included in the transportation improvement program package that is now being assembled. Because of the state and federal budget cycles, the bike path will be built 3 years from now. The expected cost of doing it then is $68,000. What is the implied interest rate for postponing the bike path?   (*Answer:* 23.84%)

14.14 The Engineering Building at State University needs $500K worth of repair and renovation. The state cannot afford these repairs in the near future. If the repairs are not performed, the building will require replacement in 8 years. The building will cost $2.8M to replace. Determine the implied interest rate for forgoing the repairs.

14.15 Runway construction at the local airport is expected to cost $35M. The runway will require maintenance costing $200K per year. At the end of each 6-year period, the runway will require resurfacing at a cost of $1750K. Assume an interest rate of 6% and a life of 30 years. How many landings and takeoffs per day are needed to justify the runway, if each landing or takeoff is worth $85 to the airport authority?   (*Answer:* 96)

14.16 A proposed steel bridge has a life of 50 years. The initial cost is $350,000, and the annual maintenance costs $12,000. The bridge deck will be resurfaced every 10 years (in years 10, 20, 30, and 40) for $90,000, and anticorrosion paint will be applied every 5 years (in years 5, 10, 15,..., 45) for $22,000. The state highway department uses an interest rate of 6%, and 800 vehicles per day will go over the bridge. What is the required benefit per vehicle to justify the bridge?

14.17　A new transit system and a new freeway are proposed to relieve congestion. Both have lives of 40 years and no salvage values. Their benefits are comparable at the specified interest rate of 10%. The local taxpayers are responsible for operating costs and the unsubsidized portion of first costs. What is the local EAC of each alternative? What is the overall EAC for each? Is the preferred alternative different for local and national interests?

Factor	Transit System	New Freeway
First cost	$240M	$400M
Federal construction subsidy	80%	90%
Annual operating cost	$80M	$70M

14.18　A state program is subsidizing first costs for alternative energy usage. Carla and Charlie are the owners of a small architectural and engineering firm (A&E firm). They have found the following data for renovating the heating system of a client's home. Each system should last 20 years. Their client's interest rate is 10%. Which system is preferred by their client? Is this economically efficient from a broader perspective?

Factor	Solar	Gas Furnace
First cost	$6000	$2000
State subsidy	40%	0%
Annual fuel cost	$40	$400

14.19　Exhibit 14.7's demand curve describes how users value picnics, fishing, boating, and camping at small state parks. If admission fees are $2, estimate the annual receipts for admission to a proposed park. Estimate the total annual benefit and the consumers' surplus for having another park.　(*Answer:* consumers' surplus = $175K)

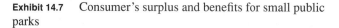

Exhibit 14.7　Consumer's surplus and benefits for small public parks

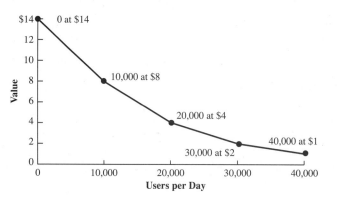

14.20   A new toll bridge is to be constructed over the Green River at a cost of $120M. The bridge requires maintenance costing $400K annually over its 50-year life. Every 10 years, the bridge will require repainting at a cost of $1M. The value to motorists using this bridge is estimated to be $1.60 per trip. If the interest rate is 10% and 25,000 vehicles per day travel over the bridge, how much toll should be charged for each crossing? Rounding that toll up to the nearest nickel, what is the consumers' surplus?

14.21   A dam may be built at a cost of $10M to eliminate the periodic flooding that Lowville experiences. That flooding averages a cost of $300K per year. Additional expenses can be incurred to divert water to another area for irrigation, to add electric power generation, and to permit recreation. Which alternative is best? Since political support is crucial, allocate the costs to the uses. The interest rate is 5% and the horizon is 50 years. Calculate a B/C ratio for your recommended alternative.

	Irrigation	Recreation	Electricity
First cost	$ 4M	$200K	$3M
Annual cost	100K	30K	500K
Annual benefit	450K	50K	850K

    (*Answer:* B/C = 1.05)

14.22   An engineer failed to allow for a windstorm during construction of a building. Collapse occurred while a crew was trying to add bracing to the structural frame. No one was injured, but one worker was killed. Compensation to the worker's family for lost income will be made using a 7% interest rate. The worker was 25 and had a salary of $22,000. Assume retirement at age 65 and an annual $1000 salary increase. What is the present value of the family's lost income? Is this a realistic estimate of the value of a human life?

14.23   A young man, 30 years of age, was killed while working as a lineman for the local electric authority. His annual salary was $27,500. The electric authority also contributed to health insurance coverage and retirement at a cost of 12% of the annual salary. His salary was projected to increase annually by $1500, and the lineman was expected to retire at age 55. What amount should the estate receive for lost income and benefits, using an interest rate of 8%?

14.24   A proposal for dividing a four-lane highway is being considered. The cost is $6M for the roadbed, $3M every 15 years for pavement, and $125K every year for maintenance. The time savings to travelers are worth $300K per year. Using the NHTSA standard, how many lives must be saved each year to justify the highway? Assume that *i* is 5% and *N* is 60 years.   (*Answer:* .76 lives per year)

14.25 How many lives per year must be saved to justify a bypass that will cost $12.5M to construct and $135K per year to maintain? Time savings for motorists using the bypass are valued at $100K annually. The city uses 7% to evaluate capital projects and the bypass is expected to handle traffic for 40 years.

14.26 The state Department of Transportation (DOT) is considering a proposal for an additional turn lane at a busy intersection. The added maintenance costs will be $3000 annually for the next 10 years, and the time savings to motorists are valued at $4000 per year. The improved traffic flow is projected to save .8 lives per year. Using an interest rate of 12% and the NHTSA standard, how much should DOT be willing to pay for the improvement?

14.27 Construct a cost-effectiveness measure to rank the following park-acquisition alternatives. Assume that the state's interest rate is 8% and that the horizon is 30 years. If a budget of about $20 million is established, what is the implicit annual value of each citizen served? Which parks should be created?

Park Site	First Cost (M)	Cost/Year (M)	Users/Year
A	$4.4	$.12	30,000
B	5.3	.22	50,000
C	3.5	.32	25,000
D	5.9	.31	40,000
E	4.2	.24	15,000
F	3.1	.18	20,000
G	4.7	.20	25,000
H	3.9	.16	35,000

(*Answer:* $20.85 per year)

14.28 The federal government is considering investing $400M in an East Coast light rail system. The construction of the stations will receive 58% of the budget. The stations range in size from small to very large. In order to keep "pork-barrel" decisions to a minimum, each possible station site has been assigned a number. Using the data presented below, determine which stations should be constructed.

Station	Construction Cost (M)	Passengers Served
1	$35	230,000
2	25	120,000
3	68	544,000
4	72	390,000
5	47	302,500
6	52	354,000
7	19	140,750
8	8	55,250

APPENDIX 14A   VALUE OF B/C AND IRR

For mutually exclusive alternatives, both B/C and IRR techniques require incremental analysis. However, the difficulties in inferring the relative attractiveness of two proposals based on their B/C ratios are particularly acute, as shown in Example 14A.1.

Example 14A.1   Comparing IRRs and B/C Ratios

In this example, the interest rate is 10% and proposals A and B are mutually exclusive. Compare their ratios and their IRRs for the total projects, and for the correct incremental comparison.

	Proposal A	Proposal B
First cost	$10 M	$ 12M
Annual benefit	$ 1.3M	$ 8M
Annual cost	$ 0	$5.8M
Life (years)	20	20

Solution

	Proposal A	Proposal B
$B/C_{10\%}$	$11.07/10 = 1.107$	$68.11/(12 + 49.38) = 1.1097$
IRR	11.54%	17.62%
$B - C_{10\%}$	$1.07M	$6.73M

Incremental Analysis for B−A	
First cost	$2.0M
Annual benefit	6.7M
Annual cost	5.8M
$B/C_{10\%}$	$55.34M/(2M + 49.38M) = 1.077$
i_{B-A}	45.0%
$B - C_{10\%}$	$5.66M

Even though all of the B/C ratios are close to 1.1 at $i = 10\%$, the IRR for proposal B, at 17.6%, is much better than the 11.5% IRR for proposal A. All three incremental values confirm that B is the best choice. The incremental PW of $5.66M and the incremental IRR (i_{B-A}) of 45% emphasize how much better B is, while the incremental ratio of 1.077 does not.

  As emphasized in Chapter 10, the easiest way to make a mutually exclusive comparison is to rely on the present values of $B - C = \$1.07M$ for A and $6.73M for B. Proposal B's value is $5.66M higher.

PROBLEMS

14A.1   Calculate incremental B/C ratios and IRRs to compare the alternatives presented in Problem 14.7.

14A.2   Calculate incremental B/C ratios and IRRs to compare the alternatives in Problem 14.17.

14A.3   Calculate incremental B/C ratios and IRRs to compare the alternatives in Problem 14.18.

CHAPTER 15 INFLATION

The Situation and the Solution

Over time the amounts of many cash flows increase due to inflation. At the same time, the value that can be purchased by a dollar, a yen, or a peso is changing. These changes in value complicate engineering economic decision making.

The solution is to apply the tool of geometric gradients, since inflation and deflation are often modeled as constant rates rather than as constant amounts. As in Chapter 4, geometric gradients are usually modeled using spreadsheets. Appendix 15A describes a more mathematical, formula-based approach to inflation.

Chapter Objectives

After you have read and studied the sections of this chapter, you should be able to:

Section 15.1
Define inflation, deflation, and differential inflation.

Section 15.2
Estimate inflation rates using inflation indexes.

Section 15.3
Identify common terminology and assumptions for inflation.

Section 15.4
Solve for PW or EAC when including inflation.

Section 15.5
Include leases, insurance payments, and other prepaid expenses under inflation.

Section 15.6
Include loan payments, depreciation, and income taxes under inflation.

Section 15.7
Combine four common sources of geometric gradients.

Appendix 15A
Calculate an equivalent discount rate for use in the engineering economy factors.

Key Words and Concepts

Inflation   A decrease in the value of the dollar (or other monetary unit); or, equivalently, the increase in the general level of prices of purchased items.

Deflation   An increase in the value of the dollar (or other monetary unit); or, equivalently, the decrease in the general level of prices of purchased items.

Differential inflation   For an item, a price that is changing at a different rate than are prices in the economy as a whole.

Geometric gradient   A cash flow that increases or decreases at a constant rate (see also Chapter 4).

Consumer price index (CPI)   The best-known government measure of inflation, which is the relative composite price of a "market basket" of goods and services.

Producer price indexes (PPIs)   Government measures of inflation for different industries.

15.1 DEFINING INFLATION, DEFLATION, AND DIFFERENTIAL INFLATION

Inflation applies to prices in general, which is the economy as a whole. Different inflation rates can also be linked to specific items. If the price of an item is inflating, then more dollars (pesos, yen, etc.) are needed to buy it now than in the past. If an economy is **inflationary**, then the prices of most or all items are increasing. **Deflation** is rarer, but it does occur. This term applies when prices are falling rather than increasing, as in inflation.

> **Inflation** is a decrease in the value of the dollar (or other monetary unit); or, equivalently, increases in the general level of prices of purchased items. **Deflation** is the much rarer opposite condition.

For example, a consulting engineering firm now pays $20,000 per year for accounting services. Suppose that these expenses are expected to increase by 6% a year for the next 15 years. Over this same 15 years, the inflation rate is expected to be 4% per year. The 4% inflation rate is not linked to any specific item; it is for the whole economy. The inflation rate for the specific item, accounting services, is 6%.

The ANSI standard symbol is f for the rate of inflation (or deflation) in the general level of prices. That is for the 4% that applies to the whole economy.

Differential inflation for an item means that its price is changing at a different rate than are prices in the economy.

Because these two inflation rates are different, accounting services are subject to **differential inflation**. This is another way of describing inflation and it is detailed later in this section. The differential inflation rate can be treated as a geometric gradient or, as shown in Section 15.4, the item and general inflation rates can be considered, together with the interest rate, to calculate the PW and the EAC for the cost of accounting services.

The Basic Geometric Formula Revisited.   Inflation rates are usually estimated as having a constant *rate* of change. This type of cash flow is a **geometric gradient** (also known as an *escalating series*). With inflation, the price increase in year 8 is linked to both the price in year 7 and the inflation rate. It is not linked to a constant amount of a price increase, which would be an arithmetic gradient.

A **geometric gradient** is a cash flow that increases or decreases at a constant rate.

An engineer's annual raise provides another example. When this year's raise is given, it will be based on many factors (the engineer's performance, the firm's profitability, the rate of inflation, and the current market for that engineering specialty). However, it will almost certainly be thought of by the firm and the engineer as a percentage increase over the current salary. This is a geometric gradient, and it is *compound growth* (See Chapter 2).

To calculate cash flows with inflation, the geometric gradient formulas developed in Chapter 4 are used. Equations 4.1 and 4.2 are repeated here as Equations 15.1 and 15.2. These were based on the basic compound-interest equation, where the only change from the *(F/P,i,N)* factor is that the value at the end of the first year, P_1, is the base. Thus, the $20,000 cost of accounting services is assumed to occur *at the end of the first year*, rather than at time 0. In Section 15.7 these formulas are elaborated to include multiple geometric gradients.

The formula for the value in year *t* is Equation 15.1, where *g* is the rate of the geometric gradient. This formula is easy to use, but restating it recursively makes the constant rate of change even clearer (see Equation 15.2). This recursive format is easier to use in manual and computerized tables.

$$F_t = P_1(1 + g)^{(t-1)} \qquad (15.1)$$
$$F_t = F_{t-1}(1 + g) \qquad (15.2)$$

Differential Inflation.   When the inflation rate, f, is 5%, most goods cost 5% more each year. Some items, such as labor or energy costs, might be inflating at a higher or lower rate. If labor's inflation rate is 6%, then it would have a differential inflation rate of 1% over the economy's 5% inflation rate. If energy's inflation rate is 4%, then it would have a differential inflation rate of -1% over the economy's 5% inflation rate. (Note: Equation 15.5 in Section 15.3 describes another approach for calculating differential inflation rates, which yields very similar answers.)

An item's **differential inflation** rate equals the item's inflation rate minus the economy's inflation rate.

An item's **differential inflation** rate equals the item's inflation rate minus the economy's inflation rate, f. The differential inflation rate is positive if an item's rate of price increases exceeds the economy's rate of price increases.

The differential inflation rate is often estimated directly, rather than estimating the two component rates separately. It may be easier and more accurate to estimate that labor costs will go up 1% faster than inflation due to a union contract, for instance, than to estimate the economy's inflation rate. Thus, the differential inflation rate is usually more stable or less volatile than the two inflation rates on which it is based. For most items, the differential inflation rate is 0%.

Example 15.1 shows how to calculate a cash flow's inflation-adjusted value, given a differential inflation rate. Example 15.2 illustrates a negative differential inflation rate. It also shows that if the differential inflation rate is 0%, then a cash flow's inflation-adjusted value matches its value before adjusting for inflation. *This implies that inflation can be ignored if the differential inflation rate is 0%, which it is for many (and often most) items.*

Example 15.1 Inflation-Adjusted Cost of Accounting Services

An engineering consulting firm purchases accounting services for $20,000 per year. These costs are expected to inflate at 6% per year, while the economy's inflation rate, f, is 4%. What is the inflation-adjusted cost of these services in each of the next 5 years?

Solution
The first step is to calculate the differential rate of inflation for accounting services. That equals the item's inflation rate minus the economy's rate, or $6\% - 4\% = 2\%$. The cash flow for year 1 is $20,000; Equation 15.2 is applied to calculate the inflation-adjusted cash flows for years 2 to 5.

$$CF_2 = 20,000 \cdot 1.02 = \$20,400$$

$$CF_3 = 20,400 \cdot 1.02 = \$20,808$$

$$CF_4 = 20,808 \cdot 1.02 = \$21,224$$

$$CF_5 = 21,224 \cdot 1.02 = \$21,649$$

These values represent the true cost of the accounting services. They are becoming slightly more expensive each year, since they are inflating 2% faster than the economy is. These cash flows are stated in the same dollars as CF_1, that is, in year-one dollars.

Example 15.2 Differential Inflation Rate Is Less than or Equal to 0

As in Example 15.1, the accounting services purchased by the engineering consulting firm cost $20,000 per year, and the economy's inflation rate, f, is 4% per year. Suppose these services inflate at 4% per year instead. What is the inflation-adjusted cost for each of the next 5 years? What are these inflation-adjusted costs if productivity improvements hold the fee constant at $20,000 per year?

Solution
When the item inflation rate equals 4%: The first step is to calculate the differential rate of inflation for the accounting services. That equals the item's inflation rate minus the economy's rate, or $4\% - 4\% = 0\%$. The cash flow for year 1 is $20,000; Equation 15.2 is applied to calculate the inflation-adjusted cash flows for years 2 to 5. However, since $g = 0\%$, these values all equal $20,000.

When improved productivity holds the fee constant: The item inflation is 0%, since the fees are not changing. The differential inflation rate is -4%, which is found by subtracting the economy's rate of 4% from the item's rate of 0%. The cash flow for year 1 is $20,000; Equation 15.2 is applied to calculate the inflation-adjusted cash flows for years 2 to 5. Again these cash flows are stated in the same year-one dollars as CF_1.

$$CF_2 = 20,000 \cdot .96 = \$19,200$$
$$CF_3 = 19,200 \cdot .96 = \$18,432$$
$$CF_4 = 18,432 \cdot .96 = \$17,695$$
$$CF_5 = 17,695 \cdot .96 = \$16,987$$

15.2 MEASURING INFLATION WITH INDEXES

Consumer Price Index (CPI).   Inflation is measured by the federal government. The measures allow the government to validly report family income, the cost of programs, gross national product, etc. The best-known measure of inflation, in the general level of prices (f), is the **consumer price index (CPI)**. This index measures the current cost of a "market basket" of family purchases for food, transportation, housing, etc., and divides it by the cost in a specified base year. An index of 143 means that current prices are 1.43, or 143%, of the prices in the base year. In other words, 43% more dollars must be paid now than in the base year to buy the same market basket of goods and services. Periodically, the government shifts to a new base year. The values over several decades that are shown in Exhibit 15.1 are relative to the base of 100 in early 1983.

The **consumer price index (CPI)** measures the yearly cost of a market basket of purchases as compared with the cost in a base year.

Examples 15.3 and 15.4 illustrate the use of Exhibit 15.1. Basically, a price in any year is divided by the index for that year, and the price in any other year is then found by multiplying that first result by the index for that other year.

Example 15.3   Buying Power of Family Income

Carol and George are both engineers. In 1984 their combined income was $64,500. How much does their income need to be in 1989 to have the same purchasing power?

Solution
From Exhibit 15.1, the indexes for 1984 and 1989 are, respectively, 105.3 and 126.1. Dividing $64,500 by the 1984 index states their income in base-year dollars. Multiplying this value by the 1989 index states the required income in 1989 dollars.

$$\text{Income}_{1989} = (\text{Income}_{1984}/\text{Index}_{1984})\text{Index}_{1989}$$
$$= (64,500/105.3)126.1 = \$77,241$$

Example 15.4   First-Class Stamps and Inflation

The cost of a 1-ounce letter was 15¢ in 1980 and 29¢ in 1992. Did the real cost of mailing a letter really increase?

Solution

This is the same as asking if the costs of the stamps had increased faster than inflation. From Exhibit 15.1, the indexes for 1980 and 1992 are 86.3 and 141.9, respectively. If stamps had increased at the inflation rate of the CPI, the cost in 1992 would be calculated as follows:

$$\text{Stamp}_{1992} = (15/86.3)141.9 = 24.7¢$$

Since stamps were 4.3¢ more expensive than this in 1992, their cost had increased faster than inflation.

Exhibit 15.1   Inflation and the CPI

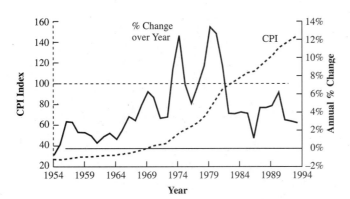

Year	Index	Year	Index	Year	Index	Year	Index
1954	26.7	1964	31.2	1974	51.9	1984	105.3
1955	26.8	1965	31.8	1975	55.5	1985	109.3
1956	27.6	1966	32.9	1976	58.2	1986	110.5
1957	28.4	1967	33.9	1977	62.1	1987	115.4
1958	28.9	1968	35.5	1978	67.7	1988	120.5
1959	29.4	1969	37.7	1979	76.7	1989	126.1
1960	29.8	1970	39.8	1980	86.3	1990	133.8
1961	30.0	1971	41.1	1981	94.0	1991	137.9
1962	30.4	1972	42.5	1982	97.6	1992	141.9
1963	30.9	1973	46.2	1983	101.3	1993	145.8

These CPI indexes are for December of each year (*Monthly Labor Review*). The base index of 100 is in early 1983.

Annual Inflation Rate.   Exhibit 15.1 can be used to calculate the annual inflation rate. That rate is simply the change in the index for the year divided by the index at the year's beginning. For example, the inflation rate in 1988 (an election year) is calculated as (120.5 − 115.4)/115.4 or 4.4%. Exhibit 15.2 shows the annual inflation rates for the last 10 years.

These inflation rates can change significantly from year to year. In 1964 the rate was 1%; 3 years later in 1967 it had tripled to 3%, and then doubled again by 1969 to 6.2%. From 1978 to 1979 it increased from 9% to 13.3%. The only deflation rate was in 1954, $-.7\%$.

Exhibit 15.2   Annual Inflation Rates

Year	Index	Inflation Rate
1984	105.3	3.949%
1985	109.3	3.799
1986	110.5	1.098
1987	115.4	4.434
1988	120.5	4.419
1989	126.1	4.647
1990	133.8	6.106
1991	137.9	3.064
1992	141.9	2.901
1993	145.8	2.748

To calculate the inflation rate over more than one year requires the use of Equation 15.3. Equation 15.3 modifies Equation 15.1 by introducing a base year of T and by using t rather than $t - 1$ as a power. When described by the inflation index in Exhibit 15.1, inflation is occuring in every year. Equations 15.1 and 15.2 were written to describe a geometric gradient, where the first year's cash flow is given, and inflation occurs only in subsequent years. Example 15.5 applies Equation 15.3 to the CPI and to the stamps. Note: when Equation 15.3 is applied to the general level of price changes, g is replaced by f.

$$F_{T+t} = F_T(1 + g)^t \tag{15.3}$$

Example 15.5   Average Inflation Rates for Stamps and the CPI

The cost of a 1-ounce letter was 15¢ in 1980 and 29¢ in 1992. What is the average annual inflation rate in cost of a 1-ounce letter? Compare this with the inflation rate f, measured by the CPI.

Solution
The inflation rate for the economy, f, is calculated first, using the CPI values. From Exhibit 15.1 and Example 15.4, the indexes for 1980 and 1992 are 86.3 and 141.9, respectively. In this case, the period is 12 years. The inflation rate for the economy is calculated using Equation 15.3.

$$141.9 = 86.3(1 + f)^{12}, \text{or}$$
$$f = (141.9/86.3)^{.08333} - 1 = 4.23\% \text{ for the economy, or CPI}$$

The same calculation for the stamps leads to a higher result, since we know the stamps have inflated faster than inflation (see Example 15.4).

$$29¢ = 15¢(1 + g)^{12} \text{ or}$$
$$g = (.29/.15)^{.08333} - 1 = 5.65\% \text{ for the stamps}$$

Producer Price Indexes.   Industries, like consumers, are subject to inflation. However, they purchase different goods and services than do consumers, and so the CPI is not the firm's best measure of inflation. The **producer price indexes (PPIs)** are developed for specific industries, some of which are included in Exhibit 15.3. These indexes are correlated with the CPI, but they are more specific and thus more accurate—for the industry they apply to. Similar indexes are published by state and private organizations (see Chapter 16 on cost estimating). For example, if you need to estimate the productivity and labor rates of carpenters for a specific metropolitan area, there is likely to be a published index.

Producer price indexes (PPIs) are a set of measures of the yearly cost of purchases by firms as compared with costs in a base year.

Exhibit 15.3   Producer price indexes

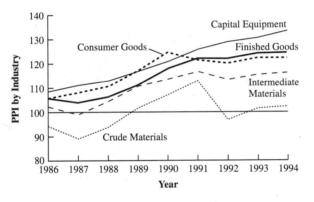

15.3 INFLATION ASSUMPTIONS AND TERMINOLOGY

Differential Inflation.   Section 15.1 defined the differential inflation rate as the item's inflation rate minus the economy's inflation rate. Usually, an item's differential inflation rate is 0% (within the limits of our ability to forecast the future), and inflation can be ignored for that item (see Example 15.2). There may be a larger number of dollars shown on invoices and checks, but each dollar is worth less.

For example, when estimating future costs of leased space, labor, and raw materials, most firms will assume a differential inflation rate of 0%. There is inflation in prices, and the rate may be uneven; however, on average it is about the same as the average rate of inflation in the economy.

Some items—such as energy prices, health care, and land values—may have differential inflation rates. In fact, some items, such as electronics and computers, may actually be declining in price. (A 486 IBM-compatible computer costs $500 less in 1994 than in 1993, even though the inflation rate has been about 5%.)

Inflation rates for the economy and for individual items are hard to estimate accurately. However, there is one class of items with a known inflation rate of 0%. Most loans and fixed financial obligations have fixed terms. Even if inflation changes the value of the payments, the payment amounts do not change, and there is a 0% inflation rate for those items. Depreciation and its associated income tax deductions (see Chapters 12 and 13) also have a 0% inflation rate. With fixed payments and deductions, the differential inflation rate is the negative of the economy's inflation rate (equals $0\% - f$).

Inflation Terminology.   When a problem says maintenance costs will be $1500 per year for 10 years, have those costs been adjusted for inflation? In most cases, a constant amount is a clue that the answer is yes, the costs have been adjusted. This is a cost with a differential inflation rate of 0%, so that inflation can be ignored.

Suppose those same maintenance costs are described as $1500 the first year, with yearly increases of 3%. Then we need to know more to decide. The 3% may be an estimate of the inflation rate or of the differential inflation rate for maintenance costs. Or it may be an estimate of the rate at which the physical amount of maintenance is increasing as equipment gets older. Or the 3% may be a combination of increasing amounts of maintenance and changing prices for that maintenance.

Certain words tell us whether the values have been adjusted for inflation. The terms *constant-value*, *year-0*, and *1994* dollars adjust for inflation in the item's price and the dollar's value, so that prices are stated in constant value terms.

Then or *nominal* dollars are written on a check in the year 199x. These numeric values have been inflated to accommodate inflation in the cost of an item, but they do not account for changes in the value of the dollar.

(I recommend that the language *current dollars* be avoided, as it is not clear whether *current now* or *current then* or *current at that time* is intended.)

To summarize, we can divide the terminology into two columns, depending on whether or not cash flow values have been adjusted for changes in the value of the dollar:

Adjusted	Not Adjusted
Constant-value	Then
Real	Nominal
Year-0	Tomorrow's
1994 or 1999	
Today's	
~~Current (now)~~	~~Current (then)~~

The constant-value dollar used most frequently in this chapter is the year-one dollar. This is the first year's cost for energy, labor, and raw materials and the first year's revenue received from selling products and services. Unless stated otherwise, first costs, salvage values, and other cash flows are also assumed to be stated in year-one dollars.

Matching Interest Rates to Inflation Assumptions. In this book, the interest rate for the time value of money is stated in terms of constant-value dollars. The assumption is that the interest rate is a real interest rate. Thus, if an item's cost is stated in constant-value terms, then inflation can be ignored.

This text's approach matches a common practice of estimating costs in constant-value terms. For example, maintenance costs will be $1500 per year and the salvage value will be 15% of the first cost. In this case, the assumption is that the inflation rates for the maintenance costs and for the used object match the economy's inflation rate. To be consistent, a real interest rate is used.

Another approach is to use a market interest rate. When a bank charges 9% interest on a loan and the inflation rate is 5%, then the bank is earning a real interest rate of about 4%. The market interest rate of 9% includes both the time value of money and an estimate of current and predicted inflation.

For exact calculation of real and market interest rates, Equation 15.4 is used.

$$(1 + \text{Market rate}) = (1 + i)(1 + f) \tag{15.4}$$

Thus, with a market rate of 9% and an inflation rate of 5%, the real interest rate is 3.81% {equals $[(1+ \text{market rate}) / (1 + f)] - 1$ or $1.09/1.05 - 1$}. While it is possible to correctly evaluate projects using market interest rates, usually it is easier with the real interest rate, i. Many cash flows are estimated as constant in real terms; that is, their differential inflation rate is 0%. If a market interest rate is used with these estimates, then inflation has been adjusted for—twice, which is once too often, so the results are wrong. Examples 15.7–15.11 in Section 15.4 illustrate the use of these inflation assumptions.

Accuracy of Inflation Estimates. Inflation rates can be calculated for historical periods with reasonable accuracy—to 2 or even 3 significant digits. In Example 15.5 the inflation rate was calculated to be 4.23%. However, predictions into the future are far less accurate. It is more realistic to estimate inflation rates to 1 or perhaps 2 significant digits.

Often, the approach for estimating inflation rates is first to estimate f, the economy's rate. Then the differential inflation rates for other items are identified as being a percent or two or more higher or lower. When estimates are constructed in this fashion, then a differential inflation rate of 2% or -3% is the natural result.

It is also possible to use Equation 15.5 and the exact theoretical relationship between the different inflation rates to calculate a differential inflation rate. This is similar to the calculation above for the real interest rate, 3.81%, when there is a market rate, 9%, that must be adjusted for inflation, 5%. Example 15.6 illustrates the calculation of an exact historical differential inflation rate.

$$1 + \text{Differential inflation rate} = \frac{1 + \text{Item's inflation rate}}{1 + f} \tag{15.5}$$

Example 15.6   Differential Inflation Rate for First-Class Stamps

The cost of a 1-ounce letter was 15¢ in 1980 and 29¢ in 1992. What was the differential rate of inflation in first-class postage?

Solution

In Example 15.5, the inflation rate for the economy over the 12 years was calculated to be 4.23%. The inflation rate for the same period for first-class stamps was calculated to be 5.65%.

As defined in Section 15.1, the differential inflation rate equals the arithmetic difference between these, or 1.42%. For projecting into the future, this would be rounded to the more realistic accuracy of 1.4%.

Theoretically, the exact differential inflation rate is calculated by using Equation 15.5.

$$1 + \text{Differential inflation rate} = 1.0565/1.0423$$

$$= 1.0136$$

Thus, the exact differential inflation rate is 1.36%. In this case, this rounds to the same 2-digit accuracy of 1.4% for a projected differential inflation rate.

The exact differential inflation rate of 1.36% is given in Example 15.6. In estimating future cash flows, that formula's extra precision is misleading. Realistically, there is little difference between a 1.4% and a 1.6% differential inflation rate. In this case, the highest-quality inflation estimate would state that first-class postage will inflate at a rate 1.4% above the economy's inflation rate—whatever f turns out to be.

15.4 SOLVING FOR PW OR EAC WHEN INCLUDING INFLATION

Chapter 4 presented spreadsheets to calculate economic equivalence for cash flows with geometric gradients, such as inflation. The examples in this chapter will also be solved using spreadsheets. To expand the number of different software packages and examples presented in this text, these spreadsheets have been produced using the Table feature of WordPerfect™ 6.0. Similar features are or will be available in other word processing packages. In spreadsheets the variables are defined in a data block (see Chapter 4). In this case the data block items are placed above the columns where they are used.

The financial functions used in these spreadsheets are summarized in Exhibit 15.4. These functions are very similar to the Lotus 1-2-3, Quattro Pro, and Excel functions presented in Chapter 4. The parameters are sometimes in a different order. Like those spreadsheet functions presented earlier, the Type parameter is optional to indicate beginning- and end-of-period cash flows. There are also functions for straight-line, double declining balance, and sum-of-the-years'-digits depreciation.

The major difference between these functions and the earlier ones is the sign convention. In WordPerfect, the formula for calculating the present worth of five end-of-period cash flows of $1000 each at 6% is:

PV(6%,1000,5,0), which results in $4112.4.

The signs of the PW and the $1000 cash flows are the same rather than opposite, as in the functions of Excel and Quattro Pro. Notice that even though there is no future end-of-horizon (FV) cash flow in this example, it is still necessary to enter a 0 for the fourth parameter. This is not an optional parameter.

Exhibit 15.4   WordPerfect 6.0 engineering economy functions

Block Functions	Annuity or Investment Functions
NPV(list, rate%)	PV(rate%, payment, periods, FV[, Type])
IRR(list, rate%)	PMT(rate%, PV, periods, FV[, Type])
	FV(rate%, PV, payment, periods[, Type])
	RATE(PV, payment, periods, FV[, Type])
	TERM(rate%, PV, payment, FV[, Type])

Finding the economic value of cash flows affected by inflation is like any other problem in this text—once the cash flows have been stated in constant-value terms. However, the answer can vary slightly, depending on how that constant-value calculation is made, as shown in Examples 15.7–15.9, which revisit Example 15.1.

Example 15.7 uses the 2% differential inflation rate calculated by subtracting f, the 4% CPI, from the item's inflation rate of 6%. Example 15.8 uses a 1.92% differential inflation rate calculated using the division approach of Equation 15.4. Finally, Example 15.9 shows that using the two rates explicitly is equivalent to the 1.92% rate used in Example 15.8.

Example 15.7   PW with a Differential Inflation Rate of 2%

In Example 15.1, the inflation-adjusted cost of accounting services was calculated for 5 years. These costs begin at $20,000 per year; they inflate at 6% per year; and the economy's inflation rate is 4%. Thus, the differential inflation rate was 2% per year. If the engineering firm's interest rate is a real 8%, calculate the PW of these services over the next 5 years.

Solution
In Example 15.1, the 2% differential inflation rate and Equation 15.2 were used to calculate the inflation-adjusted (or constant-value dollar) cash flows. Each year's cash flow is simply (1 + absolute address of 2%) or 1.02 times the previous year's cash flow. Those values, shown in column B of Exhibit 15.5, are simply multiplied by the $(P/F, 8\%, t)$ values in column C. The results, shown in column D, are then summed to calculate the PW. If an EAW were needed, then the PW would be multiplied by $(A/P, 8\%, 5)$.

Rather than using columns C and D, the PW could be calculated with a NPV(list,rate%) function from the values in column B. The cash flows and the PW are stated in the same year-one dollars as CF_1.

Exhibit 15.5   PW of accounting services with 2% differential inflation

A	B	C	D
	Differential Rate 2%	Interest Rate 8%	
t	Constant $ Cash Flow	$(P/F, 8\%, t)$	PW_t
1	−$20,000	.9259	−$18,519
2	−20,400	.8573	−17,489
3	−20,808	.7938	−16,517
4	−21,224	.7350	−15,600
5	−21,649	.6806	−14,734
		Total PW	−$82,861

Example 15.8   PW with Differential Inflation from Equation 15.4

Using the data from Example 15.7 (and Example 15.1) and Equation 15.5, calculate the differential inflation rate and the PW of the accounting services over the next 5 years.

Solution
Equation 15.5 divides the factor for inflation in accounting services (1.06) by $(1 + f)$, which is 1.04, rather than using the 2% difference between the inflation rates.

$$\text{Differential inflation rate} = (1.06/1.04) - 1 = 1.92\%$$

As in Example 15.1, the differential inflation rate, now 1.92%, and Equation 15.2 are used to calculate the inflation-adjusted cash flows that are shown in column B of Exhibit 15.6. For example, the constant-value dollar cash flow value for year 2 is (1 + absolute address of 1.92%) or 1.0192 times −$20,000. Then the value for each later year is 1.0192 times each earlier year.

    Those values, shown in column B, are simply multiplied by the $(P/F, 8\%, t)$ values in column C. The result shown in column D is then summed to compute the PW. Columns C and D could be replaced with a NPV(list,rate%) function. If an EAW were needed, then the PW would be multiplied by $(A/P, 8\%, 5)$.

    With the different value for the differential inflation rate, of course, a different PW is calculated. However, the new PW of −$82,738 differs from the one calculated in Example 15.7 of −$82,861 by only .15%. The cash flows and the PW are stated in the same year-one dollars as CF_1.

Exhibit 15.6   PW of accounting services with
1.92% differential inflation

A	B	C	D
	Differential Rate 1.92%	Interest Rate 8%	
t	Constant $ Cash Flow	$(P/F, 8\%, t)$	PW_t
1	-$20,000	.9259	-$18,518
2	-20,384	.8573	-17,475
3	-20,775	.7938	-16,491
4	-21,174	.7350	-15,563
5	-21,581	.6806	-14,688
		Total PW	-$82,738

Rather than using the differential inflation rate, it is possible to use the two inflation rates—for the item and the economy (f). Equation 15.1 or 15.2 is used to calculate the nominal-dollar cash flows using the inflation rate for the item, such as accounting services, energy costs, or computer costs.

Then, to compute the inflation-adjusted, constant-value dollars, or real cash flows, Equation 15.6 is used. This equation recognizes that the higher the inflation rate for the economy (f), the less each of the nominal dollars is really worth. This equation is used in Example 15.9. This equation assumes that the real and nominal values for CF_1 are equal, so year-one dollars are the assumed base.

$$\text{Real } CF_t = \text{Nominal } CF_t/(1 + f)^{t-1} \qquad (15.6)$$

Example 15.9   PW with Two Inflation Rates

In Examples 15.7 and 15.8, the PW of the cost of accounting services was calculated using differential inflation rates. Calculate the PW using the 6% inflation rate for the services and the 4% inflation rate for the economy (f).

Solution
The first step here is to apply Equation 15.1 or 15.2 to calculate the nominal-dollar cash flows shown in column B of Exhibit 15.7. These are calculated at the 6% inflation rate for accounting services. Then Equation 15.5 and f, the economy's inflation rate of 4%, are used to calculate the real, or constant-value, cash flows shown in column C. Column B can be calculated recursively, that is

$$\text{Nominal } CF_t = \text{Nominal } CF_{t-1}(1 + \text{absolute address of 6\%}).$$

However, column C cannot be; it must use Equation 15.6.

Those column C values are simply multiplied by the $(P/F, 8\%, t)$ values in column D. The result is shown and then summed in column E. If an EAW were needed, then the PW would be multiplied by $(A/P, 8\%, 5)$. As in the previous examples, columns D and E could be replaced by the NPV(list,rate%) function.

Except for rounding of the differential inflation rate, this approach is identical mathematically to that used in Example 15.8. Here, the $5 difference in the PW is due to using a differential inflation rate of 1.92% (in Example 15.8), rather than 1.9231%, which is implicitly used here. As before, the constant-value dollar cash flows and the PW are measured in year-one dollars.

Exhibit 15.7   PW of accounting services with separate inflation rates

A	B	C	D	E
	Item Rate	f	Interest Rate	
	6%	4%	8%	
t	Nominal $ Cash Flow	Constant $ Cash Flow	$(P/F, 8\%, t)$	PW$_t$
1	−$20,000	−$20,000	.9259	−$18,518
2	−21,200	−20,385	.8573	−17,476
3	−22,472	−20,777	.7938	−16,493
4	−23,820	−21,176	.7350	−15,564
5	−25,250	−21,583	.6806	−14,689
			Total PW	−$82,743

Examples 15.10 and 15.11 are more realistic problems. They have multiple cash flows, and only some of the cash flows are affected by differential inflation. Later sections explicitly deal with inflation and leases or prepaid expenses (Section 15.5) and taxes or loans (Section 15.6).

Example 15.10   EAC with Multiple Differential Inflation Rates

Process Engineering Design is evaluating whether or not to upgrade to a more highly computerized design approach. The change will increase the annual costs of upgrading software, training, and replacing computers by $25K. The change also requires a time 0 expense of $50K for new software licenses and training.

The continuing computer costs are expected to fall by 5% per year, while f is 4% per year. The new design approach is expected to save $35K per year in engineering labor costs. Those costs have a differential inflation rate of 2% per year.

If the firm uses a real interest rate of 10%, what is the EAW of the change? Use a time horizon of 5 years.

Solution

The first step here is to calculate the differential inflation rates. The rate for engineering labor is given as 2%. The differential inflation rate for computer costs is found by subtracting f (4%) from the inflation rate for computers (−5%) to get −9%. The year-one cash flows of $25K and $35K and the time 0 cost of $50K are all assumed to be stated in year-one dollars.

In column B of Exhibit 15.8, the constant-value dollar cash flows for computing costs are found by multiplying the previous year's value by (1 + absolute address of −9%) or .91. Similarly, in column C the constant-value dollar cash flows for labor savings are found by multiplying the previous year's value by (1 + absolute address of 4%) or 1.04.

Column D includes the time 0 expense, as well as the total from the two annual costs. Column E is the PW of each cash flow at an interest rate of 10%. Multiplying the total PW, $12,061, by $(A/P, .1, 5)$ (equals .2638) gives the EAW of $3182.

Exhibit 15.8 Multiple inflation rates for Process Engineering Design

A	B	C	D	E
Rate	−9%	4%		10%
t	Computing Constant $ Cash Flow	Engineering Labor Constant $ Cash Flow	Total Constant $ Cash Flow	$(P/F, 10\%, t)$ PW_t
0			−$50,000	−$50,000
1	−$25,000	$35,000	10,000	9091
2	−22,750	36,400	13,650	11,281
3	−20,703	37,856	17,153	12,888
4	−18,839	39,370	20,531	14,023
5	−17,144	40,945	23,801	14,779
			Total PW	$12,061

Example 15.11 PW for a Dam

A dam will supply annual recreation benefits of $500K for 250 users per day and 200 days per year at $10 per user-day. However, its main purpose is to generate electricity. It will save $3 million in fuel costs for Arctic Power & Light. The dam will cost $400K annually for labor to run and maintain, and $40 million to build.

Calculate the PW using a real interest rate of 4% and a 25-year life. The differential interest rates are 0% for recreation, 2% for fuel cost, and 1% for labor. There is a 5% inflation rate for the economy (f).

Solution

In this case, f is irrelevant, since the real interest rate and each of the differential inflation rates are given. The annual expense for labor is shown in column B of Exhibit 15.9, where a 1% differential inflation rate is used to calculate the constant-value dollar cash flows. In column C the 2% differential inflation rate for energy savings is used to calculate those constant-value dollar cash flows. Note that all cash flows and equivalent worths in the table are stated in $1000s, and the initial data is assumed to be year-one dollars.

Column D includes the time 0 expense and the constant $500K per year in recreational benefits, as well as the total from the two annual costs. Column E is the PW of each entry in column D at an interest rate of 5%. Column E could be omitted and a NPV(list,rate%) function used instead.

Exhibit 15.9   PW of a dam

A	B	C	D	E
Rate	1%	2%		5%
t	Labor Constant $ Cash Flow	Energy Constant $ Cash Flow	Total Constant $ Cash Flow	PW$_t$
0			−$40,000K	−$40,000K
1	−$400K	$3000K +500	3100K	2952K
2	−404K	3060K +500	3156K	2863K
3	−408K	3121K	3213K	2776K
4	−412K	3184K	3272K	2691K
5	−416K	3247K	3331K	2610K
6	−420K	3312K	3392K	2531K
7	−425K	3378K	3454K	2455K
8	−429K	3446K	3517K	2381K
9	−433K	3515K	3582K	2309K
10	−437K	3585K	3648K	2239K
11	−442K	3657K	3715K	2172K
12	−446K	3730K	3784K	2107K
13	−451K	3805K	3854K	2044K
14	−455K	3881K	3926K	1983K
15	−460K	3958K	3999K	1923K
16	−464K	4038K	4073K	1866K
17	−469K	4118K	4149K	1810K
18	−474K	4201K	4227K	1756K
19	−478K	4285K	4306K	1704K
20	−483K	4370K	4387K	1653K
21	−488K	4458K	4470K	1604K
22	−493K	4547K	4554K	1557K
23	−498K	4638K	4640K	1511K
24	−503K	4731K	4728K	1466K
25	−508K	4825K	4817K	1423K
			Total PW	$12,386K

15.5 LEASES AND OTHER PREPAID EXPENSES

Lease, insurance, and subscription payments are typically made at the beginning of the period, rather than the end. These costs inflate in exactly the same fashion as do other costs; however, because of their difference in timing, Equation 15.2, $F_t = F_{t-1}(1 + g)$, works properly, while Equations 15.1 and 15.6 do not. For these prepaid expenses, the first cash flow, P_0, is typically at time 0 and stated in year-0 dollars, rather than the end of the first period, as was P_1. To account for this difference in timing, Equation 15.7 modifies Equation 15.1 by using t rather than $t - 1$ as the exponent.

$$\text{\Large$\ast$}F_t = P_0(1 + g)^t \qquad\qquad (15.7)$$

Example 15.12 illustrates the application of inflation with lease payments. Example 15.13 includes both insurance payments and end-of-period energy costs.

Example 15.12 Valuing Lease Payments

TomKat Engineering has grown to the size that its two principals, Katherine and Tom, want to maintain. They are considering a long-term lease for their office space. To maintain flexibility during the firm's growth period, all leases have been for a year, and the company has moved or expanded its space four times. But now, the firm may be able to save money by leasing for a longer period.

The firm's lease will cost $120,000 to extend for another year. Each year, this cost has been increasing by f, the inflation rate, 5%. The firm's landlord is also willing to write a lease for $120,000 per year for a 3-year term. In present value terms, how much do they save by making the longer-term commitment if the firm's discount rate is 15%? (Assume that this is a real interest rate.)

Solution

As shown in Exhibit 15.10, lease payments are prepaid expenses. These payments would normally be made monthly, but annual payments have been assumed to simplify the calculations. The exhibit includes both constant-value year-0 dollars and then-dollar values. The calculations use the constant-value numbers, which match the assumption that the interest rate is a real rate and not a market rate.

Exhibit 15.10   TomKat lease payments in constant and then dollars

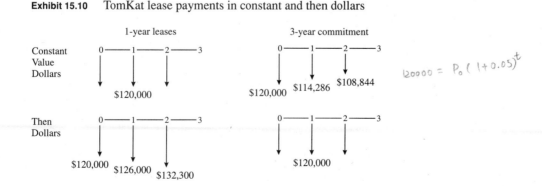

$$120000 = P_0(1 + 0.05)^t$$

If an annual lease is used for each year, then there is no differential inflation, and f, the 5% inflation rate, can be ignored.

$$P_{\text{annual lease}} = -120,000[1 + (P/A, .15, 2)]$$
$$= (-120,000)(1 + 1.626) = -\$315,120$$

Note: the $-\$315,120$ is stated in year-0 dollars.

If the 3-year lease is used, then differential inflation exists. The lease payments are fixed at $120,000 per year, but the payments for years 2 and 3 are made in "cheaper" or less valuable dollars. First, we convert these to year-0 dollars at the ends of years 1 and 2, which adjust for 5% inflation.

$$\text{Lease}_2 = -120,000(P/F, .05, 1) = -120,000/1.05 = -\$114,286$$
$$\text{Lease}_3 = -120,000(P/F, .05, 2) = -114,286/1.05^2 = -\$108,844$$

Then we find the present worth of all three cash flows.

$$\text{PW(Lease}_1) = -120,000(P/F, .15, 0) = -120,000/1 = -\$120,000$$
$$\text{PW(Lease}_2) = -114,286(P/F, .15, 1) = -114,286/1.15 = -\$99,379$$
$$\text{PW(Lease}_3) = -108,844(P/F, .15, 2) = -108,844/1.15^2 = -\$82,302$$
$$P_{\text{3-year lease}} = -\$301,681, \text{ which is \$13,440 cheaper in year-0 dollars.}$$

Example 15.13   Insurance for Copper River Valves

Copper River Valves is evaluating whether to add a new line of valves designed for more demanding and hazardous applications. Most of the annual costs are for labor, purchased parts, and materials. These begin at $25K in year 1, and they have no differential inflation rate. The liability insurance premiums are expected to increase at a differential inflation rate of 6% from an initial cost of $5K. The energy costs will decrease at a differential inflation rate of -2% from an initial cost of $7K. The inflation rate, f, is 3%.

For planning purposes, the demand is assumed to be constant in volume. The annual income is $90K in year 1, and there is no differential inflation rate for revenue. Capital equipment costing $200K is required. The equipment will have no salvage value after 10 years. What is the EAW of this line of valves if i equals 10%?

Solution

Assume the initial insurance cost ($5000) and the first cost ($200K) are stated in year-0 dollars. Convert these to year-one dollars by multiplying by $(1 + f)$ or 1.03. The time-0 cash flow stated in year-one dollars equals $-5150 - 200K(1.03)$, or $-\$211,150$. The constant-value dollar cost for the liability insurance is computed in column B of Exhibit 15.11, using the 6% differential inflation rate. Each year's constant-value cash flow is 1.06 times the previous year's.

Column C computes the constant-value dollar cost for energy, using the -2% differential inflation rate. Column D for the total constant-value dollar cash flow includes the time 0 cost to purchase the equipment. Years 1–10 in column D include the $65K in net annual income plus the values for that year from columns B and C.

The PW is $109,599, and the EAW is found by multiplying this value by $(A/P, 10\%, 10)$ or .1627 to get $\underline{\$17,832}$. These values are stated in year-one dollars like the $25K in annual costs and $90K in annual income.

Exhibit 15.11 Insurance for Copper River Valves

A	B	C	D	E
Rate	6%	−2%		10%
t	**Insurance Constant $ Cash Flow**	**Energy Constant $ Cash Flow**	**Total Constant $ Cash Flow**	**PW_t**
0		65000 − 7000 − 845 → $211,150	−$211,150	−$211,150
1	−$5150	−$7000	52,541	47,765
2	−5459	−6860	52,353	43,267
3	−5787	−6723	52,143	39,176
4	−6134	−6588	51,910	35,455
5	−6502	−6457	51,652	32,072
6	−6892	−6327	51,367	28,995
7	−7305	−6201	51,055	26,199
8	−7744	−6077	50,715	23,659
9	−8208	−5955	50,344	21,351
10	−8701	−5836	59,164	22,810
			Total PW	$109,599

15.6 DEPRECIATION AND LOAN PAYMENTS

Two categories of costs have cash flows that do not change with inflation—tax reductions due to depreciation and payments on loans or bonds with fixed interest rates.

Depreciation is computed on the initial cost of the capital equipment. Thus, higher rates of inflation decrease the value of the fixed depreciation deduction. Since the depreciation deduction is fixed by the cost of the purchase, its inflation rate is 0%; that is, it does not change as prices change. Thus, its differential inflation rate equals $-f$, the negative of the inflation rate for the economy. This rate is larger than most differential inflation rates, but its effect is limited by tax "sharing" between the firm and the government. Example 15.14 illustrates this effect.

Loans or bonds with fixed interest rates have payments that do not change with inflation. Making payments on a fixed-rate mortgage, for example, is much easier if annual salary raises keep up with inflation, since the house payment then becomes a smaller fraction of the nominal and real monthly incomes (see Example 15.15). Of course, the expected inflation rate is part of the process by which the interest rate for the loan is determined.

Firms also borrow money for which the interest payments are tax deductible. Example 15.16 illustrates the tax effects for depreciation and interest payments. The effect of inflation on the real cost of making payments is usually the opposite of inflation's effect on the real value of the depreciation deduction. Thus, inflation tends to matter more if one of fixed-loan payments or depreciation, but not both, occurs on a project.

Example 15.14   Depreciation and Inflation

Copper River Valves is evaluating some manufacturing equipment whose first cost is $250K. The operating costs are $45K per year in year-0 dollars, and the salvage value is $0 after 10 years. The firm's real after-tax interest rate is 9%, and the firm's pretax profits average $40 million per year. Inflation is expected to average 7% over the next 10 years. What is the EAC of this equipment? How much higher or lower is this than the value that ignores inflation?

Solution

Since the operating costs are stated in year-0 dollars, the first cost will be as well. Then, which values are affected by f, the inflation rate? The first cost is at time 0, so it is not affected. However, the depreciation deduction is affected, and it has a differential inflation rate of $-f$ or -7%. The annual $45K in operating cost is best assumed to have no differential inflation. Thus, the easiest way to include inflation is to convert the depreciation deduction from nominal to year-0 dollars.

From Exhibit 12.7, manufacturing equipment has a 7-year recovery period for MACRS. The percentages from Exhibit 12.9 are multiplied by the first cost and entered in column B of Exhibit 15.12. Column C uses the differential inflation rate to state this tax deduction in year-0 dollars. (Since the first cost is at time 0 and the tax deduction is at the end of the first year, Equation 15.3 is used.)

Column D shows the taxable income in year-0 dollars. This is computed by subtracting the column C values from the operations cost of $-\$45,000$ that is shown in the data block for column D. Column E computes the tax savings, or negative tax, from the costs of this equipment by multiplying each year's value in column D times the tax rate shown in the data block for column E. From Chapter 13 that tax rate is 35% with $40M in pre-tax profits.

Column F computes the after-tax cash flow (ATCF) by subtracting the negative tax in column E from the operations cost of $-\$45,000$ that is shown in the data block for column D. Column G uses the interest rate of 9% to compute the PW of each ATCF. This column could be omitted and the function NPV(list,rate%) used instead.

In year-0 dollars the PW is $-\$385,940$, and the EAC is found by multiplying this value by $(A/P, 9\%, 10)$, or .1668, to get $64,375. To calculate the value ignoring inflation's impact on depreciation, the easiest approach is to change the differential inflation rate for the MACRS deduction to 0%, as shown in Exhibit 15.13.

At $-\$372,684$, the PW is better (less negative) when ignoring inflation, since the MACRS deduction is larger in year-0 dollars. The EAC is $62,164, which is $2211 or 3.4% lower. While this difference is small, in some industries profits are less than 10% of costs, so that this difference could amount to half of the potential profits linked to this machinery.

Exhibit 15.12 Depreciation and inflation at Copper River Valves

A	B	C	D	E	F	G
	$250,000 First Cost	−7% Differential Inflation Rate	−$45,000 Operations Cost	35% Tax Rate		9% Interest Rate
t	MACRS Deduction Nominal $	MACRS Deduction Constant $	Taxable Income	Tax	ATCF	PW$_t$
0					−$250,000	−$250,000
1	$35,725 $\;^{3525(0.93)^1}$	$33,224	−$78,224	−$27,378	−17,622	−16,167
2	61,225 $\;^{61225(0.93)^2}$	52,954	−97,954	−34,284	−10,716	−9020
3	43,725	35,171	−80,171	−28,060	−16,940	−13,081
4	31,225	23,358	−68,358	−23,925	−21,075	−14,930
5	22,325	15,531	−60,531	−21,186	−23,814	−15,478
6	22,300	14,428	−59,428	−20,800	−24,200	−14,430
7	22,325	13,433	−58,433	−20,452	−24,548	−13,429
8	11,150	6239	−51,239	−17,934	−27,066	−13,584
9			−45,000	−15,750	−29,250	−13,468
10			−45,000	−15,750	−29,250	−12,356
					Total PW	−$385,940

Exhibit 15.13 Ignoring inflation's impact on depreciation at Copper River Valves

A	B	C	D	E	F	G
	$250,000 First Cost	0% Differential Inflation Rate	−$45,000 Operations Cost	35% Tax Rate		9% Interest Rate
t	MACRS Deduction Nominal $	MACRS Deduction Constant $	Taxable Income	Tax	ATCF	PW$_t$
0					−$250,000	−$250,000
1	$35,725	$35,725	−$ 80,725	−$28,254	−16,746	−15,364
2	61,225	61,225	−106,225	−37,179	−7821	−6583
3	43,725	43,725	−88,725	−31,054	−13,946	−10,769
4	31,225	31,225	−76,225	−26,679	−18,321	−12,979
5	22,325	22,325	−67,325	−23,564	−21,436	−13,932
6	22,300	22,300	−67,300	−23,555	−21,445	−12,787
7	22,325	22,325	−67,325	−23,564	−21,436	−11,726
8	11,150	11,150	−56,150	−19,653	−25,348	−12,721
9	0	0	−45,000	−15,750	−29,250	−13,468
10			−45,000	−15,750	−29,250	−12,356
					Total PW	−$372,684

Example 15.15   Mortgage Interest and Inflation

What is the constant-dollar after-tax cost (the IRR) of a mortgage loan? The loan is for $100K at 9%, with 30 equal annual payments. Assume inflation is 4% and that the taxpayer itemizes and is in the 28% tax bracket.

Solution

Exhibit 15.14   After-tax mortgage loan rate considering inflation

A	B	C	D	E	F
9% Loan's Interest Rate	$9734 Loan Payment	$100,000 First Cost	28% Tax Rate		4% f
t	Interest Portion of Loan	Loan Balance at End of Year	Tax Savings	ATCF in Nominal $	ATCF in Year-0 $
0		$100,000		$100,000	$100,000
1	$9000	99,266	$2520	−7214	−6936
2	8934	98,467	2502	−7232	−6687
3	8862	97,595	2481	−7252	−6447
4	8784	96,645	2459	−7274	−6218
5	8698	95,609	2435	−7298	−5999
6	8605	94,481	2409	−7324	−5788
7	8503	93,250	2381	−7353	−5587
8	8393	91,909	2350	−7384	−5395
9	8272	90,447	2316	−7418	−5211
10	8140	88,854	2279	−7454	−5036
11	7997	87,117	2239	−7495	−4868
12	7841	85,224	2195	−7538	−4708
13	7670	83,161	2148	−7586	−4556
14	7484	80,911	2096	−7638	−4411
15	7282	78,460	2039	−7695	−4273
16	7061	75,788	1977	−7756	−4141
17	6821	72,875	1910	−7824	−4017
18	6559	69,700	1836	−7897	−3898
19	6273	66,239	1756	−7977	−3786
20	5962	62,467	1669	−8064	−3680
21	5622	58,356	1574	−8159	−3581
22	5252	53,874	1471	−8263	−3487
23	4849	48,989	1358	−8376	−3398
24	4409	43,664	1235	−8499	−3316
25	3930	37,860	1100	−8633	−3238
26	3407	31,534	954	−8780	−3167
27	2838	24,639	795	−8939	−3100
28	2217	17,123	621	−9113	−3039
29	1541	8930	431	−9302	−2983
30	804	0	225	−9509	−2932
				IRR	2.38%

Handwritten annotations: "9000 × 0.28" above the Tax Savings column near row 0–1, and "9000 −2520" near the ATCF Nominal column for row 1.

This is a simplification of a typical mortgage, which has 360 monthly rather than 30 annual payments. The most intuitive way to solve this problem is to calculate interest payments and tax savings in nominal dollars. Then nominal after-tax cash flows are converted to year-0 dollars.

In Exhibit 15.14, columns B and C summarize the amortization schedule over the 30 years. In column B, each year's interest payment is 9% of the previous year's loan balance. The annual payment of $9734, shown in the data block for column B, is computed using PMT(rate%, PV, periods, FV) or PMT(A1, C1, 30, 0). Each year's principal payment is $9734 minus the year's interest payment. This is the yearly change in the loan balances shown in column C.

In column D, each year's tax savings are computed as 28% of the year's interest payment. This is subtracted from the annual payment of $9734 to calculate column E, the after-tax cash flows in nominal dollars.

In column F, each of these values is converted to year-0 dollars using the 4% value of f. This is the first point in this example where inflation is considered. An answer in year-0 dollars is desired (this is when the $100,000 is received), and the first payment is made a full year later, at the end of the year. Thus, the conversion formula for column F is

$$\text{ATCF}_{\text{year-0 \$}} = \text{Nominal ATCF}/(1 + f)^t$$

Finally, the IRR is computed using IRR(list,rate%), where the rate% is a guess to start the solution process. Note that this interest rate is only 2.38%! The dramatic drop from 9% before tax deductions and inflation is because the tax deduction is based on the full interest payment, so that the home owner really only pays about $(1 - 28\%)$ times 9%, or 6.48%, interest after taxes. Then the 4% inflation rate reduces the true cost of this to about 2.48%.

To calculate this rate more exactly and to match the results of Exhibit 15.14, Equation 15.5 is used to divide 1.0648 by 1.04 to get 2.38%.

Example 15.16   Depreciation, Interest, and Inflation

Assume that the Copper River Valves' manufacturing equipment evaluated in Example 15.14 will be financed with a 5-year, 11% loan that has equal annual payments. What is the EAC of this equipment? Check this answer by computing the PW of the loan separately and combining that PW with the PW from Example 15.14. That equipment had a first cost of $250K and annual operating costs of $45K—both in year-0 dollars.

Solution
The first step is to calculate the annual payment.

$$\text{Loan payment} = 250K(A/P, 11\%, 5) = 250K \cdot .2706 = \$67.65K$$

This value is subject to a differential inflation rate of -7%, like the MACRS tax deduction. Note that if inflation and the tax deduction on the interest were not decreasing the true cost of the payments, a firm with an interest rate of 9% would *not* borrow money at 11%.

Exhibit 15.15 calculates columns B through H in nominal dollars, then converts them to constant-value dollars in column I. Columns C and D are calculated as they were in Exhibit 15.14. For each year in column F, the taxable income equals the cash flow for operating costs, minus the MACRS deduction and minus the interest portion of the loan payment. For each year in column G, the after-tax cash flow (ATCF) equals the cash flows for operating costs minus the loan payment (these loan payments occur only in years 1 through 5).

Exhibit 15.15   Interest, depreciation, and inflation at Copper River Valves

A	B	C	D	E	F	G	H	I	J
	$250,000 First Cost	11% Loan's Interest Rate	$67,643 Loan Payment	−$45,000		35% Tax Rate		7% *f*	9% Interest Rate
t	MACRS Deduction Nominal $	Interest Portion of Loan	Principal Portion of Loan	Operating Cost in Nominal $	Taxable Income in Nominal $	Tax in Nominal $	ATCF in Nominal $	ATCF in Year-0 $	PW*t*
0			$250,000						$ 0
1	$35,725	$27,500	209,857	−$48,150	−$111,375	−$38,981	−$76,811	−$71,786	−65,859
2	61,225	23,084	165,299	−51,521	−135,830	−47,540	−71,623	−62,558	−52,654
3	43,725	18,183	115,839	−55,127	−117,035	−40,962	−81,807	−66,779	−51,566
4	31,225	12,742	60,939	−58,986	−102,953	−36,034	−90,595	−69,114	−48,962
5	22,325	6703	0	−63,115	−92,143	−32,250	−98,507	−70,234	−45,648
6	22,300			−67,533	−89,833	−31,442	−36,091	−24,049	−14,340
7	22,325			−72,260	−94,585	−33,105	−39,155	−24,384	−13,339
8	11,150			−77,318	−88,468	−30,964	−46,354	−26,979	−13,540
9	0			−82,731	−82,731	−28,956	−53,775	−29,250	−13,468
10				−88,522	−88,522	−30,983	−57,539	−29,250	−12,356
								Total PW	−$331,730

From Exhibit 15.15, the PW has changed to −$331,730 with the loan from −$385,940 in Example 15.14. This $54,210 difference occurs because the cost of the loan is less than the firm's interest rate, which is 9% after taxes. In fact, as detailed in Exhibit 15.16, the true cost of the loan is about 0%. The ATCFs for the loan payments in year-0 dollars add to about $250,000. As stated in Example 15.15, the loan's cost can be approximated as (pretax rate)(1 − the tax rate) minus *f*, the inflation rate. This equals:

$$\text{After-tax interest rate} \approx .09(1 - .35) - 7\% = .15\% \approx 0\%$$

Exhibit 15.16 computes the PW of the loan using nominal dollars for the interest portion of the loan payment and for the resulting tax savings. These are converted in column F to year-0 dollars by dividing the nominal values by $(1 + f)^t$. This is the most accurate way to compute the PW of the loan.

The resulting value of $49,024, however, does not equal the value of $54,210 calculated above. The reason is that the calculations in Exhibit 15.12 used a differential inflation rate of −7% for the MACRS and interest deductions. At this rate, the calculation of year-0 dollars is done using a factor of $.93^t$. Exhibits 15.15 and 15.16 instead leave the calculation of year-0 dollars until the last step, when a divisor of 1.07^t is used to account for the impact of inflation. This would correspond to a multiplier of $.9346^t$ instead of $.93^t$.

These differences are small and much less than the uncertainty in the estimated future values of inflation. Theoretically, the approach shown in Exhibits 15.15 and 15.16 is more correct.

Exhibit 15.16   After-tax PW of loan for Copper River Valves

A	B	C	D	E	F
11% Loan's Interest Rate	$67,643 Loan Payment	$250,000 First Cost	35% Tax Rate		7% *f*
t	Interest Portion of Loan	Loan Balance at End of Year	Tax Savings	ATCF in Nominal $	ATCF in Year-0 $
0		$250,000		$250,000	$250,000
1	$27,500	209,857	$9625	−58,018	−54,222
2	23,084	165,299	8080	−59,563	−52,025
3	18,183	115,839	6364	−61,279	−50,022
4	12,742	60,939	4460	−63,183	−48,202
5	6703	0	2346	−65,296	−46,555
				PW at 9%	$49,024

15.7 INFLATION AND OTHER GEOMETRIC GRADIENTS

Chapter 4 included a formal mathematical model for geometric gradients. This section incorporates into that model either the differential inflation rate or the two inflation rates—one for the item and f for the economy.

The Four Geometric Gradients.   Future costs and revenues are often estimated as changing at the same *rate* as in the past. Common examples include the cost of supplying more power to a growing city, the labor cost to run an assembly line, and the cost to repay a loan during an inflationary period. Respectively, these are geometric gradients for the amount of power, labor's unit cost, and the dollar's value. To compare two cash flows we have to find present or annual worths using compound interest, which is a fourth geometric gradient.

This book defines four geometric gradients:

$v \equiv$ rate of change in the volume of the item

$g \equiv$ rate of inflation or deflation in the item's price

$f \equiv$ rate of inflation or deflation in the economy, as measured by the value of a dollar (ANSI standard symbol).

$i \equiv$ interest rate for money's time value (ANSI standard)

The first two of these, v and g, are combined into a single g in the ANSI standard. That g measures a constant rate of change in cash flows, but it cannot be used to find differential inflation rates, unless v equals 0.

Obviously, a single problem could involve all of these gradients. The effect of each gradient is more easily seen if they are considered individually. Positive rates for some gradients increase the present worth, while positive rates for other gradients decrease the present worth.

Consider an electric power problem, where:

Increases in the number of kWhs used increase future costs.

Increases in the cost per kWh increase future costs.

Inflation in the economy decreases the value of those costs.

The interest rate discounts these costs to their present values.

This can be expressed mathematically, as shown in Exhibit 15.17. Exhibit 15.17 is somewhat intimidating, but it is really only describing the spreadsheets that have already been done in this chapter.

Exhibit 15.17   Mathematical factors for the four geometric gradients

Period	Base Amount	v Volume Change	g Price Change	f Inflation	Constant-Value Cash Flow	i PW Factor	PW
1	100	1.0	1.0	1.0	100	$1/(1+i)^1$	$100/(1+i)$
2	100	$1+v$	$1+g$	$\dfrac{1}{(1+f)}$	$\dfrac{100(1+v)(1+g)}{(1+f)}$	$\dfrac{1}{(1+i)^2}$	$\dfrac{100(1+v)(1+g)}{(1+f)(1+i)^2}$
$\vdots$	$\vdots$	$\vdots$	$\vdots$	$\vdots$	$\vdots$	$\vdots$	$\vdots$
t	100	$(1+v)^{(t-1)}$	$(1+g)^{(t-1)}$	$\dfrac{1}{(1+f)^{(t-1)}}$	$\dfrac{100[(1+v)(1+g)]^{(t-1)}}{(1+f)^{(t-1)}}$	$\dfrac{1}{(1+i)^t}$	$\dfrac{100[(1+v)(1+g)]^{(t-1)}}{(1+f)^{(t-1)}(1+i)^t}$

Example 15.17   Evaluating the Cost of Pumping Sewage

Greenacres is a rapidly growing suburb of Brownsky, a major Sun Belt city. Greenacres pays Brownsky to treat and dispose of its sewage. Greenacres must pump the sewage up 200 feet, from near river level to the top of the bluff where Brownsky sits. The electrical cost this year (year 1) will be $1.25 million. Greenacres is expected to grow at 7%. The estimated inflation rate in the cost per kWh is 5%. The economy is inflating at 3%. Greenacres' interest rate is 8%. What is the present worth of the pumping cost for year 10?

Solution

Exhibit 15.18 Spreadsheet for Greenacres sewage pumping (Example 15.17)

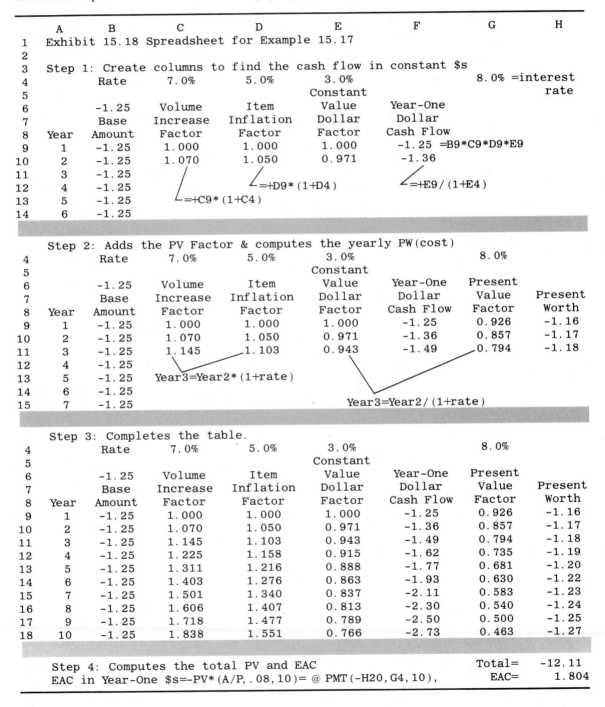

	A	B	C	D	E	F	G	H
1	Exhibit 15.18 Spreadsheet for Example 15.17							
2								
3	Step 1: Create columns to find the cash flow in constant $s							
4		Rate	7.0%	5.0%	3.0%		8.0% =interest	
5					Constant			rate
6		-1.25	Volume	Item	Value	Year-One		
7		Base	Increase	Inflation	Dollar	Dollar		
8	Year	Amount	Factor	Factor	Factor	Cash Flow		
9	1	-1.25	1.000	1.000	1.000	-1.25 =B9*C9*D9*E9		
10	2	-1.25	1.070	1.050	0.971	-1.36		
11	3	-1.25						
12	4	-1.25		=+D9*(1+D4)	=+E9/(1+E4)			
13	5	-1.25	=+C9*(1+C4)					
14	6	-1.25						

Step 2: Adds the PV Factor & computes the yearly PW(cost)

	A	B	C	D	E	F	G	H
4		Rate	7.0%	5.0%	3.0%		8.0%	
5					Constant			
6		-1.25	Volume	Item	Value	Year-One	Present	
7		Base	Increase	Inflation	Dollar	Dollar	Value	Present
8	Year	Amount	Factor	Factor	Factor	Cash Flow	Factor	Worth
9	1	-1.25	1.000	1.000	1.000	-1.25	0.926	-1.16
10	2	-1.25	1.070	1.050	0.971	-1.36	0.857	-1.17
11	3	-1.25	1.145	1.103	0.943	-1.49	0.794	-1.18
12	4	-1.25						
13	5	-1.25	Year3=Year2*(1+rate)					
14	6	-1.25						
15	7	-1.25			Year3=Year2/(1+rate)			

Step 3: Completes the table.

	A	B	C	D	E	F	G	H
4		Rate	7.0%	5.0%	3.0%		8.0%	
5					Constant			
6		-1.25	Volume	Item	Value	Year-One	Present	
7		Base	Increase	Inflation	Dollar	Dollar	Value	Present
8	Year	Amount	Factor	Factor	Factor	Cash Flow	Factor	Worth
9	1	-1.25	1.000	1.000	1.000	-1.25	0.926	-1.16
10	2	-1.25	1.070	1.050	0.971	-1.36	0.857	-1.17
11	3	-1.25	1.145	1.103	0.943	-1.49	0.794	-1.18
12	4	-1.25	1.225	1.158	0.915	-1.62	0.735	-1.19
13	5	-1.25	1.311	1.216	0.888	-1.77	0.681	-1.20
14	6	-1.25	1.403	1.276	0.863	-1.93	0.630	-1.22
15	7	-1.25	1.501	1.340	0.837	-2.11	0.583	-1.23
16	8	-1.25	1.606	1.407	0.813	-2.30	0.540	-1.24
17	9	-1.25	1.718	1.477	0.789	-2.50	0.500	-1.25
18	10	-1.25	1.838	1.551	0.766	-2.73	0.463	-1.27

Step 4: Computes the total PV and EAC   Total=   -12.11
EAC in Year-One $s=-PV*(A/P,.08,10)= @ PMT(-H20,G4,10),   EAC=   1.804

The cost in year 10 occurs 9 years later than the $1.25M cost in year 1. Each year, a kilowatt-hour of electricity costs 5% more and each year 7% more kWhs are used. Thus, the nominal-dollar cost of the pumping is $(1.25 \cdot 10^6) \cdot (1.05^9) \cdot (1.07^9)$, which equals 3.57M nominal dollars. On the other hand, since f is 3%, the real value of that $3.57M is only 2.73M year-one dollars (equals $3.57/1.03^9$). Finally, the present worth of that $2.73M equals 1.27M year-one dollars (equals $2.73/1.08^{10}$). This problem is presented as a spreadsheet in Exhibit 15.18, where the PW and EAC are calculated for the 10-year period.

15.8 SUMMARY

This chapter has focused on inflation, which is one source of geometric gradients—cash flows modeled by a constant rate of change. The chapter has concluded with a mathematical model of four geometric gradients. The first geometric gradient is due to changes in a physical volume, such as the number of items manufactured and sold each year or the amount of electicity generated each year. The second geometric gradient is due to changes in the price of an item, such as labor costs, oil prices, or computer prices. The third geometric gradient is due to inflation (f), which changes the value of the dollars that are received and spent in future years. The fourth geometric gradient is the interest rate used to calculate the time value of money.

The two geometric gradients for inflation (in the price of an item and in the value of a dollar) are often combined into a single rate, which is called the differential inflation rate for that item. Because of the difficulty in estimating the inflation rate for the value of the dollar, which in turn influences the inflation rates for all items, this differential inflation rate may be estimated more accurately than the other two inflation rates.

Whether a particular problem has one or four kinds of geometric gradients, and whether they influence one or several sources of cash flows (energy costs, labor costs, loan repayments, etc.), geometric gradients are arithmetically cumbersome. Thus, this chapter has relied heavily on spreadsheets.

This chapter focused on inflation, while simpler geometric gradients were covered in Chapter 4, for several reasons. First, inflation occurs often and it must be considered in many economic problems. Second, a proper understanding of differential inflation often limits the consideration of inflation to recognizing that the differential inflation rate is 0% and that inflation can be ignored. Third, since there are one to three inflation rates that can be stated for each set of cash flows—with different rates for computers, labor costs, energy costs, etc.—it is easy to get them confused unless inflation is well understood.

REFERENCES

Buck, James R., and Chan S. Park, *Inflation and Its Impact on Investment Decisions*, 1984, Industrial Engineering and Management Press (IIE).

Jones, Byron W., *Inflation in Engineering Economic Analysis*, 1982, Wiley.

U.S. Department of Labor, *Monthly Labor Review*.

PROBLEMS

15.1   Sam receives a starting salary offer of $30,000 for year 1. If inflation is 4% each year, what must his salary be to have the same purchasing power in year 10? Year 20? Year 30? Year 40?   (*Answer:* Salary$_{10}$ = $42,699)

15.2   Assume that Sam (in Problem 15.1) receives an annual 5% raise. How much more, in year-one dollars, is his salary in year 10? Year 20? Year 30? Year 40?   (*Answer:* $2698 in year 10)

15.3   Due to competition from a new plastic, revenues for the mainstay product of Toys-R-Plastic are declining by 5% per year. Revenues will be $2M for this year. The product will be discontinued at the end of year 6. If the firm's interest rate is 10%, calculate the PW of the revenue stream.

15.4   The cost of machinery used to assemble widgets has kept pace with inflation. Using Exhibit 15.1, calculate the cost, in 1992 dollars, for a machine that cost $48K in 1976.   (*Answer:* $117.0K)

15.5   The price of widgets ($3800 in 1992) has kept pace with inflation. Using Exhibit 15.1, calculate the cost in 1954.

15.6   Using Exhibit 15.1 calculate the annual inflation rates and the average for the following decades:
       a.   1974–1984   (*Answer:* 7.33%)
       b.   1964–1974

15.7   Using Exhibit 15.1 calculate the annual inflation rates and the average for the following decades:
       a.   1970–1980
       b.   1983–1993

15.8   Inflation is 4%. If $1000 is invested in an account paying 6% compounded semiannually, what is the year-0 dollar value of the account at the end of the 5 years?

15.9   Assume that your salary is $35,000 in 1995, and $60,000 in 2035. If inflation has averaged 2% per year, what is the real or differential inflation rate of salary increases?   (*Answer:* –.6%)

15.10 In the 1920 Sears Roebuck catalog, an oak chest of drawers cost $8 plus freight. In 1990 this same chest of drawers, in good condition, cost $1200. If the average rate of inflation over the 70-year period has been 3%, what has been the average yearly rate of appreciation, adjusted for inflation?

15.11 The ABC Block Company anticipates receiving $4000 per year from its investments (with a differential inflation rate of 0%) over the next 5 years. If ABC's interest rate is 8% and the inflation rate is 3%, determine the present value of the cash flows.

15.12   Assume inflation is 5% per year. What is the price tag in 6 years for an item that has a differential inflation rate of 4% and that costs $400 today?   (*Answer:* $671)

15.13   The expected rise in prices due to inflation over the next 6 years is expected to be 30%. Determine the average annual inflation rate over the 6-year period.

15.14   Explain how high inflation in a booming real estate market can benefit an engineer who sells a home 5 years after he buys it.

15.15   Felix Jones, a recent engineering graduate, expects a starting salary of $35,000 per year. His future employer has averaged 5% per year in salary increases for the last several years. If inflation is estimated to be 4% per year for the next 3 years, how much, in year-one dollars, will Felix be earning each year? What is the differential inflation rate for Felix's salary?

15.16   The price of a National Computer 686VX computer is presently $2200. If deflation of 2% per quarter is expected on this computer, what will its price be in nominal dollars at the end of 1 year? If inflation is 4.5% per year, what will the price be in year-0 dollars?
       (*Answer:* $1942 year-0 dollars)

15.17   You place $4000 into an account paying 8% compounded annually. Inflation is 5% during each of the next 3 years. What is the account's value at the end of the 3 years in year-0 dollars?

15.18   Using Exhibit 15.1, how much money in 1990 would be equivalent to $2000 in 1967?

15.19   Two series of indexes from the *Historical Statistics of the United States* (shown below) must be combined with Exhibit 15.1 to construct a long-term measure of inflation. From 1779 to 1990, what has the inflation rate averaged?

1910 = 100 (base)							
Year	1779	1785	1800	1803	1830	1850	1864	1880
Index	226	92	129	118	91	84	193	100

1926 = 100 (base)							
Year	1890	1910	1920	1921	1932	1940	1950	1954
Index	56	90	154	98	65	79	162	181

15.20   The Widget Company will need a certain type of milling machine for its new assembly line. The machine presently costs $85,000. It has had a differential inflation rate of 2%. Widget will not need to purchase the machine for 2 years. If inflation is expected to be 4% per year during those 2 years, determine the nominal price of the machine. What is the present worth of the machinery if the market rate of interest for Widget is 9%?

15.21   If $10,000 is deposited in a 5% savings account and inflation is 3%, what is the value of the account at the end of year 20 in year-0 dollars? If the time value of money is 4%, what is the present worth?

15.22   The cost of garbage pickup in Green Valley is $4,500,000 for year 1. Estimate the cost in year 3 in year-one dollars and in nominal dollars. The population is increasing at 6%, the real cost per ton is increasing at 5%, and the inflation rate is estimated at 4%.   (*Answer:* $5.57M in nominal $s)

15.23   If residents of Green Valley (see Problem 15.22) reduce their individual trash generation by 2% per year, by how much will the cost in year 3 have decreased (in year-one dollars)?

15.24   If the interest rate is 7%, what is the present worth in year-one dollars of the following:
   a.   The cost in Problem 15.22   (*Answer:* $4.21M)
   b.   The savings in Problem 15.23   (*Answer:* $166.6K)

15.25   Redo Problem 15.22 for $N = 10$ and $N = 25$.

15.26   Redo Problem 15.23 for $N = 10$ and $N = 25$.

15.27   Redo Problem 15.24 for $N = 10$ and $N = 25$.

15.28   Redo Example 15.12 assuming that the 15% interest rate is a market interest rate.

15.29   Generate a different solution for Example 15.17 by assuming that inflation starts at time 0.

15.30   Your beginning salary is $30,000. You deposit 10% each year in a savings account that earns 6% interest. Your salary increases by 5% per year and inflation is 3% per year. What value does your savings account show after 40 years? What is the value in year-one dollars?
   (*Answer:* $973,719 and $307,455)

15.31   The market for widgets is increasing by 15% per year from current profits of $200,000. Investing in a design change will allow the profit per widget to stay steady; otherwise they will drop by 3% per year. If inflation in the economy is 2%, what is the present worth in year-one dollars of the savings over the next 5 years? 10 years? The interest rate is 10%.

15.32   Bob has lost his job and had to move back in with his mother. She agreed to let Bob have his old room back on the condition that he pay her $1000 rent per year, and an additional $1000 every other year to pay for her biannual jaunt to Florida. Since he is down on his luck, she will allow him to pay his rent at the end of the year. If inflation is 6% and Bob's interest rate is 15%, how much is the present cost (in year-one dollars) for a 5-year contract? (Trips are in years 2 and 4.)

15.33   Enrollment at City University is increasing by 3% per year, its cost per credit hour is increasing by 8% per year, and state funds are decreasing by 4% per year. State funds currently pay half of the costs for City U., while tuition pays the rest. What annual increase in tuition is required?

15.34   A 30-year mortgage for $100,000 has been issued. The interest rate is 10% and payments are made annually. If your time value of money is 12%, what is the PW of the payments in year-one dollars if inflation is 0%? 3%? 6%? 9%?   (*Answer: $PW_{3\%} = -\$70,156$*)

15.35   A homeowner is considering an upgrade from a fuel-oil–based furnace to a natural gas unit. The investment in the fixed equipment, such as a new boiler, will be $2500 installed. The cost of the natural gas will average $60 per month over the year, instead of the $145 per month that the fuel oil costs. If funds cost 9% per year and differential inflation in fossil fuels will be 3% per year, how long will it take to recover the initial investment? (Solve on a monthly basis.)

15.36   John earns a salary of $40,000 per year, and he expects to receive increases at a rate of 4% per year for the next 30 years. He is purchasing a home for $80,000 at 10.5% for 30 years (under a special veterans preference loan with 0% down). He expects the home to appreciate at a rate of 6% per year. He will also save 10% of his gross salary in savings certificates that earn 5% per year. Assume that his payments and deposits are made annually. If inflation is assumed to have a constant 5% rate, what is the value (in year-one dollars) of each of John's two investments at the end of the 30-year period?

15.37   In Problem 15.36, how much does the value of each investment change if the inflation and appreciation rates are halved? Doubled?

15.38   In Problem 15.34, assume the house is owned by an engineer whose effective tax rate for state and federal income taxes is 30%. What is the after-tax IRR for the mortgage if inflation is 0%? 3%? 6%? 9%?

15.39   (Refer to Problem 13.7.) LargeTech Manufacturing has obtained estimates for a new semiconductor product. The machinery costs $800K, and it will have no salvage value after 8 years. LargeTech is in the highest tax bracket. Income starts at $200K per year and increases by $100K per year, except that it falls by $300K per year in years 7 and 8. Expenses start at $225K per year and increase by $50K per year. Revenues and expenses have no differential inflation. If inflation is 5%, what are the before-tax cash flows and taxable income in year-0 dollars under
a.   Straight-line depreciation?
b.   SOYD depreciation?
c.   Double-declining balance depreciation?

d. MACRS?

e. Each method if the machine's salvage value is $40,000?

15.40 (Refer to Problem 13.9.) An engineering and construction management firm with a taxable income of $1M is opening a branch of-
fice. Nearly $250,000 worth of furniture will be depreciated under
MACRS. Inflation is expected to be 6%.

a. What are the after-tax cash flows for the first 5 years in year-0
dollars?

b. If the office is closed early in year 6 and the furniture is sold to a
second-hand office furniture store for 10% of its initial cost, what
tax is paid in year-0 dollars?

15.41 (Refer to Problem 13.12.) The R&D lab of BigTech Manufacturing
will purchase a $1.8 million process simulator. It will be replaced at
the end of year 5 by a newer model. Inflation will be 4% over the 5
years. Use MACRS and a tax rate of 40%. The simulator's salvage
value is $.5M.

a. What is the after-tax cash flow in year 5 in year-0 dollars due to
the simulator?

b. What is the equipment's after-tax EAC in year-0 dollars over the
5 years if the interest rate is 10%?

15.42 (Refer to Problem 13.13.) Find the EAC in year-0 dollars after taxes
for a computer that costs $15,000 to purchase and $3500 per year
to run with a 0% differential inflation rate, and that has a salvage
value of $1000 in year-0 dollars at the end of year 7. Inflation is ex-
pected to be 5% per year. The company's profit exceeds $20 million
per year, and it is purchasing $850,000 worth of capital assets this
year. The company uses an interest rate of 12%.

15.43 Assume that the computer in Problem 15.42 is financed with a 4-
year loan that has constant total payments calculated at 11%. Find the
EAC in year-0 dollars after taxes.

15.44 (Refer to Problem 13.30.) We-Clean-U Inc. expects to receive
$42,000 each year for 15 years from the sale of its newest soap,
OnGuard. There will be an initial investment in new equipment of
$150,000. The expenses of manufacturing and selling the soap will
be $17,500 per year. Expenses and receipts have no differential infla-
tion beyond the CPI of 7%. Using MACRS depreciation, a marginal
tax rate of 42%, and an interest rate of 12%, determine the EAW in
year-0 dollars of the product. How different is this from the $225 that
was calculated by ignoring inflation?

15.45 Assume the equipment in Problem 15.44 is financed with a 13% loan
that will be repaid with five equal annual payments. Determine the
EAW in year-0 dollars of the product. How different is this from the
value in Problem 15.44?

15.46   (Refer to Problem 13.33.) If the simulator in Problem 15.41 is purchased on a 3-year loan, what is the EAC in year-0 dollars? The loan is for 80% of the cost, the interest rate is 12%, and the payments are uniform.

15.47   (Refer to Problem 13.34.) The simulator in Problem 15.41 will save $600,000 per year in year-0 dollars in experimental costs.
   a.   What is the after-tax IRR?
   b.   What is the after-tax IRR if the loan in Problem 15.46 is used to buy the equipment?
   c.   Graph the IRR for percentages financed from 0% to 90%.

APPENDIX 15A   FORMULAS BASED ON THE EQUIVALENT DISCOUNT RATE

This appendix develops formulas and an equivalent discount rate that can sometimes replace spreadsheets. This adds inflation in the item's price and in the economy (f) to the formulas developed in Appendix 4A. This could be described as using finesse to replace brute force.

The underlying basis for this approach is that the factors for geometric gradients for changing volume, changing prices, inflation, and the time value of money have the same form. Each factor relies on $(1 + \text{rate})^{\text{year}}$. Thus, for some problems, these geometric rates can be combined into an equivalent discount rate. This transformation is most useful (1) if financial calculators rather than computerized spreadsheets are used, or (2) for "what-if" analyses in which manipulating equations is often easier than manipulating tables.

Consider the question of showing the dependence of Example 15.17's present worth (see Exhibit 15.18) on the assumed rate of volume change. Since the number of kWhs is the first column computed, each new value for the rate of volume change requires recomputing the entire table.

However, this table is based on the assumption of a constant base amount that has grown at a constant rate, with constant rates of change for the unit price of electricity and for inflation. These are the assumptions that permit the development of an equivalent discount rate. As a result, the present worth of the tth cash flow is

$$\frac{(1 + \Delta\text{Volume})^{t-1}(1 + \Delta\text{ItemPrice})^{t-1}}{(1 + \Delta\text{Inflation})^{t-1}(1 + i)^t} \cdot \text{CashFlowYear}_1$$

We could rewrite this using the notation of Exhibit 15.17 as

$$\frac{(1 + v)^{t-1}(1 + g)^{t-1}}{(1 + f)^{t-1}(1 + i)^t} \cdot \text{BaseAmount}$$

This can be rewritten by defining a rate, r, that accounts for all but the time value of money factor, i. This rate is the geometric gradient for the cash flow values stated in year-0 dollars. Equation 15A.1 is used to calculate r.

$$(1 + r) = (1 + v)(1 + g)/(1 + f) \qquad (15\text{A}.1)$$

Once r has been defined, the equations for present worth can be developed as they were in Appendix 4A. The Equations 4A.3, 4A.4, and 4A.5 are repeated here, with g replaced by r, as 15A.2, 15A.3, and 15A.4, respectively. As before, A_1 is the cash flow in year 1.

If $r < i$

$$PW = [A_1/(1 + r)] \cdot (P/A, x, N) \qquad (15A.2)$$
$$\text{where } (1 + x) = (1 + i)/(1 + r)$$

If $r < i$

$$PW = [A_1/(1 + r)] \cdot N \qquad (15A.3)$$

If $r > i$, then the x as calculated above would be negative. This might occur when the volume is increasing at a high rate or when the item has a high rate of differential inflation. Instead, a future worth factor can be used to calculate a present worth.

$$PW = [A_1/(1 + i)](F/A, x, N) \qquad (15A.4)$$
$$\text{where } (1 + x) = (1 + r)/(1 + i)$$

It may be less obvious than looking at the first few rows of a spreadsheet, but these equations make the same assumptions as the tables that were developed in the main body of this chapter. The cash flow in year 1 is the base value, and all later cash flows are derived by multiplying that base value by a series of geometric gradients.

Example 15A.1   Greenacres' Sewage Pumping (Revisited)

Calculate the present worth for Example 15.17 to match Exhibit 15.18.

Solution
Now

$$(1 + r) = 1.07 \cdot 1.05/1.03 = 1.090777$$

and $r = 9.0777\%$. Since this exceeds $i = 8\%$, we let

$$(1 + x) = (1 + r)/(1 + i) = 1.090777/1.08$$

and $x = 0.9978\%$. Now

$$PW = [1.25M/1.08](F/A, .9978\%, 10) = 1.1574 \cdot 10.4612 = \$12.11M$$

which is the same answer as before.

If we wanted to analyze the dependence of the PW on any of the four geometric rates, this equation is much easier to use than the full table of Exhibit 15.18.

Example 15A.2   Heavy Metal Casting

Assume that Heavy Metal Manufacturing is producing a lead casting for shielding of nuclear reactors. This casting is replaced on a 5-year schedule. Very few reactors are being built or retired; thus, there is essentially no change expected in the annual volume. However, the price Heavy Metal can charge is expected to increase at 3% per year, while inflation is expected to be 7% per year. What is the difference between the present worth of sales calculated with the exact inflation rates and the present worth calculated using the approximate rate of differential inflation? Heavy Metal's sales in the first year will be $850,000; its interest rate for the time value of money is 10%; and it believes a 20-year time horizon is appropriate.

Solution

The exact calculation lets $(1 + r) = 1.03/1.07$; thus, $r = -3.74\%$, which is less than 10%. Therefore, Equation 15A.2 is used.

$$(1 + x) = (1 + r)/(1 + i), \text{ so } x = 14.27\%, \text{ and}$$
$$PW = \$850,000(P/A, 14.27\%, 20) = \$5.542M.$$

The approximate calculation uses -4%, which equals $3\% - 7\%$, as the rate of differential inflation. This is still less than i (all negative values for r are less than i), so Equation 15A.2 is still used:

$$(1 + x) = (1 + r)/(1 + i), \text{ so } x = 14.58\%, \text{ and}$$
$$PW = 850,000(P/A, 14.58\%, 20) = \$5.446M.$$

The difference is $96,000, or less than 2% of the true value. Given the probable accuracy of the inflation estimates, the approximation is probably accurate enough.

PROBLEMS

15A.1   Smallville is suffering a 1% annual loss of population and property values. Even so, Smallville must maintain its tax collections at a constant value of $3.2 million. If the inflation rate is 4.5%, what inflation rate in taxes for the remaining taxpayers is required for Smallville to attain its goal? (While Smallville uses a rate of 6% for the time value of money, that rate is irrelevant to this problem.)

15A.2   Redo Problem 15.22, using an equivalent discount rate.

15A.3   Redo Problem 15.27, using an equivalent discount rate.

15A.4   Redo Problem 15.31, using an equivalent discount rate.

15A.5   Redo Problem 15.34, using an equivalent discount rate.

15A.6   Redo Example 15.15, using an equivalent discount rate.

PART FIVE   DECISION-MAKING TOOLS

Cost estimating, sensitivity analysis, probability theory, and multiobjective techniques extend the complexity of the problems that engineering economy can solve. Each tool is the basis for more advanced courses on mathematical models for decision making. However, substantial realism in modeling and depth in analysis can be added easily.

The cash flows in homework problems and examples have been supplied in each chapter; however, in the real world estimating cash flows is often necessary (Chapter 16). The benefits of an engineering project often span a decade or longer, and even the cost of construction must be estimated. Thus, it is necessary to use *sensitivity analysis* to show how recommended decisions depend on uncertainties in the assumed or estimated data (Chapter 17).

Probability theory explicitly models uncertainty about cash flows with expected values to describe economic returns, and with standard deviations to describe risks (Chapter 18). *Multiobjective techniques* balance cash flow measures and objectives that cannot be directly linked to cash flows, such as flexibility or technological learning (Chapter 19).

CHAPTER 16 ESTIMATING CASH FLOWS

The Situation and the Solution

Cash flows for real-world problems must be estimated—they are not specified in a problem statement.

The solution is to use a method that best balances the cost of estimating and its accuracy—for the problem being analyzed. Possible methods include analogies, cost-estimating relationships such as learning curves and parametric costing, and detailed engineering data with bids and quotes.

Chapter Objectives

After you have read and studied the sections of this chapter, you should be able to:

Section 16.1
Distinguish between the costs *of making* a decision and the costs *determined by* that decision.

Section 16.2
Define the stages of a project's life cycle and identify the costs occurring in each.

Section 16.3
Match the methods for estimating cash flows to the stages where each is used—based on the methods' cost and accuracy.

Section 16.4
Define the two common design criteria and write functional or performance specifications.

Section 16.5
Correctly model the base case for evaluating alternatives that change current operations.

Section 16.6
Use indexes to adjust for inflation and productivity changes.

Section 16.7
Estimate the costs of facilities using capacity functions.

Section 16.8
Estimate sales using growth or S-curves.

Section 16.9
Using learning curves, estimate the labor content of the ith unit from the first unit's labor content.

Section 16.10
Estimate costs using factor estimates.

Section 16.11
Correct for possible inaccuracies in accounting data to be used for cost estimating.

Key Words and Concepts

Design to cost   A method that matches the first cost to a defined budget.

Minimizing life-cycle cost   A method that emphasizes the time value of money.

Functional specifications   These define both what is to be accomplished and how that functionality is to be achieved.

Performance specifications   These define only the required level of performance, and they leave the definition of how it is to be achieved to the bidders.

Cost estimating is required for engineering economy. Future and current costs and benefits must be estimated to provide the basic data to be analyzed. Flawed data cannot reliably support decision making. Any engineering economic analysis is only as good as the estimates it is based on.

This chapter develops models for the following cash flow estimation problems:

Estimating the future demand for a new product using **growth curves** or **S-curves.** Examples include the demand for telephones, radios, TVs, color TV, and cable TV.

Estimating the cost of a new facility with a **size index** or by *scaling* the known cost of a similar facility using a **capacity function.** Examples include a petroleum refinery, an office building, and an athletic stadium.

Estimating the cost of the 75th jet or 8th ship using the cost of the first one and **learning curves**, which reflect the predictable decrease in unit costs as experience accumulates.

Estimating a job's cost using a **factored estimate** based on a materials list, labor hours (e.g., for electricians and laborers), and equipment hours (e.g., for cranes and cement mixers). This is used for new buildings, remodeling assembly lines, and new product prototypes.

16.1 THE IMPACT OF EARLY PROJECT DECISIONS

Penny-wise and Pound-foolish.   Early decisions have a large impact on later costs. Benjamin Franklin's words about being "penny-wise and pound-foolish" are particularly applicable. Skimping on up-front planning and design costs saves "pennies" when compared with the costs of redesigns and modifications. Example 16.1 is only one of the many projects that could be used to substantiate this phrase.

Why is there never enough time to do it right, but always enough time to do it over?

Example 16.1   Anchorage Performing Arts Center

In the late 70s and early 80s, Alaska's construction industry boomed as the state funneled tax money from Prudhoe Bay oil revenues to many municipal projects. Anchorage (with slightly over half of the state's population) built a convention center, a sports arena, a new library, and a new performing arts center.

The performing arts center was plagued with problems in planning and budgeting. The arts communities identified ambitious goals for the three theatres, and few financial constraints were imposed on the architect early in the project.

Guess how severe the problems in construction and operation were?

Solution

The problems were very severe. The original architect was fired during construction for designing a facility that met the goals of the arts communities—but that the city could not afford to build. The redesign of a facility that was under construction significantly increased the cost of what was finally built.

The project's final construction cost was $63 million for a facility with 3295 seats. If money's time value is assumed to be 7% and each seat is used three times per week, then the investment cost is $8.58 per seat per use, plus the costs to operate and maintain it and to pay the performers. Decisions today have little impact on these costs; they were largely determined by the conceptual and preliminary design of the building.

There are also substantial technical problems with the facility. These include avalanches of snow off the stainless-steel roof onto pedestrians, leaks in a roofing system never before used in Alaska, and poor handicapped access that required redesign and modification. The high cost to operate and the extensive technical problems have led to continuing arguments over whether the city should mothball the facility, arguments that began only 4 years after it opened.

Quantifying the Impact of Early Decisions.   Exhibit 16.1 summarizes data from the acquisition of military weapon systems. This graph and similar ones developed for civilian construction projects emphasize that most costs are determined during the planning stage.

Exhibit 16.1's first three milestones represent a detailed breakdown of a project's planning stage into conceptual design, preliminary systems design, and detailed systems design. Even though the majority of the costs are incurred

Exhibit 16.1   Determined- vs. spent-to-date funds over a project's life

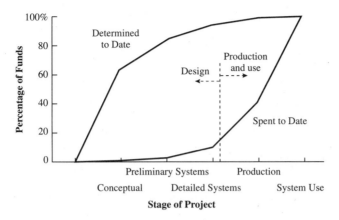

during system operation, the operating costs are largely determined during these three steps of the planning stage.

Most of the costs are defined by the specifications developed during conceptual design. How fast, how far, and how high must the jet fly? What ordnance must the tank fire, carry, and withstand? Or, at the personal level, should your transportation between your home and your office be a bicycle, public transportation, a used "commuter" auto, or a sports car?

Since so much of the system's costs are defined at the conceptual and preliminary systems design stages, it is penny-wise and pound-foolish to skimp on the planning steps.

16.2 CASH FLOW ESTIMATING AND LIFE-CYCLE STAGES

Hidden Costs.   Many costs and consequences are hidden at the time of decision. The analogy to an iceberg shown in Exhibit 16.2 is useful; often 90% of the costs occur after acquisition. For example, consider a public project funded by a bond, such as a school, library, or performing arts center. The debate is over the cost of construction, the incremental property tax paid to repay the bond, and the need for the facility. Often, the cost for schoolteachers, librarians, heat and light, etc., to provide the services *far* exceeds the cost of construction. Yet, these "hidden" costs receive far less attention.

Costs during the Project Life Cycle.   The four stages of the project life cycle are (1) planning, (2) growth/construction, (3) operation/production, and (4) retirement/disposal. The later stages include many of the hidden costs shown in Exhibit 16.2.

To ensure that cash flows are not overlooked, they are categorized. Sometimes, categorizing costs as recurring or nonrecurring is useful, but more often, simple checklists like the ones shown in Example 16.2 and the following are used:

Exhibit 16.2   Iceberg of hidden costs

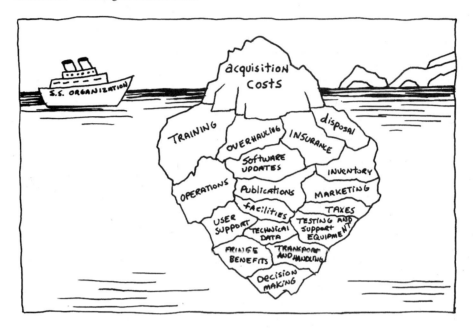

Working capital

Training

Public relations

Accounting and auditing

General management and administration

Contract services

Software design, development, and maintenance

Security

Insurance and workers' compensation

Taxes

Fringe benefits

Capital recovery and return requirements

Design and engineering

These lists of costs are often supplemented with lists of <u>tangible</u> and intangible <u>benefits</u>. For example, reducing the time spent waiting for the doctor or improving patient morale might be part of an economic analysis within a hospital. Similarly, reducing both the number of defective parts and the lead time before shipping an order would be intangible benefits within a manufacturing environment.

Exhibit 16.3 defines several cost groupings for manufacturing. Several are important standard categories in which accountants report data on historical performance. This historical data is often the basis for estimating future costs and performance.

Exhibit 16.3   Price of a manufactured product

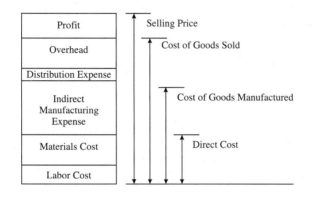

Example 16.2   Estimating Checklist for an Oil Company

A firm wants to ensure that engineers do not forget some important costs when analyzing oil exploration and development projects in Alaska. What's the best approach?

Solution
One firm uses a more complete version of the checklist shown in Exhibit 16.4. Its version has about 15 more categories and short comment lines with each category.

Exhibit 16.4   Estimating checklist for an oil company

☐ Project Execution Plan	☐ Code Requirements	☐ Communications Equipment
☐ Process Flow Diagram	☐ Permitting Requirements	☐ Furniture and Fixtures
☐ Plot Plan	☐ Safety Requirements	☐ Geotechnical Requirements
☐ Construction Drawings	☐ Gravel Requirements	☐ Functional Checkout
☐ Schedule Constraints	☐ Demolition	☐ Utility Tie-Ins
☐ Equipment List	☐ Foundation Requirements	☐ Operating Fluids
☐ Transportation and Freight	☐ Painting and Coating	☐ Labor Productivity
☐ Spare Parts	☐ Insulation	☐ Payroll Benefits and Burdens
☐ Vendor Representatives	☐ Fireproofing	☐ Cleanup
☐ Purchasing Plan	☐ Fire Protection	☐ Scaffolding
☐ Contracting Plan	☐ Instrumentation	☐ Material Handling
☐ Documentation and As-Builts	☐ Fabsite Leases and Fees	☐ Temporary Facilities
☐ Vendor Shop Visits	☐ Fees	☐ Job-Site Utilities
☐ Expediting	☐ Taxes	☐ Security
☐ Quality Assurance	☐ Inflation and Escalation	☐ Allocated Costs
☐ Other Costs	☐ Risk Analysis	☐ Contingency

16.3 CASH FLOW ESTIMATING STANDARDS

Stages of Cash Flow Estimating.   Engineering economy is mostly concerned with decisions made during steps in the planning stage—conceptual design, preliminary systems design, and detailed systems design. Once a project is operating, engineering economy's role is to evaluate possible modifications and recommend the financially attractive point for project termination.

Estimating is essentially forecasting, and there are many techniques from statistics that are used. There are also techniques that have been specifically developed for estimating cash flows.

Conceptual Design.   Little data exists when a project is an idea or a concept, so the methods are simple. In a simple parametric model, a building's projected size is multiplied by a typical cost per square foot to derive an estimated construction cost. Other cost estimates may be derived by analogy from a project that is similar in type and in size.

Two common techniques are used to adjust for differences between the project being estimated and the project with known data. First, indexes are used to adjust for inflation, gains or losses in productivity, project location, and project size (see Section 16.6). Second, a capacity function model can be used to adjust for scale differences if the projects have different capacities.

For example, a 100,000-square-foot warehouse costs less than twice what a warehouse of 50,000 square feet would cost. The capacity function model uses a scale coefficient to adjust for the lower cost per square foot of the larger building (see Section 16.7).

Preliminary Systems Design.   At this stage, more detail about the project is available. There may be computer simulations of the system, and there is some ability to break the project down into its individual parts. Parametric models are likely to be used, but at a finer level of detail. For example, a building might be divided into hallways, offices, meeting rooms, and mechanical/plumbing areas, and then the relevant cost per square foot for each kind of space would be applied.

Final Systems Design.   At this stage, detailed data on the project will be available. There may be bids and quotes, and there will be accurate design data to match with standard cost factors or distributions.

An organization does not need to depend only on internal resources in constructing these estimates. There are many external sources that acquire and summarize cost data; examples include *Walker's Building Estimator's Reference Book, Richardson's Standards,* and *Means Cost Data.* However, specific in-house historical data that represents similar projects in the same locality, with similar-quality labor and management, similar weather, etc., is generally better than generalized data.

Cost Estimate Definitions and Accuracy.   The American Association of Cost Engineers (now AACE International) defines three kinds of estimates, which are done in order as a project progresses [Humphreys and Wellman]. The first is an

order-of-magnitude estimate, the second a **budget estimate**, and the third a **definitive estimate**. Even when these estimates are *done properly*, actual costs rarely match the estimates. The reasonable range for actual costs around the estimated value is:

Estimate Type	Actual vs. Estimate
Order-of-magnitude	−30% to +50%
Budget	−15% to +30%
Definitive	−5% to +15%

If a well-prepared budget estimate is $1 million, then the actual costs should fall within a range of $850,000 to $1.3 million.

These estimates are typically produced sequentially. An order-of-magnitude estimate would be appropriate early in a feasibility study; a budget estimate would be produced after a firm definition of the project's scope was developed; and the definitive estimate might require that the design be at least 65% complete.

The methods and detail for more accurate estimates take longer and cost more, as shown in Exhibit 16.5. At the same time, detailed planning with definitive estimates can allow contractors or bidders to bid lower due to smaller contingency factors. Remember that these planning and estimating costs, even when they are high, are still much smaller than the costs for construction and operation. Also, note that overruns are often larger than underruns. Murphy's Law ("What can go wrong, will go wrong") does play some role in this.

Exhibit 16.5   Accuracy vs. cost of doing estimate

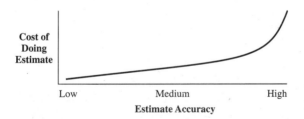

Revenues and benefits are usually harder to estimate accurately than costs, so the percentages cited by Humphreys and Wellman are better than can be expected for revenues and benefits. It is also hard to decide if the benefits of more planning and estimating exceed the benefits and risks of getting started so that the facility can operate as soon as possible.

16.4 DESIGN CRITERIA AND SPECIFICATIONS

Design Criteria.   There are two approaches to designing products and facilities. They are distinguished by the flexibility of the initial budget for the first costs.

Design to cost ensures that first costs match a budget, while **minimizing life-cycle costs** maximizes present worth.

Design to cost emphasizes first-cost expenditures that must fit within a fixed budget. Features, square footage, etc., are increased or decreased to keep first costs within the limit.

Minimizing life-cycle costs or maximizing PW allows for incurring extra first costs that are justified because they reduce future expenses. This is the approach that engineering economy emphasizes.

Many public projects are designed to cost. A congressional budget item or a local bond issue may identify the project and a budget for capital costs when the project is still at the conceptual design stage. From that point on, features may be added or subtracted to ensure that the first cost remains targeted on the identified value. Similarly, in a firm, if adjusting the budget for the first cost is difficult, then design to cost can be used to build the best possible project—within the budget. This approach increases the need for accurate estimates.

The shortcoming of this approach is obvious from an engineering economy standpoint. For example, reducing the insulation in a building reduces construction costs, but increases heating and cooling bills. The proper decision is made by comparing cash flows at different insulation thicknesses, and choosing the thickness that minimizes the equivalent annual costs for construction and for operation, which is minimizing life-cycle costs.

Many federal projects (such as highways, transit systems, and weapon systems) have been designed to cost [Michaels and Wood]. First costs were minimized, but the costs to operate, maintain, and replace early soared. So the U.S. government developed policies that promoted minimizing life-cycle costs. Some of these policies are not stated in dollar terms. For example, electronic systems are supposed to be designed in a modular fashion. This simplifies repair and facilitates later upgrades.

Once a project has been approved and is underway, the focus often becomes design to cost. The project management team is focused on completing the project—on time, under budget, and within specifications.

Specifying Performance.   The accuracy of cost estimates is also influenced by how project specifications are defined. **Functional specifications** define specifically how to complete a project or manufacture a product, while **performance specifications** simply define what a project or product must do.

Functional specifications focus on how to do something, while **performance specifications** focus on what something must do.

An engineering team that writes functional specifications is assumed to be able to specify the best possible way to complete the project or build the product. However, hiring another firm to build the product or complete the project is often done because the hired firm has more expertise in that technology.

When writing performance specifications, the engineering team must be sure that they are realistic. Otherwise the project cannot be completed, nor the product manufactured. However, the team assigned with meeting the specifications is responsible for devising the best way to achieve that level of performance while minimizing costs.

As the advantages of performance specifications have become clearer in recent years, earlier trends toward functional specifications have been reversed. Exhibit 16.6 contrasts both types of military specifications to illustrate this difference.

Another example of a preference for performance specifications is shown by engineers who work for owners or for design firms that are careful not to specify *how* a contractor shall perform the work—that is the contractor's area of expertise.

Exhibit 16.6   Performance vs. functional specification

Performance Specification Example. Excerpts from the one-page U.S. Army Signal Corps Specification #486 for a Heavier-than-Air Flying Machine, December 23, 1907:

> The flying machine will be accepted only after a successful trial flight...packed for transportation in army wagons. It should be capable of being assembled and put in operating condition in about one hour...to carry two persons having a combined weight of about 350 pounds, also sufficient fuel for a flight of 125 miles...have a speed of at least forty miles per hour in still air...a trial endurance flight will be required of at least one hour...include the instruction of two men in the handling and operation of this flying machine.

Functional Specification Example. Examples from Jack Anderson's column of April 16, 1986 (hopefully since simplified):

> GI cocoa: 20 pages (example: "When washed with petroleum ether, not less than 98% by weight shall pass through a U.S. Standard No. 200 sieve.")

> GI chewing gum: 17 pages

> Olives: 17 pages and 10 sizes

> Doughnuts (example: when "cut vertically or horizontally with a sharp knife, shall not be greasy over one-eighth inch in depth at any place on the cut surface," and frosted doughnuts must retain their glaze when "subjected to ordinary shocks of transportation.")

16.5 MODELING THE BASE CASE

Many applications of engineering economy compare a new alternative with a current mode of operation. Often this includes a bad assumption. The key questions are: What happens if the current mode of operation continues? Does the current level of sales continue or does it fall?

Often, firms incorrectly assume that continuing the current operating mode causes the future to look like the present. However, many manufacturing and service operations have found that they must improve what they provide their customers or the firms will die.

The comparison is not before and after the project, but rather with and without the project.

A worldwide competitive environment requires continuous improvement for competitive survival. For example, in many production situations the cost of robots exceeds the costs of having the jobs performed by people. The key difference may be in consistency and quality, which may have financial benefits that are difficult to quantify. These financial benefits may be translated into choosing the appropriate base case, as in Example 16.3.

Example 16.3   Justifying Robots for Welding and Painting

HouseHold Appliances is considering replacing people with robots for the welding and painting operations on its washers and dryers. Many of its competitors have already done so, but HouseHold is a small firm with a low production volume.

The proposed revision of its plant will cost $900,000. Since new technicians would have to be hired, the annual savings in labor costs are limited and power costs for the plant will go up. The annual savings in operations and maintenance costs are expected to be only $50,000. If the facility will last 10 years and have no salvage value, should it be built? Assume that sales of this line currently bring in a revenue of $400,000 per year. HouseHold uses an interest rate of 12%.

Solution

If the sales revenue of $400,000 per year can be maintained without the project, it is not justified. In other words, a constant demand for the product implies that the robot project is not justified.

$$PW = -900,000 + 50,000(P/A, .12, 10) = -\$617,490$$

However, if we assume that revenue will decline as the gap between our quality and our competitor's quality increases, then a different answer may emerge. Basically, declining demand may require process improvement. For simplicity, assume that the volume of production does not decline, but that the price HouseHold can charge does decline. If revenue declines by $35,000 per year (less than 10%), the answer is: do it.

$$PW = -900,000 + 50,000(P/A, .12, 10) + 35,000(P/G, .12, 10)$$
$$= -617,490 + 35,000(20.254) = \$91,400$$

16.6 USING INDEXES FOR AN ORDER-OF-MAGNITUDE ESTIMATE

Indexes of inflation were developed in Chapter 15. These can be combined with indexes for productivity, geographical location, and facility size to create rough or order-of-magnitude estimates. Indexes are the essence of the historical analogy approach, and they are often included in other approaches as well. Index data can be purchased in text or database formats (in *Dodge, Means, Richardson's,* and *Walker's*), or an organization can develop its own.

An example productivity index would adjust for the lower productivity of labor under prolonged overtime conditions. For example, 2 hours of overtime per day for 5-day weeks is estimated to result in a lost efficiency of 2–8%, and 4 overtime hours per day for 7-day weeks is estimated to result in a lost efficiency of 28% [*Walker's*].

Exhibits 16.7–16.9 summarize a few cost and productivity indexes used to estimate the cost of constructing facilities [U.S. Air Force]. The cost growth factor (Exhibit 16.7) accounts for inflation. Area cost factors (Exhibit 16.8) account for location-specific factors, which include weather, seismic codes, mobilization, climate, availability and productivity of labor, and overhead and profit for the contractor. The size-adjustment factor (Exhibit 16.9) accounts for the difference in square footage between the database of historical costs and the project being estimated. For the graph in Exhibit 16.9 the ratio between the size of the estimated facility and the typical size (table in Exhibit 16.9) is the *x*-axis value. Then, the corresponding cost index is the *y*-axis value. These indexes are applied in Example 16.4.

Exhibit 16.7 Cost growth factor

Year	January	April	July	October
1988	.938	.947	.955	.964
1989	.973	.982	.991	1.000
1990	1.009	1.018	1.026	1.034
1991	1.043	1.051	1.059	1.064
1992	1.073	1.081	1.087	1.091
1993	1.099	1.106	1.112	1.116
1994	1.125	1.131	1.138	1.142
1995	1.151	1.157	1.164	1.168
1996	1.177	1.184	1.191	1.195
1997	1.204	1.211	1.218	1.223

Exhibit 16.8 Area cost factors

Location	Factor[1]	Location	Factor
Huntsville, AL	.88	Denver, CO	.97
Aleutians, AK	3.48	Delaware	.99
Anchorage, AK	1.79	Cape Kennedy, FL	.91
Phoenix, AZ	.98	Tampa, FL	.78
Little Rock, AR	.87	Atlanta, GA	.86
Los Angeles, CA	1.19	Macon, GA	.80
Travis AFB, CA	1.22	Honolulu, HI	1.33

[1] 1.00 equals a 144-city average (3 cities each in 48 states).

Exhibit 16.9 Size-adjustment factors and chart

Facility	Typical Size	$/ft^2 (FY90)[1]
Outpatient clinic	30,000	$144
Dental clinic	15,000	155
Administrative	25,000	82
Barracks	40,000	72
Messhalls	16,000	157
Community fire station	3,500	98

[1] FY90 costs are at the $1.00 point in October 1989.

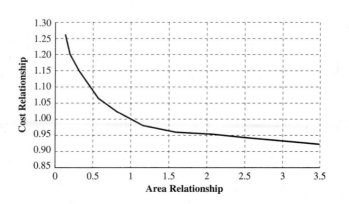

Example 16.4   Estimate Messhall Cost

Estimate the cost of a 12,000-square-foot messhall to be constructed in Shemya, Alaska, during summer 1994.

Solution
The midpoint of the construction season is July 1994, so the cost growth or inflation factor is 1.138 (from Exhibit 16.7). Since Shemya is part of the Aleutian Island chain, the location factor is 3.48 (from Exhibit 16.8). Finally, the base unit cost for a messhall is $157 per square foot (from Exhibit 16.9). The messhall is 75% of the typical size, so the cost relationship factor (from the chart in Exhibit 16.9) is 1.03. The estimated cost is determined by multiplying these factors together:

$$\text{Estimate} = (157)(12{,}000)1.03 \cdot 3.48 \cdot 1.138$$
$$= \$7{,}685{,}000$$

16.7 USING CAPACITY FUNCTIONS FOR ORDER-OF-MAGNITUDE ESTIMATES

The capacity function model uses a capacity parameter to estimate a facility's cost. Larger facilities cost more, but usually there are economies of scale in their construction. Equation 16.1 is sometimes called the 60% rule because the capacity exponent is so often close to 60%.

$$C_x = C_k(S_x/S_k)^n \tag{16.1}$$

where

$C_x \equiv$ Estimated cost of facility of size S_x
$C_k \equiv$ Known cost of facility of size S_k
$n \equiv$ Capacity exponent (normally < 1).

To improve the model's accuracy, these exponents are identified for specific facilities, such as refineries or power generation, perhaps with a size parameter specified. Exhibit 16.10 is an abbreviated list of these exponents, which are applied in Examples 16.5 and 16.6.

Exhibit 16.10   Capacity exponents

Exponent	Facility	Exponent	Facility
.70	LP gas recovery in refineries	.65	Oxygen plants
.73	Polymerization, small plants	.65	Styrene plants
.91	Polymerization, large plants	.98	Ammonia, and nitric acid or urea
.61	Steam generation, large, 200 psi	.75	Chlorine plants, electrolytic
.81	Steam generation, large, 1000 psi	.57	Refineries, small
.88	Power generation, 2000–20,000 kW	.67	Refineries, large
.50	Power generation, oil field, 20–200 kW	.55	Hydrogen sulfide removal
.64	Sulfur from H_2S		

Source: Humphreys and Wellman, Table 2.1.

There are limits to how different the sizes of the estimated and known facilities can be. Generally, the ratio should be less than 2, and it should never be greater than 5 [Humphreys and Wellman]. Similarly, due to changes in technology, the gap in timing should be no more than 5 years.

Example 16.5   Adjusting to the Market

Your engineering group has just finished the preliminary design of a small refinery. The estimated cost is $19 million for a capacity of 2 million barrels per year. Your management has asked you for order-of-magnitude estimates for facilities with capacities of 1 million and 3 million barrels per year. The managers also want a comparison of cost per barrel of capacity.

Solution
From Exhibit 16.10, the coefficient is .57 for small refineries.

$$C_x = C_k(S_x/S_k)^n$$
$$C_1 = C_2(1/2)^{.57} = 19M(.6736) = \$12.8M, \text{ or } \$13M$$
$$C_3 = C_2(3/2)^{.57} = 19M(1.26) = \$23.9M, \text{ or } \$24M$$

In tabular form, this is:

Capacity (millions of barrels)	1	2	3
Cost	$13M	$19M	$24M
Cost/Bbl Capacity	$12.8	$9.50	$8.0

Example 16.6   A New Power Plant

The community of Upper Snowshoe has expanded its winter tourism very rapidly. Three years ago, it built a new 10,000-kW power plant for $4.5M. The local construction cost index was 135 at that time, and it has risen to 157 now. In current dollars, what is the estimated cost of a 16,000-kW power plant?

Solution
The first step is to convert the $4.5M into current dollars by multiplying by 157/135, which results in $5.23M as the estimated current cost of a 10,000-kW power plant.

Then the capacity function is used to scale this to the cost of a 16,000-kW plant. From Exhibit 16.10, the capacity coefficient is .88.

$$C_x = C_k(S_x/S_k)^n$$
$$C_{16,000} = C_{10,000}(16,000/10,000)^{.88} = 5.23M(1.512)$$
$$= \$7.91M, \text{ or } \$8M$$

16.8 USING GROWTH CURVES

Demand or sales, and thus revenues, are often estimated with *growth* or *S-curves*. These models, as shown in Exhibits 16.11 and 16.12, assume that demand is initially low. After product refinement and market awareness, sales grow rapidly until the volume approaches a limit and growth tapers off.

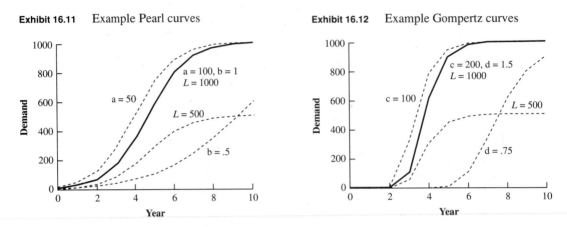

Exhibit 16.11   Example Pearl curves

Exhibit 16.12   Example Gompertz curves

There are several forms of these models, two of which are presented here. The first of these, the Pearl curve (Equation 16.2), is symmetrical about its point of inflection. The second, the Gompertz curve (Equation 16.3), does not have this property. The choice between these two equations should be based on fitting past sales in the industry being analyzed. Problem 16.20 illustrates that the curves differ when fit to the same first few points. Example curves are graphed in Exhibits 16.11 and 16.12.

These curves are defined here so that demand (D) is a function of time (t) and the upper limit (L). These curves can also be formulated for cumulative sales or even for technological performance, such as the speed performance for jet aircraft.

Pearl Curve $\qquad\qquad\quad$ ✳ $\quad D = L/(1 + ae^{-bt})\qquad\qquad\qquad$ (16.2)

Gompertz Curve $\qquad\quad$ ✳ $\quad D = Le^{-ce^{-dt}}\qquad\qquad\qquad\quad$ (16.3)

Example 16.7 illustrates use of a Pearl curve with given a and b. These curves can be linearized, as shown in Equations 16.2′ and 16.3′. Regression can be used to estimate the a, b, c, and d coefficients with three or more periods of data (beyond this text). Or, as shown in Example 16.8, two data points can be used to estimate the coefficients. However, as discussed in Eschenbach and Geistauts, the limit L cannot reliably be estimated from the data used to estimate a, b, c, and d.

Linearized Pearl Curve $\qquad$ $\ln[(L - D)/D] = \ln a - bt\qquad\qquad$ (16.2′)

Linearized Gompertz Curve $\quad$ $\ln[\ln(L/D)] = \ln c - dt\qquad\qquad$ (16.3′)

Example 16.7   ElectroChip Market Growth

ElectroChip is developing a chip that will convert old IBM PC clones into home-security monitors. It expects the chips to sell for $50 each the first year and to drop in price by $5 per year. It believes that the upper limit for annual demand is 1,000,000 chips. Assume that a Pearl curve with a = 10,000 and b = 3 applies. What is the expected demand in each of the next 6 years? What is the expected revenue?

Solution

$$D = L/(1 + ae^{-bt}) = 1,000,000/(1 + 10,000e^{-3t})$$

Year	1	2	3	4	5	6
Demand (D)	2005	38,778	447,608	942,114	996,950	999,848
Revenue (R)	$.1M	$1.7M	$17.9M	$33.0M	$29.9M	$25.0M

By multiplying the estimated sales in each year by $50, $45, $40, $35, $30, and $25, respectively, the revenue data shown in the table can be generated and then graphed, as shown in Exhibit 16.13.

Exhibit 16.13   Pearl curve for ElectroChip market growth

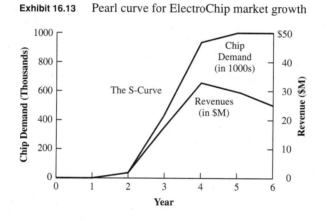

Example 16.8   BigState MediState Demand

BigState has instituted an incentive for hospitals to file their claims for reimbursement under the MediState system in a specific electronic format. Assume that the number of hospitals is steady at 1000, and that 5 and 70 hospitals switched in the first and second years, respectively. In order to plan for the upgrade of the state's computer network, you must estimate the number of hospitals that will switch in each of the next 4 years. Assume that a Gompertz curve applies.

Solution

The limit for the number of hospitals, L, is 1000, and 75 hospitals have switched through year 2.

$$D = Le^{-ce^{-dt}}$$

Using the first 2 years of data ($t = 1$ and 2), estimates of c and d must be derived using Equation 16.3' as follows:

$$\ln[\ln(1000/5)] = 1.66739 = \ln c - d \cdot 1$$
$$\ln[\ln(1000/75)] = .95176 = \ln c - d \cdot 2$$

Subtracting the second equation from the first solves for d; then that value is substituted to find ln c and c.

$$.71563 = d \text{ and } 2.3830 = \ln c \text{ or } c = 10.8376$$

(More accurate estimates would be available from regression if there were more data.) The results are summarized in Exhibit 16.14.

Exhibit 16.14   Gompertz curve for BigState MediState demand

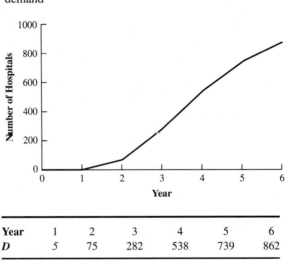

Year	1	2	3	4	5	6
D	5	75	282	538	739	862

16.9 USING LEARNING CURVES

Learning curves were developed to model the cost of producing complex systems in which operator experience, process improvements, and design changes reduce labor hours in a predictable manner. In words, the model is usually expressed as follows: For each doubling of cumulative production, the labor hours per unit fall to a specified percentage of the previous value. The lower the percentage, the faster the number of hours falls. This percentage is often 80% (see also Exhibit 16.15).

Exhibit 16.15   Example learning-curve coefficients

.65	Prototype assembly
.70	Final assembly—complex
.80	Final assembly—simple
.80	Packaging
.82	Shearing metal plates
.83	Printed circuit-board fabrication
.85	Welding shop
.95	Machine shop
.985	Grinding, chipping, blast cleaning

Source: Smith, p. 27, and Nanda and Alder, p. 30.

Once the number of labor hours is found using Equation 16.4, then the appropriate cost per hour is applied.

Let

$$H_i = \text{hours for the } i\text{th item}$$
$$\Theta = \text{rate}$$
$$H_i = H_1 \Theta^{(\ln i / \ln 2)} \qquad (16.4)$$

When this type of relationship is graphed on normal coordinates, the decline in cost is obvious, but the relationship is not (see Exhibit 16.16). On log-log paper (see Exhibit 16.17), it appears as a straight line, since doubling the number of units on the *x*-axis reduces the labor hours by a constant percentage.

Exhibit 16.16   Learning curve with 80% rate

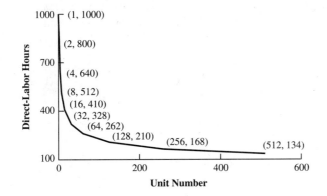

Exhibit 16.17   Log-log learning curve with 80% rate

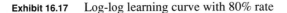
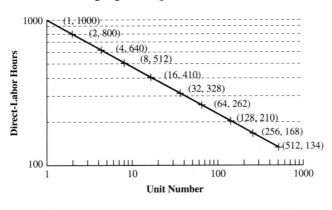

Example 16.9   Learning Curve for AirTruck

AirTruck, a manufacturer of freight aircraft, is evaluating the potential payoff of a process improvement program that should change its learning-curve percentage from .85 to .8.

The jet on which this program would start requires an estimated 8000 person-days for its first unit. Contrast the curves for person-days for the two learning-curve percentages.

Solution
The calculation of person-days is easiest if the unit numbers are doubled each time (1, 2, 4, 8, ...). Then the learning curve coefficient is simply multiplied by the person-day term to calculate the next one.

Exhibit 16.18   AirTruck labor person-days at 85% vs. 80% learning curves

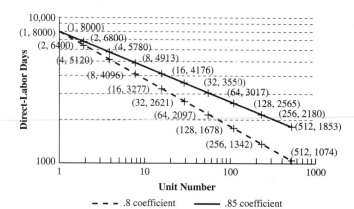

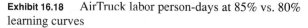

Unit Number	Existing $\Theta = .85$	Potential $\Theta = .8$
1	8000	8000
2	6800	6400
4	5780	5120
8	4913	4096
16	4176	3277
32	3550	2622
64	3018	2098
128	2565	1678
256	2180	1342
512	1853	1074

16.10 USING FACTOR ESTIMATES

Factor estimates are more detailed parametric estimates. Rather than using only total square footage or total capacity, the project being estimated is broken down into smaller pieces.

If the level of detail is only moderate, then it corresponds to estimates made at the preliminary systems design or the budget estimate stage. Example 16.10 illustrates a factor estimate with a moderate level of detail.

If the project is broken down into exhaustive detail, then it corresponds to estimates made at the detailed systems design or the definitive estimate stage. Thus, person-days will be broken down by craft and experience level, with the associated costs. The capital items will be based on bids, catalogs, and perhaps a few capacity functions applied to adjust the cost of a particular unit. Working capital will be linked to a cash flow budget.

Example 16.10 HouseHold Plant Revision

Provide a more detailed estimate of the cost of the plant revision that was estimated at $900,000 in Example 16.3. (Note: a similar, more detailed analysis of operating cost and revenue implications should also be done, although they are not included here.)

Solution
The earlier estimate was based on:

$600K	Six robots at $100,000 each
125K	Material-handling system
10K	Materials for rooms, etc.
200K	Labor for construction and training
−35K	Salvage value of old conveyor
$900K	Total

A more detailed factor estimate might be:

$225K	3 painting robots (Atlas 342s) at $75,000 each
150K	1 long travel welding robot (Atlas 358)
250K	2 welding robots (Atlas 357)
75K	Asynchronous conveyor (MagicRamp 2440)
55K	Paint-handling system (Atlas 342—38/38/38)
10K	Material for rooms, paint booths, etc.
120K	600 days at $25/hour for mechanical technicians
96K	600 days at $20/hour for training
12K	100 days at $15/hour for disassembly
−45K	Sale of old conveyor after disassembled
$948K	Total

16.11 POTENTIAL PROBLEMS WITH ACCOUNTING DATA

Cost Allocation.   One difficulty in estimating costs is that some organizational activities cannot be linked to specific projects, products, or services. For example, the receiving and shipping areas of a manufacturing plant are used by all incoming materials and all outgoing products. These materials and products differ in their weight, size, fragility, value, number of units, packaging, etc., and the cost depends on all of these factors.

Other costs, such as the organization's management, sales, and administrative expenses, are difficult to link directly to products or services. As a result, accountants have developed mechanisms for allocating overhead using burden vehicles, such as direct-labor hours. This allocation represents average, not incremental, performance.

Three common ways of allocating overhead are: direct-labor hours, direct-labor cost, and direct-materials cost. The first two differ significantly only if the cost per hour of labor differs for different products. Example 16.11 illustrates the difference between the latter two for a simple case of two products.

The principle is that the analyst must determine which indirect or overhead expenses will be changed because of an engineering project. What are the incremental cash flows? The changes in costs incurred must be estimated. Loadings, or allocations, of overhead expenses cannot be used.

This issue has become very important because in some firms, automation has reduced direct-labor content to less than 5% of the product's cost. Yet in some of these firms, the basis for allocation of overhead is still direct-labor hours or cost. Other firms are shifting to activity-based costing (ABC) where each activity is linked to specific cost drivers, and the number of dollars allocated as overhead is minimized. Exhibit 16.19 illustrates the difference between activity-based costing and traditional overhead allocations [see Tippett and Hoekstra].

Exhibit 16.19   Activity-based costing vs. traditional overhead allocation[1]

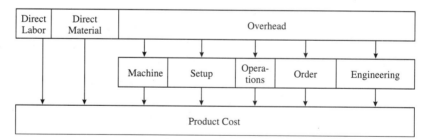

[1] Based on an example by Kim La Scola Needy.

Example 16.11   HouseHold Appliance

HouseHold does not manufacture its own motors or electronic controls. The motors for washers must be sturdier than those for dryers, and the controls for washers are also more complex.

As a result, HouseHold produces a higher fraction of the dryer's value itself, and it purchases a higher fraction of the washer's value. Using the data shown below, allocate $750,000 in overhead on the basis of (1) labor cost and (2) material cost.

	Dryers	Washers
Number per year	4200	5000
Labor cost (each)	$40	$20
Material cost (each)	$50	$75

Solution

First, the labor and material costs for the dryers, the washers, and in total are calculated.

	Dryers	Washers	Total
Number per year	4200	5000	
Labor cost (each)	$40	$20	
Material cost (each)	$50	$75	
Labor cost	$168,000	$100,000	$268,000
Material cost	$210,000	$375,000	$585,000

Then the allocated cost per labor dollar, $2.80, and the allocated cost per material dollar, $1.28, are calculated by dividing the $750K in overhead to be allocated by $268K and $585K, respectively. Finally, the $750,000 in allocated overhead is split between the two products using (1) labor costs, and (2) material costs.

	Dryers	Washers	Total
Labor cost	$168,000	$100,000	$268,000
Overhead/labor $	2.80	2.80	
Allocation by labor $	470,149	279,851	750,000
Material cost	210,000	375,000	585,000
Overhead/material $	1.28	1.28	
Allocation by material $	269,231	480,769	750,000

If labor cost is the burden vehicle, then 63% of the $750K in overhead is allocated to the dryers. If material cost is the burden vehicle, then 64% is allocated to the washers. In both cases, the $750,000 has been split between the two products. However, neither value is relevant for our purposes, since neither is an incremental cost.

Timely and Accurate Data.   Centralized accounting systems have often been accused by project managers of being too slow or being "untimely." Because engineering economy is not dealing with the daily problem of project control, this is less of an issue. However, if an organization establishes multiple files and systems so that project managers (and others) may have the timely data that they need, then the level of accuracy in one or all systems may be low. Then cost estimates will have to consider other internal data sources.

There are several cases in which data on equipment or inventory values may be questionable. When inventory is valued on a "last in, first out" basis, the remaining inventory may be valued too low. Similarly, land may be valued at its acquisition cost, and thus be significantly undervalued. Finally, capital equipment may be valued at either a low or a high value, depending on allowable depreciation techniques and company policy.

16.12 SUMMARY AND CONCLUSIONS

This chapter has detailed one of the steps in the engineering economy process originally described in Chapter 1. With these tools, you can estimate the consequences of alternatives and construct cash flow diagrams. But the results of an engineering economy study are no better than the data that is estimated and used.

Therefore, when doing a study one must identify the level of accuracy required for the estimate and the methodology that will achieve that accuracy. It requires identifying a base case, if the project is a modification of an existing system.

The tools available for estimating include statistics used for parametric forecasting, indexes of inflation and productivity, capacity functions, learning curves, and factored estimates.

REFERENCES

Barrie, Donald S., and Boyd C. Paulson, *Professional Construction Management: Including C.M., Design-Construct, and General Contracting,* 3rd ed., 1992, McGraw–Hill.

Dodge Manual for Construction Pricing and Scheduling, published annually, McGraw–Hill Information Systems.

Eschenbach, Ted G., and George A. Geistauts, "Forecasting Technological Advances," *Handbook of Engineering and Technology Management,* Dundar Kocaoglu, ed., (forthcoming) Wiley.

Humphreys, Kenneth K., and Paul Wellman, *Basic Cost Engineering,* 2nd ed., 1987, Marcel Dekker.

Means Cost Data, published annually (Building Construction, 52nd, 1994; Repair and Remodeling, 15th, 1994; Electrical, 17th, 1994; Plumbing, 17th, 1994; Mechanical, 17th, 1994; and Square Foot, 15th, 1994), Robert Snow Means Co.

Michaels, Jack V., and William P. Wood, *Design to Cost,* 1989, Wiley.

Nanda, Ravinder, and George L. Alder, eds., *Learning Curves: Theory and Applications,* 1982, Institute of Industrial Engineers.

Richardson's General Construction Estimating Standards and *Process Plant Construction Estimate Standards,* published annually, Richardson.

Smith, Jason, *Learning Curve for Cost Control,* 1989, Institute of Industrial Engineers.

Tippet, Donald D., and Peter Hoekstra, "Activity-Based Costing: A Manufacturing Management Decision-Making Aid," *Engineering Management Journal,* Volume 5, Number 2, June 1993, American Society for Engineering Management, pp. 37–42.

U.S. Air Force, *HQ USAF Annual Construction Pricing Guide for FY 90–94 Major Construction Program,* January 1988.

Walker's Building Estimator's Reference Book, published periodically, Frank R. Walker Co.

PROBLEMS

16.1　Provide a checklist for ensuring the completeness of a cost estimate for an organization you are familiar with.

16.2　Provide an example of poor up-front planning and economic analysis that led to poor performance.

16.3　Provide an example of excellent up-front planning and economic analysis that supported excellent performance.

16.4　Analyze a budget estimate that can be compared to an executed cost. Categorize the estimate and compare its accuracy with the AACE International standards. How did you account for "change orders"?

Use the U.S. Air Force data from Exhibits 16.7–16.9 to answer Problems 16.5–16.10.

16.5　An estimate for a new runway needs to be updated for a new construction starting date. When the midpoint of construction was July 1996, the estimated cost was $4.8 million. If the project is being delayed by 1 year, what is the new estimated cost?

16.6　A standard barracks plan of 40,000 square feet will be constructed in Huntsville, Alabama, and Anchorage, Alaska. What is the estimated cost for each site in July 1995?　(*Answer:* $2.95M and $6.00M)

16.7　A typically sized administration building is 25,000 square feet, and it costs $82 (FY90) per square foot. These figures were used to estimate the cost of a requested facility. However, only 75% of the funds were authorized, so the facility must be downsized. Estimate how much smaller the administration building must be.

16.8　Estimate the cost to build a 50,000-square-foot outpatient clinic in Blytheville, Arkansas, with construction beginning in the spring of 1995.　(*Answer:* $5.51M)

16.9　Estimate the cost of a 12,000-square-foot dental clinic to be built in Los Angeles, California, with a midpoint construction period of October 1997.

16.10   Estimate the cost of a 4000-square-foot community fire station to be built in Tampa, Florida, with a midpoint construction period of July 1996.

16.11   Estimate the cost of expanding a planned new school by 20,000 square feet. The appropriate capacity exponent is .66, and the budget estimate for 200,000 square feet was $15M.   (*Answer:* $1M)

16.12   The estimated cost to construct a small polymerization plant is required before your firm's management will fund any preliminary design work. Data shows that a plant twice the size of the proposed plant cost $22M to construct. Using order-of-magnitude estimating methods, determine the estimated cost of the proposed plant.

16.13   New EPA regulations have just been proposed. If they are enacted, the capacity of the hydrogen sulfide removal plant that is part of a coal facility will have to be doubled. The definitive estimate for the original removal plant is $12M. What is an order-of-magnitude estimate for the expanded plant?

16.14   Five years ago, POW Chemical built a fertilizer-production plant for $8M. The inflation rates over the last 5 years have been 3%, 5%, 3%, 6%, and 2%, respectively. In current dollars, what is the order-of-magnitude estimated cost of a plant with 50% more capacity?

16.15   Calculate and draw a Pearl curve over 10 years. Let $L = 1000$, $a = 50$, and $b = .7$.   (*Answer:* $D_1 = 39$, $D_2 = 75$)

16.16   The manufacturer of a new, improved widget must estimate sales for the first 6 years of production. The upper limit on annual demand for these widgets is believed to be 1,000,000 units. The widget will be sold for $12 the first year and will decrease in price by $.50 each year. Using a Pearl curve with $a = 100,000$ and $b = 4$, determine the expected demand for the 6-year period and determine the revenue that can be expected each year.

16.17   Calculate and draw a Gompertz curve over 5 years. Let $L = 500$, $c = 12$, and $d = .7$.   (*Answer:* $D_1 = 1$, $D_2 = 26$)

16.18   The Heritage Cable Network received commitments to carry the network's programming from 35 cable operators the first year of operation and 90 commitments the second year. The number of cable operators is expected to remain steady at 2000 for the next several years. Using a Gompertz curve, estimate the number of commitments Heritage should anticipate each year for the next 4 years. (*Answer:* $D_3 = 299$)

16.19   Mrs. Meadows Cookies provides many varieties of fresh-baked cookies and pastries for specialty shops across the country. Two years ago a new cookie, the Choco-Mucho, was introduced and recorded sales of $1.2M the first year and $1.5M the second year. The company anticipates that sales of the cookie will reach a maximum of $5M. Using a Pearl curve, estimate the revenues for the cookie for the next 4 years.

16.20 Calculate and draw growth curves over 10 years. Let $L = 1000$ and the numbers for years 1 and 2 be 5 and 50, respectively.
    a. Calculate the coefficients and draw a Pearl curve.
    b. Calculate the coefficients and draw a Gompertz curve.
    c. Contrast the two curves. Is there any basis for preferring one of them?

16.21 Determine the time required to produce the 2000th item if the first item requires 180 minutes to produce and the learning-curve percentage is 92%. (*Answer:* 72.1 minutes)

16.22 For AirTruck in Example 16.9 calculate the curve with $\Theta = .82$, which someone has suggested as a more likely outcome. (*Answer:* $D_2 = 6560$)

16.23 Plot the curves for Example 16.9 on standard x-y–coordinates rather than on log-log coordinates.

16.24 Determine the labor costs required to produce the 100th item if the first item requires 45 days to produce and the learning curve percentage is 82%. The average cost of labor is $8.50 per hour and two 9-hour shifts work each day. The average number of people working on the item is eight.

16.25 For Example 16.9 (the AirTruck learning curve), approximate the cumulative person-day savings of having $\Theta = .8$ rather than .85. Assume that 10 planes per year will be built.

16.26 For Problem 16.25, calculate the PW of the savings if each person-day costs $250, $i = 10\%$, and $N = 10$ years.

16.27 Contrast the learning-curve coefficients for simple and complex assembly operations. In which case does learning occur faster? Why is this true?

16.28 The size-adjustment graph of Exhibit 16.9 is a type of capacity function curve. Estimate the capacity function coefficient for each of the following sizes: .5, 1.5, 2, and 3.

16.29 Par Golf Inc. produces two types of golf bags: the standard and the deluxe. The total overhead cost to be allocated for the two bags is $35,000. Determine the net revenue Par can expect from the sale of each bag.
    a. Use direct-labor cost overhead allocation.
    b. Use direct-materials cost overhead allocation.

Data Item	Standard	Deluxe
Direct-labor cost	$50,000	$65,000
Direct-material cost	$40,000	$47,500
Selling price	$60	$95
Units produced (and sold)	1800	1400

CHAPTER 17 SENSITIVITY ANALYSIS

The Situation and the Solution

Unlike most end-of-chapter problems, the real world contains data that are not precisely known. For example, first costs, annual revenues, the project's life, and any salvage values are estimated.

The solution is to analyze the project under different assumptions that represent the range of reasonable estimates. Sensitivity analysis examines how the recommended decision depends on the estimated data.

Chapter Objectives

After you have read and studied the sections of this chapter, you should be able to:

Section 17.1
Describe the uses of sensitivity analysis and recognize the techniques used.

Section 17.2
Set reasonable limits of uncertainty for estimated data.

Section 17.3
Describe the techniques of sensitivity analysis.

Section 17.4
Interpret sensitivity information from spiderplots.

Section 17.5
Construct a spiderplot by hand or with a spreadsheet.

Section 17.6
Analyze problems with multiple alternatives.

Section 17.7
Conduct sensitivity analysis on the interaction of two cash flow elements.

Key Words and Concepts

Sensitivity analysis   A method that examines how a recommended decision depends on estimated cash flows.

Cash flow element   A cash flow (P or F), a set of cash flows (A or G), an interest rate (i), a horizon or life (N), a tax rate, etc.

Breakeven chart   A chart that depicts economic desirability as a function of one cash flow element.

Scenario   A representative, believable combination of cash flow elements.

Spiderplot   A graphic that depicts economic performance as a function of each variable under analysis.

Base case   The set of most likely values for the cash flow elements.

17.1 WHAT IS SENSITIVITY ANALYSIS?

Sensitivity analysis examines how uncertainty in estimated cash flows influences recommended decisions. Does using another interest rate or a shorter time horizon cause a different alternative to have the best present worth? Would a 10% savings in construction costs make an unattractive project economically attractive? Would a 20% cost overrun make the present worth negative?

Sensitivity analysis examines uncertainty in all data items that affect the problem's cash flows. These include the cash flows themselves (P, A, F, and G), the interest rate (i), the horizon (N), depreciation rules, the tax rate, sales volume, etc. As a group, these are defined as **cash flow elements**.

Sources of Uncertainty.   Cash flow elements may be uncertain for the following reasons:

1. Measurement error
2. Unclear specifications
3. A volatile future

For example, measuring current labor output and monthly labor costs is affected by the number of holidays, the amount of overtime, the number of employees, and how many paydays occurred during the month. Errors in any of these will cause measurement errors in estimating labor output and labor costs.

> **Sensitivity analysis** examines how a recommended decision depends on estimated input variable values.

> **Cash flow elements** include P, A, F, G, i, N, tax rates, etc.

Feasibility studies often have unclear specifications, but even well-defined building projects may not be completely specified. For example, specifying the size of the building, room sizes and uses, etc., largely determines the cost of the building; however, at early stages of design, specification of interior and exterior finishes is omitted, even though the choice of carpeting, tile, sheet vinyl, and other elements influence that cost.

Finally, estimates of such things as product demand, the corrosion rate in oil field piping, and the growing demand for electrical power are inherently uncertain. Each depends on a future that is only partly predictable.

Since engineering economy emphasizes the future, this uncertainty cannot be avoided. The uncertainty must be analyzed using sensitivity analysis. While this has always been true, sensitivity analysis is becoming more important (1) as technology and global competition force more rapid change, and (2) as competitors increasingly rely on more complete analysis supported by cheaper computing power.

Breakeven charts depict economic loss and profit as a function of one problem variable. **Scenarios** are representative, believable combinations of cash flow elements. **Spiderplots** depict economic performance as functions of several variables.

Techniques for Sensitivity Analysis.   Sensitivity analysis is a broad term that includes a variety of techniques. Three techniques will be detailed in this chapter. Example 17.1 and Exhibit 17.1 illustrate a **breakeven chart** or graphical breakeven analysis. The minimum value of a variable that avoids a loss is the breakeven quantity. Example 17.2 illustrates the use of **scenarios** or plausible combinations of values in sensitivity analysis. The **spiderplot** approach is developed more fully in Section 17.5 to combine, contrast, and compare the multiple breakeven charts shown in Exhibit 17.2.

Example 17.1   Buying vs. Renting

George recently graduated with a degree in aeronautical engineering from a Midwest university. He has accepted a job in Seattle, and he is deciding between renting and buying a home. George's hobby is model railroading, and he will not move again unless he leaves Seattle for a job elsewhere in the country.

If George buys the house, it will cost him about $5000 in transaction costs at time 0, and another $20,000 when he sells it. His mortgage and lease payments would be about the same, but he has estimated that the lifestyle and tax advantages of ownership are worth about $3600 per year. George's interest rate is 9%. How many years must George live in Seattle for buying to be more economically attractive?

Solution
One approach is to simply try different years. Another is to graph the present worth of buying the home as a function of the number of years it is owned. Trying N equals 1, 10, and 20 years certainly spans the likely ownership periods.

$$PW_1 = -5000 - 20,000(P/F, 9\%, 1) + 3600(P/A, 9\%, 1)$$
$$= -5000 - 20,000(.9174) + 3600(.9174) = -\$20,045$$

$$PW_{10} = -5000 - 20,000(P/F, 9\%, 10) + 3600(P/A, 9\%, 10)$$
$$= -5000 - 20,000(.4224) + 3600(6.418) = \$9657$$

$$PW_{20} = -5000 - 20,000(P/F, 9\%, 20) + 3600(P/A, 9\%, 20)$$
$$= -5000 - 20,000(.1784) + 3600(9.129) = \$24,296$$

Exhibit 17.1 Present worth of house buying vs. years of ownership

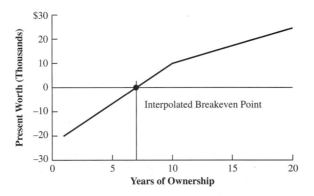

As shown in Exhibit 17.1 a linear interpolation between 1 year and 10 years suggests that 7.1 years is the breakeven point. In fact, $PW_6 = -\$776$ and $PW_7 = \$2178$, so the breakeven point is closer to 6 than to 7 years.

Example 17.2 State Park Evaluation

The state park system economically evaluates land purchases, but the state legislature sets the interest rate and the horizon that the park system is required to use. The interest rate for a benefit/cost analysis may be set for 3%, 6%, or 9%; similarly, the horizon may be set at 10, 20, or 50 years.

The land for one park will cost $2 million, and the annual benefit will be $180,000. Assume that the salvage value equals $0. Construct three scenarios using combinations of interest rates and horizons representing favorable, middle-of-the-road, and unfavorable situations. Under which scenarios should the land be acquired?

Solution
Since the benefits occur later than the costs, the project will be more attractive when lower interest rates and longer time horizons are used. The favorable scenario is 3% and 50 years; the middle-of-the-road scenario is 6% and 20 years; and the unfavorable scenario is 9% and 10 years.

$$B/C_{fav} = 180,000(P/A, 3\%, 50)/2,000,000 = 2.32$$
$$B/C_{midroad} = 180,000(P/A, 6\%, 20)/2,000,000 = 1.03$$
$$B/C_{unfav} = 180,000(P/A, 9\%, 10)/2,000,000 = .58$$

The favorable and middle-of-the-road scenarios favor the park; however, the 1.03 is a "just barely" recommendation.

Why Do Sensitivity Analysis?   The uncertainty that is present in estimated cash flows requires sensitivity analysis. How sensitivity analysis is used depends on the stage of the economic analysis. The reasons for doing sensitivity analysis include:

1. Making better decisions
2. Deciding which data estimates merit refinement
3. Focusing managerial attention on the key variables during implementation

   Example 17.1 illustrated using sensitivity analysis to make a decision. George can compare his best guess for how long he will be in Seattle with the breakeven value of just over 6 years. Similarly, the scenarios in Example 17.2 can be used to decide whether or not to purchase the parkland.

   In Example 17.1, buying a home is very important to George's finances. If he wants to improve the reliability of his analysis, which cash flow elements should he estimate more accurately? Which uncertainty is more important: (1) the number of years; (2) the 9% interest rate; (3) the $5000 cost at time 0; (4) the $20K cost at time N; or (5) the $3600-per-year value of home ownership?

   Example 17.3 and Exhibit 17.2 analyze this problem using breakeven analysis. However, the set of four breakeven charts is difficult to use. One of the weaknesses of breakeven analysis is that there is no indication of whether the breakeven values are likely to occur or are nearly impossible. Section 17.2 introduces the need to define the limits of the uncertainty for each variable.

   The second weakness of breakeven charts is that they are separate charts, so it is difficult to compare variables. As shown in Section 17.4, the spiderplot is the best technique for comparing uncertainty in different cash flow elements. During a feasibility study or the development of a cost estimate, a spiderplot can be used to show where additional analysis and data gathering are needed—before a decision is made. Finally, after the decision is made, sensitivity analysis can show which cash flow elements merit the most attention during project implementation.

   For example, a project's largest cost may be a fixed cost to purchase equipment. Some negotiation may be possible, but little change in the cost is likely. The number of dollars involved in the cost of constructing the building to house the equipment may be smaller, but construction cost may need to be the focus of managerial attention. Schedule delays in the building might be the source of variability in revenue projections, and the cost of construction might be the major source of cost uncertainty.

Example 17.3   Buying vs. Renting (Example 17.1 Revisited)

Analyze the breakeven points for George for the initial and final costs of buying and selling the home and for the annual value of owning the home in Example 17.1. Assume $N = 10$ years for the calculations of other breakeven points.

Solution

George's present worth is

$$PW = -\text{FirstCost} - \text{FinalCost} \cdot (P/F, i, N) + \text{AnnBen} \cdot (P/A, i, N)$$

where FirstCost = $5000, FinalCost = $20,000, AnnBen = $3600, i = 9%, and N = 10.

These functions are graphed in Exhibit 17.2, but the breakeven values can also be calculated by setting the PW equation equal to 0. For FirstCost, the equation is:

$$0 = -\text{FirstCost}_{BE} - 20,000 \cdot (P/F, 9\%, 10) + 3600(P/A, 9\%, 10)$$

$$\text{FirstCost}_{BE} = -20,000 \cdot .4224 + 3600 \cdot 6.418 = \$14,660 \quad (193\% \text{ increase})$$

The equations and values for FinalCost and AnnBen are:

$$\text{FinalCost}_{BE} = (-5000 + 3600 \cdot 6.418)/.4224 = \$42,860 \quad (114\% \text{ increase})$$

$$\text{AnnBen}_{BE} = (5000 + 20,000 \cdot .4224)/6.418 = \$2095 \quad (42\% \text{ decrease})$$

Exhibit 17.2   Breakeven charts for the present worth of house buying

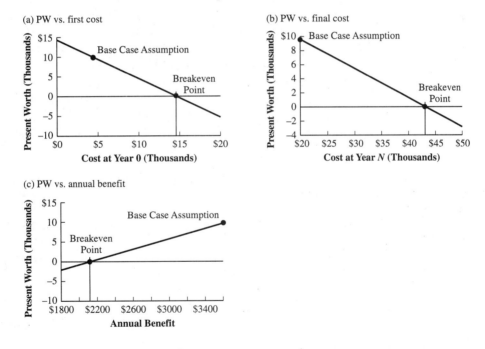

17.2 UNCERTAIN DATA AND ITS IMPACT

Defining the Limits of Uncertain Data.   The first step in sensitivity analysis is identifying the limits for each cash flow element. Some cash flows may be certain, but others are not. Exhibit 17.3 summarizes reasons why certain cash flow elements may have small or large uncertainties.

Exhibit 17.3   Possible reasons for large and small uncertainties

Cash Flow Element	Level of Uncertainty	Reason
First cost	Small	Off-the-shelf purchase
	Large	R&D required
Revenues	Small	Existing product
	Large	New product
Horizon	Large	Far in future
Salvage value	Limited	Fraction of first cost

Often, these <u>uncertainties are stated as percentages</u> of the most likely values. For example, annual operating costs or the costs of purchased equipment might be known within ±10%, or between 90% and 110%, of the most likely value. Similarly, the actual construction cost for a complex facility might be expected to be between 70% and 150% of the order-of-magnitude estimate. How important to the decision's reliability is it to reduce this uncertainty to the 85%-to-130% range of the budget estimate or to the 95%-to-115% range of the definitive estimate? (See Section 16.3 for levels of accuracy.)

Often, the limits on the uncertain data are asymmetrical. Remember that the AACE International error percentages for cost estimates, as shown in the previous paragraph, were larger above than below the most likely value. Unfortunately, this is because there seem to be more ways for things to go wrong than for things to go right. Thus, costs are more likely to go up than down; revenues are more likely to go down than up; and delays until start-up are more likely to be longer than shorter. This is illustrated in Example 17.4.

The **base case** is the set of most likely values for the cash flows, interest rates, etc.

The set of most likely values is the starting point for an analysis. It is the **base case**. However, the limits of uncertainty that have been described here are as important as the base case. In order to reduce the potential for bias, these limits should be estimated before the first present worth is calculated. If the PW is calculated first, then someone might be tempted to use limits that support the "desired" outcome.

Example 17.4   Northern Electric

Postulate and justify the limits of uncertainty for an electric power-generating facility with the following six parameters:

Building cost	$5M	Salvage value	$1M
Equipment cost	$10M	i	10%
Annual net revenue	$2.5M	N	20 years

Solution

Northern Electric's experience is that major buildings cost between 80% and 150% of the estimates used to generate the request for proposals (RFP). The major equipment consists of natural gas fired turbines, which are "off-the-shelf" items with actual costs between 90% and 110% of estimated costs. Growth in demand and associated revenues are difficult to predict, since this region has a boom/bust economy. Annual net revenues should be within 60% and 120% of the estimated quantity. The salvage value, S, has an estimated accuracy of 70% to 130%.

The appropriate interest rate for real economic returns over inflation should be between 6% and 15%. This is 60% to 150% of the base case value. (This range has been made larger than it might be, so that nonlinearities would be more apparent.) The project horizon of $N = 20$ years should be between 15 years and 30 years, which is 75% to 150% of the base case value. These uncertainties and the base case values, S, are summarized in Exhibit 17.4.

Exhibit 17.4   Uncertainties in the data for Northern Electric

Element	Lower Limit	Base Case	Upper Limit
Building cost	80%	$5M	150%
Equipment cost	90%	$10M	110%
Annual net revenue	60%	$2.5M	120%
Salvage value	70%	$1M	130%
i	60%	10%	150%
N	75%	20 years	150%

Estimating Sensitivities.   The time value of money ensures that some cash flow elements play a larger role in the calculation of PWs or EACs. The calculated economic worth depends more heavily or is more sensitive to the values of the more important cash flow elements.

A cash flow in an early year is more important than a similar cash flow in year N, since the later cash flow is discounted more heavily. A cash flow that occurs every year, such as revenues from sales or labor costs, tends to be more important than a cash flow that occurs once, such as an overhaul or a salvage value.

Similarly, when the problem has a long time horizon, then uncertainty in the interest rate is important. When the interest rate is low, then the length of the study period is important.

17.3 THE PROCESS OF SENSITIVITY ANALYSIS

Sensitivity analysis examines how a recommended decision depends on estimated cash flow elements. The first step of sensitivity analysis was discussed in Section

17.2—defining the limits of uncertainty for each cash flow, the interest rate, and the problem horizon.

The next step is to apply the techniques of sensitivity analysis to answer the questions of Section 17.1. In the real world, this is likely to call for additional data gathering and alternative generation. For example, rather than making a decision using very uncertain data, more data will be gathered. Similarly, rather than relying on an alternative that risks large losses, where possible, new alternatives will be devised.

Basic Techniques.   Basic breakeven charts and scenarios have been presented as two techniques of sensitivity analysis.

Spiderplots will be presented in Section 17.4 as the best way to emphasize the relative impact of each cash flow's uncertainty on the project's present worth, internal rate of return, or benefit/cost ratio. This technique emphasizes individual variables, as do the collective breakeven charts in Exhibit 17.2, but spiderplots make comparisons easy.

Scenarios emphasize the impact of combinations of uncertainty on the total outcome. For example, the interest rate and horizon were combined in Example 17.2 to create three scenarios for the benefit/cost ratio. While many cash flows may be different between the scenarios, there should only be two to five scenarios. More are likely to be confusing rather than helpful. Often, a most likely, a pessimistic, and an optimistic scenario are the best combination.

Section 17.5 illustrates how both of these techniques can be applied to problems with multiple alternatives. Section 17.6 illustrates how graphs with two variables can illustrate the impacts of uncertainty more completely than two-variable scenarios can.

More Advanced Techniques.   As shown in Chapter 18, uncertainty can be characterized by probability distributions. When this is done, then the tools of expected value and risk analysis are quite useful.

Specific tools for sensitivity analysis that depend on probability distributions include simulation (briefly described in Section 18.7) and stochastic sensitivity analysis. Simulation uses probability distributions for all of the cash flow elements, then solves the problem many times and analyzes the results. Stochastic sensitivity analysis uses the probability distributions for each cash flow element to draw a special spiderplot [see Eschenbach and Gimpel]. This connects uncertainty in the data with uncertainty in the outcome.

17.4 SPIDERPLOTS

Defining Spiderplots.   In Example 17.1, a graph of present worth vs. N was used to define a breakeven period of home ownership. The present worth was plotted as a function of the number of years that the home was owned. Similar curves were drawn in Exhibit 17.2 for the initial and final costs for buying and selling the home and for the annual value of home ownership. A spiderplot combines the curves with an appropriate *x*-axis.

The *x*-axis of a spiderplot measures all variables in a common unit: the percentage of each base case value. For example, consider the lower and upper limits for each cash flow element in the Northern Electric problem (Example 17.3). The horizon's lower limit, 15 years, was stated as 75% of the base case value of 20 years, and 30 years as 150% of the base case value. All variables were stated as a percentage of the base case value—building cost as a percentage of $5M, equipment cost as a percentage of $10M, etc.

Note that the dollar amounts of building cost, annual revenue, and salvage value are *not* comparable and thus a common *x*-axis in dollars would not be meaningful. Building cost is measured in year 0 dollars; annual revenues in years 1 to *N* dollars; and salvage value in year *N* dollars. These are different dollars, and they cannot be used as a common unit.

Section 17.5 explains how to construct spiderplots, while the rest of this section explains how to interpret them. Exhibit 17.5 is a spiderplot that compares the uncertainties due to all four variables in Example 17.1. These uncertainties are defined as follows. The problem statement identified a horizon of 1 to 20 years or 10% to 200% of the 10-year base case. It seems reasonable to select limits of ±10% for the first cost of buying and ± one-third for the final selling cost. Assume that George has said the annual benefit is between 80% and 150% of $3600.

On a spiderplot, there are two directions to measure uncertainty. On the *x*-axis, the potential uncertainty in the cash flow element is measured. On the *y*-axis is measured the impact of that uncertainty on the present worth, internal rate of return, or benefit/cost ratio.

In the center of the "spider" in Exhibit 17.5 is the *base case*. This is the set of estimates that would have been used, if no sensitivity analysis had been done. Each cash flow is represented by a curve between its lower and upper limits that

Exhibit 17.5   Spiderplot for buying vs. renting (based on Example 17.1)

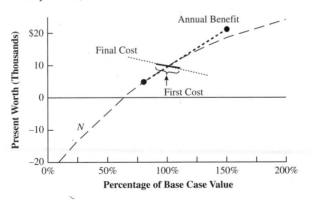

Exhibit 17.5   (*continued*)

Cash Flow Element (base case)	Change (*x*-axis)	Present Worth of Buying (*y*-axis)
Horizon (10 years)	Decreases to 1 year (10%) Increases to 20 years (200%)	Decreases to −$20,046 Increases to $24,294
First cost ($5000)	Decreases by 10% to $4500 (90%) Increases by 10% to $5500 (110%)	Increases to $10,155 Decreases to $9155
Final cost ($20,000)	Decreases by 1/3 to $13,333 (67%) Increases by 1/3 to $26,667 (133%)	Increases to $12,471 Decreases to $6839
Annual benefit ($3600)	Decreases by 20% to $2880 (80%) Increases by 50% to $5400 (150%)	Decreases to $5035 Increases to $21,207

passes through the base case. Exhibit 17.5 also contains a table that restates the graphical information.

Interpreting a Spiderplot.   A properly drawn spiderplot depicts the following three things:

1. Limits of uncertainty for each cash flow element
2. Impact of each cash flow element on the PW, EAC, or IRR
3. Identification of which cash flow elements might change the recommendation

Determining the limits of uncertainty for each cash flow was described in Section 17.2. These limits of uncertainty are stated as percentages of the base case, and they are measured along the *x*-axis. Note that extending the curve for a cash flow element beyond its proper *x*-axis limits is quite misleading. Such extensions can dramatically overstate the impact of variables on the present worth, internal rate of return, or benefit/cost ratio, which is shown as the *y*-axis coordinate. Less important variables can also be identified as crucial.

In Example 17.3, breakeven values were calculated for the initial cost to buy the home ($14,660), the final cost to sell it ($42,860), and the annual benefit of owning the home ($2095). All of these values are outside the limits identified in Exhibit 17.5. Extending each of these variables from 0% to 200% of the base case values would create a mathematical fiction. Yet analysts do make the error of defining the *x*-axis as ±50% or ±100% and then extending the curve for each variable to the minimum and maximum percentages.

Deciding whether uncertainty in a cash flow element might suggest a different recommendation is graphically shown by including a standard of comparison. A present worth of $0, a benefit/cost ratio of 1.0, and a target IRR all provide cut-off values that are lines parallel to the *x*-axis. Points below the cut-off value indicate one decision, and points above the cut-off indicate another decision.

When drawing the curve of PW vs. a variable, only two points are needed for linear relationships. For example, cash flows are multiplied by the *engineering economy factors* to compute the PW, EAW, or EAC. So those economic measures are linear functions of first costs, annual costs, arithmetic gradients for sales, etc. If a cash flow is in the numerator, then the B/C ratio is a linear function of the cash flow; if the cash flow is in the denominator, then the B/C ratio is a nonlinear function of the cash flow. All of these measures are nonlinear functions of interest rates, lives, and geometric gradients. This linearity or nonlinearity of the PW for various cash flow elements is illustrated in Examples 17.5 and 17.6. The IRR is a nonlinear function for all cash flow elements.

If a cash flow or a geometric gradient has a base case of 0, then an alternate *x*-axis must be used for that variable, since 80% or 150% of 0 is still 0.

Example 17.5 Interpreting a Spiderplot for Northern Electric

Use a spiderplot (see Exhibit 17.6) to analyze which data estimates in Example 17.4 should be refined before the final decision is made.

Exhibit 17.6   Spiderplot of PW for Northern Electric

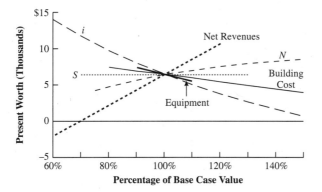

Solution
Only one cash flow element, net annual revenues, has a curve with any negative present worth values. The building's cost, the interest rate (*i*), and the horizon (*N*) have curves that approach a negative present worth. The equipment cost has little uncertainty ($\pm 10\%$) and the salvage value (*S*) does not occur until year *N*, so those two cash flow elements have little impact on the *uncertainty* in the present worth.

Example 17.6 Interpret the Spiderplot for George's Home Purchase (Based on Exhibit 17.5)

Exhibit 17.7 repeats Exhibit 17.5 for reference. Interpret the curve to decide which data estimates George should refine before making a final decision.

Exhibit 17.7   Spiderplot for George's home purchase

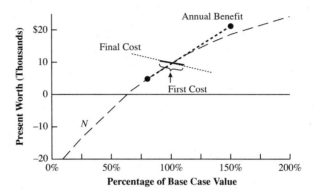

Solution

Only the curve for the ownership period (the number of years or the problem's horizon) has negative present worth values. Applying the reasonable limits of ±10% to the first cost, ±1/3 to the final cost, and -20% and +50% to the annual benefit shows that a change in one of those cash flows is not likely to change the recommended decision. George should focus on how long he will live there in making his decision.

17.5 CONSTRUCTING A SPIDERPLOT

By Hand.   Constructing a spiderplot is the type of computational and graphical task at which computers excel, yet spiderplots can be created by hand. If done by hand and if the *y*-axis is PW, EAC, or EAW, then cash flows can be plotted using one end-point calculation and the base case. The other end point is determined by linearity and its *x*-axis coordinate, as shown in Example 17.7.

Both end points should be calculated for interest rates, lives, time periods, geometric gradients, and all curves if the *y*-axis is the internal rate of return or benefit/cost ratio. These curves are nonlinear, although they sometimes approach linearity.

Example 17.7   Hand Construction of a Spiderplot

Construct a spiderplot for a problem with the following first cost, annual revenue, interest rate, life, and uncertainties. Assume the salvage value is $0. Use present worth for the *y*-axis.

Base Case	Cash Flow Element	Lower Limit	Upper Limit
−$90	First cost	95%	115%
$12	Annual revenue	60%	120%
10%	*i*	80%	120%
20	*N*	75%	150%

Solution

The equation for the base case is:

$$PW = -90 + 12(P/A, 10\%, 20) = -90 + 12(8.514) = \$12.2$$

All curves will pass through this point, since 0% change in the base case value is part of each curve.

For the first cost line, the upper limit is used. (For plotting by hand, a larger distance between the two points increases accuracy.)

$$PW_{FC\text{-}upper} = -90 \cdot 1.15 + 12(P/A, 10\%, 20) = -\$1.34$$

For the annual revenue line, the lower limit is used:

$$PW_{revenue\text{-}lower} = -90 + 12 \cdot .6(P/A, 10\%, 20) = -\$28.7$$

Since i and N are nonlinear, both end points must be calculated:

$$PW_{i\text{-}lower} = -90 + 12(P/A, 8\%, 20) = \$27.8$$
$$PW_{i\text{-}upper} = -90 + 12(P/A, 12\%, 20) = -\$.37$$
$$PW_{N\text{-}lower} = -90 + 12(P/A, 10\%, 15) = \$1.27$$
$$PW_{N\text{-}upper} = -90 + 12(P/A, 10\%, 30) = \$23.1$$

Exhibit 17.8 plots these points and illustrates that only the annual revenue has PWs significantly below \$0. All variables approach or slightly drop below a PW of \$0.

Exhibit 17.8   Hand constructed spiderplot

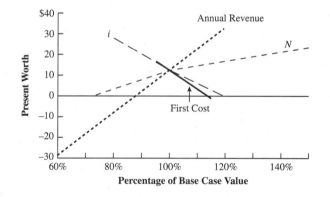

Using a Spreadsheet.   Spreadsheets are ideal for constructing spiderplots. The computer automates the hand calculations in Example 17.7, and adds powerful graphing functions.

The only major difference is that some spreadsheets, such as Lotus 1-2-3 and QuattroPro, require that multiple curves on the same *x*-axis use the same *x*-axis coordinates. Thus, each of the cash flows in Example 17.8 must have a calculation for 95% and 115%, even though those are the limits only for the first cost.

Example 17.8   Spreadsheet Construction of a Spiderplot

Use a spreadsheet to construct the spiderplot that was done by hand in Example 17.7.

Solution

Exhibit 17.9 contains a spreadsheet showing the calculations needed to construct the spiderplot, along with representative formulas. Listed below are the steps.

Exhibit 17.9   Make a spiderplot with a spreadsheet

	A	B	C	D	E
1	Base Case		Lower &	Upper Limits	
2	-90	First Cost	95%	115%	
3	12	A=revenue	60%	120%	
4	10%	i	80%	120%	
5	20	N	75%	150%	
6	12	PW Base case			
7	=$A$2 + @PV($A$3, $A$4, $A$5)				
8					
9	%	First Cost	A=revenue	i	N
10	60%		-28.7		
11	75%		-13.4		1.3
12	80%		-8.3	27.8	3.9
13	95%	16.7	7.1	15.7	10.4
14	100%	12.2	12.2	12.2	12.2
15	115%	-1.3	27.5	2.5	16.6
16	120%		32.6	-0.4	17.8
17	150%				23.1
18					
19		=+$A$2 + @PV($A$3, $A$4, $A$5*A17)			
20		=+$A$2 + @PV($A$3, $A$4*A16, $A$5)			
21		=+$A$2 + @PV($A$3*A16, $A$4, $A$5)			
22		=+$A$2*A15 + @PV($A$3, $A$4, $A$5)			
23	Entered in order from lower and upper limits				

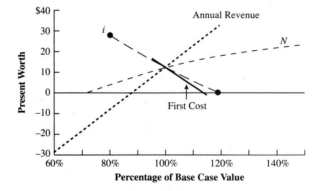

1. Create the data block of base case values and lower and upper limits.

2. Write the formula for the base case PW using absolute addresses (when the formula is copied, the addresses must not change).

3. Create the column headings for the cash flow elements to be changed and the row headings for the percentage points on the *x*-axis. These row headings are the different values listed in the lower and upper limits—listed in increasing order. Add a row for 100%—the base case.

4. Copy the formula for the base case into each cell of the top row of the table. Then edit the entry for each column, so that the cash flow element for that column is multiplied by the percentage value in column A.

   For example, the formula for the *revenue* entry in cell C10 is $A$2+@PV($A$3,$A$4*A10,$A$5). The *revenue* value in $A$4 is multiplied by cell A10, which has the 60% value.

5. Copy the top row into the rest of the table. The values in the 100% row should all match the calculation of the base case PW.

6. Delete all entries that are beyond the limits for that column. For example, in this case the first cost column should only have values between 95% and 115%. Similarly, the interest rate column should only have entries between 80% and 120%.

7. Create a graph. The type should be XY. (If a line graph is selected, the *x*-axis values will be treated as labels and will be spaced evenly.) Each column of the table is a series. These series all have the same number of cells and *include the deleted entries*.

   For this example, the series for first cost is cells B10 to B17, and the series for the *x*-axis is cells A10 to A17. Many spreadsheet packages have a group function that allows the *y*-axis values to be defined as a group.

   If PW is the *y*-axis, add a variable that equals 0 for all *x*-axis values. If the *y*-axis is a B/C ratio, set the extra variable equal to 1. If the *y*-axis is the IRR, set the extra variable equal to the minimum attractive rate of return.

   Label the graph appropriately.

Choosing a *y*-axis. Exhibit 17.10 redraws the spiderplot of Example 17.8 with four different *y*-axes. In each case, the *x*-axis coordinates are the same, but the shapes of some of the curves are different. For example, in converting from a present worth to an equivalent annual worth, an $(A/P, i, N)$ factor is used. If the cash flow element is *i* or *N*, then this factor is nonlinear and the curve's shape is changed.

However, the basic meaning of the spiderplot is not affected by the change in axis. The choice of the *y*-axis should be matched to the decision maker's preference.

Exhibit 17.10   Comparing different *y*-axes for spiderplots

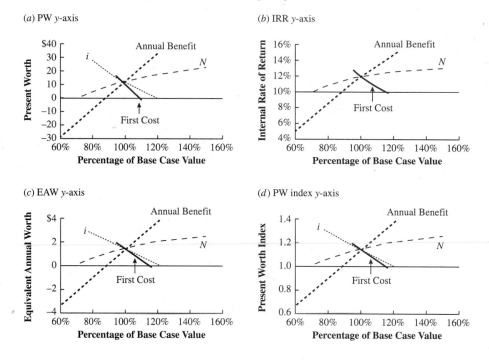

(*a*) PW *y*-axis

(*b*) IRR *y*-axis

(*c*) EAW *y*-axis

(*d*) PW index *y*-axis

17.6 MULTIPLE ALTERNATIVES

Spiderplots.   If there are two mutually exclusive alternatives, then the difference between them can be graphed as shown in Sections 17.4 and 17.5. However, if the problem is one of constrained project selection or if there are three or more mutually exclusive alternatives, then a graph similar to those shown in this section is needed.

The number of "spiders" in a spiderplot equals the number of alternatives, so to avoid information overload the maximum number of cash flow elements should be three. Example 17.9 applies this technique to mutually exclusive alternatives, and Example 17.10 applies it to constrained project selection.

Example 17.9   Multiple Mutually Exclusive Alternatives

Conspicuous Consumption Toys has a short-term (about 1 year) need for a part that will be included in this year's fad toy. The part can be purchased or it can be produced through either a capital- or a labor-intensive process.

The expected volume is 8000 units, with a lower limit of 70% and an upper limit of 120%. The purchase price for 8000 is estimated to be $50 per part ±10%. For each 10% increase or decrease in the quantity, the expected price will change by 2.5%. The price falls if the quantity increases.

The labor-intensive process will cost $40 per part for labor (±20%) and $50,000 in capital improvements, with no salvage value. The capital-intensive process will cost $400,000, with a lower limit of 95% and an upper limit of 115%. These capital expenditures also have no salvage value. The capital-intensive process also requires $5 per part in labor.

Construct a spiderplot using cost per part as the y-axis.

Solution

The base case cost per part for each alternative is found as follows:

$$\text{LaborInt} = 40 + 50,000/8000 = \$46.25$$
$$\text{CapitalInt} = 5 + 400,000/8000 = \$55$$
$$\text{Purchased} = \$50$$

The cost of the labor-intensive process has a range of ±$8 for labor productivity, based on ±20% of the $40 labor cost. The range for volume is −$1.04 to +$2.68 per part (50,000/9600 − 6.25 to 50,000/5600 − 6.25).

The capital-intensive process has limits of 95% to 115% of its $400K capital cost, or $380K to $460K. The cost per part is $47.5 to $62.5 (5 + 380/8; 5 + 460/8). Since the capital cost is assumed to equal $400K as the volume changes, the cost per part ranges from $76.4 to $46.7(5 + 400K/5600; 5 + 400K/9600).

The purchased part has a cost uncertainty of ±10%, or $45 to $55 at a volume of 8000. If the volume decreases by 30%, then the price increases by 7.5%, to $53.75. If the volume increases by 20%, then the price decreases by 5%, to $47.50.

The four sources of uncertainty produce six curves, as shown in Exhibit 17.11. The three curves for each cost as the volume changes are not independent. In all cases, the cost falls as volume increases, but it falls fastest for the capital-intensive process.

The labor-intensive process seems to be superior for most assumptions within the limits described above. Even if the capital-intensive process were slightly cheaper, it is likely that the labor-intensive process would be preferred on the basis of risk. If sales are high, the firm will do well, and it can probably afford a slightly higher cost per part. On the other hand, if sales are low, the firm will be doing poorly, and it will be additionally disadvantaged by the higher-than-anticipated costs of the capital-intensive option.

Exhibit 17.11   Conspicuous Consumption Toys—cost per part

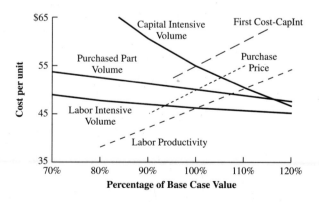

Example 17.10   Oil Company Project Selection vs. Oil Price

Six possible drilling projects are being ranked to see which ones should be funded within the available budget of $5 million to $6 million. The profitability of each well is determined by its first cost (in millions), the number of years it will last, its annualized rate of production (in barrels or bbl), the variable cost of production and transportation from that well (in dollars per barrel), and the average selling price of oil over the well's life (in dollars per barrel). Assume a base case average oil price of $20 per barrel, with lower and upper limits of ±25%. Analyze the sensitivity diagram with respect to the price of oil to see which drilling projects are the most attractive for low, base case, and high oil prices.

Project	First Cost (in $M)	Life (years)	Production Cost ($/bbl)	Annualized Volume (bbl)
A	$2	20	$ 9	60,000
B	1	10	14	40,000
C	1.5	7	8	40,000
D	2.1	25	10	50,000
E	1.6	16	9	20,000
F	2	20	5	20,000

Solution

Note that this is a spiderplot, where each of the six projects has only one curve to its spider. The basic equation for each project is:

$$0 = -\text{FirstCost} + \text{Volume} \cdot (\text{OilPrice} - \text{ProdCost}) \cdot (P/A, i, \text{Life})$$

Rewriting this as a spreadsheet function for each project's IRR, the result in Quattro Pro (for Excel, the function is RATE) is:

$$@\text{IRATE}(\text{Life}, \text{Volume} * (\text{OilPrice} - \text{ProdCost}), -\text{FirstCost})$$

Then a table of IRR values for each project at each of the three oil prices is constructed to support the graph. Exhibit 17.12 graphs the central part of the sensitivity plot (very low and very high internal rates of return are not graphed). In general, those projects with higher production costs are more sensitive to the price of oil.

Exhibit 17.12   Which oil projects should be accepted?

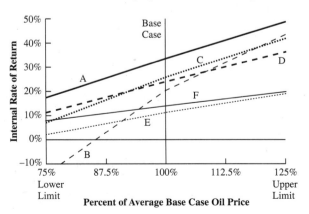

Project A is the most attractive in all three cases, and it should be funded. Project C is ranked second in the base case and third with the upper limit of oil prices, and is tied for third/fourth with the lower limit of oil prices. These together cost $3.5M.

In the base case and with the lower limit, project D is ranked third and second, respectively. Project B is ranked fourth for the base case and second with oil prices at the upper limit. Since project B is ranked last and has a very negative IRR if average prices fall to the lower limit, project D appears to be a better choice. The total recommended budget is $5.6M for projects A, C, and D.

Unfortunately, the decision to fund these projects must be made with current information. It is not possible to be certain about average oil prices over the next 25 years.

Scenarios. Exhibits 17.13a and 17.13b illustrate an effective graphic for scenarios with multiple alternatives. Grouping the present worth by scenarios (as in Exhibit 17.13a) emphasizes which project is best for each scenario. Project A is clearly the best under the bad scenario, and it is slightly better than projects B

Exhibit 17.13   Mutually exclusive projects and scenarios

(a)   Projects grouped by scenarios

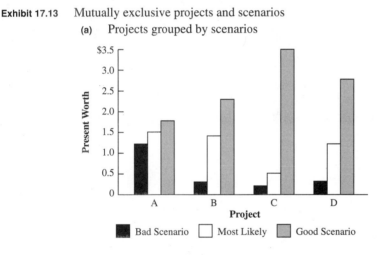

(b)   Scenarios grouped by projects

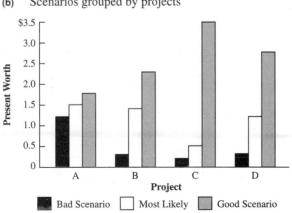

and D under the most likely scenario. Project C is the best in the good scenario, but the worst otherwise.

Grouping the present worths by project (as in Exhibit 17.13b) emphasizes that project A has the most stable performance under the different scenarios (and project C is the least stable). This grouping also allows construction of a kind of "visual average."

17.7 SENSITIVITY ANALYSIS WITH MULTIPLE VARIABLES

The spiderplots of Sections 17.3–17.6 have changed only one cash flow element at a time; however, it is possible to analyze changes in two variables at a time. Example 17.11 analyzes the "scenario" in Example 17.2 to show under what circumstances the project seems to be worth doing.

As demonstrated in Example 17.11, two kinds of graphs are possible. The first, shown in Exhibit 17.14, uses one variable for the x-axis and draws curves for different values of the other variable.

The second, shown in Exhibit 17.15, solves a breakeven equation and creates go/no go regions for all combinations of the two variables. Here, the approach is to write a cash flow equation and to set it equal to 0. Selecting the values for one cash flow element and solving for the other one defines a breakeven curve. On one side of the curve, the project is economically positive; on the other side, it is economically negative. Points close to the line are clearly within a zone of economic indifference.

When choosing which variables to pair in these graphs, several criteria are possible. The two cash flow elements may be:

1. The most important
2. Logically linked (such as two inflation rates or price and quantity sold)
3. The ones with the most uncertainty

Example 17.11   Two Variable Sensitivity Analysis (Example 17.2 Revisited)

The park acquisition problem assumed a land cost of $2 million, a salvage value of $0, and an annual benefit of $180,000. Possible values for the interest rate were 3%, 6%, and 9%; for the project life, the possible values were 10, 20, and 50 years.

Construct two sensitivity charts to show how the recommended decision depends on the choice of the interest rate and time period.

Solution
The easier graph to construct is to plot the present worth as a function of i or N. Either variable can be selected for the x-axis. Exhibit 17.14 places the horizon, N, on the x-axis and uses the values for i to define the curves. Clearly, at 3% almost all study periods support a positive recommendation, while at 6% the neutral point is about 20 years; at 9%, even longer than 50 years is required to justify the project.

The second graph is probably most easily constructed by creating a table of assumed values for N and calculated values for i that create a present worth of 0—the breakeven point. When that table is graphed, the breakeven chart in Exhibit 17.15 is created.

Exhibit 17.14   Park acquisition PW vs. *i* and *N*

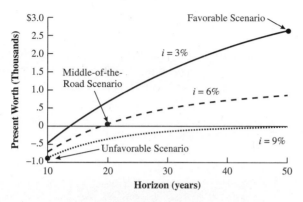

The equation is $0 = -2M + 180K(P/A, i, N)$, or in Quattro Pro spreadsheet terms: @IRATE(*N*,-2000,180). The values are as follows:

N (years)	10	15	20	25	30	35	40	45	50
i (%)	-1.9	4.0	6.4	7.5	8.1	8.5	8.7	8.8	8.9

The graph is Exhibit 17.15. Within the range of expected values for *i* and *N*, most combinations support acquisition of the parkland.

Exhibit 17.15   Park acquisition Go/No Go with *i* vs. *N*

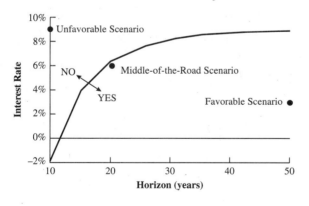

17.8 SUMMARY

Sensitivity analysis requires a sense of proportion to properly judge the significance of the results. In some cases, the apparent economic difference between the alternatives is small. The graphical and numerical techniques of this chapter calculate or display the economic breakeven point. When likely conditions are close to economic breakeven, then the irreducibles that were not included in the cash flow diagrams should dominate the decision making.

Sensitivity analysis is needed due to uncertainties in estimated data. It is used to make better decisions, to decide which data to refine, and to focus attention during implementation. The process begins by identifying the most likely values and the limits of uncertainty for each cash flow element.

Breakeven charts can be used to analyze one variable at a time, and scenarios are used to evaluate plausible combinations of variables. Spiderplots are used to compare the sensitivity of the economic measure (PW, EAC, or IRR) to different variables.

These spiderplots can be constructed by hand, but they are an ideal application of spreadsheets. Spiderplots can be applied to multiple alternatives. Two-variable graphical sensitivity analyses can be done either by plotting several curves of present worth or by defining a breakeven curve of one variable versus another variable.

REFERENCES

Eschenbach, Ted, "Chapter 3. Sensitivity Analysis," *Cases in Engineering Economy*, 1989, Wiley.

Eschenbach, Ted, "Quick Sensitivity Analysis for Feasibility Studies and Small Projects," *Proceedings of the Annual AACE Meeting*, June 1992.

Eschenbach, Ted, and Robert J. Gimpel, "Stochastic Sensitivity Analysis," *The Engineering Economist*, Volume 35, Number 4, Summer 1990, pp. 305–321.

Eschenbach, Ted, and Lisa S. McKeague, "Exposition on Using Graphs for Sensitivity Analysis," *The Engineering Economist*, Volume 34, Number 4, Summer 1989, pp. 315–333.

PROBLEMS

Note: Each problem is based on or refers to items noted in parentheses.

17.1   (Problem 6.3) What is the breakeven annual savings for the heat exchanger?   (*Answer:* $13,920)

17.2   (Problem 6.9) What is the breakeven quarterly revenue for the machine?   (*Answer:* $2054)

17.3   (Problem 8.7) What is the breakeven annual benefit for ABC's new machine?

17.4   (Problem 8.1) What is the breakeven annual time savings for the highway project?

17.5   (Problem 7.22) How does the effective annual rate depend on the amount of the down payment? Vary the down payment from 0% to 25%.

17.6   (Problem 18.9) What is the breakeven probability for the most likely annual benefit? Assume that the probability of $300K in annual benefits is three times the probability of $500K.

17.7   (Problem 14.18) What is the breakeven fuel cost for the gas furnace from the client's perspective? From a broader perspective?

17.8   (Problem 6.13) Assume that the R&D period is uncertain, rather than a known 4 years. Find the breakeven R&D period.

17.9   (Problem 5.10) What is the minimum annual revenue for breakeven? Maximum first cost? Minimum salvage value or maximum salvage cost? Construct breakeven charts for each.

17.10   (Problem 5.30) What is the minimum net annual revenue for breakeven? Maximum development expenses? Maximum salvage cost? Construct breakeven charts for each.

17.11   (Problem 11.26) Assume that the $10K-per-mile cost and gradient for pothole fixing is dependent on the technology used. What is the breakeven cost and gradient for the 8-year policy to be optimal? (*Answer:* $5300 vs. 7-year)

17.12   (Problem 13.12) Graph the EAC as a function of the salvage value, which may vary from $0 to $.6M.

17.13   (Problem 10.4) Create scenarios of useful life and MARR that favor the Smoothie and the Blaster.

17.14   (Problem 10.20) Create scenarios of useful-life and interest-rate combinations that favor each of the three alternatives.

17.15   (Problem 18.10) Create and analyze good, most likely, and bad scenarios using the given first costs and net revenues. How does this differ from the states of nature in Problem 18.10?

17.16   (Problem 18.11) Create and analyze good, most likely, and bad scenarios using the given first costs, net revenues, and project life. How does this differ from the states of nature in Problem 18.11?

17.17   (Problem 7.12) Assume that cost to proceed, initial monthly sales, and the monthly gradient can vary 20% to the good and 30% to the bad. Create bad, most likely, and good scenarios, and find the IRR for each.   (*Answer:* i_m = 10.45%, 15.23%, 21.45%)

17.18   (Problem 14.27) How sensitive is the recommendation to the state's interest rate? Does changing the rate to 4% or to 12% alter the recommended set of projects?

17.19   (Problem 10.11) Construct reasonable limits of uncertainty for the first cost, annual savings, and salvage value. Construct a spiderplot for the difference in the EAC for the two security systems.

17.20   (Problem 3.39)
    a. Construct reasonable limits of uncertainty for the annual college cost, the interest rate, and the initial year of deposit.
    b. Construct a spiderplot for the amount of the initial deposit.

17.21   (Problems 3.39 and 17.20) Construct reasonable scenarios from the limits in Problem 17.20a, and then calculate the amount of the initial deposit for each scenario.

17.22   (Problem 2.32) A certificate of deposit compounds interest annually. What amount has accumulated for retirement at age 65 if $10,000 is deposited at a certain age? Analyze the sensitivity of the final accumulation as a function of age and interest rate with a spiderplot. Assume that age 35 and an interest rate of 10% are the base case, with respective limits of 25 to 55 years and 5% to 15%.   (*Answer*: Age = $25.9K–$452.6K, i = $43.2K–$662.1K)

17.23   (Problem 19.5) How sensitive is the recommended decision to the weight placed on the EAC? Vary the weight from 0 to 10 and identify the recommended site for each weight.

17.24   (Problem 13.30) Construct a spiderplot if the annual revenues may vary by –40% to +20%, the annual expenses by –20% to +30%, and the initial investment by ±10%.

17.25   (Problem 3.22) If HiTek Manufacturing's borrowed principal were to change by ±10%, the interest rate by ±20%, and the number of years between 10 and 30, which changes would be most significant? Use a spiderplot.

17.26   (Problems 3.27 and Chapter 15) Different lottery winners have different interest rates for the time value of money.
a.   Analyze the present worth of the payments for real interest rates ranging from 5% to 12%.
b.   Repeat part a with four curves for inflation rates of 0%, 3%, 6%, and 9%. 5.36

17.27   (Problem 5.36) This problem originally asked for a breakeven salvage value (which is –$357.3K). Assume the salvage value is actually $50K. Using the following limits construct a spiderplot.

Variable	Lower Limit	Upper Limit
First cost	80%	140%
Annual expense	80%	150%
Annual benefit	60%	120%
Interest rate	80%	140%
Horizon	50%	150%
Salvage value	0%	200%

17.28   (Problem 10.25) Create a spiderplot for the EAC difference between the two bridges.

Variable	Lower Limit	Upper Limit
First cost (galvanized)	80%	140%
First cost (painted)	80%	150%
Annual operations (galvanized)	80%	120%
Annual operations (painted)	60%	140%
Interest rate	75%	125%

(*Answer:* First cost$_{galv.}$ = -$.3M to +$3.9M)

17.29 (Problem 15.30) Assume that your annual cash flow deposit, savings account rate, salary increase rate, and inflation rate could each become 10% better or 20% worse. Construct a spiderplot of the value after 40 years in year-one dollars.

17.30 (Problem 16.11) Construct a spiderplot for changes in the size of the addition ($\pm$10,000 square feet); in the capacity coefficient (ranging from .6 to .8); and in the size of the original facility ($\pm$10,000 square feet).

17.31 (Problem 16.15) Draw a spiderplot for the level of demand in year 5 (halfway through the 10-year horizon). Assume that L, a, and b each vary by $\pm$20%.

17.32 (Problem 16.20) Draw a spiderplot for the level of demand in year 5 (halfway through the 10-year horizon) when $L = 1000 \pm 20\%$, year 1's demand ranges from 4 to 10, and year 2's demand ranges from 40 to 70.
   a.   Assume a Pearl curve.
   b.   Assume a Gompertz curve.

17.33 (Problem 9.20) Analyze the sensitivity of the benefit/cost ratio rankings of the nine projects to the interest rate mandated by Congress. Use 10% to 20% as a reasonable range for the interest rate on the x-axis. Use the benefit/cost ratio as the y-axis.

17.34 (Problem 9.8) For the seven projects at National Motors' Rock Creek plant, analyze the impact of the assumed life on the recommendations. Use a common range of 80% to 150% of estimated life and a y-axis of internal rate of return.
   a.   Construct the multiple alternative sensitivity graph.
   b.   For a budget of $500,000, how different are the sets of recommended projects at the two extremes and at the base case?

17.35 (Problem 9.26) Assume that the projects of NewTech's Ceramic Division will have common errors in the estimates. In other words, if one has annual benefits that increase by 20%, then the benefits of all three will increase by 20%.
   a.   Construct a graph of the internal rate of return of each project as first costs range between 85% and 130% of the estimate.
   b.   Construct a graph of the internal rate of return of each project as annual benefits range between 75% and 110% of the estimate.

        c.   Construct a graph of the internal rate of return of each project as each life is shortened by 20% or lengthened by 100%.

        d.   Construct a spiderplot for the most important two variables in parts a, b, and c.

17.36   (Problem 9.26) Redo Problem 17.35c, assuming that each project's life is shortened by 2 years or lengthened by 5 years instead of by percentages. Use the number of years less or more as the x-axis. Do the graphs of Problems 17.35c and 17.36 have similar shapes? Do they lead to the same conclusion? Is one of them better? If so, why?

17.37   (Problems 4.8–4.10 and 15.22–15.24) The cost of garbage pickup in Green Valley will be $4.5 million this year. Estimate the cost in 10 years in year-one dollars. In base case estimates, the population is increasing by 6% annually, the cost per ton is increasing at 5% annually, the inflation rate is 4% annually, and residents will individually reduce their trash generation by 2% annually.

        a.   Construct reasonable limits of uncertainty for each geometric gradient.

        b.   Construct a spiderplot for the year-10 cost (measured in year-one dollars).

17.38   (Problem 4.20) Construct a spiderplot for the annual rate of increase in tuition at City University. The reasonable limits for enrollment increases are 1% to 4%. The cost per credit-hour taught is increasing between 4% and 10%. State funds are decreasing between 0% and 6% per year.

17.39   (Problem 14.10) Construct reasonable limits for the following variables, and then create a spiderplot: construction costs, repaving cost, road life, interest rate, and value per trip.

17.40   (Chapter 16) Construct a sensitivity graph for the cost of the 10th, the 100th, and the 1000th item on a learning curve when the first item requires 8000 person-days. Vary the learning-curve function coefficient from .8 to .85.

17.41   (Problems 2.27–2.30) Construct a multiple alternative sensitivity chart for the four loan patterns in Chapter 2 when $100,000 is borrowed, to be repaid over 5 years. Assume a base case interest rate of 11% and a range of 7% to 15%. Use the present worth of the payments as the y-axis.

17.42   (Problems 17.14 and 10.20) Divide a two-variable sensitivity chart (for useful life and interest rate) into regions that favor each of the three alternatives.

CHAPTER 18   UNCERTAINTY AND PROBABILITY

The Situation and the Solution

It is unrealistic to assume that future cash flows are known exactly. For example, estimates of market demand 5 years from now or energy costs in 20 years may vary broadly.

The solution is to explicitly consider probability distributions for cash flows, interest rates, time horizons, etc. Then measures of average or expected return and of risk can be used for economic decision making.

Chapter Objectives

After reading and studying the sections of this chapter, you should be able to:

Section 18.1
Define probability and develop a probability distribution.

Section 18.2
Calculate the expected value of a probability distribution.

Section 18.3
Use expected values to make economic decisions.

Section 18.4
Model more complex problems using economic decision trees.

Section 18.5
Measure risk using the standard deviation of a probability distribution.

Section 18.6
Understand risk/return tradeoffs in economic decision making.

Section 18.7
Construct probability distributions with multiple independent variables.

Key Words and Concepts

Probabilities   Measures of expected frequency or likelihood.

Probability distribution   Set of possible individual outcomes with their probabilities.

Expected value   The weighted average or $E(X) = \sum_i x_i \cdot P(x_i)$.

Economic decision trees   Figures describing the timing, probabilities, and consequences of sequential decisions and chance events.

Standard deviation (σ, dispersion about the expected value)   The common measure of risk. It equals $\sqrt{\{E(X^2) - [E(X)]^2\}}$, which is the square root of the average of the square minus the square of the average.

Dominated   A term that describes an alternative with a lower expected PW and a higher risk or standard deviation than a better alternative.

Dominant   A term that describes an alternative with a higher expected PW and a lower risk or standard deviation than all other alternatives.

   This chapter assumes no previous study of probability, and it focuses on using probabilities to make economic decisions. Limited space precludes including topics such as the normal, beta, and lognormal distributions. While valuable, knowledge of these distributions is not essential in modeling the consequences of each alternative and in choosing the best alternative.

18.1 PROBABILITIES

The basic measure of uncertainty is **probability**. Even without formal training in probability and statistics, most of us have developed an intuitive understanding for probabilities. That understanding may be based on a 20% chance of rain or on a 1 in 2 chance of getting heads on a coin flip.

Probability is a measure of expected frequency or likelihood.

   In fact, probabilities are derived from:

1.   Historical data
2.   Mathematical models
3.   Subjective estimates

   Probabilities of rain or of a telephone pole's or a computer's life can be derived from historical data. A coin flip or a random event's probability can be derived mathematically from a model of a *fair* or unbiased chance. The probability of when a competing firm will introduce its product can be subjectively estimated by a senior executive or by a junior engineer.

In all cases these probabilities must satisfy three axioms:

1. All probabilities are ≥ 0 ($0 \Rightarrow$ never).
2. All probabilities are ≤ 1 ($1 \Rightarrow$ always).
3. The probabilities of the mutually exclusive outcomes sum to 1.

The **probability distribution** for mutually exclusive outcomes has individual probabilities (each between 0 and 1 inclusive) that collectively sum to 1.

The set of possible individual outcomes with their probabilities is called a **probability distribution**.

Since probability is a long word, a shorter notation has been developed. The probability of event A is commonly written as $P(A)$. Similarly, the probability that the present worth is greater than $875 is written as $P(\text{PW} > 875)$. Examples 18.1, 18.2, and 18.3 illustrate the historical, mathematical, and subjective bases for probability in more detail, using three different probability distributions.

Example 18.1   Cost of Construction

A building is scheduled for late fall construction. Based on historical records, the chance of sunny weather is 20% and the chance of an early winter is 30%. With sunny weather construction will cost $225,000. If it is an early winter, the building will cost $295,000. If the weather is between the two extremes, construction will cost $250,000. What is the probability distribution for the construction cost?

Solution

Three *states of nature* have been identified, along with the first cost of each. However, only two probabilities are given. The third axiom, that the probabilities of the mutually exclusive outcomes sum to 1, is used to find the P for the intermediate weather. Specifically, this must equal $1 - .2 - .3$, which equals .5 or 50%. Mathematically, the paired probabilities and first costs form the probability distribution, but the meaning is clearer if the states of nature are also given.

State of Nature	First Cost	Probability
Sunny	$225,000	.2
Intermediate	$250,000	.5
Winter	$295,000	.3

Example 18.2   Life of a Crash Absorber

The Crooked Mountain Expressway has many bridge abutments with crash-absorbing barriers. Each year, one-third of these barriers are damaged in collisions and must be replaced or repaired. There is no apparent pattern to which bridge abutments are hit; rather, it appears to be random. If a crash absorber survives 5 years, it is replaced under regular maintenance. What is the probability distribution for the life of a crash absorber?

Solution

This mathematical model assumes randomness of collisions over the years. Thus, if a crash absorber survives to the beginning of a year, there is a one-third chance of being hit during the year. For example, a crash absorber can be hit in year 2 only if it was not hit in year 1, and the one-third chance of a hit comes true in year 2. Thus, two equations form the probability model:

$$P(\text{hit in year } t) = P(\text{not hit by year } t - 1) \cdot (1/3)$$

$$P(\text{not hit by year } t) = P(\text{not hit by year } t - 1) - P(\text{hit in year } t)$$

So for the first year, the $P(\text{hit}) = 1/3$ and $P(\text{not hit yet}) = 1 - 1/3 = 2/3$. For the second year, the $P(\text{hit}) = (2/3) \cdot (1/3) = 2/9$, and the $P(\text{not hit yet}) = 2/3 - 2/9 = 4/9$. This continues, as tabulated below, for 5 years. Finally, the probability that it was not hit at all is found by subtracting the probabilities that it is hit from 1 (axiom 3). The distribution is shown graphically in Exhibit 18.1. The two 5-year outcomes could be combined ($P = .1975$).

Year	P(hit)	P(not hit yet)
1	.333	.667
2	.222	.444
3	.148	.296
4	.0988	.1975
5	.0658	.1317
5	.1317	Replaced anyway

Exhibit 18.1 Probability density function for crash absorber life

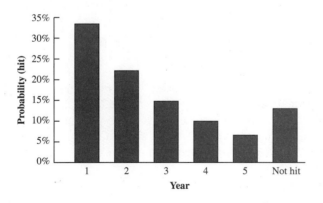

Example 18.3   Micro Pizza Heater: Market Demand

A factory renovation is needed to build a compact microwave with a new shape, which will be called the Micro Pizza Heater. The low sales-volume prediction (20,000 heaters per year) has a subjectively estimated probability of 30%. The most likely market prediction is 30,000 units sold per year. The optimistic market prediction (30,000 sold the first year, with annual increases of 5000) has a subjectively estimated probability of 10%. In all cases the factory equipment and the market will last 5 years. The net revenue will be $10 per microwave. What is the probability distribution for the net revenue?

Solution
Three *states of nature* are identified and linked with annual cash flows. However, no most-likely probability is given. Because the probabilities sum to 1, the P(selling 30,000/year) must equal $1-.3-.1$, which equals .6, or 60%. The probability distribution for net revenue is a 5-year projection, as follows:

| | State | | |
Year	Pessimistic $P = .3$	Intermediate $P = .6$	Optimistic $P = .1$
1	$200,000	$300,000	$300,000
2	200,000	300,000	350,000
3	200,000	300,000	400,000
4	200,000	300,000	450,000
5	200,000	300,000	500,000

18.2 COMPUTING EXPECTED VALUES

Expected value or mean is a weighted average. The weights are probabilities.

The **expected value, E,** is the most often used measure of a probability distribution. Each possible value is weighted by multiplying the value by its probability; these products are then summed. Since the probabilities are weights that sum to 1, the expected value is a weighted average. Thus, for Example 18.1, the average first cost is found by multiplying each first cost by its probability:

$$E_{first\ cost} = .2(225,000) + .5(250,000) + .3(295,000)$$
$$= \$258,500$$

As shown by this example, the expected value may not even be a listed possibility. The expected value is "close" to the intermediate value. Here the expected value exceeds the intermediate value, because the pessimistic value is both further away from the intermediate value and more likely than the optimistic value. The value of $258,500 should be used to compute expected present worths and to evaluate the project.

Example 18.4   Expected Life for Crash Absorber (Example 18.2 continued)

Compute the expected life for the crash absorber in Example 18.2.

Solution
Assume that crashes and replacements occur at the end of the year. The probability distribution in Example 18.2 can be used, with one minor change. For computing the average

life, there is no difference between being hit in year 5 and being a scheduled replacement. Thus, those two probabilities are summed in the next-to-last line of the accompanying table.

To find the expected life, each year is multiplied by its probability. Then these products are summed.

Year	P(year)	Product
1	.333	.33
2	.222	.44
3	.148	.44
4	.0988	.40
5	.1975	.99
	Expected life	2.60 years

Example 18.5. Micro Pizza Heater: Expected Market Demand (Example 18.3 continued)

The company uses a minimum attractive rate of return of 12%. What is the expected value for the equivalent annual revenue?

Solution
We begin by computing the equivalent annual revenue for each state of nature. The pessimistic and most likely states have uniform flows of $200,000 and $300,000, respectively. The optimistic state's equivalent annual revenue is found as follows:

$$EAW = 300,000 + 50,000(A/G, .12, 5)$$
$$= 300,000 + 50,000(1.7746) = \$388,730$$

The expected equivalent annual revenue is:

$$E_{annual\ revenue} = .3(200,000) + .6(300,000) + .1(388,730)$$
$$= \$278,873$$

Equation 18.1 is the general formula for calculating expected values, to be used whenever a set of discrete possibilities has been identified. This weighted average simply takes each value of x and multiplies it by its probability. A similar formula relying on integration is used for continuous distributions. Note that the simpler formula built into most calculators assumes that each probability equals 1 divided by the number of data points, which is often untrue for models of economic decisions.

$$E = \sum_i x_i \cdot P(x_i) \qquad (18.1)$$

This formula can also be directly applied to present worths and equivalent annual costs to quantify the expected worth (cost) of an alternative.

18.3 CHOOSING ALTERNATIVES USING EXPECTED VALUES

So far we have computed expected values only for a single cash flow element. The best choice among alternatives has the maximum expected worth or the minimum expected cost. We calculate each alternative's expected present worth and choose the largest. Or we calculate the expected equivalent annual cost for each alternative and choose the lowest.

① The first step is calculating the expected value of each probabilistic cash flow element. For example, if there are probability distributions for the first cost, annual revenues, salvage value, and cost of an overhaul, then four expected values are calculated. The second step is to find the PW, EAC, or IRR.

As detailed in Section 18.7, this substitution of expected values in PW or EAC calculations can be inappropriate or approximate. For example, if there is a probability distribution on N or i, then this answer is only approximate (see Example 18.16 in Section 18.7). Or a more complete analysis that includes risk (see Sections 18.5 and 18.6) may be planned. In this case, the complete distribution of the present worth is needed—not just its expected value.

Examples 18.6 (typical for feasibility studies) and 18.7 (typical for insurance or risk avoidance) illustrate the principle of maximizing expected value. Maximizing expected value is not the *only* criterion used; however, it is clearly the most commonly used and most important criterion. More complete discussions of decision making under uncertainty and risk include other criteria [see, for example, Buck].

Example 18.6   New Product Development

ElectroToys has a new product idea, but the first cost of developing the product is uncertain, as is the annual net revenue. Assume that the first cost occurs at time 0, and that the net annual revenues begin at the end of year 1. ElectroToys uses an interest rate of 12% and new toys have a life of 5 years. Should the product be developed if the probabilities are as shown?

First Cost	P	Net Revenue	P
$20,000	.2	−$ 5,000	.3
30,000	.5	10,000	.5
50,000	.3	30,000	.2

Solution
The first step is to identify the alternatives, which in this case are go and no go. The second step is to calculate the expected values for the first cost and the net revenue.

$$E_{\text{first cost}} = .2(20{,}000) + .5(30{,}000) + .3(50{,}000)$$
$$= 4000 + 15{,}000 + 15{,}000 = \$34{,}000$$

$$E_{\text{annual net revenue}} = .3(-5000) + .5(10{,}000) + .2(30{,}000)$$
$$= -1500 + 5000 + 6000 = \$9500$$

The third step is to calculate the expected present worth of the new product.

$$E_{PW} = -34,000 + 9500(P/A, 12\%, 5) = \$245$$

Since this slightly exceeds a PW of 0, this product can be expected to earn just over the 12% minimum requirement. The expected present worth of rejecting the new product is $0.

Example 18.7 Flood Damage Protection

The community of Lowville has a problem with flooding from a nearby river. Building a levee will reduce both the probability of a flood and its consequences. Higher levees cost more, but they are less likely to be overtopped, and less water floods the town if they are overtopped. Which of the following levees minimizes the expected total annual cost to Lowville? Lowville uses an interest rate of 6% for flood protection projects, and all of the levees should last 50 years.

Levee Height	First Cost	P(flood) for Each Year	Damages if Flood Occurs
None	$ 0	.1	$500,000
2 meters	600,000	.05	125,000
4 meters	650,000	.01	50,000
6 meters	750,000	.001	5000

Solution
Calculating the EAC of the first cost merely requires multiplying the first cost by $(A/P, .06, 50)$. Calculating the annual expected flood damage for each alternative is done by multiplying the P(flood) times the damages if a flood happens. For example, the expected annual flood damage with no levee is $.1 \cdot 500,000$ or $50,000. Then the EAC of the first cost and the expected annual flood damage are summed. The 4-meter levee is somewhat cheaper than the 2-meter levee.

Levee Height	EAC of First Cost	Expected Annual Flood Damages	Total Expected EAC
None	$ 0	$50,000 _{125K (0.05)}	$50,000
2 meters	38,067	6250	44,317
4 meters	41,239	500	41,739
6 meters	47,583	5	47,588

18.4 ECONOMIC DECISION TREES

Engineering economy problems often involve sequences of decisions that may be intermixed with chance occurrences. Rather than using tables, we use **economic**

Economic decision trees describe the timing, probabilities, and consequences of sequential decisions and chance events.

decision trees to describe the problem's structure. <u>Each node of the tree represents a decision point or a chance occurrence. The branches that radiate from a decision node are the alternatives, and the branches that radiate from a chance node are the states of nature.</u>

A simple example is shown in Exhibit 18.2 for insuring a car against collision damage. The first node is a decision node with two alternatives for the next year—spending $300 to buy collision insurance with a $500 deductible and self-insuring for property damage. After the decision is made, then there are chance nodes for the different severities of an accident (if any). Each branch from a chance node has a probability between 0 and 1. The probabilities for the branches from a chance node sum to 1, since some outcome must occur (one chance branch must be selected).

Exhibit 18.2   Example decision tree—auto insurance

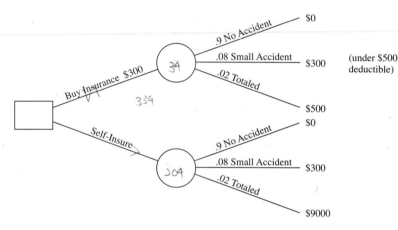

Decision trees are used in a two-stage process. In the first stage, their construction graphically organizes and structures the problem, as shown in Exhibit 18.2. Starting with the chance occurrence or the decision that occurs first, nodes and branches are added until the problem is completely described. At the same time, probabilities are entered on the chance branches and any cash flows are entered on both the chance and alternative branches.

If a decision must be made before the outcome of any chance event is known, then the first node is a decision node. For example, in Exhibit 18.2, the first event is a decision between two alternatives—buying insurance and self-insuring. There is no cost at the decision stage for self-insuring, but buying insurance costs $300.

Since our driving habits are likely to be the same with and without insurance, the accident probabilities are the same with and without insurance. The 100% chance of something happening this year is assumed to be divided into a 90% chance of no accident, an 8% chance of a small accident (at $300, which is less than the deductible), and a 2% chance of totaling the $9000 vehicle.

When drawing a decision tree, it is useful to use <u>boxes or diamonds for decision nodes and circles for chance nodes. It is essential that the probabilities of</u>

the branches from a chance node satisfy all three probability axioms (including summing to 1).

Once the tree is complete, the second stage begins. Expected value analysis is used to roll back the values and to choose the best alternatives. Here the principle of maximizing present or equivalent annual worth is applied.

For Exhibit 18.2, the expected values of the costs of an accident with and without insurance equals:

$$EV_{\text{accident w/ins.}} = .9 \cdot 0 + .08 \cdot 300 + .02 \cdot 500 = \$34$$
$$EV_{\text{accident w/o ins.}} = .9 \cdot 0 + .08 \cdot 300 + .02 \cdot 9000 = \$204$$

As shown in Exhibit 18.3, these expected costs can be entered into the chance nodes. Then these expected costs are combined with the cost for each decision to see that self-insuring costs $204 per year, or $130 less than the $334 cost for buying insurance and paying the deductible.

Exhibit 18.3   Rolling back a decision tree—auto insurance

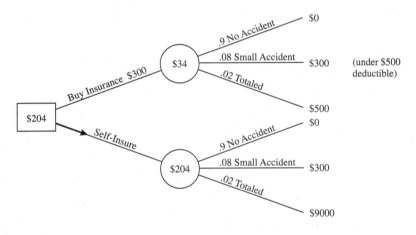

This is not surprising, since the costs of operating the insurance company ensure that the expected value of the payouts must be less than the premiums collected. This is also an example of when *expected values alone do not determine the decision.* Buying insurance has an expected cost that is $130 per year higher, but that insurance limits the maximum loss to $500 rather than $9000. The $130 may be worth spending to avoid that risk. (If there is a loan on the car, the lender is likely to require the insurance.)

Sequential Decisions.   ElectroToys' new product development (Example 18.6) illustrates that new and better alternatives can be evaluated by decision trees. If the net revenue is negative, then ElectroToys is likely to discontinue production after the first year. There is still a loss, but this new alternative will save $5000 per year for four years. This alternative represents adding another decision node after demand for the product is known. Example 18.8 details this.

More generally, in later years it is possible to expand, contract, or modify operations to increase revenues and decrease costs. Some alternatives have more flexibility for these changes, and those alternatives will often be far more valuable than a simple expected value analysis might suggest. A detailed decision tree that shows how the alternatives' performance can be improved will also allow the calculation of more accurate expected values.

Example 18.8   ElectroToys May Discontinue Production

ElectroToys' potential new product has an expected first cost for development of $34,000. If net revenues are negative, the product will be discontinued after the first year. Calculate the expected present worth of the revenues and of the product. How much does the optional early termination improve the product's PW?

Solution

The decision tree for the revenues is shown in Exhibit 18.4. In finding the expected net revenue, it is not possible to apply an *average* life, since the number of years is different for the early termination option. The early termination has a probability of .3 and a loss of $5K for 1 year, and the other two outcomes require a (P/A) factor for 5 years.

$$E_{\text{net revenue}} = .3(-5K)/1.12 + [.5(10K) + .2(30K)](P/A, .12, 5)$$
$$= -1339 + 11K \cdot 3.605 = \$38,316$$

Subtracting the expected first cost of $34K leaves an expected profit of $4316, which has a present worth that is $4071 higher than the $245 for accepting the toy without the early termination option.

Exhibit 18.4   A partial decision tree—ElectroToys

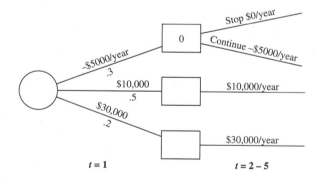

Cash Flows that Occur over Time.   For engineering economy problems it is also useful to line up events and actions by time periods, so that a labeling of the cash flow timings can be aligned with the tree. Example 18.10 does the rollback for the decision tree that is developed in Example 18.9.

Example 18.9   MedEQuipt New Product Development Tree

MedEQuipt has identified a need for a new blood analysis test unit. However, it will be used only in some infertility clinics, so the potential market is relatively small. The first decision between alternatives is whether to (1) forgo the project, (2) begin normal development, or (3) accelerate development. If MedEQuipt forgoes the project, no costs or profits result. Normal development will take a year, cost $450,000, and almost certainly lead to a successful project. Thus, there is a 95% chance of meeting a market need that should produce profits of $100,000 per year for 10 years. These revenues begin in the second year, and they are assumed to occur at the end of each year.

The accelerated program skips some steps, so it only costs $250,000. Since it is done earlier, there are operating profits of $50,000 for the first year. Since a larger market share is established, the profits for the 10 years beginning in year 2 are larger, $125,000 per year. Unfortunately, the accelerated program has a 40% chance of failure. Assume the costs occur at time 0, and that the actions of MedEQuipt's competitors will close the market niche if the development alternative is unsuccessful. Describe this problem using a decision tree.

Solution
The decision has three alternatives. Forgoing development has a sure outcome of $0, each of the other two has costs and leads to a single chance node. While the chance branches represent success or failure, the probabilities and consequences following normal and accelerated development differ. The decision tree in Exhibit 18.5 has three branches from the decision node 1 (one no go and two go), and each of the two go branches have chance nodes (nodes 2 and 3) with two branches apiece.

Exhibit 18.5   MedEQuipt decision tree

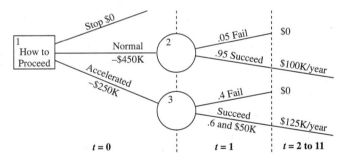

Notice that years are included for each cash flow, since the value of the cash flows depends on when they occur (if they occur at all). Notice also that the probabilities from a chance node must sum to 1 (see the axioms in Section 18.1).

Example 18.10   Calculating EVs for MedEQuipt Decision Tree

If MedEQuipt (in Example 18.9) uses a 12% minimum attractive rate of return and maximizes present worth, which alternative should be adopted?

Solution

Expected value analysis, or rolling back, begins at the nodes furthest out in Exhibit 18.5. In this case, these are the nodes for the expected return from normal development (node 2) and from accelerated development (node 3). The results are summarized in Exhibit 18.6 and explained in the following text.

Exhibit 18.6   Completed MedEQuipt decision tree

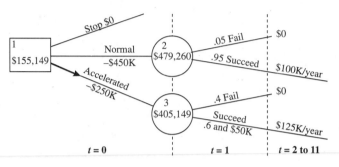

The expected value for normal development is calculated as a present value, as follows:

$$E_{\text{normal returns}} = .05(0) + .95(100, 000)(P/A, .12, 10)(P/F, .12, 1)$$
$$= 0 + .95(504,484) = \$479,260$$

If MedEQuipt were to select the branch for normal development, costs of $450,000 at time 0 are incurred and the annual returns have an expected present worth of $479,260. The expected net present worth for normal development is $479,260 minus $450,000, or $29,260.

If MedEQuipt were to select accelerated development, the probabilities differ and the potential $50,000 in revenue for year 1 must be included, but the principles are identical. To simplify the formulas, the present worth of the returns are first calculated:

$$PW_{\text{accelerated returns}} = (P/F, .12, 1)[50,000 + 125,000(P/A, .12, 10)]$$
$$= (50,000 + 706,278)/1.12 = \$675,248$$

Now it is easy to calculate the expected value for node 3:

$$E_{\text{accelerated returns}} = .4(0) + .6(675,248) = \$405,149$$

If MedEQuipt selects accelerated development, costs of $250,000 at time 0 are incurred and returns have an expected present worth of $405,149. The expected net present worth for accelerated development is $405,149 minus $250,000, or $155,149.

Since accelerating is the best alternative, this last value is entered into node 1. Notice that the branch for the accelerated branch is bold, since that is the best decision. Chance branches are not bold, since it is not possible to control the states of nature.

18.5 RISK

In an intuitive sense, risk is the chance of getting an outcome other than the expected value. Quantitatively, risk can be described as the probability of certain outcomes, such as the chance of a loss. In Exhibit 18.2, for collision insurance on an auto, self-insurance risks a $9000 loss that has a 2% probability. For a given expected return, firms prefer to minimize risk, where possible.

Usually risk is measured as the amount of dispersion about the expected value. While there are a variety of measures of risk, the most common measure is the **standard deviation** (σ) of the outcomes. The standard deviation begins by calculating the dispersion—the difference between the possible outcomes and the average, $X - \mu$. This is squared for each outcome, so that both positive and negative deviations receive positive weights. The expected value, or weighted average, of these deviations is then calculated. Finally, the square root is found. Thus, the result is the square root of the weighted average of the squared deviations from the mean, or $\sqrt{E[(X - \mu)^2]}$.

Standard deviation is the common measure of risk.

Now this is not the formula for the standard deviation built into most calculators. One difference is that most calculator formulas fit equally likely data points from a randomly drawn sample, so that 1 divided by the number of data points becomes the probability for all. Here we will use a weighted average for the squared deviations since the outcomes are not equally likely.

The **standard deviation** (σ) is the square root of *the average of the squares minus the square of the average*, or $\sqrt{E(X^2) - [E(X)]^2}$.

The second difference is that for calculations (by hand or the calculator), it is easier to use the following equivalent formula:

$$\text{Standard deviation } (\sigma) = \sqrt{E(X^2) - [E(X)]^2} \qquad (18.2)$$

$$= \sqrt{\sum_i x_i^2 \cdot P(x_i) - \left[\sum_i x_i \cdot P(x_i)\right]^2} \qquad (18.2')$$

In words, this is the square root of the difference between the average of the squares and the square of the average. The mathematics that prove that these two formulas are equivalent to $\sqrt{E[(X - \mu)^2]}$ can be found in virtually any probability and statistics text. Omitting the square root or squaring the standard deviation calculates the *variance*. The standard deviation is preferred since it is measured in the same units as the expected value. The variance is measured in "squared dollars"—whatever they are.

The variance and standard deviation are both based on an averaging of squared deviations from the mean. Consequently, negative values are impossible, and they are clear indicators of arithmetic mistakes. The variance and the standard deviation equal 0 only when there is a single certain value, which becomes the expected value. Otherwise, the variance and standard deviation are positive.

The calculation of a standard deviation by itself is only a descriptive statistic of questionable value. However, as shown in the next section on risk/return trade-offs, when the standard deviation of each alternative is calculated and these are compared, then it is useful. But first, some examples of calculating the standard deviation.

Tables are used for calculating $E(X)$ and $E(X^2)$ in Examples 18.12 and 18.13. These are easily constructed as spreadsheets or by using the table arithmetic available in some word processors.

Example 18.11   Standard Deviation for Cost of Construction
(Example 18.1 continued)

The potential costs of construction (first costs) under the three states of nature were $225,000, $250,000, and $295,000, with probabilities of .2, .5, and .3, respectively. At the beginning of Section 18.2, the expected value or mean of this distribution was calculated to be $258,500. What is the standard deviation of the construction cost?

Solution
First, each first cost is squared to calculate the average of the squared first costs, $E(\text{first cost}^2)$. Each squared value is weighted by its probability, so the calculation is as follows:

$$E_{\text{first cost}^2} = .2(225,000^2) + .5(250,000^2) + .3(295,000^2)$$
$$= 6.74825 \cdot 10^{10}$$

Then the standard deviation can be found, as follows:

$$\sigma_{\text{first cost}} = \sqrt{6.74825 \cdot 10^{10} - 258,500^2} = \sqrt{660,250,000}$$
$$= \$25,695$$

Example 18.12   Standard Deviation for Crash Absorber Life
(Examples 18.2 and 18.4 continued)

In Example 18.4, the expected value was calculated to be 2.60 years. What is the standard deviation?

Solution
Once the weighted average of the squared years has been calculated to be 9.07 (see accompanying table), then the calculation of the standard deviation is straightforward.

Year^2	$P(\text{year})$	Product
1	.333	.33
4	.222	.89
9	.148	1.33
16	.0988	1.58
25	.1975	4.94
	E_{year^2}	9.07

$$\sigma_{\text{year}} = \sqrt{9.07 - 2.60^2} = 1.52 \text{ years}$$

Example 18.13   Standard Deviation for Micro Pizza Heater's Market Demand
(Examples 18.3 and 18.5 continued)

In Example 18.5 the $E_{\text{annual revenue}}$ was calculated to be $278,873. What is the standard deviation of that annual revenue?

Solution

Once the expected value of the squared equivalent annual revenue has been calculated to be 81,111,101,290 (see accompanying table), the standard deviation is calculated as follows:

$$\sigma = \sqrt{81,111,101,290 - 278,873^2} = \$57,801.$$

Equivalent Annual Revenue	(Equivalent Annual Revenue)^2	Probability	Product
$200,000	40,000,000,000	.3	12,000,000,000
300,000	90,000,000,000	.6	54,000,000,000
388,730	151,111,012,900	.1	15,111,101,290
	Expected value [Equivalent annual revenue]^2		81,111,101,290

18.6 RISK/RETURN TRADEOFFS

Alternatives with higher risk are generally accepted only when higher returns are also expected. These risks may be that a project or an investment loses money, or it may involve bankruptcy, a firing, or death. It is clear that most people and most businesses are risk-averse.

Buying Insurance.   One example of a risk/return tradeoff can be found in the insurance business. An individual or a firm buys insurance to minimize or eliminate the risk of certain kinds of disaster. Owners of homes and vehicles buy liability, fire, theft, and/or collision insurance. Businesses buy liability, fire, and key person insurance. In every case, the costs of operation for the insurance companies and their stockholder dividends ensure that what is paid out is less than what is taken in. Often, less than half of the premiums received are paid out in claims. Thus, for you or the insured firm, the expected value of what is received may only be half of what is paid in.

Nevertheless, buying insurance can be an intelligent decision. The death of a key individual at a critical time can cause a firm to go bankrupt. Similarly, a fire can bankrupt a firm or a family. The small certain cost of insurance is often better than a possible catastrophic uninsured loss.

Balancing Risks and Returns.   Sometimes the potential loss is not catastrophic, and increases in risk must be balanced with increases in expected return. Are the increases commensurate? Basically, firms want to accept projects that have (1) a large expected value for the present worth or equivalent annual worth, and (2) a small standard deviation.

The starting point is to calculate the expected values and standard deviations for each alternative. Occasionally an alternative will be **dominant**, as it will have both a higher expected value and a lower risk than any other alternatives. More often one of the alternatives will be **dominated;** that is, at least one other alter-

A **dominated** alternative is worse than another alternative because it has a lower expected PW and a higher risk or standard deviation. A **dominant** alternative has a higher expected PW and a lower risk or standard deviation than all other alternatives.

native will have a better expected value and a lower risk. However, as shown in Example 18.14, sometimes each alternative is best—for some level of risk.

Example 18.14   Calculating Risk/Return Tradeoffs for MedEQuipt (Examples 18.9 and 18.10 continued)

If MedEQuipt (in Examples 18.9 and 18.10) maximizes present worth and minimizes risk, which alternative should be adopted?

Solution

The decision tree with expected values can be used here as well (Exhibit 18.7 repeats 18.6, with standard deviations also shown).

Stop alternative. This alternative's expected value and standard deviation are both $0.

Normal development. The expected value for normal development was calculated as a present value as follows:

$$E_{\text{normal returns}} = .05(0) + .95(100,000)(P/A, .12, 10)(P/F, .12, 1)$$
$$= 0 + .95(504,484) = \$479,260.$$

$$E_{\text{PW normal}} = \$479,260 - \$450,000, \text{ or } \$29,260.$$

To calculate the standard deviation, the first step is to identify the possible outcomes and their probabilities. That is the probability distribution for the PW under normal development.

$29260 = 0.05 \, (-450) + 0.95 \, x$

Probabilities	Normal Present Worth
.05	−$450,000
.95	54,484

This probability distribution still has an expected value of $29,260. The second step is to calculate the expected value of the squares and the standard deviation.

$$E(\text{PW}^2) = .05 \cdot (-450,000)^2 + .95 \cdot 54,484^2 = 1.2945 \cdot 10^{10}$$

$$\sigma_{\text{PW}} = \sqrt{1.2945 \cdot 10^{10} - 29,260^2} = \$109,949$$

Exhibit 18.7   MedEQuipt decision tree with standard deviations

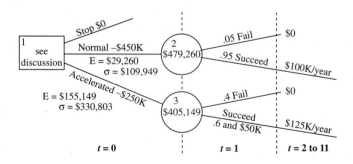

Accelerated Development. The expected present worth under accelerated development was calculated as follows:

$$PW_{\text{accelerated returns}} = (P/F,.12,1)[50{,}000 + 125{,}000(P/A,.12,10)]$$
$$= (50{,}000 + 706{,}278)/1.12 = \$675{,}248$$

$$E_{\text{accelerated returns}} = .4(0) + .6(675{,}248) = \$405{,}149.$$
$$E_{\text{PW}} = \$405{,}149 - \$250{,}000 = \$155{,}149.$$

To calculate the standard deviation, the first step is to identify the possible outcomes and their probabilities. That is the probability distribution for the PW under accelerated development.

Probabilities	Accelerated PW
.4	−$250,000
.6	425,248

This probability distribution still has an expected value of $155,149. The second step is to calculate the expected value of the squares and the standard deviation.

$$E(PW^2) = .4 \cdot (-250{,}000)^2 + .6 \cdot 425{,}248^2 = 1.3350 \cdot 10^{11}$$

$$\sigma_{\text{PW}} = \sqrt{1.3350 \cdot 10^{11} - 155{,}149^2} = \$330{,}803$$

Comparing the Alternatives. As we can see in the following table, none of the three alternatives is dominant. When the three are compared, an increasing level of return is associated with more risk.

Alternative	EV_PW	σ_PW
Stop	$ 0	$ 0
Normal development	29,260	109,949
Accelerated development	155,149	330,803

In this case, tripling the risk increases the return more than fivefold, so it is likely that accelerated development is still the best alternative. However, a rigorous treatment of the tradeoff question is beyond the scope of this text.

Efficient Frontier for Risk/Return Tradeoffs.   Dealing with tradeoffs where no alternative is dominant has been the focus of much research in engineering economy and finance. Before we very briefly discuss some of the possibilities, let us consider Exhibit 18.8, which shows a conceptual model of the tradeoffs between risk and average return. Since expected return is on the y-axis and the standard deviation is on the x axis, a dominant alternative is one that is plotted above and to the left of another alternative. Thus, alternative B dominates alternative D, as alternative B has both a higher expected return and a lower standard deviation.

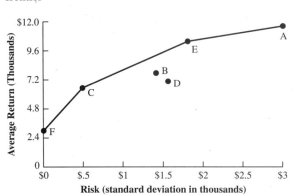

Exhibit 18.8   Risk/return tradeoff and the efficient frontier

Point F in Exhibit 18.8 represents the risk-free rate of return on government securities, where the standard deviation is $0. This risk-free rate is available for investment of any temporarily excess funds the firm has.

Alternatives F, C, E, and A are not dominated. Moreover, if linear combinations of these alternatives could be formed, such as in a stock portfolio, then the line that connects alternatives F, C, E, and A would form an *efficient frontier* for the feasible region of (risk, return) combinations. Alternative B and combinations such as F and A or D and E are within the linear envelope of these points.

Approaches to Risk/Return Tradeoffs.   Choosing among alternatives A, C, E, and F in a rigorous way is beyond us at this point. However, there are at least three useful approaches. The first treats risk and return as multiple criteria, and it requires the tools of Chapter 19. The second approach is that of capital asset pricing theory [see Bussey and Eschenbach or Levy and Sarnat].

The third approach is utility theory. This can be illustrated by asking whether you would rather receive a sure $1 million in a lottery or have a 50% chance of receiving $5 million. Even though a 50% chance of having $5 million has an expected value of $2.5 million, virtually all of us would choose the sure $1 million. This can be explained by noting that the utility of the first $1 million to us is larger than the utility of the next $4 million.

While the second and third approaches are theoretically preferable, there are substantial problems in applying them to the typical engineering economy problem. Consequently, many decisions are made using unquantified tradeoffs between risk and return. The expected value of each alternative is calculated, along with a measure of risk, such as the standard deviation. Then a decision maker chooses the maximum expected value within the level of risk considered acceptable for that return. Alternatively, the expected value and standard deviation are considered using the tools of Chapter 19.

18.7 PROBABILITY DISTRIBUTIONS FOR PW AND SIMULATION

So far, this chapter has considered probability distributions, expected values, and standard deviations for one variable at a time. The approach can be extended to find the expected present worth or expected equivalent annual cost for several independent variables. If many variables or possible values must be considered, then simulation (described at the end of this section) is a more appropriate approach.

Probability Distributions with Multiple Independent Variables.   Under the assumption of independent probability distributions with a few values for a few variables, the different combinations can be evaluated. Each combination has a probability, a value for each variable, and a total economic worth. Not only does this approach support expected value decision making, but, as shown in Example 18.15, it also provides a probability distribution function for present worth. This probability distribution can be used to calculate the risk associated with an alternative.

In Section 18.3, expected values for each cash flow element were calculated. In Examples 18.6 and 18.7, the alternative with the best expected present worth or equivalent annual cost was selected. Those expected values were exact since the probabilistic cash flows did not occur within an engineering economy factor. However, the probability distributions were not derived, so it was not possible to calculate a measure of risk.

Example 18.15   Probability Distribution for PW for ElectroToys' New Product Development (Example 18.6 continued)

ElectroToys' new product idea has an uncertain first cost for development and uncertain annual net revenues, as shown in the accompanying table. Assume that the first cost occurs at time 0, and that the net annual revenues begin at the end of year 1. ElectroToys uses an interest rate of 12%, and new toys have a life of 5 years. What is the probability distribution for the present worth if the product is developed and if it is maintained for the full 5 years?

First Cost	P	Net Revenue	P
$20,000	.2	−$ 5,000	.3
30,000	.5	10,000	.5
50,000	.3	30,000	.2

Solution
The first step is to identify the nine combinations of first cost and net revenue. Each first cost is paired with each net revenue. Exhibit 18.9 summarizes this, along with the probability of each combination. Based on the assumption of statistical independence, the probability of the first cost is multiplied by the probability of the net revenue to find the probability of the combination. For example, the probability of the high first cost and the low revenue is $.3 \cdot .3 = .09$.

The last column of the table in Exhibit 18.9 is included to show that the original expected value of $245 that was calculated in Example 18.6 is indeed exact. These calculations are only valid if the probability distributions for the first cost and the net revenue are independent. The graph in Exhibit 18.9 lists the present worths in increasing order to make the probability distribution clearer.

Exhibit 18.9 shows the advantage of the greater detail in the probability distribution. The decision tree of Exhibit 18.4 correctly increased this example's PW by canceling the project after 1 year if net annual returns were negative, which had a probability of .3. Exhibit 18.9 shows that the probability of a negative PW is much higher at .45—or nearly 1 chance in 2. The .15 probability that a high first cost and an intermediate annual return will lead to a PW of −$13,952 cannot be avoided. Once the high first cost has occurred, it is a sunk cost. On the other hand, once the negative annual returns have been observed for year 1, they can be avoided for years 2 through 5.

Exhibit 18.9   Probability distribution of PW with two independent variables (ElectroToys' product development)

First Cost (Probability)	Annual Net Return (P)	Joint P	Present Worth	P·PW
$20,000 (.2)	$ −5000 (.3)	.06	−$38,024	−$2281
20,000 (.2)	10,000 (.5)	.10	16,048	1605
20,000 (.2)	30,000 (.2)	.04	88,143	3526
30,000 (.5)	−5000 (.3)	.15	−48,024	−7204
30,000 (.5)	10,000 (.5)	.25	6048	1512
30,000 (.5)	30,000 (.2)	.10	78,143	7814
50,000 (.3)	−5000 (.3)	.09	−68,024	−6122
50,000 (.3)	10,000 (.5)	.15	−13,952	−2093
50,000 (.3)	30,000 (.2)	.06	58,143	3489
			Expected value	$245

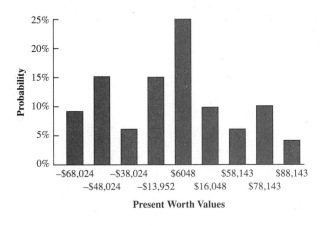

Present Worth Values

Example 18.16   Risk/Return for ElectroToys' New Product Development (Examples 18.6 and 18.15 continued)

In Example 18.6, the conclusion was that ElectroToys should develop the product since the expected present worth was $245 at its MARR of 12%. This value was confirmed by the more complete analysis shown in Example 18.15.

However, if risk is considered, is this still a clear choice?

Solution

The answer is no. Four of the combinations tabulated in Example 18.15 have negative present worths, and their combined probability is

$$.06 + .15 + .09 + .15 = .45$$

To calculate the standard deviation, we change the last column of the table in Exhibit 18.9. In Exhibit 18.10, that column is the joint probability times the square of the present worth. The result is that the average of the squares, $E(PW^2)$, equals 2,037,483,015.

Then the standard deviation of the present worth is calculated as

$$\sigma_{PW} = \sqrt{2,037,483,015 - 245^2} = \$45,138$$

The value of this project is unclear. Does this level of risk exceed the risk of most projects that ElectroToys undertakes? We cannot know without more data.

Exhibit 18.10 Standard deviation of PW for ElectroToys (Example 18.6)

First Cost (Probability)	Annual Net Return (P)	Joint P	Present Worth	P·PW²
$20,000 (.2)	$ −5000 (.3)	.06	−$38,024	86,749,475
20,000 (.2)	10,000 (.5)	.10	16,048	25,753,830
20,000 (.2)	30,000 (.2)	.04	88,143	310,767,538
30,000 (.5)	−5000 (.3)	.15	−48,024	345,945,686
30,000 (.5)	10,000 (.5)	.25	6048	9,144,576
30,000 (.5)	30,000 (.2)	.10	78,143	610,632,845
50,000 (.3)	−5000 (.3)	.09	−68,024	416,453,812
50,000 (.3)	10,000 (.5)	.15	−13,952	29,198,746
50,000 (.3)	30,000 (.2)	.06	58,143	202,836,507
			Expected Value	2,037,483,015

***N* and *i* Need Complete Distributions for Exact Answers.** In Section 18.3, expected values were calculated for each cash flow element, then those values were used to calculate the expected present worth or expected equivalent annual cost.

Example 18.17 shows that this answer is not exact if N is the probabilistic cash flow element. A similar example could be constructed for i. The expected value operation is exact for those cash flow elements falling outside of the engineering economy factors. For example, present worth is a linear function of first cost or salvage value. However, it is not linear in i, N, or geometric gradients.

In Example 18.17, the uncertainty in N is even symmetric. However, the chance of a shorter life has more impact on the present worth than the chance of a longer life. This is due to *discounting* by the interest rate to find the time value of money.

Example 18.17   Uncertain *N* Requires Probability Distribution for PW for Exact Answer

A project has a first cost of $22,000, no salvage value, and a net annual benefit of $4000. The firm uses an interest rate of 12%. The benefits may last 5, 10, or 15 years, with respective probabilities of 20%, 60%, and 20%. What is the expected value of the present worth?

Solution

An approximate answer can be obtained by using the expected value for the number of years, 10 years. This is the best single estimate of the project's life. Since 5 and 15 years are equidistant from 10 years and they have the same probability, the probability distribution is symmetric, and the mean (average) is the point of symmetry. Mathematically, the average life of 10 years equals $.2 \cdot 5 + .6 \cdot 10 + .2 \cdot 15$. The present worth for this life is:

$$PW = -22,000 + 4000(P/A, 12\%, 10) = \$601$$

The exact answer is obtained by finding the present worth for each possible life, then finding the expected value, as shown in Exhibit 18.11.

Exhibit 18.11   Exact expected value computations with uncertain *N*

N	*P*	PW	*P*·PW	*P*·PW²
5	.2	−$7581	−$1516	11,494,312
10	.6	601	361	216,721
15	.2	5243	1049	5,497,810
		Expected values	−$106	17,208,843

$$\sigma_{PW} = \sqrt{17,208,843 - (-106^2)} = \$4147$$

The exact value of the expected PW is negative, and it indicates a different recommendation than the approximate value does; that is, that the project should not be undertaken. Notice also that the standard deviation of the PW is relatively large.

However, it should be noted that the approximation is not that far off. The $707 difference in calculated PWs is small relative to the uncertainty in the original values.

Continuous and Discrete Probability Distributions.   If distinct possibilities, such as optimistic, most likely, and pessimistic, have been identified and discrete probabilities assigned, the concepts in this chapter can be directly applied. However, some situations are better modeled with continuous probability distributions. Examples include time to complete a project, demand for a product, and cost of a construction project. Loosely put, the distinction is measuring rather than counting. There are families of probability distributions, such as the normal, the beta, and the lognormal, that can be applied (through integration rather than summation). However, this requires a deeper understanding of probability than this chapter assumes.

Another possibility is simply using a discrete approximation as a substitute for a continuous distribution. Then the question becomes how many categories or points to use. This chapter has typically used three estimated probabilities to describe uncertainty. While estimates that are derived from historical data may merit finer detail, that is certainly not true for subjectively estimated data. Much of the data in engineering economy, at least when generating minimum and maximum likely values, is highly subjective. Thus, you should not estimate probabilities for seven points, which implicitly suggest an unrealistically high state of knowledge and precision.

However, use of three- or five-point estimates (see Example 18.18), rather than a single point is highly recommended. The use of a single number suggests that the number is known with certainty, such as a count of 150,000 vehicles per year. The economic analysis that results, in turn, suggests that the project's present worth is far more certain than it really is. The use of multiple-point estimates also allows explicit calculations of a project's risk or uncertainty to accurately reflect significant probabilities that traffic counts may be as low as 114,600 or as high as 185,400.

In some cases, the project's technical feasibility may not work for some possible values. For example, the maximum traffic load might be 175,000 vehicles per year in Example 18.18. Then the economic analysis must consider the impact of this limit.

Example 18.18 Approximating a Continuous Probability Distribution with a Discrete Distribution

The annual traffic on a proposed new highway is uncertain. It is predicted to follow a normal distribution, with a mean of 150,000 vehicles per year and a standard deviation of 20,000 vehicles per year. How can this be approximated using three discrete values?

Solution
As shown in Exhibit 18.12, this normal distribution is continuous, with the values between 100,000 and 200,000 having the largest probabilities. Note that normal distributions are symmetric about their mean. They have 68% of their probability within ± 1 standard deviation of the mean, 95% within ± 1.96 standard deviations, and 99.7% within ± 3 standard deviations.

Exhibit 18.12   Normal distribution for predicted traffic volumes

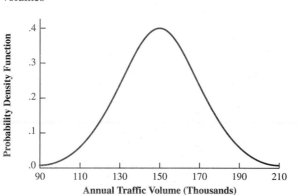

Assuming that the mean has a probability of 68% and that the remaining 32% is split evenly between the mean ±1.77 standard deviations leads to a discrete approximation with the same mean and standard deviation as that normal distribution. In this case, the result is:

$$\text{mean} = 150,000 \text{ has a } P = .68$$
$$\text{mean} - 1.77 \cdot \sigma = 150,000 - 35,400$$
$$= 114,600 \text{ has a } P = .16$$
$$\text{mean} + 1.77 \cdot \sigma = 150,000 + 35,400$$
$$= 185,400 \text{ has a } P = .16$$

Simulation.   If there are many cash flow elements that are uncertain, then computerized simulation is likely to be the best approach. Once a probability distribution is identified for each variable within our economic model, then numerous trials can be run to generate a distribution of the possible outcomes. This is usually computerized, and it is called simulation. Exhibit 18.13 illustrates simulation for a simple problem, where the first cost, annual return, and life all have their own probability density functions defined. In this case, 1000 iterations were run. For each iteration, three values were randomly selected based on each probability distribution, one each for first cost, annual return, and life. Then these were combined with the known discount rate of 8% to calculate the present worth. The 1000 values of present worth were then combined to produce Exhibit 18.13.

Exhibit 18.13   Simulation of present worth

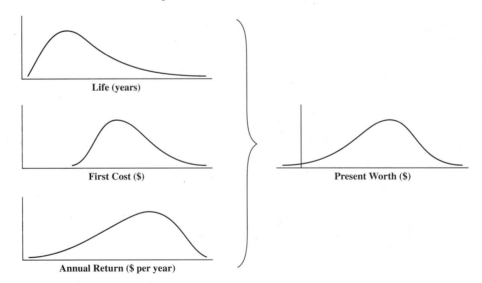

There are simulation packages such as @Risk® that can be used with spreadsheets. These simulation packages have two components. The first is a set of probability distribution functions. These functions are used to specify which cash flows are uncertain, and which probability distributions should be used. For example, a machine's life might be specified to follow a normal distribution, while the cost to build it might follow a lognormal distribution. The second component in simulation packages is the commands to specify which output values should be summarized, how many iterations should be run, etc.

If only a few variables and potential values must be considered, then the enumerative approach of this chapter can be used to produce the same results (any differences are the result of random sampling within simulation).

More complex calculations are possible. See, for example, the discussion of stochastic dominance based on means and variances in [Park and Sharp-Bette]. Also, the question of approaches for analyzing risk is discussed in [Cooper and Chapman].

18.8 SUMMARY

This chapter has briefly reviewed the application of probability to engineering economy problems. It has not attempted to rigorously develop the probabilistic concepts—leaving that for a course in probability and statistics. Rather, this chapter has described how to use the twin concepts of expected return and of risk as measured by the standard deviation.

This chapter has shown that the goal is to maximize expected return and to minimize the standard deviation. However, they both tend to increase together. Consequently, there is a tradeoff between low returns with low risk and high returns with higher risks.

This development has been applied both to simple problems with only three states of nature and to more complex problems with sequences of chance and decision nodes. These more complex problems are best developed using economic decision trees.

In closing, uncertainty and risk characterize the real world. Only in a textbook are the values known with certainty. Thus, these tools are absolutely essential for the effective application of engineering economy in the real world.

REFERENCES

Buck, James R., *Economic Risk Decisions in Engineering and Management,* 1989, Iowa State University Press.

Bussey, Lynn E., and Ted G. Eschenbach, *The Economic Analysis of Industrial Projects,* 2nd ed., 1992, Prentice-Hall.

Cooper, Dale, and Chris Chapman, *Risk Analysis for Large Projects*, 1987, Wiley.

Levy, Haim, and Marshall Sarnat, *Capital Investment and Financial Decisions,* 4th ed., 1990, Prentice-Hall International.

Park, Chan S., and Gunter P. Sharp-Bette, *Advanced Engineering Economics*, 1990, John Wiley.

PROBLEMS

18.1   Annual savings in labor costs due to an automation project have a most likely value of $35,000. The high estimate of $45,000 has a probability of .1, and the low estimate of $30,000 has a probability of .2. What is the probability distribution and the expected value for the annual savings?   (*Answer:* $35K)

18.2   Over the last 10 years, the hurdle or discount rate for projects from the Advanced Materials Division of SuperTech has been 15% three times, 25% twice, and 20% the rest of the time. There is no recognizable pattern. What is the probability distribution and the expected value for next year's discount rate?

18.3   A railroad between Fairbanks and the Brooks Range in Alaska will have a most likely construction cost of $900,000 per kilometer. If there is more permafrost than estimated (probability of .3), then the cost will increase by 40%. If there is less permafrost than estimated (probability of .2), then the cost will decrease by 25%. What is the probability distribution and the expected value for the first cost per kilometer?   (*Answer:* $963K/km)

18.4   Determine the average or expected grade if the instructor assigns 15% As, 35% Bs, 25% Cs, and 15% Ds. (Base your answer on a four-point scale, where A = 4.)

18.5   The financial vice president for the Memphis Mudcats Baseball Club is determining the expected revenues for this year. History indicates that the revenues depend on how the team finishes in the six-team Southeastern League. Determine the expected revenues for the ball club.

League Standing	Probability	Total Revenues[1]
1st	.30	$1,000,000
2nd	.20	850,000
3rd	.18	600,000
4th	.15	
5th	.10	

[1]Total revenues are $450K for 4th or lower.

18.6   The construction time for a highway project depends on weather conditions. It is expected to take 240 days if the weather is dry and the temperature is hot. If the weather is damp and cool, the project is expected to take 320 days. Otherwise, it is expected to take 275 days. The historical data suggests that the probability of cool, damp weather is 35% and that of dry, hot weather is 40%. What is the project's probability distribution and expected completion time?

18.7   You recently received a traffic ticket for a serious moving violation. If you have an accident or receive another moving violation within the next 3 years, you will become part of the "assigned risk" pool and you will pay an extra $500 per year for insurance. If the probability of an accident or moving violation is 20% per year, what is the probability distribution of your "extra" insurance payments over the next 4 years? Assume that insurance is purchased annually, and that violations register at the end of the year—just in time to affect next year's insurance premium.

18.8 If your interest rate is 8%, what is the expected value of the present worth of the "extra" insurance payments in Problem 18.7? (*Answer:* $440.6)

18.9 MoreTech uses a discount rate of 15% to evaluate engineering projects. Should the following project be undertaken if its life is 10 years and it has no salvage value?

First Cost	P	Net Revenue	P
$200,000	.1	$ 70,000	.3
300,000	.6	80,000	.6
500,000	.3	100,000	.1

(*Answer:* $46.5K, yes)

18.10 DOTPUFF (Department of Transportation, Public Utilities, and Fly Fishing) uses a discount rate of 10% to evaluate engineering projects. Should the following project be undertaken if its life is 10 years and it has no salvage value?

First Cost	P	Annual Benefit	P
$1,200,000	.1	$300,000	.3
1,800,000	.6	400,000	.6
2,700,000	.3	500,000	.1

18.11 Joe Doe, project engineer, has estimated the benefits and costs of a new product line, along with the possible states of nature and their probabilities. Using Joe's information, determine the expected values for each cash flow element. Find the expected PW at 8% if the product line's life is 10 years.

Element	P = .25	P = .50	P = .10	P = .15
First cost	$100,000	$125,000	$130,000	$150,000
Annual benefits	15,000	20,000	30,000	45,000
Annual costs	5000	4000	6000	3500
Salvage value	10,000	10,000	15,000	20,000

(*Answer:* $10.89K)

18.12   Determine the net present value of the following alternative. Use an interest rate of 8% and a life of 6 years.

Element	P = .10	P = .20	P = .30	P = .40
First cost	$10,000	$12,000	$15,000	$16,000
Annual benefits	4000	6000	8000	9500
Annual costs	2000	2000	3500	4500
Salvage value	500	750	'1000	1200

18.13   The ABC Company has just installed a heat exchanger at a cost of $80,000. Given the following estimates and probabilities for the yearly savings and useful life and assuming that the heat exchanger has no salvage value at the end of its useful life, determine the expected rate of return.

Savings per Year	Probability	Useful Life (Years)	Probability
$18,000	.15	12	1/6
20,000	.75	5	2/3
22,000	.10	4	1/6

18.14   The XYZ Company has just installed a new automated production line for $100,000. Estimates and probabilities for savings per year, useful life, and salvage value are given below. XYZ pays taxes at an effective 27% marginal rate. Determine the expected EAW of this purchase if the after-tax MARR is 7%.

Element	P = .30	P = .65	P = .05
Savings per year	$18,000	$22,000	$24,000
Useful life (years)	9	10	16
Salvage value	$10,000	$15,000	$17,000

a.   XYZ uses straight-line depreciation with a depreciable life of 10 years, and a depreciable salvage value of $10,000.
b.   XYZ uses MACRS depreciation.

18.15   The city water authority must choose among three options for a levee used to control local flooding. Regardless of which option is chosen, the levee will be replaced in 10 years. The first option is to completely rebuild the levee at a cost of $150,000. If completely rebuilt, the levee's chance of failure is 10% over the next 10 years. The second option is to repair the levee at a cost of $35,000. If this option is chosen, the levee will require another repair job in 5 years at a present cost of $27,750. If the levee is repaired, the chance of failure is estimated to be 28%. The third option is to do nothing to the levee. If nothing is done to the levee, the chance of failure is estimated to be 60%. Weighted by flood size and timing over the 10 years, the expected present cost of a levee failure is $400,000. Which option should be chosen?

18.16 In Problem 18.9, what if the project's life may be 5, 10, or 15 years with equal probabilities? What is the approximate expected value of the project's present worth? What is the exact value?   (*Answer:* exact = $24.4K)

18.17 Firerock Tire Company, a very large, profitable company, may install automated production equipment that will cost $110,000 and offer substantial savings in labor costs and increase productivity. The exact savings are not known, but the plant engineer has made the estimates tabulated below. If Firerock's after-tax MARR is 8%, determine the proposal's PW.

a. The equipment has a declared salvage value of $9200 and a depreciable life of 8 years. Firerock uses the sum-of-the-years' digits method for all depreciation.

b. Firerock uses MACRS depreciation.

Element	$P = .09$	$P = .16$	$P = .50$	$P = .16$	$P = .09$
Annual savings	$38,000	$45,000	$60,000	$68,000	$72,000
Annual cost	$20,000	$22,000	$26,000	$28,000	$32,000
Salvage value	$10,000	$12,000	$15,000	$18,000	$20,000
Useful life (years)	8	10	12	14	16

18.18 A new wood-forming machine will cost Wood-Chuckers $35,000. The machine is expected to last 4 years and have no salvage value. Wood-Chuckers' MARR is 12%. Using the information shown below, determine the risk associated with the purchase.

P	.25	.30	.45
Annual savings	$10,000	$11,000	$13,000

18.19 Bellville Electric Company may invest in new power-switching equipment. The net present worth of the three possible outcomes are $3430, $5870, and $6210. The probabilities of each outcome are .50, .30, and .20, respectively. Calculate the expected return and risk associated with this proposal.   (*Answer:* E_{PW} = $4718, σ_{PW} = $1293)

18.20 Flextire Manufacturing is considering two mutually exclusive proposals. Each will cost $80,000 and will last 6 years. Cash flows and estimated probabilities are presented below. Based on a MARR of 10%, determine the expected return and risk associated with each proposal. If Flextire is risk-averse, which proposal might it choose?

Proposal A		Proposal B	
Benefits per year	**Probability**	**Benefits per year**	**Probability**
$18,000	.25	$17,500	.40
20,000	.60	20,500	.35
23,000	.15	23,000	.25

18.21   What is your risk associated with Problem 18.8?

18.22   What is MoreTech's risk associated with Problem 18.9?   (*Answer:* $\sigma_{PW} = \$110.6K$)

18.23   What is DOTPUFF's risk associated with Problem 18.10?

18.24   The values given below are present worths. What decision(s) should be made and what is the expected value?

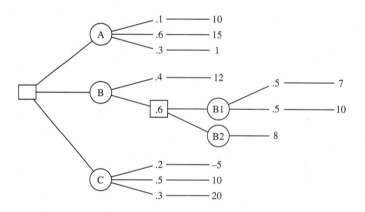

(*Answer:* $E_A = \$10.3$ is best)

18.25   What are the risks of actions A, B/B1, B/B2, and C in Problem 18.24? Graph the four actions on a risk/return plot. Are any dominated?   (*Answer:* $\sigma_A = \$6.24$, A dominates C, B/Bl has best risk/return tradeoff)

18.26   Use an economic decision tree to describe a problem that you are familiar with.

18.27   A new product's chief uncertainty is its annual net revenue. So far, $23K has been spent on development, but an additional $30K is required to license a patent. The firm's interest rate is 10%. What is the expected PW for deciding whether or not to proceed? Use an economic decision tree.

	State		
	Bad	**OK**	**Great**
Probability	.4	.5	.1
Net revenue	−$3K	$10K	$25K
Life (years)	5	5	10

18.28 In Problem 18.27, how much is it worth to the firm to terminate the product after 1 year if the net revenues are negative? Use an economic decision tree.

18.29 In Problem 18.27, describe the risk to the firm in proceeding with the product.
   a. Find both the probability of a negative PW and the standard deviation of the present worth.
   b. How much do the answers change if the early termination in Problem 18.28 is allowed?

18.30 Use an economic decision tree to find the probability distribution and the expected PW for an assembly line modification. The first cost is $80K, and its salvage value is $0. The firm's interest rate is 9%. The savings shown below depend on whether the assembly line runs one, two, or three shifts, and on whether the product is made for 3 or 5 years.

Savings per Year	Probability	Useful Life (Years)	Probability
$15,000	.2	3	.6
30,000	.5	5	.4
45,000	.3		

18.31 In Problem 18.30, how much is it worth to the firm to be able to extend the product's life by 3 years, at a cost of $50K, at the end of the product's initial useful life? Use an economic decision tree.

18.32 In Problem 18.30, describe the risk to the firm in proceeding.
   a. Find both the probability of a negative PW and the standard deviation of the present worth.
   b. How much do the answers change if the possible life extension in Problem 18.31 is allowed?

CHAPTER 19   MULTIPLE OBJECTIVES

The Situation and the Solution

Some consequences are difficult to state in terms of cash flows. Saving a life, improving product quality, being the market leader, reducing variability, and maintaining flexibility for an uncertain future are all linked to future cash flows. These are objectives on which alternatives must be evaluated.

Rarely does a single alternative achieve the best outcome for all objectives; thus, tools for balancing multiple objectives are required. Graphical techniques can be used to summarize evaluations, additive models can be used to weigh objectives, and hierarchical trees can be used to structure more complex problems.

Chapter Objectives

After you have read and studied the sections of this chapter, you should be able to:

Section 19.1
Define and identify objectives and attributes.

Section 19.2
Describe the process of multiattribute evaluation.

Section 19.3
Describe the rules for selecting attributes.

Section 19.4
Provide examples of evaluating alternatives on the attributes.

Section 19.5
Summarize graphical techniques.

Section 19.6
Construct numerical scales for performance evaluations.

Section 19.7
Create a single value for each alternative through the use of additive weighting.

Section 19.8*
Create a hierarchy of attributes and weight the attributes and the outcomes.

Key Words and Concepts

Objective A goal or desired outcome, such as maximizing present worth, minimizing risk, or maximizing quality.

Attribute A measure of achievement of an objective, such as present worth, the IRR's standard deviation, or the number of defects per million units.

Dominant A term that describes an alternative that is as good or better than all others on all attributes and better on at least one attribute.

Tangible Outcome or consequence that can be measured, but not in dollars.

Intangible Outcome or consequence that is a yes/no or subjectively estimated.

Mutually exclusive attributes These measure different or independent aspects of a problem.

Collectively exhaustive attributes Attributes that include all important aspects of a problem.

Dominated A term that describes an alternative that is inferior to another alternative on all attributes.

Decision rule A standard for selecting an alternative, such as dominance, satisficing, or maximizing a weighted sum.

Satisficing A decision rule that emphasizes meeting the minimal requirements on all attributes—that is, being satisfactory.

19.1 MULTIPLE ATTRIBUTES

Definitions and Tradeoffs. Real projects have multiple **objectives**. Each alternative is assessed by an **attribute** for each objective. If an alternative is as good or better on all attributes and better on at least one, then it is the best choice; it is a **dominant** alternative. This almost never happens. Usually, each alternative is highly rated on some, but not all, attributes. Choosing an alternative requires tradeoffs.

An **objective** is a desired outcome; an **attribute** is a measure of an objective's achievement; and a **dominant** alternative is as good or better on all attributes and better on at least one.

*The example in section 19.8 is based on material that was developed by Drs. Nancy Mills and Paul McCright.

Some example tradeoffs include balancing reliability improvement vs. cost reduction, automation vs. employment, risk vs. average return, fast but uneven growth vs. steady growth, and pollution control vs. economic efficiency.

Example 19.1 details a set of objectives, attributes, and tradeoffs. This example will be developed further in later sections.

Example 19.1   Harold's First Engineering Job—Objectives and Attributes

Harold will be graduating soon with a B.S. in metallurgical engineering. Describe what his objectives and their attributes might be as he considers possible job offers.

Solution
His objectives might be (1) to maximize his salary, (2) to maximize his professional growth, and (3) to maximize his "fun out of life."

Attributes that could be used to measure or describe his achievement of each objective could be (1) salary, (2) promotion outlook or quality of intern/training programs, and (3) his city's lifestyle rating.

Tradeoffs are likely. He may have to accept a lower initial salary if he accepts the job with the most potential. And the job with the best location for enjoying life may not be best on the other two attributes.

Attribute Categories   An alternative's consequences often include three categories of attributes:

1. Cash flows that are measured in dollars.
2. Tangibles that are quantified, but not in dollars.
3. Intangibles that are subjectively estimated.

Attributes that are measured in dollars are evaluated using the time value of money (see Chapters 1–17). Possible measures include present worth, equivalent annual cost, internal rate of return, and benefit/cost ratios. If probability distributions are given, then expected values are used (see Chapter 18).

The second category of attributes includes decibels of noise, lives saved, days from order to shipment, probability of failure, or number of defects per million—all of which can be numerically quantified, but not in dollars. These measurable attributes are called **tangibles**. Risk, when measured by the standard deviation (see Chapter 18), is a tangible.

Tangibles are attributes that are numerically quantified, but not in dollars.

The third category includes those attributes that cannot be quantified, such as fit with the organization's strategic mission or capabilities, potential for follow-on projects, unquantified risk of major problems, consequences of failure, and even the analyst's or decision maker's "gut feel" about the project. These attributes must be subjectively estimated or even reduced to a simple yes or no answer, and they are defined as **intangibles** or **irreducibles**.

Intangibles or **irreducibles** are attributes that cannot be measured; they must be subjectively estimated or answered yes/no.

Example 19.2 illustrates these three categories. The objectives of the solution to this example are important so often that they are illustrated in diagrams like Exhibit 19.1 [Park and Thuesen].

Exhibit 19.1   Three common objectives

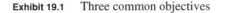

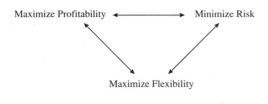

Maximize Profitability ⟷ Minimize Risk

Maximize Flexibility

Example 19.2   Flexible Automation vs. Hard Automation vs. People

Manufacturing MicroPizzaHeaters (a microwave oven with a new shape) requires a new assembly line. Three alternatives have been identified. One emphasizes robots, which can be reprogrammed for other tasks; the second emphasizes specially designed machines that operate very efficiently; and the third emphasizes people. Identify an objective for each category of attributes.

Solution
Profitability is an objective that would be measured in dollars using present worth or equivalent annual costs.

The alternative's risks would be measured in the standard deviation of the present worth or equivalent annual cost.

The flexibility of each alternative would be subjectively estimated, since it is impossible to measure. The flexibility of people to learn new tasks exceeds that of robots, and both have far more flexibility than machines designed to perform one task efficiently.

Selecting Multiple Objectives   As soon as objectives are expanded beyond maximizing economic value, the following questions arise. How are attributes chosen? Whose subjective judgments are used for intangible attributes? How are alternatives to be judged (acceptable/not acceptable, rank from best to worst, or numerically rated on a scale from 1 to 10)?

When selecting objectives or attributes, they should be as independent (or **mutually exclusive**) as possible and include as many of the important problem aspects as possible (or **collectively exhaustive**). It is critical that dollars, tangibles, and intangibles all be appropriately considered. The intangibles should include positive answers to two ethical questions: What is fair to all concerned? and, What is the right thing to do?

While most engineers consider subjective data less desirable than numerical data, most accept that subjective data is better than no data at all. If subjective judgments are made hastily, then the dangers of low validity are acute. However, subjective estimates made after careful analysis merit full credibility.

Example 19.3 illustrates why it is financially important to include tangible and intangible attributes that are difficult to quantify in dollar terms. Ignoring substantial benefits like these and failing to invest in better technology can ensure that your customers move to competing firms. On the other hand, poorly selected, poorly executed, and/or poorly timed projects have bankrupted firms. This chapter

Mutually exclusive attributes measure different or independent problem aspects, and attributes that are **collectively exhaustive** include all important problem aspects.

[see also Canada & Sullivan and Clemen] provides the tools to evaluate projects, so that the best ones are implemented.

Example 19.3   Multiple Objectives in Manufacturing

Often, the justification for using robots and computer-aided design, drafting, and manufacturing (CAD/CAM) systems relies on objectives that include those listed in Exhibit 19.2.

What other objectives might be important? Which of the potential payoffs are measured in dollars, tangibles, or intangibles? How are these objectives linked to improved cash flows for the firm?

Exhibit 19.2   Potential payoffs from implementing computer-integrated manufacturing

Reductions	
5–50%	Personnel costs
15–30%	Engineering design costs
30–60%	Work-in-process inventory
75–95%	Transaction volume for control
30–95%	Lead times
Gains	
40–70%	Production rate
200–300%	Capital equipment up-time
200–300%	Product quality
300–3500%	Engineering productivity

Solution

Unlisted objectives include using less floor space, improving product or firm reputation, developing knowledge about new technology, raising morale among engineers (for CAD), and increasing flexibility.

Only two categories in Exhibit 19.2, personnel and engineering design costs, are measured in dollars. The other payoffs in Exhibit 19.2 and the unlisted objective of using less floor space are tangibles that can be measured. The objectives of improving product or firm reputation, developing knowledge about new technology, raising morale among engineers, and increasing flexibility are usually treated as intangibles.

These tangible and intangible objectives are linked to the firm's bottom line, even though they are not measured in dollars. Using less floor space means faster movement of parts and people and better communications. Reducing work-in-process inventory reduces the need for working capital, the amount of space required, the risk of obsolescence, and the amount of rework. Reducing the transaction volume reduces the time spent on and the number of personnel required for processing paper, computer messages, etc.

Reducing lead times allows faster response and increases delivery-estimate reliability, both of which increase sales. If lead time can be reduced enough, then a make-to-order manufacturing policy with possible customization for each customer can replace a policy of make-to-stock and deliver from stock to customer.

Increasing the production rate increases revenues and material costs without increasing labor or capital costs. A 1% production rate increase may increase profits by 5%. Increasing the time that capital equipment is up or available for use increases production and may remove the need for additional machines.

Increasing product quality reduces rework or scrap costs for defective products, improves customer satisfaction, and may be necessary to remain in business. In a CAM system, for example, automated quality inspection may be 97% accurate, whereas human inspection under optimal conditions may be only 78% accurate.

Increasing engineering productivity reduces the number of engineers required and supports the design of better products that can be produced more cost-effectively. Knowledge about new technology can be used to define and evaluate future projects. Higher morale will increase efforts and creativity. An improved product or firm reputation increases sales. Larger profits *will* result as new manufacturing systems are designed to increase flexibility to achieve economies of scope and to respond to an uncertain future.

New manufacturing technologies can be evaluated with multiattribute techniques, but creative alternatives are sometimes better. For example, an "expensive" robot was being economically analyzed for the task of drilling a series of holes. There was a frame to guide the human operator, but too many holes were skipped during an 8-hour shift. A new alternative involving little expense solved the problem and the expensive robot project was shelved. Painting squares in bright colors around the guide holes on the frame proved much cheaper.

Multiple objectives exist for all large-scale projects. As illustrated in Exhibit 19.3 for manufacturing projects, larger projects have more impact on the firm, they depend more heavily on intangibles, and judgment plays a larger role. The larger the project, the more important the tools of this chapter. Similarly, deciding whether to proceed with a new dam with flood control, electrical power generation, and recreational benefits requires a multiple-objective approach, whereas selecting the roofing material for the new city hall may only need an EAC analysis.

Exhibit 19.3 Need for multiattribute evaluation for CIM

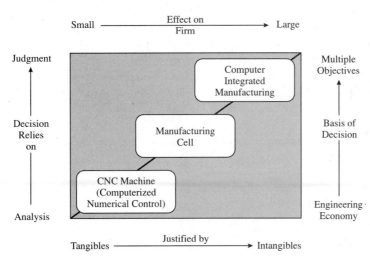

Summary.   This section has defined the basic terms; identified the attribute categories *dollar*, *tangible*, and *intangible*; and established the importance of considering multiple objectives. The next section details the *process* of multiattribute decision making, including the example of a graduating engineer who must accept a job offer. Later sections will detail each of these steps.

19.2 THE PROCESS OF EVALUATING MULTIPLE OBJECTIVES

The steps for evaluating multiple objectives are:

1.   Identify the objectives, attributes, and alternatives.
2.   Evaluate the alternatives on each attribute.
3.   Summarize the results and eliminate any dominated or unsatisfactory alternatives.
4.   Construct numerical scales for the evaluations.
5.   Rate the importance of the attributes.
6.   Use an additive weighted model to select the best alternative.

These steps are sequential, but they are repeated when new information or alternatives must be considered. Steps 5 and 6 can also be accomplished by constructing a hierarchy of attributes, as described in Section 19.8.

Eliminating Dominated and Unsatisfactory Alternatives.   Some problems, such as selecting a building site or a new car, have *too many* alternatives. Even though most projects do not have *that many* alternatives, decision makers or analysts still eliminate alternatives before the final set is evaluated. When many alternatives exist, it is often possible to simplify the problem in step 3 by eliminating **dominated** or unsatisfactory alternatives. An alternative that is inferior to another alternative on all attributes cannot be the best choice; neither can an alternative that fails to meet a constraint be the best choice.

A **dominated** alternative can be eliminated because another alternative is better on all attributes.

Example 19.4 revisits Example 19.1—Harold's first engineering job—to illustrate the process and definitions for a simple problem. Note that his best choice is not the best on *any* attribute. Rather, it achieves the best balance among all objectives. Also note that he eliminated a fourth alternative as dominated, since its salary, promotion outlook, and "city" attributes were all less attractive than San Francisco's. A fifth alternative was eliminated as unsatisfactory because it failed to meet a minimum salary constraint he had set.

Example 19.4   Harold's First Engineering Job—the Multiattribute Process (Based on Example 19.1)

Harold will be graduating soon with a B.S. in metallurgical engineering. He has been job hunting, and he has five job offers. Evaluate the five alternatives using multiattribute decision making.

Solution

1. The attributes he is considering are salary, promotion outlook, and city (see Example 19.1). The alternatives are National Motors in Detroit, American Containerships in San Francisco, National Aerospace in Seattle, Southern Oil in Galveston, and Midwest Forgings in Chicago.

2. On a scale of 1 to 10 (1 is bad; 10 is excellent), he has evaluated the alternatives on all three attributes. (Note: salaries have been adjusted for the local cost of living, since San Francisco is much more expensive and Seattle is somewhat more expensive than the national average.)

3. The Galveston job was dominated by the San Francisco job, since its salary, promotion outlook, and city ratings were 7, 8, and 7, respectively. Note that a dominated alternative can match, but not beat, the alternative it is dominated by on some attributes. The offered salary in Chicago was too low to meet Harold's minimal standard or constraint. The ratings for the remaining three alternatives are summarized below.

Attribute	Detroit	San Francisco	Seattle
Salary	10	7	9
Promotion outlook	5	10	9
City	3	8	6

4. Harold constructed his scales by judgment and assigned them numeric values in step 2. (This step has already been done.)

5. Harold rated the importance of the attributes by placing 60% of the weight on the salary and splitting the other 40% between the other two criteria.

Attribute	Weight
Salary	.60
Promotion outlook	.20
City	.20

6. For an additive model, Harold simply multiplies each performance evaluation by the attribute weight. In other words, each salary evaluation receives a weight of 60%, each evaluation of the promotion outlook receives a weight of 20%, and each evaluation of the city also receives a weight of 20%. These are added together to calculate a total value. Since the Seattle job has the highest additive weighting, it is the best fit for Harold's requirements.

$$\text{Detroit} = .6(10) + .2(5) + .2(3) = 7.6$$
$$\text{San Francisco} = .6(7) + .2(10) + .2(8) = 7.8$$
$$\text{Seattle} = .6(9) + .2(9) + .2(6) = 8.4$$

Decision Rules for Choosing the Best Alternative.   Example 19.4 can also be used to illustrate three possible **decision rules**: dominance, **satisficing**, and maximizing a weighted sum. Dominance is rarely useful in choosing the best alternative, because rarely is one alternative the best on all attributes. The term *satisficing* was developed to describe the behavior of managers and firms. It was found that often firms did not search for the best possible solution, but rather the first satisfactory solution was implemented.

These criteria consider the needs, wants, and "ignorables" that make up a problem. The *needs* define the satisfactory alternatives (those that "satisfice"). Performance of the alternatives on the *wants* attributes determines if any alternative dominates (the alternative is better on all wants). After prioritizing the importance of the wants and evaluating each alternative, then a weighted sum model combines all of the data into *one* number for each alternative. The *ignorables* are less important wants that are unlikely to influence the final choice.

None of the three alternatives listed in Example 19.4 are dominant. Detroit is better than San Francisco and Seattle on salary. San Francisco is better than Detroit and Seattle on promotion outlook and city. Seattle is better than San Francisco on salary and better than Detroit on promotion outlook and city. The fourth alternative, Galveston, was dominated by San Francisco.

In Example 19.4, the three listed alternatives are all satisfactory, since each would be acceptable if considered alone. There was a fifth alternative, which Harold eliminated as unsatisfactory due to a low salary. This alternative did not *satisfice*.

Given the variety of attributes and their unequal importance, the additive model is most useful. Choosing the largest value of 8.4 (Seattle), 7.8 (San Francisco), and 7.6 (Detroit) works. Keep in mind that some of the data may be *soft*, or *approximate*, that small differences are not significant, and that sensitivity analysis is warranted. An organized step-by-step evaluation procedure forces Harold to clarify what is important to him and to communicate that to others, such as his family.

19.3 IDENTIFYING THE ATTRIBUTES

The first step in multiple-objective problems is to identify the attributes that will be considered. This step should consider the following guidelines.

1. Attributes with the same value for every alternative can be ignored.
2. The attributes should be as independent as possible. Should any attributes that are not mutually exclusive be deleted?
3. The set of attributes should be complete. Should any be added or subdivided so that the set is collectively exhaustive?
4. The performance evaluations of the alternatives on each attribute should be meaningful.
5. The number of attributes to be compared should not be too large, or the problem becomes unwieldly. A common guideline is to have fewer than seven.

These guidelines will in themselves involve tradeoffs. For example, subdividing or adding attributes makes each attribute more specific and easier to evaluate, while also decreasing the independence of the attributes. An attribute may be included in the early stages simply to spark the creative process for an alternative that truly excels for that attribute. Example 19.5 applies these guidelines to Harold's first engineering job.

Example 19.5   Harold's First Engineering Job—Choosing the Attributes (Based on Example 19.4)

Apply the guidelines for attribute selection to Example 19.4.

Solution
The guidelines can be applied as follows.

1.   The potential attribute "job uses and builds on B.S. degree" should not be added since all of the existing alternatives require a B.S. in metallurgical engineering— unless another job possibility that does not require this degree must be considered.

2.   The second attribute, promotion outlook, should be adjusted, since it is at least partly a salary attribute.

3.   Perhaps a better attribute than promotion outlook would be how interesting the work is. This would also seem to make the set of attributes more exhaustive. The prospect for future promotion-based salary increases could be included in the salary attribute.
    The city attribute could be subdivided into the categories of climate, job prospects for fiancée, distance to family, atmosphere, etc., or one of the published ratings of livability could be used. The salary attribute could be decomposed into basic salary, benefits, bonuses, and cost of living (and perhaps expected increases). These subdivisions ensure that we consider each attribute more completely, but without the tools of Section 19.8, subdividing the salary attribute is more confusing than helpful.

4.   Harold was able to rate each attribute originally. Subdividing the city attribute could permit more accurate evaluations of the three alternatives on each new attribute. At the same time, the task of assigning priorities to the larger number of attributes would become more difficult.

5.   Three attributes are certainly not too many, and they seem to include the most important aspects of the problem.

These guidelines do not lead to single correct answers. Whether an attribute should be subdivided, added, or eliminated calls for the exercise of judgment. In particular, in selecting your first engineering job the attributes may be (or may have been) vastly or slightly different.

Example 19.6 applies the guidelines of attribute selection to the objectives of Exhibit 19.1.

Example 19.6   Examining Profitability, Risk, and Flexibility as Attributes (Based on Exhibit 19.1)

Exhibit 19.1 presented the three objectives of profitability, risk, and flexibility. These are examples of attributes that are, respectively, cash flows, tangibles, and intangibles. Discuss why these are often three of the most important attributes.

Solution

Maximizing profitability using PW, EAC, and IRR measures has been the focus of Chapters 1–17, so it will not be discussed further here.

Minimizing risk or variability was discussed extensively in Chapter 18. The common measure was the standard deviation of the PW, EAC, or IRR value used to measure profitability. This quantitative measure evaluates the outcomes of an alternative under an assumed set of probability distributions and states of nature.

Maximizing flexibility must be defined with care or it will overlap with the attribute of minimizing risk. The focus of this attribute must be an alternative's ability to accommodate unforeseen circumstances that were not part of the probability distributions used in evaluating risk.

For example, consider a building that could be built with a foundation and first three floors now, but would accommodate another two floors later. Building all five floors at once provides economies of scale, but unused space is very expensive. Building in two stages allows flexibility in timing and in the exact design of the upper two floors. A probability distribution for when the space will be needed and profitability estimates for the cost of building in stages are required to calculate the risk for each alternative. The flexibility measure would cover the ability to do it later or earlier or in a different way than originally anticipated.

19.4 EVALUATING THE ALTERNATIVES

Evaluating alternatives requires the acquisition of data and the exercise of judgment. For Harold's first job, the results have already been presented in Examples 19.1, 19.4, and 19.5. This section simply further develops (in Example 19.7) the staged building construction concept introduced in Example 19.6.

Summary tables, such as in Exhibit 19.4, are a significant decision-making aid as compared to paragraphs of discussion. This tabular "score card" simply reports the rating for each alternative on a variety of attributes—using either numbers and/or words. The score card succinctly summarizes any prior discussion and analysis. The evaluations are easy to interpret and to see all at once, and the score card visually supports the tradeoffs required in multiattribute decision making.

Example 19.7   LowTech Manufacturing's Staged Construction

LowTech Manufacturing is building a new office complex next to its manufacturing plant. There are two plans. The first is a five-floor complex, where only three floors are needed for the first 3 to 10 years. The top two floors would remain empty until needed by LowTech. The second is a three-story building whose foundation and first three floors would accommodate adding another two floors later.

Building in two stages reduces the costs now, but total costs will increase substantially. Building in two stages allows flexibility in timing and in the exact design of the upper two floors. A probability distribution for when the space will be needed and data for the profitability estimates are summarized in the table below.

To simplify the calculations, your boss has said to assume that the benefits of having the new space are $1M per floor for each year the floor is needed. LowTech management has identified an interest rate of 10% and an analysis period of 20 years.

	Cash flows		Expansion Built/Needed (year)	P(year)
	One Stage	**Two Stages**		
			3	.2
First cost	$20M	Year 0: $15M	5	.5
		Year 3, 5, or 10: $12M	10	.3
Salvage value	$5M	$6M		

Solution
First, find the PW for both alternatives with each timing of the second stage. With three floors in use, the benefits are $3M per year for the first 3, 5, or 10 years. With five floors in use, the benefits are $5M per year for the last 17, 15, or 10 years. The $E(PW)$ for each alternative is the weighted average (using probabilities of .2, .5, and .3) for using five floors in 3, 5, or 10 years.

$$PW_{\text{one stage, yr. 3}} = -20M + 5M(P/F, 10\%, 20) + 3M(P/A, 10\%, 3)$$
$$+ 5M(P/A, 10\%, 17)(P/F, 10\%, 3) = \$18.34M$$

$$PW_{\text{one stage, yr. 5}} = -20M + 5M(P/F, 10\%, 20) + 3M(P/A, 10\%, 5)$$
$$+ 5M(P/A, 10\%, 15)(P/F, 10\%, 5) = \$15.73M$$

$$PW_{\text{one stage, yr. 10}} = -20M + 5M(P/F, 10\%, 20) + 3M(P/A, 10\%, 10)$$
$$+ 5M(P/A, 10\%, 10)(P/F, 10\%, 10) = \$11.02M$$

$$E(PW_{\text{one stage}}) = .2(18.34M) + .5(15.73M) + .3(11.02M) = \$14.84M$$

$$PW_{\text{two stages, yr. 3}} = -15M + 6M(P/F, 10\%, 20) + 3M(P/A, 10\%, 3)$$
$$+ 5M(P/A, 10\%, 17)(P/F, 10\%, 3) - 12M(P/F, 10\%, 3) = \$14.47M$$

$$PW_{\text{two stages, yr. 5}} = -15M + 6M(P/F, 10\%, 20) + 3M(P/A, 10\%, 5)$$
$$+ 5M(P/A, 10\%, 15)(P/F, 10\%, 5) - 12M(P/F, 10\%, 5) = \$13.43M$$

$$PW_{\text{two stages, yr. 10}} = -15M + 6M(P/F, 10\%, 20) + 3M(P/A, 10\%, 10)$$
$$+ 5M(P/A, 10\%, 10)(P/F, 10\%, 10) - 12M(P/F, 10\%, 10) = \$11.54M$$

$$E(PW_{\text{two stages}}) = .2(14.47M) + .5(13.43M) + .3(11.54M) = \$13.07M$$

The standard deviation of the present worth for each alternative is found using the following table.

$$\sigma_{\text{one stage}} = \sqrt{E(\text{PW}^2_{\text{one}}) - E^2(\text{PW}_{\text{one}})}$$

$$= \sqrt{227.4 - 14.84^2} = \$2.7\text{M}$$

$$\sigma_{\text{two stage}} = \sqrt{E(\text{PW}^2_{\text{two}}) - E^2(\text{PW}_{\text{two}})}$$

$$= \sqrt{172 - 13.07^2} = \$1.07\text{M}$$

N	P	$\text{PW}_{\text{one stage}}$	$\text{PW}^2_{\text{one stage}}$	$P \cdot \text{PW}^2_{\text{one stage}}$	$\text{PW}_{\text{two stages}}$	$\text{PW}^2_{\text{two stages}}$	$P \cdot \text{PW}^2_{\text{two stages}}$
3	.2	\$18.34M	\$336.36	67.27	\$14.47M	\$209.38	41.88
5	.5	15.73M	247.43	123.72	13.43M	180.36	90.18
10	.3	11.02M	121.44	36.43	11.54M	133.17	39.95
	$E(\text{PW}) = \$14.84\text{M}$			$E(\text{PW}^2) = 227.42$	$E(\text{PW}) = \$13.07\text{M}$		$E(\text{PW}^2) = 172.01$

Exhibit 19.4 Summary table for Example 19.7: LowTech Manufacturing office complex

Attributes	One Stage	Two Stages
Profitability—$E(\text{PW})$	\$14.84M	\$13.07M
Risk—σ_{PW}	2.69M	1.07M
Flexibility	Below average	Above average

19.5 GRAPHICAL TECHNIQUES

Even if more sophisticated techniques will be used, simple graphical techniques illustrate the potential tradeoffs and provide a framework for checking the data and thoroughly understanding it before more mathematical and less intuitive calculations are made.

Shaded Circles and Squares. Exhibit 19.4 can be converted into the shaded-square table of Exhibit 19.5. The squares for quantitative attributes, such as present worth and the standard deviation, cannot have the numerical accuracy of tables like that in Exhibit 19.4. Nevertheless, the squares ease the task of "combining" attributes.

Each attribute must be translated into a scale—from poor to excellent or 0 to 10—and the scale values must be defined in terms of shaded squares. For the

Exhibit 19.5 Shaded circle diagram for Example 19.7: LowTech Manufacturing office complex

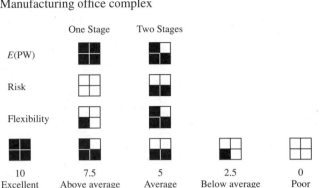

standard deviation in Exhibit 19.4, higher values are worse, so they are evaluated as "below average." Note: if multicolor printing is available to you, then red for good and black for bad can be used, as in *Consumer Reports*. Any alternative that is dominant will be red, and dominated alternatives will tend to be black. *Consumer Reports* provides a fine example of one advantage of shaded-circle diagrams. Suppose the analyst has data on 10 to 50 attributes, which is being provided to tens or thousands of decision makers. Any decision maker can individually select attributes (less than seven) to emphasize by highlighting columns.

Polar Graph. Exhibit 19.6 is a polar graph, where each arrow drawn from the center corresponds to an attribute. The scale can be poor to excellent or 0 to 10, but *worst* should be at the center and *best* at the arrow's end. Any attribute that can be measured or subjectively rated can be included, and different attributes can have different scales.

Exhibit 19.6 Polar graph for Example 19.7: LowTech Manufacturing office complex

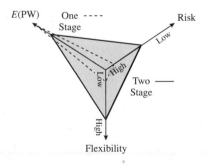

After the values are plotted on each arrow, those points are connected to form polyhedrons for each alternative. The larger the polyhedron, the better the

alternative. Strengths and weaknesses of all alternatives on all attributes are shown on a single graph.

The overall desirability of each alternative may be only in rough proportion to the areas of their respective polyhedrons, since axes ordering does influence the areas. When constructing polar graphs at least 3 attributes must be used (a practical upper limit is probably 8 to 10), and the spacing of the axes is usually symmetrical, although 45° and 30° spacing may be used.

Another constraint on polar graphs is the number of alternatives that can be shown. Even three alternatives begins to be somewhat crowded and five or more are almost impossible to read. Polar graphs have been drawn where the length of the arrows or the angles around each arrow are proportional to the attribute's importance. These refinements do not seem to work well. Polar graphs are best used as summaries of the unweighted performance of two or three alternatives on less than seven attributes.

19.6 NUMERICAL SCALES FOR EVALUATION

The conversion of *verbal* ratings, such as excellent or more than average, or numerical values to or from a standardized scale is assisted by simple graphs such as those shown in Exhibits 19.7 to 19.10. These graphs are commonly called "scoring functions." While other scales are possible, a common choice is 0 to 10, with 0 = worst and 10 = best.

Numerical Variables.   Exhibit 19.7 presents the scoring function for the E(PW) in Example 19.7. In this case, the worst case is assumed to be a PW of 0, and the the best case is assumed to be the maximum of the two alternatives. Intermediate values are found by linear interpolation. In Exhibit 19.7, the two-stage score is $10 \cdot 13.07M/14.84M$, or 8.81. Exhibit 19.7 is an example of an increasing scoring function, where larger values are better. For increasing scoring functions, the scaled values are found using Equation 19.1.

$$\text{Score}(x) = 10(x - \text{Worst value})/(\text{Best value} - \text{Worst value}) \qquad (19.1)$$

Exhibit 19.7   Conversion of E(PW) to a numerical scale

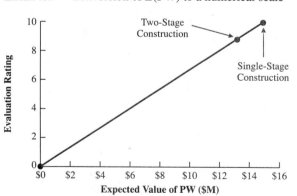

Costs and standard deviations require decreasing scoring functions, such as shown in Exhibit 19.8, since smaller values are better. Equation 19.1 could be used, but to avoid negative values for the numerator and denominator, the interpolating scoring equation (Equation 19.2) can be used instead.

$$\text{Score}(x) = 10(\text{Worst value} - x)/(\text{Worst value} - \text{Best value}) \qquad (19.2)$$

For Example 19.7, the standard deviations for the present worth were, respectively, $2.69M and $1.07M for one-stage and two-stage construction. Zero is the best possible standard deviation, and $2.69M is the worst for this case. The scaled value for $\sigma_{\text{PW}_{\text{two stage}}}$ is found as follows:

$$\text{Score}(1.07) = 10(2.69 - 1.07)/(2.69 - 0) = 6.02.$$

In some problems, it is better to use the standard deviation divided by the expected value as the measure of risk. The reason is that PWs that range between $100M and $104M and between $11 and $15M each have standard deviations of about $1M, but the relative risk is far larger when the standard deviation is 10% of the expected value than it is when it is 1% of the expected value.

Exhibit 19.8   Conversion of standard deviations to numerical scale

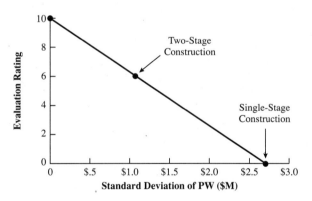

Choosing the Best Case and the Worst Case.   Usually it is better to use reasonable limits or zero as the best and worst cases. In Exhibit 19.7, an $E(\text{PW})$ equal to $0 was the worst case. In Exhibit 19.8, a standard deviation equal to $0 was the best case.

However, judgment may suggest that $0 is not a good best or worst case. Consider the simple problem of defining a scale for an initial salary offer, as shown in Exhibit 19.9. In both cases, the best salary is $35,000, but the worst salary could be assumed to be $0 or $22,000. Since a starting salary of less than $22,000 is (hopefully) unrealistic, that value is a better choice to define the worst case.

Exhibit 19.9   Conversion of salaries to numerical scale

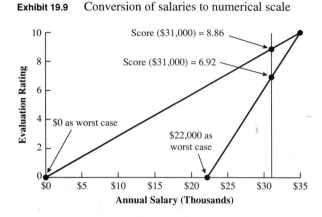

The importance of the assumed worst salary can be illustrated by considering the scores for a $31,000 salary. If $0 is the worst case, then the score for $31,000 − 10(31,000/35,000) = 8.86. If $22,000 is the worst case, then the score or scaled value for $31,000 = 10(31,000 − 22,000)/(35,000 − 22,000), or 6.92.

If five alternatives are considered on an attribute, then the worst of the five might be the 0 and the best of the five might be 10 (see Example 19.8). However, if only two alternatives are considered, then scoring the better outcome of the two as a 10 and the other as a 0 on each attribute probably overstates the differences between the alternatives.

It is better if all attributes are rated over the same ranges. In Example 19.8, the choice of $0 for best or $800 for best may depend on whether other attributes have ratings of 0 to 10 or if ratings more typically have a range of 5, such as ratings between 5 and 10. It is important that about the same range of values be used for each attribute. This is why numerical attributes are converted to a numerical 10-point scale. Otherwise, the weights or priorities that will be assigned in Section 19.7 will be misleading.

Example 19.8   Rating a Computer's Cost

Susan is choosing a computer, and she has narrowed the possibilities to five choices. She will be considering cost, a variety of performance measures, and dealer and manufacturer reliability, but first she wants to rate the cost performance on a scale of 0 to 10. If $0 is the best cost, what are the scaled values for the five computers? If the $800 cost of Mr. Value is the best, what are the scaled values for the five computers?

Cost	Computer
$ 800	Mr. Value
1100	Samsonic
1200	Graphic Power
1800	BBC (Big Blue Computers)
2400	Pear Computing

Solution

For example, substitute the $1100 cost of Samsonic in Equation 19.2.

$$\text{Score}(1100) = 10(2400 - 1100)/(2400 - \text{best})$$

If $0 is the best, then

$$\text{Score}(1100) = 10(2400 - 1100)/(2400 - 0) = 5.42.$$

If $800 is the best, then

$$\text{Score}(1100) = 10(2400 - 1100)/(2400 - 800) = 8.125.$$

The other results are summarized below.

Computer	Cost	$0 best	$800 best
Mr. Value	$ 800	6.67	10
Samsonic	1100	5.42	8.125
Graphic Power	1200	5	7.5
BBC (Big Blue Computers)	1800	2.5	3.75
Pear Computing	2400	0	0

Verbal Variables. Many variables, such as flexibility in Example 19.7, are subjectively estimated or rated. In most cases, a simple linear scale to convert from words to ratings is appropriate. For example, Exhibit 19.10 converts the five verbal ratings ranging from poor to excellent to the values (0, 2.5, 5, 7.5, 10). This linear conversion is the norm; however, nonlinear relationships can also be used [Keeney], especially when above-average and below-average ratings have significantly different consequences.

Exhibit 19.10   Numerical scale from verbal evaluation (Example 19.7)

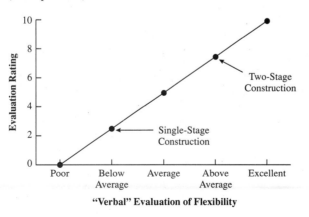

"Verbal" Evaluation of Flexibility

Missing Values. Incomplete data is a common problem—especially in feasibility studies. Assigning an "average" rating or making a "blue sky" or "pulled out of the air" guess are the two recommended solutions. Assigning a 0 or worst rating is the wrong approach. Assuming the worst case, rather than making a "blue sky" guess, distorts the final evaluation by the maximum amount. The value that is used should be as close as possible to the final one that would be assigned after gathering more data.

19.7 ADDITIVE MODELS

Examples 19.1, 19.4, and 19.5 (about Harold's job choice) introduced a simple *additive model*. The overall scores for Detroit, Seattle, and San Francisco were calculated by *adding* together, for the three attributes, the product of that job's performance evaluation and the attribute's weighted importance. This additive model is simple, easy to understand, and powerful.

The additive model has two major advantages over graphical techniques. First, the additive model includes *weighted importance ratings* that can recognize the higher priority or importance of some attributes. Second, by adding together the results for different attributes, a strength on one attribute can compensate for weakness on another. Because a strength can compensate for a weakness, the additive model is sometimes called a *compensatory model*.

This section presents:

1. Direct assignment of weights that add to 100%.
2. Subjective assignment of importance ratings.
3. Tabular additive models.
4. Graphical additive models.

Whether direct assignment of weights or subjective assignment of importance ratings is used, the resulting weights for the attributes apply to all alternatives and the resulting weights sum to 100%.

Direct Assignment of Weights.   In Example 19.5, weights of 60% for salary, 20% for promotion outlook, and 20% for the city were directly assigned by Harold.

This approach is easy if the number of attributes to be prioritized is not too large. Even more important, the approach is natural and intuitive. If there are many attributes, then the approach of subjectively assigning importance ratings is better. Changing the weight on one attribute implies that others must be changed as well, since the weights must sum to 100%

Directly assigning weights is sometimes described as allocating 100 points. For problems with more than four attributes, ensuring that the total points sum to 100 may interfere with the more subjective judgment process. Extra thought on substance is helpful, but time spent on arithmetic is wasted.

Subjective Assignment of Importance Ratings.   An additive model can be created by assigning subjective importance ratings to each attribute. These importance ratings are then totaled, and each is divided by the sum to create the additive weight.

For convenience, these importance ratings are defined on the same 10-point scale used to evaluate an alternative on an attribute. Thus, an attribute that is rated a 10 is one of the most important, and an attribute that is rated a 2 has little importance.

In Example 19.7, three attributes were identified to evaluate the choice between one-stage and two-stage construction of LowTech Manufacturing's office complex. Suppose that the expected value of the present worth is rated as a 10. Suppose that risk as measured by the standard deviation is rated as a 7. Finally, suppose that flexibility is rated as a 5. Then Exhibit 19.11 summarizes the calculation of the weights, I (for importance).

Exhibit 19.11   Subjective importance ratings

Attribute	Importance Rating	Additive Weight (I)
E(PW)	10	10/22 = .4545
σ_{PW}	7	7/22 = .3182
Flexibility	5	5/22 = .2273
Total	22	1.00

More than one attribute can receive the same rating. For example, if two attributes were tied as most important, then each would be rated as a 10. The key is that the difference between two ratings represents the decision maker's or the analyst's best judgment of their relative importance.

These additive weights are the decimal fractions or percentage weights that are placed on each attribute. Furthermore, the sum of all the weights equals 1 (or 100%). If using more than one rater or judge is desirable, then their individual ratings could simply be averaged for each attribute.

It seems best to allow the analyst and the decision maker to determine the importance rating, and let a computer or a calculator do the arithmetic. In any case, these additive weights become multipliers for the ratings on each attribute.

The subjective assignment of importance ratings has at least two advantages. First, it allows for at least a rough measure of the relative difference between the importance of any two attributes. A rating of 10 is five times more important than a 2 rating. Second, if reconsideration shows that a rating is wrong, only that rating needs to be changed, and then the computer will automatically recalculate all weights.

Tabular Additive Models.   Exhibit 19.12 combines the weights from Exhibit 19.11 with the evaluations of each attribute on the 10-point numerical scale. This is an example of a tabular additive model. It succinctly summarizes which attributes are most important, the performance of each alternative on each attribute, and the totals to be used to compare the alternatives.

Exhibit 19.12   Tabular additive model

Attributes	Importance (I) (weight)	One-Stage Construction		Two-Stage Construction	
		Rating (R)	I·R	Rating (R)	I·R
E(PW)	.4545	10	4.55	8.8	4.00
σ_{PW}	.3182	0	.00	6.0	1.91
Flexibility	.2273	2.5	.57	7.5	1.70
Total	1.000		5.12		7.61

Since the additive weights sum to 1 (or 100%) and the best equals 10 for all attributes, the weighted sum implies how close this alternative comes to the ideal. In fact, the total divided by 10 equals the fraction of the ideal [Baird].

Graphical Additive Models.   Exhibit 19.13 is a stacked bar chart, which is the graphical analog of the data shown in Exhibit 19.12. The total heights of the stacked bars correspond to the 5.12 and 7.62 total scores for one- and two-stage construction in Example 19.7. The value for profitability is the expected value of the PW; for risk, it's the standard deviation. These and the values for flexibility are shown for each alternative. The height of each subbox is a value in the y-axis series, which is the importance of the attribute (I) times the rating for that attribute (R). For one-stage construction, the expected value contributes 4.55 to thc total, the standard deviation contributes 0 to the total, and flexibility contributes .57 to the total.

Exhibit 19.13   Stacked bar chart for weighted evaluation

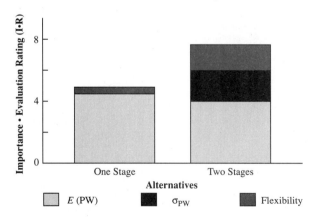

Using a spreadsheet simplifies construction of this stacked bar chart. The attributes of expected value, standard deviation, and flexibility are defined as the x-axis series. Then the I · R values for one-stage construction are defined as the

first *y*-axis series. Two-stage construction is defined as the second series, and other alternatives would be added as needed.

Closing Comment on Additive Models.   These tables and graphs are simple and effective ways to select the best alternative. They also support sensitivity analysis and reconsidering assigned values, so they are useful for planning as well as for decision making.

There are numerous techniques for deriving the weights for the criteria. Several of these depend on pairwise comparisons. Rather than directly assigning an importance rating, each pair of attributes is compared using a graphical, verbal, or numeric scale. The triangular matrix of comparisons is then processed by some technique to derive the weights. One of the simpler techniques uses regression to derive the weights, and one of the most common, the Analytic Hierarchy Process (AHP), uses left eigenvalues and eigenvectors. As shown in [Saaty], AHP supports measures of consistency that can be used to judge the quality of the subjective judgments.

19.8 HIERARCHICAL ATTRIBUTES AND OBJECTIVES

Exhibit 19.14 shows a more complex and more complete breakdown for the decision of which job to select. There are six criteria (*job characteristics,* work *environment,* ability to have *impacts on, contributions to,* personal *growth and development* through, and *overall satisfaction*). All but the overall satisfaction criteria are subdivided further. Finally, two career paths are the two alternatives— working as an engineer in industry or as an engineering professor.

This particular example [Mills and McCright] has been simplified slightly in that subjective assignments of evaluations have been made to each criterion and subcriterion. These ratings use the 0 to 10 scale and they have been written on the branches and to the right of the alternatives in Exhibit 19.14. In this case, there are no ratings below 4, because less important criteria or subcriteria are simply omitted.

These ratings are converted into weights by summing the weights for each node and dividing by the total, as was done for the additive model in Exhibit 19.11.

The weights shown on the criteria and subcriteria branches in Exhibit 19.15 are found by summing the importance ratings out of each node and then dividing each rating by that sum. The ratings of the criteria sum to 35, so the weight on job characteristics is 10/35 (or .286).

At the subcriteria level, consider the node for growth and development. There are three branches (lifelong learning, professional societies, and rewards of research) with ratings of 10, 5, and 8, respectively. Since the sum is 23, the weights are, respectively, 10/23, 5/23, and 8/23.

The columns for industry and professor in Exhibit 19.15 include the evaluation of each alternative on each subcriterion and the contribution of that to the total for the alternative. For example, the 6 rating for financial rewards for a professor is multiplied by the weight for the job characteristic criterion (.286) and the financial rewards subcriterion (.170) to get .292 (which equals $6 \cdot .286 \cdot .170$).

Exhibit 19.14   Decision hierarchy for engineering in industry vs. academia

Criteria	Alternatives	
└──Subcriteria	**Engineer in Industry**	**Engineering Professor**

10 Job Characteristics
- 8 Financial rewards ──── 10 ──── 6
- 10 Opportunities (demand) ── 8 ──── 10
- 4 Security ──── 6 ──── 10
- 6 Status/prestige ──── 6 ──── 10
- 8 Variety of work ──── 10 ──── 7
- 6 Freedom/flexibility ──── 3 ──── 8
- 5 Practical applications ──── 8 ──── 2

5 Environment
- 8 Immediate work ──── 5 ──── 7
- 4 Surroundings ──── 5 ──── 9

4 Impacts on
- 10 Directions of discipline ── 0 ──── 7
- 8 Lives of others ──── 4 ──── 10

4 Contributions to
- 6 Industry ──── 6 ──── 6
- 10 Society ──── 4 ──── 8

7 Growth and Development
- 10 Lifelong learning ──── 4 ──── 8
- 5 Professional societies ──── 4 ──── 8
- 8 Rewards of research ──── 2 ──── 10

5 Overall satisfaction ──── 6 ──── 8

Exhibit 19.15 Decision hierarchy with weights and additive totals for engineering in industry vs. academia

| Criteria | Alternatives | | | |
| | Engineer in Industry | | Engineering Professor | |
Subcriteria	Rating	Weight	Rating	Weight
.286 Job Characteristics				
.170 Financial rewards	10	.486	6	.292
.213 Opportunities (demand)	8	.487	10	.609
.085 Security	6	.146	10	.243
.128 Status/prestige	6	.220	10	.366
.170 Variety of work	10	.486	7	.340
.128 Freedom/flexibility	3	.110	8	.293
.106 Practical applications	8	.243	2	.061
		2.178		2.204
.143 Environment				
.667 Immediate work	5	.477	7	.668
.333 Surroundings	5	.238	9	.429
		.715		1.096
.114 Impacts on				
.556 Directions of discipline	0	0	7	.444
.444 Lives of others	4	.202	10	.506
		.202		.950
.114 Contributions to				
.375 Industry	6	.257	6	.257
.625 Society	4	.285	8	.570
		.542		.827
.200 Growth and Development				
.435 Lifelong learning	4	.348	8	.696
.217 Professional societies	4	.174	8	.347
.348 Rewards of research	2	.139	10	.696
		.660		1.739
.143 Overall satisfaction	6	.858	8	1.144
		5.155		7.960

547

This type of hierarchy greatly increases the flexibility of the additive model. Exhibits 19.14 and 19.15 have 16 subcriteria and 1 nonsubdivided criterion, for a total of 17 identified objectives. This is far in excess of the seven that has been suggested as an upper limit for the number of attributes. However, by subdividing the problem, a more complete and a more accurate analysis is possible.

The final result of the professor's job being better (at 7.960, versus 5.155 for industry) depends on the relative importance that is placed on the various criteria. In this case, growth and development is the second most important criterion, and the professor's job is far better on this criterion.

19.9 SUMMARY

This chapter has focused on incorporating the calculations of discounted cash flows into a broader perspective that includes objectives that cannot be expressed in dollars. This framework can include minimizing risk, maximizing flexibility, or maximizing quality, as well as maximizing present worth.

The attributes that are selected to measure the achievement of the multiple objectives should be distinct from each other, and as a set they should completely cover relevant objectives. Once the attributes have been selected and each alternative has been evaluated on each attribute, then the information must be appropriately summarized to support decision making.

The summary can be a table, which allows for the inclusion of exact numeric values; but the summary is often more effective if it is graphical instead. Shaded-square diagrams and polar graphs were presented as possible graphical techniques.

Weighted additive models allow for a prioritization of the multiple objectives, so that more important objectives carry more weight in the analysis. These compensatory models produce a single number for each alternative, so that the alternative with the best or highest total can be selected.

For more complex problems, a decision hierarchy can be used to extend the techniques of a simple additive table. This hierarchy allows the definition of objectives or criteria and subobjectives or subcriteria.

REFERENCES

Baird, Bruce F., *Managerial Decisions Under Uncertainty*, 1989, Wiley, Chapter 13.

Canada, John R., and William G. Sullivan, *Economic and Multiattribute Evaluation of Advanced Manufacturing Systems*, 1989, Prentice-Hall.

Clemen, Robert P. *Making Hard Decisions: An Introduction to Decision Analysis*, 1991, Duxbury.

Harmon, Roy L., and Leroy D. Peterson, *Reinventing the Factory: Productivity Breakthroughs in Manufacturing Today*, 1990, Free Press (Arthur Anderson cases in appendix).

Keeney, Ralph L., "A Decision Analysis with Multiple Objectives: The Mexico City Airport," *Bell Journal of Economics and Management Science*, Volume 4, Number 1, Spring 1973, pp. 101–117.

Mills, Nancy R., and Paul R. McCright, "Choosing the Ph.D. Path: A Multi-Criteria Model for Career Decisions," *Journal of Engineering Education*, Volume 82, Number 2, April 1993, pp. 109–117.

Noori, Hamid, *Managing the Dynamics of New Technology: Issues in Manufacturing Management*, 1990, Prentice-Hall.

Park, Chan S., and Gerald J. Thuesen, "Combining the Concepts of Uncertainty Resolution and Project Balance for Capital Allocation Decisions," *The Engineering Economist*, Volume 24, Number 2, Winter 1979, pp. 109–127.

Saaty, Thomas L., *The Analytic Hierarchy Process*, 1980, McGraw-Hill.

PROBLEMS

Data for Problems 19.1–19.5: National Motors is planning to build a new manufacturing plant. Data on the five sites still under consideration is shown below. Note that the Detroit site reuses some existing National Motors property, while the Big City site entitles the corporation to several tax breaks for locating in a "depressed" locale. Thus these two sites are appreciably cheaper than the others.

Site	EAC ($M)	P(Strikes)	Ease of Expansion
Detroit	$12M	.05	Average
Green Field	22M	.02	Very high
Happy Valley	20M	.01	High
Big City	16M	.03	Low
Coastal View	21M	.04	Very low

19.1 Are any of National Motors' sites dominated? Is any site dominant? (*Answer:* Coastal View dominated)

19.2 Convert the National Motors summary table into a shaded-circle or shaded-square diagram.

19.3 Draw a polar graph for National Motors.

19.4 Using a 0 = worst and 10 = best scale and linear scoring functions, convert the tabulated values into scores. For the probability of a strike, assume that the best possible is $P(\text{strike}) = 0$.

19.5 Assume that EAC has an importance of 10, $P(\text{strike})$ has an importance of 7, and ease of expansion has an importance of 4. What are the weights for each category? Which site appears to be the best?   (*Answer:* Detroit, 5.71)

Data for Problems 19.6–19.10: A company called 2020 Labs is a commercial R&D firm that specializes in developing new technology, which it then licenses to major corporations. Its Ceramics group is deciding which of four projects to attack next (identified by the major potential application). Only one project can be selected, due to manpower considerations. Like other firms, 2020 Labs maximizes present worth and minimizes risk. However, 2020 Labs also places a high value on projects that provide a large intellectual challenge, since it stimulates both productivity and morale in its employees.

Project Application	E(PW)	σ_{PW}	Challenge
Auto	$20M	$15M	Low
Aircraft	15M	10M	High
Refining	10M	4M	Average
Spacecraft	5M	3M	Very high

19.6   Are any of the Ceramics group's projects dominated? Are any dominant?

19.7   Convert the Ceramics group's summary table into a shaded-circle or shaded-square diagram.

19.8   Draw a polar graph for 2020 Lab's Ceramics group.

19.9   Develop scoring functions for 2020 Lab's Ceramics group for the $E(PW)$, for the standard deviation of the PW, and for technical challenge. (*Answer:* $R_{E(PW)=15} = 7.5$, $R_{risk=10} = 3.3$)

19.10 Construct an additive model for the Ceramics group if maximizing the present worth has an importance of 10, minimizing risk has an importance of 5, and maximizing technical challenge has an importance of 8. What are the weights for each category? Which project is rated highest?

19.11 Jane Smith is considering the following five job offers as graduation nears:

Firm	Salary	Location	Recreation*	Stability*
A	$33K	NW	5	4
B	37K	NE	8	4
C	28K	SE	3	6
D	31K	SW	3	9
E	35K	SE	9	5

* Scale: 0 = worst; 10 = Best

Jane rates the Southeast above all other locations. She prefers the other regions in the following order: Southwest, Northwest, and Northeast. Construct numerical scoring functions for salary and location. Determine which offer best fits her requirements if she assigns the following weights to her objectives: 60% to salary, 20% to location, 10% to stability, and 10% to recreational opportunities.

19.12 Mrs. Meadows Cookies wishes to maximize profit, minimize risk, and increase market share. The firm assigns weights of 70% to profit, 18% to risk, and 12% to market potential. Construct numerical scoring functions and an additive model for this problem. Which one of the following four alternatives is the best choice as a new cookie?

Brand	PW (Profit)	σ (Profit)	Share Potential
Lotsa-Chips	$10M	$6M	High
Choca-Macho	14M	7M	Low
Coco-Macs	12M	4M	Medium
Lemon Dreams	13M	5M	Medium

19.13 Draw a decision hierarchy for a significant decision with economic consequences that you must make. Examples include choosing among job offers, between graduate school and immediate employment, between a summer job and traveling in Europe, etc.

19.14 Assign importance rankings, evaluate alternatives, and select the best alternative for the decision hierarchy in Problem 19.13.

19.15 NewTech wants to use an additive model for choosing its best approach to a particular marketing/R&D problem. In particular, its management has decided that the expected value should receive twice the weight as the standard deviation of the present worth. What is the score of each alternative, and which should be chosen? The probabilities for low, medium, and high sales are, respectively, .3, .5, and .2.

	Present Worth ($M)		
Choice	Low	Medium	High
A	$5	$7	$10
B	3	7	18

19.16 Find a published example of an additive model. Critique the model's objectives. Are they mutually exclusive and collectively exhaustive? Do the ratings seem to be biased? Is the conclusion reasonable?

APPENDIX

END-OF-PERIOD COMPOUND INTEREST TABLES

Tables Included in Appendix

.25%	.5%	.75%	1%	1.5%	2%
3%	4%	5%	6%	7%	8%
9%	10%	11%	12%	13%	14%
15%	16%	17%	18%	19%	20%
25%	30%	35%	40%	50%	60%
70%	80%	90%	100%		

Limits of the Factors

Factor	When $N = \infty$	When $i = 0$
$(F/P, i, N)$	∞	1
$(P/F, i, N)$	0	1
$(A/P, i, N)$	i	$1/N$
$(P/A, i, N)$	$1/i$	N
$(A/F, i, N)$	0	$1/N$
$(F/A, i, N)$	∞	N
$(P/G, i, N)$	$1/i^2$	$N(N-1)/2$
$(A/G, i, N)$	$1/i$	$(N-1)/2$

	Single Payment		Uniform Payment Series				Arithmetic Gradient		
N	Compound Amount Factor F/P	Present Worth Factor P/F	Capital Recovery Factor A/P	Present Worth Factor P/A	Sinking Fund Factor A/F	Compound Amount Factor F/A	Present Worth Factor P/G	Uniform Payment Factor A/G	N
1	1.003	.9975	1.0025	.998	1.0000	1.000	0	0	1
2	1.005	.9950	.5019	1.993	.4994	2.003	.995	.499	2
3	1.008	.9925	.3350	2.985	.3325	3.008	2.980	.998	3
4	1.010	.9901	.2516	3.975	.2491	4.015	5.950	1.497	4
5	1.013	.9876	.2015	4.963	.1990	5.025	9.901	1.995	5
6	1.015	.9851	.1681	5.948	.1656	6.038	14.826	2.493	6
7	1.018	.9827	.1443	6.931	.1418	7.053	20.722	2.990	7
8	1.020	.9802	.1264	7.911	.1239	8.070	27.584	3.487	8
9	1.023	.9778	.1125	8.889	.1100	9.091	35.406	3.983	9
10	1.025	.9753	.1014	9.864	.0989	10.113	44.184	4.479	10
11	1.028	.9729	.0923	10.837	.0898	11.139	53.913	4.975	11
12	1.030	.9705	.0847	11.807	.0822	12.166	64.589	5.470	12
13	1.033	.9681	.0783	12.775	.0758	13.197	76.205	5.965	13
14	1.036	.9656	.0728	13.741	.0703	14.230	88.759	6.459	14
15	1.038	.9632	.0680	14.704	.0655	15.265	102.244	6.953	15
16	1.041	.9608	.0638	15.665	.0613	16.304	116.657	7.447	16
17	1.043	.9584	.0602	16.623	.0577	17.344	131.992	7.940	17
18	1.046	.9561	.0569	17.580	.0544	18.388	148.245	8.433	18
19	1.049	.9537	.0540	18.533	.0515	19.434	165.411	8.925	19
20	1.051	.9513	.0513	19.484	.0488	20.482	183.485	9.417	20
21	1.054	.9489	.0489	20.433	.0464	21.533	202.463	9.908	21
22	1.056	.9466	.0468	21.380	.0443	22.587	222.341	10.400	22
23	1.059	.9442	.0448	22.324	.0423	23.644	243.113	10.890	23
24	1.062	.9418	.0430	23.266	.0405	24.703	264.775	11.380	24
25	1.064	.9395	.0413	24.205	.0388	25.765	287.323	11.870	25
26	1.067	.9371	.0398	25.143	.0373	26.829	310.752	12.360	26
27	1.070	.9348	.0383	26.077	.0358	27.896	335.057	12.849	27
28	1.072	.9325	.0370	27.010	.0345	28.966	360.233	13.337	28
29	1.075	.9301	.0358	27.940	.0333	30.038	386.278	13.825	29
30	1.078	.9278	.0346	28.868	.0321	31.113	413.185	14.313	30
31	1.080	.9255	.0336	29.793	.0311	32.191	440.950	14.800	31
32	1.083	.9232	.0326	30.717	.0301	33.272	469.570	15.287	32
33	1.086	.9209	.0316	31.638	.0291	34.355	499.039	15.774	33
34	1.089	.9186	.0307	32.556	.0282	35.441	529.353	16.260	34
35	1.091	.9163	.0299	33.472	.0274	36.529	560.508	16.745	35
36	1.094	.9140	.0291	34.386	.0266	37.621	592.499	17.231	36
40	1.105	.9050	.0263	38.020	.0238	42.013	728.740	19.167	40
48	1.127	.8871	.0221	45.179	.0196	50.931	1040.055	23.021	48
50	1.133	.8826	.0213	46.946	.0188	53.189	1125.777	23.980	50
60	1.162	.8609	.0180	55.652	.0155	64.647	1600	28.751	60
70	1.191	.8396	.0156	64.144	.0131	76.394	2148	33.481	70
72	1.197	.8355	.0152	65.817	.0127	78.779	2266	34.422	72
80	1.221	.8189	.0138	72.426	.0113	88.439	2764	38.169	80
90	1.252	.7987	.0124	80.504	.00992	100.8	3447	42.816	90
100	1.284	.7790	.0113	88.382	.00881	113.4	4191	47.422	100
120	1.349	.7411	.00966	103.6	.00716	139.7	5852	56.508	120
240	1.821	.5492	.00555	180.3	.00305	328.3	19399	107.586	240
360	2.457	.4070	.00422	237.2	.00172	582.7	36264	152.890	360
480	3.315	.3016	.00358	279.3	.00108	926.1	53821	192.670	480
∞	∞	0	.0025	400	0	∞	160000	400	∞

	Single Payment		Uniform Payment Series				Arithmetic Gradient		
N	Compound Amount Factor F/P	Present Worth Factor P/F	Capital Recovery Factor A/P	Present Worth Factor P/A	Sinking Fund Factor A/F	Compound Amount Factor F/A	Present Worth Factor P/G	Uniform Payment Factor A/G	N
1	1.005	.9950	1.0050	.995	1.0000	1.000	0	0	1
2	1.010	.9901	.5038	1.985	.4988	2.005	.990	.499	2
3	1.015	.9851	.3367	2.970	.3317	3.015	2.960	.997	3
4	1.020	.9802	.2531	3.950	.2481	4.030	5.901	1.494	4
5	1.025	.9754	.2030	4.926	.1980	5.050	9.803	1.990	5
6	1.030	.9705	.1696	5.896	.1646	6.076	14.655	2.485	6
7	1.036	.9657	.1457	6.862	.1407	7.106	20.449	2.980	7
8	1.041	.9609	.1278	7.823	.1228	8.141	27.176	3.474	8
9	1.046	.9561	.1139	8.779	.1089	9.182	34.824	3.967	9
10	1.051	.9513	.1028	9.730	.0978	10.228	43.386	4.459	10
11	1.056	.9466	.0937	10.677	.0887	11.279	52.853	4.950	11
12	1.062	.9419	.0861	11.619	.0811	12.336	63.214	5.441	12
13	1.067	.9372	.0796	12.556	.0746	13.397	74.460	5.930	13
14	1.072	.9326	.0741	13.489	.0691	14.464	86.583	6.419	14
15	1.078	.9279	.0694	14.417	.0644	15.537	99.574	6.907	15
16	1.083	.9233	.0652	15.340	.0602	16.614	113.424	7.394	16
17	1.088	.9187	.0615	16.259	.0565	17.697	128.123	7.880	17
18	1.094	.9141	.0582	17.173	.0532	18.786	143.663	8.366	18
19	1.099	.9096	.0553	18.082	.0503	19.880	160.036	8.850	19
20	1.105	.9051	.0527	18.987	.0477	20.979	177.232	9.334	20
21	1.110	.9006	.0503	19.888	.0453	22.084	195.243	9.817	21
22	1.116	.8961	.0481	20.784	.0431	23.194	214.061	10.299	22
23	1.122	.8916	.0461	21.676	.0411	24.310	233.677	10.781	23
24	1.127	.8872	.0443	22.563	.0393	25.432	254.082	11.261	24
25	1.133	.8828	.0427	23.446	.0377	26.559	275.269	11.741	25
26	1.138	.8784	.0411	24.324	.0361	27.692	297.228	12.220	26
27	1.144	.8740	.0397	25.198	.0347	28.830	319.952	12.698	27
28	1.150	.8697	.0384	26.068	.0334	29.975	343.433	13.175	28
29	1.156	.8653	.0371	26.933	.0321	31.124	367.663	13.651	29
30	1.161	.8610	.0360	27.794	.0310	32.280	392.632	14.126	30
31	1.167	.8567	.0349	28.651	.0299	33.441	418.335	14.601	31
32	1.173	.8525	.0339	29.503	.0289	34.609	444.762	15.075	32
33	1.179	.8482	.0329	30.352	.0279	35.782	471.906	15.548	33
34	1.185	.8440	.0321	31.196	.0271	36.961	499.758	16.020	34
35	1.191	.8398	.0312	32.035	.0262	38.145	528.312	16.492	35
36	1.197	.8356	.0304	32.871	.0254	39.336	557.560	16.962	36
40	1.221	.8191	.0276	36.172	.0226	44.159	681.335	18.836	40
48	1.270	.7871	.0235	42.580	.0185	54.098	959.919	22.544	48
50	1.283	.7793	.0227	44.143	.0177	56.645	1035.697	23.462	50
60	1.349	.7414	.0193	51.726	.0143	69.770	1449	28.006	60
70	1.418	.7053	.0170	58.939	.0120	83.566	1914	32.468	70
72	1.432	.6983	.0166	60.340	.0116	86.409	2012	33.350	72
80	1.490	.6710	.0152	65.802	.0102	98.068	2425	36.847	80
90	1.567	.6383	.0138	72.331	.00883	113.3	2976	41.145	90
100	1.647	.6073	.0127	78.543	.00773	129.3	3563	45.361	100
120	1.819	.5496	.0111	90.073	.00610	163.9	4824	53.551	120
240	3.310	.3021	.00716	139.6	.00216	462.0	13416	96.113	240
360	6.023	.1660	.00600	166.8	.00100	1004.5	21403	128.324	360
480	10.957	.0913	.00550	181.7	.00050	1991.5	27588	151.795	480
∞	∞	0	.005	200.000	0	∞	40000	200	∞

	Single Payment		Uniform Payment Series				Arithmetic Gradient		
N	Compound Amount Factor F/P	Present Worth Factor P/F	Capital Recovery Factor A/P	Present Worth Factor P/A	Sinking Fund Factor A/F	Compound Amount Factor F/A	Present Worth Factor P/G	Uniform Payment Factor A/G	N
1	1.008	.9926	1.0075	.993	1.0000	1.000	0	0	1
2	1.015	.9852	.5056	1.978	.4981	2.008	.985	.498	2
3	1.023	.9778	.3383	2.956	.3308	3.023	2.941	.995	3
4	1.030	.9706	.2547	3.926	.2472	4.045	5.852	1.491	4
5	1.038	.9633	.2045	4.889	.1970	5.076	9.706	1.985	5
6	1.046	.9562	.1711	5.846	.1636	6.114	14.487	2.478	6
7	1.054	.9490	.1472	6.795	.1397	7.159	20.181	2.970	7
8	1.062	.9420	.1293	7.737	.1218	8.213	26.775	3.461	8
9	1.070	.9350	.1153	8.672	.1078	9.275	34.254	3.950	9
10	1.078	.9280	.1042	9.600	.0967	10.344	42.606	4.438	10
11	1.086	.9211	.0951	10.521	.0876	11.422	51.817	4.925	11
12	1.094	.9142	.0875	11.435	.0800	12.508	61.874	5.411	12
13	1.102	.9074	.0810	12.342	.0735	13.601	72.763	5.895	13
14	1.110	.9007	.0755	13.243	.0680	14.703	84.472	6.379	14
15	1.119	.8940	.0707	14.137	.0632	15.814	96.988	6.861	15
16	1.127	.8873	.0666	15.024	.0591	16.932	110.297	7.341	16
17	1.135	.8807	.0629	15.905	.0554	18.059	124.389	7.821	17
18	1.144	.8742	.0596	16.779	.0521	19.195	139.249	8.299	18
19	1.153	.8676	.0567	17.647	.0492	20.339	154.867	8.776	19
20	1.161	.8612	.0540	18.508	.0465	21.491	171.230	9.252	20
21	1.170	.8548	.0516	19.363	.0441	22.652	188.325	9.726	21
22	1.179	.8484	.0495	20.211	.0420	23.822	206.142	10.199	22
23	1.188	.8421	.0475	21.053	.0400	25.001	224.668	10.671	23
24	1.196	.8358	.0457	21.889	.0382	26.188	243.892	11.142	24
25	1.205	.8296	.0440	22.719	.0365	27.385	263.803	11.612	25
26	1.214	.8234	.0425	23.542	.0350	28.590	284.389	12.080	26
27	1.224	.8173	.0411	24.359	.0336	29.805	305.639	12.547	27
28	1.233	.8112	.0397	25.171	.0322	31.028	327.542	13.013	28
29	1.242	.8052	.0385	25.976	.0310	32.261	350.087	13.477	29
30	1.251	.7992	.0373	26.775	.0298	33.503	373.263	13.941	30
31	1.261	.7932	.0363	27.568	.0288	34.754	397.060	14.403	31
32	1.270	.7873	.0353	28.356	.0278	36.015	421.468	14.864	32
33	1.280	.7815	.0343	29.137	.0268	37.285	446.475	15.323	33
34	1.289	.7757	.0334	29.913	.0259	38.565	472.071	15.782	34
35	1.299	.7699	.0326	30.683	.0251	39.854	498.247	16.239	35
36	1.309	.7641	.0318	31.447	.0243	41.153	524.992	16.695	36
40	1.348	.7416	.0290	34.447	.0215	46.446	637.469	18.506	40
48	1.431	.6986	.0249	40.185	.0174	57.521	886.840	22.069	48
50	1.453	.6883	.0241	41.566	.0166	60.394	953.849	22.948	50
60	1.566	.6387	.0208	48.173	.0133	75.424	1314	27.266	60
70	1.687	.5927	.0184	54.305	.0109	91.620	1709	31.463	70
72	1.713	.5839	.0180	55.477	.0105	95.007	1791	32.288	72
80	1.818	.5500	.0167	59.994	.00917	109.1	2132	35.539	80
90	1.959	.5104	.0153	65.275	.00782	127.9	2578	39.495	90
100	2.111	.4737	.0143	70.175	.00675	148.1	3041	43.331	100
120	2.451	.4079	.0127	78.942	.00517	193.5	3999	50.652	120
240	6.009	.1664	.00900	111.145	.001497	668	9494	85.421	240
360	14.73	.06789	.00805	124.282	.000546	1831	13312	107.114	360
480	36.11	.02769	.00771	129.641	.000214	4681	15513	119.662	480
∞	∞	0	.0075	133.333	0	∞	17778	133.333	∞

	Single Payment		Uniform Payment Series				Arithmetic Gradient		
N	Compound Amount Factor F/P	Present Worth Factor P/F	Capital Recovery Factor A/P	Present Worth Factor P/A	Sinking Fund Factor A/F	Compound Amount Factor F/A	Present Worth Factor P/G	Uniform Payment Factor A/G	N
1	1.010	.9901	1.0100	.990	1.0000	1.000	0	0	1
2	1.020	.9803	.5075	1.970	.4975	2.010	.980	.498	2
3	1.030	.9706	.3400	2.941	.3300	3.030	2.921	.993	3
4	1.041	.9610	.2563	3.902	.2463	4.060	5.804	1.488	4
5	1.051	.9515	.2060	4.853	.1960	5.101	9.610	1.980	5
6	1.062	.9420	.1725	5.795	.1625	6.152	14.321	2.471	6
7	1.072	.9327	.1486	6.728	.1386	7.214	19.917	2.960	7
8	1.083	.9235	.1307	7.652	.1207	8.286	26.381	3.448	8
9	1.094	.9143	.1167	8.566	.1067	9.369	33.696	3.934	9
10	1.105	.9053	.1056	9.471	.0956	10.462	41.843	4.418	10
11	1.116	.8963	.0965	10.368	.0865	11.567	50.807	4.901	11
12	1.127	.8874	.0888	11.255	.0788	12.683	60.569	5.381	12
13	1.138	.8787	.0824	12.134	.0724	13.809	71.113	5.861	13
14	1.149	.8700	.0769	13.004	.0669	14.947	82.422	6.338	14
15	1.161	.8613	.0721	13.865	.0621	16.097	94.481	6.814	15
16	1.173	.8528	.0679	14.718	.0579	17.258	107.273	7.289	16
17	1.184	.8444	.0643	15.562	.0543	18.430	120.783	7.761	17
18	1.196	.8360	.0610	16.398	.0510	19.615	134.996	8.232	18
19	1.208	.8277	.0581	17.226	.0481	20.811	149.895	8.702	19
20	1.220	.8195	.0554	18.046	.0454	22.019	165.466	9.169	20
21	1.232	.8114	.0530	18.857	.0430	23.239	181.695	9.635	21
22	1.245	.8034	.0509	19.660	.0409	24.472	198.566	10.100	22
23	1.257	.7954	.0489	20.456	.0389	25.716	216.066	10.563	23
24	1.270	.7876	.0471	21.243	.0371	26.973	234.180	11.024	24
25	1.282	.7798	.0454	22.023	.0354	28.243	252.894	11.483	25
26	1.295	.7720	.0439	22.795	.0339	29.526	272.196	11.941	26
27	1.308	.7644	.0424	23.560	.0324	30.821	292.070	12.397	27
28	1.321	.7568	.0411	24.316	.0311	32.129	312.505	12.852	28
29	1.335	.7493	.0399	25.066	.0299	33.450	333.486	13.304	29
30	1.348	.7419	.0387	25.808	.0287	34.785	355.002	13.756	30
31	1.361	.7346	.0377	26.542	.0277	36.133	377.039	14.205	31
32	1.375	.7273	.0367	27.270	.0267	37.494	399.586	14.653	32
33	1.389	.7201	.0357	27.990	.0257	38.869	422.629	15.099	33
34	1.403	.7130	.0348	28.703	.0248	40.258	446.157	15.544	34
35	1.417	.7059	.0340	29.409	.0240	41.660	470.158	15.987	35
36	1.431	.6989	.0332	30.108	.0232	43.077	494.621	16.428	36
40	1.489	.6717	.0305	32.835	.0205	48.886	596.856	18.178	40
48	1.612	.6203	.0263	37.974	.0163	61.223	820.146	21.598	48
50	1.645	.6080	.0255	39.196	.0155	64.463	879.418	22.436	50
60	1.817	.5504	.0222	44.955	.0122	81.670	1192.806	26.533	60
70	2.007	.4983	.0199	50.169	.00993	100.7	1528.647	30.470	70
72	2.047	.4885	.0196	51.150	.00955	104.7	1597.867	31.239	72
80	2.217	.4511	.0182	54.888	.00822	121.7	1879.877	34.249	80
90	2.449	.4084	.0169	59.161	.00690	144.9	2240.567	37.872	90
100	2.705	.3697	.0159	63.029	.00587	170.5	2605.776	41.343	100
120	3.300	.3030	.0143	69.701	.00435	230.0	3334.115	47.835	120
240	10.89	.09181	.0110	90.819	.00101	989	6878.602	75.739	240
360	35.95	.02782	.0103	97.218	.000286	3495	8720.432	89.699	360
480	118.65	.00843	.0101	99.157	.000085	11765	9511.158	95.920	480
∞	∞	0	.01	100	0	∞	10000	100	∞

	Single Payment		Uniform Payment Series				Arithmetic Gradient		
N	Compound Amount Factor F/P	Present Worth Factor P/F	Capital Recovery Factor A/P	Present Worth Factor P/A	Sinking Fund Factor A/F	Compound Amount Factor F/A	Present Worth Factor P/G	Uniform Payment Factor A/G	N
1	1.015	.9852	1.0150	.985	1.0000	1.000	0	0	1
2	1.030	.9707	.5113	1.956	.4963	2.015	.971	.496	2
3	1.046	.9563	.3434	2.912	.3284	3.045	2.883	.990	3
4	1.061	.9422	.2594	3.854	.2444	4.091	5.710	1.481	4
5	1.077	.9283	.2091	4.783	.1941	5.152	9.423	1.970	5
6	1.093	.9145	.1755	5.697	.1605	6.230	13.996	2.457	6
7	1.110	.9010	.1516	6.598	.1366	7.323	19.402	2.940	7
8	1.126	.8877	.1336	7.486	.1186	8.433	25.616	3.422	8
9	1.143	.8746	.1196	8.361	.1046	9.559	32.612	3.901	9
10	1.161	.8617	.1084	9.222	.0934	10.703	40.367	4.377	10
11	1.178	.8489	.0993	10.071	.0843	11.863	48.857	4.851	11
12	1.196	.8364	.0917	10.908	.0767	13.041	58.057	5.323	12
13	1.214	.8240	.0852	11.732	.0702	14.237	67.945	5.792	13
14	1.232	.8118	.0797	12.543	.0647	15.450	78.499	6.258	14
15	1.250	.7999	.0749	13.343	.0599	16.682	89.697	6.722	15
16	1.269	.7880	.0708	14.131	.0558	17.932	101.518	7.184	16
17	1.288	.7764	.0671	14.908	.0521	19.201	113.940	7.643	17
18	1.307	.7649	.0638	15.673	.0488	20.489	126.943	8.100	18
19	1.327	.7536	.0609	16.426	.0459	21.797	140.508	8.554	19
20	1.347	.7425	.0582	17.169	.0432	23.124	154.615	9.006	20
21	1.367	.7315	.0559	17.900	.0409	24.471	169.245	9.455	21
22	1.388	.7207	.0537	18.621	.0387	25.838	184.380	9.902	22
23	1.408	.7100	.0517	19.331	.0367	27.225	200.001	10.346	23
24	1.430	.6995	.0499	20.030	.0349	28.634	216.090	10.788	24
25	1.451	.6892	.0483	20.720	.0333	30.063	232.631	11.228	25
26	1.473	.6790	.0467	21.399	.0317	31.514	249.607	11.665	26
27	1.495	.6690	.0453	22.068	.0303	32.987	267.000	12.099	27
28	1.517	.6591	.0440	22.727	.0290	34.481	284.796	12.531	28
29	1.540	.6494	.0428	23.376	.0278	35.999	302.978	12.961	29
30	1.563	.6398	.0416	24.016	.0266	37.539	321.531	13.388	30
31	1.587	.6303	.0406	24.646	.0256	39.102	340.440	13.813	31
32	1.610	.6210	.0396	25.267	.0246	40.688	359.691	14.236	32
33	1.634	.6118	.0386	25.879	.0236	42.299	379.269	14.656	33
34	1.659	.6028	.0378	26.482	.0228	43.933	399.161	15.073	34
35	1.684	.5939	.0369	27.076	.0219	45.592	419.352	15.488	35
36	1.709	.5851	.0362	27.661	.0212	47.276	439.830	15.901	36
40	1.814	.5513	.0334	29.916	.0184	54.268	524.357	17.528	40
48	2.043	.4894	.0294	34.043	.0144	69.565	703.546	20.667	48
50	2.105	.4750	.0286	35.000	.0136	73.683	749.964	21.428	50
60	2.443	.4093	.0254	39.380	.0104	96.215	988.167	25.093	60
70	2.835	.3527	.0232	43.155	.00817	122.4	1231.166	28.529	70
72	2.921	.3423	.0228	43.845	.00781	128.1	1279.794	29.189	72
80	3.291	.3039	.0215	46.407	.00655	152.7	1473.074	31.742	80
90	3.819	.2619	.0203	49.210	.00532	187.9	1709.544	34.740	90
100	4.432	.2256	.0194	51.625	.00437	228.8	1937.451	37.530	100
120	5.969	.1675	.0180	55.498	.00302	331.3	2359.711	42.519	120
240	35.63	.02806	.0154	64.796	.000433	2309	3870.691	59.737	240
360	212.70	.00470	.0151	66.353	.000071	14114	4310.716	64.966	360
480	1269.70	.00079	.0150	66.614	.000012	84580	4415.741	66.288	480
∞	∞	0	.015	66.667	0	∞	4444.444	66.667	∞

End-of-Period Compound Interest Factors

	Single Payment		Uniform Payment Series				Arithmetic Gradient		
	Compound Amount Factor	Present Worth Factor	Capital Recovery Factor	Present Worth Factor	Sinking Fund Factor	Compound Amount Factor	Present Worth Factor	Uniform Payment Factor	
N	F/P	P/F	A/P	P/A	A/F	F/A	P/G	A/G	N
1	1.020	.9804	1.0200	.980	1.0000	1.000	0	0	1
2	1.040	.9612	.5150	1.942	.4950	2.020	.961	.495	2
3	1.061	.9423	.3468	2.884	.3268	3.060	2.846	.987	3
4	1.082	.9238	.2626	3.808	.2426	4.122	5.617	1.475	4
5	1.104	.9057	.2122	4.713	.1922	5.204	9.240	1.960	5
6	1.126	.8880	.1785	5.601	.1585	6.308	13.680	2.442	6
7	1.149	.8706	.1545	6.472	.1345	7.434	18.903	2.921	7
8	1.172	.8535	.1365	7.325	.1165	8.583	24.878	3.396	8
9	1.195	.8368	.1225	8.162	.1025	9.755	31.572	3.868	9
10	1.219	.8203	.1113	8.983	.0913	10.950	38.955	4.337	10
11	1.243	.8043	.1022	9.787	.0822	12.169	46.998	4.802	11
12	1.268	.7885	.0946	10.575	.0746	13.412	55.671	5.264	12
13	1.294	.7730	.0881	11.348	.0681	14.680	64.948	5.723	13
14	1.319	.7579	.0826	12.106	.0626	15.974	74.800	6.179	14
15	1.346	.7430	.0778	12.849	.0578	17.293	85.202	6.631	15
16	1.373	.7284	.0737	13.578	.0537	18.639	96.129	7.080	16
17	1.400	.7142	.0700	14.292	.0500	20.012	107.555	7.526	17
18	1.428	.7002	.0667	14.992	.0467	21.412	119.458	7.968	18
19	1.457	.6864	.0638	15.678	.0438	22.841	131.814	8.407	19
20	1.486	.6730	.0612	16.351	.0412	24.297	144.600	8.843	20
21	1.516	.6598	.0588	17.011	.0388	25.783	157.796	9.276	21
22	1.546	.6468	.0566	17.658	.0366	27.299	171.379	9.705	22
23	1.577	.6342	.0547	18.292	.0347	28.845	185.331	10.132	23
24	1.608	.6217	.0529	18.914	.0329	30.422	199.630	10.555	24
25	1.641	.6095	.0512	19.523	.0312	32.030	214.259	10.974	25
26	1.673	.5976	.0497	20.121	.0297	33.671	229.199	11.391	26
27	1.707	.5859	.0483	20.707	.0283	35.344	244.431	11.804	27
28	1.741	.5744	.0470	21.281	.0270	37.051	259.939	12.214	28
29	1.776	.5631	.0458	21.844	.0258	38.792	275.706	12.621	29
30	1.811	.5521	.0446	22.396	.0246	40.568	291.716	13.025	30
31	1.848	.5412	.0436	22.938	.0236	42.379	307.954	13.426	31
32	1.885	.5306	.0426	23.468	.0226	44.227	324.403	13.823	32
33	1.922	.5202	.0417	23.989	.0217	46.112	341.051	14.217	33
34	1.961	.5100	.0408	24.499	.0208	48.034	357.882	14.608	34
35	2.000	.5000	.0400	24.999	.0200	49.994	374.883	14.996	35
36	2.040	.4902	.0392	25.489	.0192	51.994	392.040	15.381	36
40	2.208	.4529	.0366	27.355	.0166	60.402	461.993	16.889	40
48	2.587	.3865	.0326	30.673	.0126	79.354	605.966	19.756	48
50	2.692	.3715	.0318	31.424	.0118	84.579	642.361	20.442	50
60	3.281	.3048	.0288	34.761	.00877	114.1	823.698	23.696	60
70	4.000	.2500	.0267	37.499	.00667	150.0	999.834	26.663	70
72	4.161	.2403	.0263	37.984	.00633	158.1	1034.056	27.223	72
80	4.875	.2051	.0252	39.745	.00516	193.8	1166.787	29.357	80
90	5.943	.1683	.0240	41.587	.00405	247.2	1322.170	31.793	90
100	7.245	.1380	.0232	43.098	.00320	312.2	1464.753	33.986	100
120	10.765	.0929	.0220	45.355	.00205	488.3	1710.416	37.711	120
240	115.889	.00863	.0202	49.569	.00017	5744	2374.880	47.911	240
360	1248	.00080	.0200	49.960	.0000160	62328	2483.568	49.711	360
480	13430	.00007	.0200	49.996	.0000015	671460	2498.027	49.964	480
∞	∞	0	.02	50	0	∞	2500	50	∞

	Single Payment		Uniform Payment Series				Arithmetic Gradient		
	Compound Amount Factor	Present Worth Factor	Capital Recovery Factor	Present Worth Factor	Sinking Fund Factor	Compound Amount Factor	Present Worth Factor	Uniform Payment Factor	
N	F/P	P/F	A/P	P/A	A/F	F/A	P/G	A/G	N
1	1.030	.9709	1.0300	.971	1.0000	1.000	0	0	1
2	1.061	.9426	.5226	1.913	.4926	2.030	.943	.493	2
3	1.093	.9151	.3535	2.829	.3235	3.091	2.773	.980	3
4	1.126	.8885	.2690	3.717	.2390	4.184	5.438	1.463	4
5	1.159	.8626	.2184	4.580	.1884	5.309	8.889	1.941	5
6	1.194	.8375	.1846	5.417	.1546	6.468	13.076	2.414	6
7	1.230	.8131	.1605	6.230	.1305	7.662	17.955	2.882	7
8	1.267	.7894	.1425	7.020	.1125	8.892	23.481	3.345	8
9	1.305	.7664	.1284	7.786	.0984	10.159	29.612	3.803	9
10	1.344	.7441	.1172	8.530	.0872	11.464	36.309	4.256	10
11	1.384	.7224	.1081	9.253	.0781	12.808	43.533	4.705	11
12	1.426	.7014	.1005	9.954	.0705	14.192	51.248	5.148	12
13	1.469	.6810	.0940	10.635	.0640	15.618	59.420	5.587	13
14	1.513	.6611	.0885	11.296	.0585	17.086	68.014	6.021	14
15	1.558	.6419	.0838	11.938	.0538	18.599	77.000	6.450	15
16	1.605	.6232	.0796	12.561	.0496	20.157	86.348	6.874	16
17	1.653	.6050	.0760	13.166	.0460	21.762	96.028	7.294	17
18	1.702	.5874	.0727	13.754	.0427	23.414	106.014	7.708	18
19	1.754	.5703	.0698	14.324	.0398	25.117	116.279	8.118	19
20	1.806	.5537	.0672	14.877	.0372	26.870	126.799	8.523	20
21	1.860	.5375	.0649	15.415	.0349	28.676	137.550	8.923	21
22	1.916	.5219	.0627	15.937	.0327	30.537	148.509	9.319	22
23	1.974	.5067	.0608	16.444	.0308	32.453	159.657	9.709	23
24	2.033	.4919	.0590	16.936	.0290	34.426	170.971	10.095	24
25	2.094	.4776	.0574	17.413	.0274	36.459	182.434	10.477	25
26	2.157	.4637	.0559	17.877	.0259	38.553	194.026	10.853	26
27	2.221	.4502	.0546	18.327	.0246	40.710	205.731	11.226	27
28	2.288	.4371	.0533	18.764	.0233	42.931	217.532	11.593	28
29	2.357	.4243	.0521	19.188	.0221	45.219	229.414	11.956	29
30	2.427	.4120	.0510	19.600	.0210	47.575	241.361	12.314	30
31	2.500	.4000	.0500	20.000	.0200	50.003	253.361	12.668	31
32	2.575	.3883	.0490	20.389	.0190	52.503	265.399	13.017	32
33	2.652	.3770	.0482	20.766	.0182	55.078	277.464	13.362	33
34	2.732	.3660	.0473	21.132	.0173	57.730	289.544	13.702	34
35	2.814	.3554	.0465	21.487	.0165	60.462	301.627	14.037	35
40	3.262	.3066	.0433	23.115	.0133	75.401	361.750	15.650	40
45	3.782	.2644	.0408	24.519	.0108	92.720	420.632	17.156	45
50	4.384	.2281	.0389	25.730	.00887	112.8	477.480	18.558	50
55	5.082	.1968	.0373	26.774	.00735	136.1	531.741	19.860	55
60	5.892	.1697	.0361	27.676	.00613	163.1	583.053	21.067	60
65	6.830	.1464	.0351	28.453	.00515	194.3	631.201	22.184	65
70	7.918	.1263	.0343	29.123	.00434	230.6	676.087	23.215	70
75	9.179	.1089	.0337	29.702	.00367	272.6	717.698	24.163	75
80	10.641	.0940	.0331	30.201	.00311	321.4	756.087	25.035	80
85	12.336	.0811	.0326	30.631	.00265	377.9	791.353	25.835	85
90	14.300	.0699	.0323	31.002	.00226	443.3	823.630	26.567	90
95	16.578	.0603	.0319	31.323	.00193	519.3	853.074	27.235	95
100	19.219	.0520	.0316	31.599	.00165	607.3	879.854	27.844	100
∞	∞	0	.03	33.333	0	∞	1111.111	33.333	∞

End-of-Period Compound Interest Factors

	Single Payment		Uniform Payment Series				Arithmetic Gradient		
N	Compound Amount Factor F/P	Present Worth Factor P/F	Capital Recovery Factor A/P	Present Worth Factor P/A	Sinking Fund Factor A/F	Compound Amount Factor F/A	Present Worth Factor P/G	Uniform Payment Factor A/G	N
1	1.040	.9615	1.0400	.962	1.0000	1.000	0	0	1
2	1.082	.9246	.5302	1.886	.4902	2.040	.925	.490	2
3	1.125	.8890	.3603	2.775	.3203	3.122	2.703	.974	3
4	1.170	.8548	.2755	3.630	.2355	4.246	5.267	1.451	4
5	1.217	.8219	.2246	4.452	.1846	5.416	8.555	1.922	5
6	1.265	.7903	.1908	5.242	.1508	6.633	12.506	2.386	6
7	1.316	.7599	.1666	6.002	.1266	7.898	17.066	2.843	7
8	1.369	.7307	.1485	6.733	.1085	9.214	22.181	3.294	8
9	1.423	.7026	.1345	7.435	.0945	10.583	27.801	3.739	9
10	1.480	.6756	.1233	8.111	.0833	12.006	33.881	4.177	10
11	1.539	.6496	.1141	8.760	.0741	13.486	40.377	4.609	11
12	1.601	.6246	.1066	9.385	.0666	15.026	47.248	5.034	12
13	1.665	.6006	.1001	9.986	.0601	16.627	54.455	5.453	13
14	1.732	.5775	.0947	10.563	.0547	18.292	61.962	5.866	14
15	1.801	.5553	.0899	11.118	.0499	20.024	69.735	6.272	15
16	1.873	.5339	.0858	11.652	.0458	21.825	77.744	6.672	16
17	1.948	.5134	.0822	12.166	.0422	23.698	85.958	7.066	17
18	2.026	.4936	.0790	12.659	.0390	25.645	94.350	7.453	18
19	2.107	.4746	.0761	13.134	.0361	27.671	102.893	7.834	19
20	2.191	.4564	.0736	13.590	.0336	29.778	111.565	8.209	20
21	2.279	.4388	.0713	14.029	.0313	31.969	120.341	8.578	21
22	2.370	.4220	.0692	14.451	.0292	34.248	129.202	8.941	22
23	2.465	.4057	.0673	14.857	.0273	36.618	138.128	9.297	23
24	2.563	.3901	.0656	15.247	.0256	39.083	147.101	9.648	24
25	2.666	.3751	.0640	15.622	.0240	41.646	156.104	9.993	25
26	2.772	.3607	.0626	15.983	.0226	44.312	165.121	10.331	26
27	2.883	.3468	.0612	16.330	.0212	47.084	174.138	10.664	27
28	2.999	.3335	.0600	16.663	.0200	49.968	183.142	10.991	28
29	3.119	.3207	.0589	16.984	.0189	52.966	192.121	11.312	29
30	3.243	.3083	.0578	17.292	.0178	56.085	201.062	11.627	30
31	3.373	.2965	.0569	17.588	.0169	59.328	209.956	11.937	31
32	3.508	.2851	.0559	17.874	.0159	62.701	218.792	12.241	32
33	3.648	.2741	.0551	18.148	.0151	66.210	227.563	12.540	33
34	3.794	.2636	.0543	18.411	.0143	69.858	236.261	12.832	34
35	3.946	.2534	.0536	18.665	.0136	73.652	244.877	13.120	35
40	4.801	.2083	.0505	19.793	.0105	95.026	286.530	14.477	40
45	5.841	.1712	.0483	20.720	.00826	121.0	325.403	15.705	45
50	7.107	.1407	.0466	21.482	.00655	152.7	361.164	16.812	50
55	8.646	.1157	.0452	22.109	.00523	191.2	393.689	17.807	55
60	10.520	.0951	.0442	22.623	.00420	238.0	422.997	18.697	60
65	12.799	.0781	.0434	23.047	.00339	295.0	449.201	19.491	65
70	15.572	.0642	.0427	23.395	.00275	364.3	472.479	20.196	70
75	18.945	.0528	.0422	23.680	.00223	448.6	493.041	20.821	75
80	23.050	.0434	.0418	23.915	.00181	551.2	511.116	21.372	80
85	28.044	.0357	.0415	24.109	.00148	676.1	526.938	21.857	85
90	34.119	.0293	.0412	24.267	.00121	828.0	540.737	22.283	90
95	41.511	.0241	.0410	24.398	.000987	1013	552.731	22.655	95
100	50.505	.0198	.0408	24.505	.000808	1238	563.125	22.980	100
∞	∞	0	.04	25	0	∞	625	25	∞

	Single Payment		Uniform Payment Series				Arithmetic Gradient		
N	Compound Amount Factor F/P	Present Worth Factor P/F	Capital Recovery Factor A/P	Present Worth Factor P/A	Sinking Fund Factor A/F	Compound Amount Factor F/A	Present Worth Factor P/G	Uniform Payment Factor A/G	N
1	1.050	.9524	1.0500	.952	1.0000	1.	0	0	1
2	1.103	.9070	.5378	1.859	.4878	2.050	.907	.488	2
3	1.158	.8638	.3672	2.723	.3172	3.153	2.635	.967	3
4	1.216	.8227	.2820	3.546	.2320	4.310	5.103	1.439	4
5	1.276	.7835	.2310	4.329	.1810	5.526	8.237	1.903	5
6	1.340	.7462	.1970	5.076	.1470	6.802	11.968	2.358	6
7	1.407	.7107	.1728	5.786	.1228	8.142	16.232	2.805	7
8	1.477	.6768	.1547	6.463	.1047	9.549	20.970	3.245	8
9	1.551	.6446	.1407	7.108	.0907	11.027	26.127	3.676	9
10	1.629	.6139	.1295	7.722	.0795	12.578	31.652	4.099	10
11	1.710	.5847	.1204	8.306	.0704	14.207	37.499	4.514	11
12	1.796	.5568	.1128	8.863	.0628	15.917	43.624	4.922	12
13	1.886	.5303	.1065	9.394	.0565	17.713	49.988	5.322	13
14	1.980	.5051	.1010	9.899	.0510	19.599	56.554	5.713	14
15	2.079	.4810	.0963	10.380	.0463	21.579	63.288	6.097	15
16	2.183	.4581	.0923	10.838	.0423	23.657	70.160	6.474	16
17	2.292	.4363	.0887	11.274	.0387	25.840	77.140	6.842	17
18	2.407	.4155	.0855	11.690	.0355	28.132	84.204	7.203	18
19	2.527	.3957	.0827	12.085	.0327	30.539	91.328	7.557	19
20	2.653	.3769	.0802	12.462	.0302	33.066	98.488	7.903	20
21	2.786	.3589	.0780	12.821	.0280	35.719	105.667	8.242	21
22	2.925	.3418	.0760	13.163	.0260	38.505	112.846	8.573	22
23	3.072	.3256	.0741	13.489	.0241	41.430	120.009	8.897	23
24	3.225	.3101	.0725	13.799	.0225	44.502	127.140	9.214	24
25	3.386	.2953	.0710	14.094	.0210	47.727	134.228	9.524	25
26	3.556	.2812	.0696	14.375	.0196	51.113	141.259	9.827	26
27	3.733	.2678	.0683	14.643	.0183	54.669	148.223	10.122	27
28	3.920	.2551	.0671	14.898	.0171	58.403	155.110	10.411	28
29	4.116	.2429	.0660	15.141	.0160	62.323	161.913	10.694	29
30	4.322	.2314	.0651	15.372	.0151	66.439	168.623	10.969	30
31	4.538	.2204	.0641	15.593	.0141	70.761	175.233	11.238	31
32	4.765	.2099	.0633	15.803	.0133	75.299	181.739	11.501	32
33	5.003	.1999	.0625	16.003	.0125	80.064	188.135	11.757	33
34	5.253	.1904	.0618	16.193	.0118	85.067	194.417	12.006	34
35	5.516	.1813	.0611	16.374	.0111	90.320	200.581	12.250	35
40	7.040	.1420	.0583	17.159	.00828	120.8	229.545	13.377	40
45	8.985	.1113	.0563	17.774	.00626	159.7	255.315	14.364	45
50	11.467	.0872	.0548	18.256	.00478	209.3	277.915	15.223	50
55	14.636	.0683	.0537	18.633	.00367	272.7	297.510	15.966	55
60	18.679	.0535	.0528	18.929	.00283	353.6	314.343	16.606	60
65	23.840	.0419	.0522	19.161	.00219	456.8	328.691	17.154	65
70	30.426	.0329	.0517	19.343	.00170	588.5	340.841	17.621	70
75	38.833	.0258	.0513	19.485	.00132	756.7	351.072	18.018	75
80	49.561	.0202	.0510	19.596	.00103	971.2	359.646	18.353	80
85	63.254	.0158	.0508	19.684	.000803	1245	366.801	18.635	85
90	80.730	.0124	.0506	19.752	.000627	1595	372.749	18.871	90
95	103.0	.00971	.0505	19.806	.000490	2041	377.677	19.069	95
100	131.5	.00760	.0504	19.848	.000383	2610	381.749	19.234	100
∞	∞	0	.05	20	0	∞	400	20	∞

	Single Payment		Uniform Payment Series				Arithmetic Gradient		
	Compound Amount Factor	Present Worth Factor	Capital Recovery Factor	Present Worth Factor	Sinking Fund Factor	Compound Amount Factor	Present Worth Factor	Uniform Payment Factor	
N	F/P	P/F	A/P	P/A	A/F	F/A	P/G	A/G	N
1	1.060	.9434	1.0600	.943	1.0000	1.000	0	0	1
2	1.124	.8900	.5454	1.833	.4854	2.060	.890	.485	2
3	1.191	.8396	.3741	2.673	.3141	3.184	2.569	.961	3
4	1.262	.7921	.2886	3.465	.2286	4.375	4.946	1.427	4
5	1.338	.7473	.2374	4.212	.1774	5.637	7.935	1.884	5
6	1.419	.7050	.2034	4.917	.1434	6.975	11.459	2.330	6
7	1.504	.6651	.1791	5.582	.1191	8.394	15.450	2.768	7
8	1.594	.6274	.1610	6.210	.1010	9.897	19.842	3.195	8
9	1.689	.5919	.1470	6.802	.0870	11.491	24.577	3.613	9
10	1.791	.5584	.1359	7.360	.0759	13.181	29.602	4.022	10
11	1.898	.5268	.1268	7.887	.0668	14.972	34.870	4.421	11
12	2.012	.4970	.1193	8.384	.0593	16.870	40.337	4.811	12
13	2.133	.4688	.1130	8.853	.0530	18.882	45.963	5.192	13
14	2.261	.4423	.1076	9.295	.0476	21.015	51.713	5.564	14
15	2.397	.4173	.1030	9.712	.0430	23.276	57.555	5.926	15
16	2.540	.3936	.0990	10.106	.0390	25.673	63.459	6.279	16
17	2.693	.3714	.0954	10.477	.0354	28.213	69.401	6.624	17
18	2.854	.3503	.0924	10.828	.0324	30.906	75.357	6.960	18
19	3.026	.3305	.0896	11.158	.0296	33.760	81.306	7.287	19
20	3.207	.3118	.0872	11.470	.0272	36.786	87.230	7.605	20
21	3.400	.2942	.0850	11.764	.0250	39.993	93.114	7.915	21
22	3.604	.2775	.0830	12.042	.0230	43.392	98.941	8.217	22
23	3.820	.2618	.0813	12.303	.0213	46.996	104.701	8.510	23
24	4.049	.2470	.0797	12.550	.0197	50.816	110.381	8.795	24
25	4.292	.2330	.0782	12.783	.0182	54.865	115.973	9.072	25
26	4.549	.2198	.0769	13.003	.0169	59.156	121.468	9.341	26
27	4.822	.2074	.0757	13.211	.0157	63.706	126.860	9.603	27
28	5.112	.1956	.0746	13.406	.0146	68.528	132.142	9.857	28
29	5.418	.1846	.0736	13.591	.0136	73.640	137.310	10.103	29
30	5.743	.1741	.0726	13.765	.0126	79.058	142.359	10.342	30
31	6.088	.1643	.0718	13.929	.0118	84.802	147.286	10.574	31
32	6.453	.1550	.0710	14.084	.0110	90.890	152.090	10.799	32
33	6.841	.1462	.0703	14.230	.0103	97.343	156.768	11.017	33
34	7.251	.1379	.0696	14.368	.00960	104.2	161.319	11.228	34
35	7.686	.1301	.0690	14.498	.00897	111.4	165.743	11.432	35
40	10.286	.0972	.0665	15.046	.00646	154.8	185.957	12.359	40
45	13.765	.0727	.0647	15.456	.00470	212.7	203.110	13.141	45
50	18.420	.0543	.0634	15.762	.00344	290.3	217.457	13.796	50
55	24.650	.0406	.0625	15.991	.00254	394.2	229.322	14.341	55
60	32.988	.0303	.0619	16.161	.00188	533.1	239.043	14.791	60
65	44.145	.0227	.0614	16.289	.00139	719.1	246.945	15.160	65
70	59.076	.0169	.0610	16.385	.00103	967.9	253.327	15.461	70
75	79.057	.0126	.0608	16.456	.000769	1301	258.453	15.706	75
80	105.8	.00945	.0606	16.509	.000573	1747	262.549	15.903	80
85	141.6	.00706	.0604	16.549	.000427	2343	265.810	16.062	85
90	189.5	.00528	.0603	16.579	.000318	3141	268.395	16.189	90
95	253.5	.00394	.0602	16.601	.000238	4209	270.437	16.290	95
100	339.3	.00295	.0602	16.618	.000177	5638	272.047	16.371	100
∞	∞	0	.06	16.667	0	∞	277.778	16.667	∞

	Single Payment		Uniform Payment Series				Arithmetic Gradient		
N	Compound Amount Factor F/P	Present Worth Factor P/F	Capital Recovery Factor A/P	Present Worth Factor P/A	Sinking Fund Factor A/F	Compound Amount Factor F/A	Present Worth Factor P/G	Uniform Payment Factor A/G	N
1	1.070	.9346	1.0700	.935	1.0000	1.000	0	0	1
2	1.145	.8734	.5531	1.808	.4831	2.070	.873	.483	2
3	1.225	.8163	.3811	2.624	.3111	3.215	2.506	.955	3
4	1.311	.7629	.2952	3.387	.2252	4.440	4.795	1.416	4
5	1.403	.7130	.2439	4.100	.1739	5.751	7.647	1.865	5
6	1.501	.6663	.2098	4.767	.1398	7.153	10.978	2.303	6
7	1.606	.6227	.1856	5.389	.1156	8.654	14.715	2.730	7
8	1.718	.5820	.1675	5.971	.0975	10.260	18.789	3.147	8
9	1.838	.5439	.1535	6.515	.0835	11.978	23.140	3.552	9
10	1.967	.5083	.1424	7.024	.0724	13.816	27.716	3.946	10
11	2.105	.4751	.1334	7.499	.0634	15.784	32.466	4.330	11
12	2.252	.4440	.1259	7.943	.0559	17.888	37.351	4.703	12
13	2.410	.4150	.1197	8.358	.0497	20.141	42.330	5.065	13
14	2.579	.3878	.1143	8.745	.0443	22.550	47.372	5.417	14
15	2.759	.3624	.1098	9.108	.0398	25.129	52.446	5.758	15
16	2.952	.3387	.1059	9.447	.0359	27.888	57.527	6.090	16
17	3.159	.3166	.1024	9.763	.0324	30.840	62.592	6.411	17
18	3.380	.2959	.0994	10.059	.0294	33.999	67.622	6.722	18
19	3.617	.2765	.0968	10.336	.0268	37.379	72.599	7.024	19
20	3.870	.2584	.0944	10.594	.0244	40.995	77.509	7.316	20
21	4.141	.2415	.0923	10.836	.0223	44.865	82.339	7.599	21
22	4.430	.2257	.0904	11.061	.0204	49.006	87.079	7.872	22
23	4.741	.2109	.0887	11.272	.0187	53.436	91.720	8.137	23
24	5.072	.1971	.0872	11.469	.0172	58.177	96.255	8.392	24
25	5.427	.1842	.0858	11.654	.0158	63.249	100.676	8.639	25
26	5.807	.1722	.0846	11.826	.0146	68.676	104.981	8.877	26
27	6.214	.1609	.0834	11.987	.0134	74.484	109.166	9.107	27
28	6.649	.1504	.0824	12.137	.0124	80.698	113.226	9.329	28
29	7.114	.1406	.0814	12.278	.0114	87.347	117.162	9.543	29
30	7.612	.1314	.0806	12.409	.0106	94.461	120.972	9.749	30
31	8.145	.1228	.0798	12.532	.00980	102.1	124.655	9.947	31
32	8.715	.1147	.0791	12.647	.00907	110.2	128.212	10.138	32
33	9.325	.1072	.0784	12.754	.00841	118.9	131.643	10.322	33
34	9.978	.1002	.0778	12.854	.00780	128.3	134.951	10.499	34
35	10.677	.0937	.0772	12.948	.00723	138.2	138.135	10.669	35
40	14.974	.0668	.0750	13.332	.00501	199.6	152.293	11.423	40
45	21.002	.0476	.0735	13.606	.00350	285.7	163.756	12.036	45
50	29.457	.0339	.0725	13.801	.00246	406.5	172.905	12.529	50
55	41.315	.0242	.0717	13.940	.00174	575.9	180.124	12.921	55
60	57.946	.0173	.0712	14.039	.00123	813.5	185.768	13.232	60
65	81.273	.0123	.0709	14.110	.000872	1147	190.145	13.476	65
70	114.0	.00877	.0706	14.160	.000620	1614	193.519	13.666	70
75	159.9	.00625	.0704	14.196	.000441	2270	196.104	13.814	75
80	224.2	.00446	.0703	14.222	.000314	3189	198.075	13.927	80
85	314.5	.00318	.0702	14.240	.000223	4479	199.572	14.015	85
90	441.1	.00227	.0702	14.253	.000159	6287	200.704	14.081	90
95	618.7	.00162	.0701	14.263	.000113	8824	201.558	14.132	95
100	867.7	.00115	.0701	14.269	.0000808	12382	202.200	14.170	100
∞	∞	0	.07	14.286	0	∞	204.082	14.286	∞

	Single Payment		Uniform Payment Series				Arithmetic Gradient		
N	Compound Amount Factor F/P	Present Worth Factor P/F	Capital Recovery Factor A/P	Present Worth Factor P/A	Sinking Fund Factor A/F	Compound Amount Factor F/A	Present Worth Factor P/G	Uniform Payment Factor A/G	N
1	1.080	.9259	1.0800	.926	1.0000	1.000	0	0	1
2	1.166	.8573	.5608	1.783	.4808	2.080	.857	.481	2
3	1.260	.7938	.3880	2.577	.3080	3.246	2.445	.949	3
4	1.360	.7350	.3019	3.312	.2219	4.506	4.650	1.404	4
5	1.469	.6806	.2505	3.993	.1705	5.867	7.372	1.846	5
6	1.587	.6302	.2163	4.623	.1363	7.336	10.523	2.276	6
7	1.714	.5835	.1921	5.206	.1121	8.923	14.024	2.694	7
8	1.851	.5403	.1740	5.747	.0940	10.637	17.806	3.099	8
9	1.999	.5002	.1601	6.247	.0801	12.488	21.808	3.491	9
10	2.159	.4632	.1490	6.710	.0690	14.487	25.977	3.871	10
11	2.332	.4289	.1401	7.139	.0601	16.645	30.266	4.240	11
12	2.518	.3971	.1327	7.536	.0527	18.977	34.634	4.596	12
13	2.720	.3677	.1265	7.904	.0465	21.495	39.046	4.940	13
14	2.937	.3405	.1213	8.244	.0413	24.215	43.472	5.273	14
15	3.172	.3152	.1168	8.559	.0368	27.152	47.886	5.594	15
16	3.426	.2919	.1130	8.851	.0330	30.324	52.264	5.905	16
17	3.700	.2703	.1096	9.122	.0296	33.750	56.588	6.204	17
18	3.996	.2502	.1067	9.372	.0267	37.450	60.843	6.492	18
19	4.316	.2317	.1041	9.604	.0241	41.446	65.013	6.770	19
20	4.661	.2145	.1019	9.818	.0219	45.762	69.090	7.037	20
21	5.034	.1987	.0998	10.017	.0198	50.423	73.063	7.294	21
22	5.437	.1839	.0980	10.201	.0180	55.457	76.926	7.541	22
23	5.871	.1703	.0964	10.371	.0164	60.893	80.673	7.779	23
24	6.341	.1577	.0950	10.529	.0150	66.765	84.300	8.007	24
25	6.848	.1460	.0937	10.675	.0137	73.106	87.804	8.225	25
26	7.396	.1352	.0925	10.810	.0125	79.954	91.184	8.435	26
27	7.988	.1252	.0914	10.935	.0114	87.351	94.439	8.636	27
28	8.627	.1159	.0905	11.051	.0105	95.339	97.569	8.829	28
29	9.317	.1073	.0896	11.158	.00962	104.0	100.574	9.013	29
30	10.063	.0994	.0888	11.258	.00883	113.3	103.456	9.190	30
31	10.868	.0920	.0881	11.350	.00811	123.3	106.216	9.358	31
32	11.737	.0852	.0875	11.435	.00745	134.2	108.857	9.520	32
33	12.676	.0789	.0869	11.514	.00685	146.0	111.382	9.674	33
34	13.690	.0730	.0863	11.587	.00630	158.6	113.792	9.821	34
35	14.785	.0676	.0858	11.655	.00580	172.3	116.092	9.961	35
40	21.725	.0460	.0839	11.925	.00386	259.1	126.042	10.570	40
45	31.920	.0313	.0826	12.108	.00259	386.5	133.733	11.045	45
50	46.902	.0213	.0817	12.233	.00174	573.8	139.593	11.411	50
55	68.914	.0145	.0812	12.319	.00118	848.9	144.006	11.690	55
60	101.3	.00988	.0808	12.377	.000798	1253	147.300	11.902	60
65	148.8	.00672	.0805	12.416	.000541	1847	149.739	12.060	65
70	218.6	.00457	.0804	12.443	.000368	2720	151.533	12.178	70
75	321.2	.00311	.0802	12.461	.000250	4003	152.845	12.266	75
80	472.0	.00212	.0802	12.474	.000170	5887	153.800	12.330	80
85	693.5	.00144	.0801	12.482	.000116	8656	154.492	12.377	85
90	1019	.000981	.0801	12.488	.0000786	12724	154.993	12.412	90
95	1497	.000668	.0801	12.492	.0000535	18702	155.352	12.437	95
100	2200	.000455	.0800	12.494	.0000364	27485	155.611	12.455	100
∞	∞	0	.08	12.5	0	∞	156.250	12.5	∞

	Single Payment		Uniform Payment Series				Arithmetic Gradient		
N	Compound Amount Factor F/P	Present Worth Factor P/F	Capital Recovery Factor A/P	Present Worth Factor P/A	Sinking Fund Factor A/F	Compound Amount Factor F/A	Present Worth Factor P/G	Uniform Payment Factor A/G	N
1	1.090	.9174	1.0900	.917	1.0000	1.000	0	0	1
2	1.188	.8417	.5685	1.759	.4785	2.090	.842	.478	2
3	1.295	.7722	.3951	2.531	.3051	3.278	2.386	.943	3
4	1.412	.7084	.3087	3.240	.2187	4.573	4.511	1.393	4
5	1.539	.6499	.2571	3.890	.1671	5.985	7.111	1.828	5
6	1.677	.5963	.2229	4.486	.1329	7.523	10.092	2.250	6
7	1.828	.5470	.1987	5.033	.1087	9.200	13.375	2.657	7
8	1.993	.5019	.1807	5.535	.0907	11.028	16.888	3.051	8
9	2.172	.4604	.1668	5.995	.0768	13.021	20.571	3.431	9
10	2.367	.4224	.1558	6.418	.0658	15.193	24.373	3.798	10
11	2.580	.3875	.1469	6.805	.0569	17.560	28.248	4.151	11
12	2.813	.3555	.1397	7.161	.0497	20.141	32.159	4.491	12
13	3.066	.3262	.1336	7.487	.0436	22.953	36.073	4.818	13
14	3.342	.2992	.1284	7.786	.0384	26.019	39.963	5.133	14
15	3.642	.2745	.1241	8.061	.0341	29.361	43.807	5.435	15
16	3.970	.2519	.1203	8.313	.0303	33.003	47.585	5.724	16
17	4.328	.2311	.1170	8.544	.0270	36.974	51.282	6.002	17
18	4.717	.2120	.1142	8.756	.0242	41.301	54.886	6.269	18
19	5.142	.1945	.1117	8.950	.0217	46.018	58.387	6.524	19
20	5.604	.1784	.1095	9.129	.0195	51.160	61.777	6.767	20
21	6.109	.1637	.1076	9.292	.0176	56.765	65.051	7.001	21
22	6.659	.1502	.1059	9.442	.0159	62.873	68.205	7.223	22
23	7.258	.1378	.1044	9.580	.0144	69.532	71.236	7.436	23
24	7.911	.1264	.1030	9.707	.0130	76.790	74.143	7.638	24
25	8.623	.1160	.1018	9.823	.0118	84.701	76.926	7.832	25
26	9.399	.1064	.1007	9.929	.01072	93.3	79.586	8.016	26
27	10.245	.0976	.0997	10.027	.00973	102.7	82.124	8.191	27
28	11.167	.0895	.0989	10.116	.00885	113.0	84.542	8.357	28
29	12.172	.0822	.0981	10.198	.00806	124.1	86.842	8.515	29
30	13.268	.0754	.0973	10.274	.00734	136.3	89.028	8.666	30
31	14.462	.0691	.0967	10.343	.00669	149.6	91.102	8.808	31
32	15.763	.0634	.0961	10.406	.00610	164.0	93.069	8.944	32
33	17.182	.0582	.0956	10.464	.00556	179.8	94.931	9.072	33
34	18.728	.0534	.0951	10.518	.00508	197.0	96.693	9.193	34
35	20.414	.0490	.0946	10.567	.00464	215.7	98.359	9.308	35
40	31.409	.0318	.0930	10.757	.00296	337.9	105.376	9.796	40
45	48.327	.0207	.0919	10.881	.00190	525.9	110.556	10.160	45
50	74.358	.0134	.0912	10.962	.00123	815.1	114.325	10.430	50
55	114.4	.00874	.0908	11.014	.00079	1260	117.036	10.626	55
60	176.0	.00568	.0905	11.048	.00051	1945	118.968	10.768	60
65	270.8	.00369	.0903	11.070	.000334	2998	120.334	10.870	65
70	416.7	.00240	.0902	11.084	.000216	4619	121.294	10.943	70
75	641.2	.00156	.0901	11.094	.000141	7113	121.965	10.994	75
80	986.6	.00101	.0901	11.100	.0000913	10951	122.431	11.030	80
85	1518	.000659	.0901	11.104	.0000593	16855	122.753	11.055	85
90	2336	.000428	.0900	11.106	.0000386	25939	122.976	11.073	90
95	3593	.000278	.0900	11.108	.0000251	39917	123.129	11.085	95
100	5529	.000181	.0900	11.109	.0000163	61423	123.234	11.093	100
∞	∞	0	.09	11.111	0	∞	123.457	11.111	∞

	Single Payment		Uniform Payment Series				Arithmetic Gradient		
	Compound Amount Factor	Present Worth Factor	Capital Recovery Factor	Present Worth Factor	Sinking Fund Factor	Compound Amount Factor	Present Worth Factor	Uniform Payment Factor	
N	F/P	P/F	A/P	P/A	A/F	F/A	P/G	A/G	N
1	1.100	.9091	1.1000	.909	1.0000	1.000	0	0	1
2	1.210	.8264	.5762	1.736	.4762	2.100	.826	.476	2
3	1.331	.7513	.4021	2.487	.3021	3.310	2.329	.937	3
4	1.464	.6830	.3155	3.170	.2155	4.641	4.378	1.381	4
5	1.611	.6209	.2638	3.791	.1638	6.105	6.862	1.810	5
6	1.772	.5645	.2296	4.355	.1296	7.716	9.684	2.224	6
7	1.949	.5132	.2054	4.868	.1054	9.487	12.763	2.622	7
8	2.144	.4665	.1874	5.335	.0874	11.436	16.029	3.004	8
9	2.358	.4241	.1736	5.759	.0736	13.579	19.421	3.372	9
10	2.594	.3855	.1627	6.145	.0627	15.937	22.891	3.725	10
11	2.853	.3505	.1540	6.495	.0540	18.531	26.396	4.064	11
12	3.138	.3186	.1468	6.814	.0468	21.384	29.901	4.388	12
13	3.452	.2897	.1408	7.103	.0408	24.523	33.377	4.699	13
14	3.797	.2633	.1357	7.367	.0357	27.975	36.800	4.996	14
15	4.177	.2394	.1315	7.606	.0315	31.772	40.152	5.279	15
16	4.595	.2176	.1278	7.824	.0278	35.950	43.416	5.549	16
17	5.054	.1978	.1247	8.022	.0247	40.545	46.582	5.807	17
18	5.560	.1799	.1219	8.201	.0219	45.599	49.640	6.053	18
19	6.116	.1635	.1195	8.365	.0195	51.159	52.583	6.286	19
20	6.727	.1486	.1175	8.514	.0175	57.275	55.407	6.508	20
21	7.400	.1351	.1156	8.649	.0156	64.002	58.110	6.719	21
22	8.140	.1228	.1140	8.772	.0140	71.403	60.689	6.919	22
23	8.954	.1117	.1126	8.883	.0126	79.543	63.146	7.108	23
24	9.850	.1015	.1113	8.985	.0113	88.497	65.481	7.288	24
25	10.835	.0923	.1102	9.077	.0102	98.347	67.696	7.458	25
26	11.918	.0839	.1092	9.161	.00916	109.2	69.794	7.619	26
27	13.110	.0763	.1083	9.237	.00826	121.1	71.777	7.770	27
28	14.421	.0693	.1075	9.307	.00745	134.2	73.650	7.914	28
29	15.863	.0630	.1067	9.370	.00673	148.6	75.415	8.049	29
30	17.449	.0573	.1061	9.427	.00608	164.5	77.077	8.176	30
31	19.194	.0521	.1055	9.479	.00550	181.9	78.640	8.296	31
32	21.114	.0474	.1050	9.526	.00497	201.1	80.108	8.409	32
33	23.225	.0431	.1045	9.569	.00450	222.3	81.486	8.515	33
34	25.548	.0391	.1041	9.609	.00407	245.5	82.777	8.615	34
35	28.102	.0356	.1037	9.644	.00369	271.0	83.987	8.709	35
40	45.259	.0221	.1023	9.779	.00226	442.6	88.953	9.096	40
45	72.890	.0137	.1014	9.863	.00139	718.9	92.454	9.374	45
50	117.4	.00852	.1009	9.915	.00086	1164	94.889	9.570	50
55	189.1	.00529	.1005	9.947	.00053	1881	96.562	9.708	55
60	304.5	.00328	.1003	9.967	.00033	3035	97.701	9.802	60
65	490.4	.00204	.1002	9.980	.00020	4894	98.471	9.867	65
70	789.7	.00127	.1001	9.987	.00013	7887	98.987	9.911	70
75	1272	.00079	.1001	9.992	.000079	12709	99.332	9.941	75
80	2048	.00049	.1000	9.995	.000049	20474	99.561	9.961	80
85	3299	.00030	.1000	9.997	.000030	32980	99.712	9.974	85
90	5313	.00019	.1000	9.998	.000019	53120	99.812	9.983	90
95	8557	.00012	.1000	9.999	.000012	85557	99.877	9.989	95
100	13781	.00007	.1000	9.999	.000007	137796	99.920	9.993	100
∞	∞	0	.1	10	0	∞	100	10	∞

End-of-Period Compound Interest Factors

	Single Payment		Uniform Payment Series				Arithmetic Gradient		
	Compound Amount Factor	Present Worth Factor	Capital Recovery Factor	Present Worth Factor	Sinking Fund Factor	Compound Amount Factor	Present Worth Factor	Uniform Payment Factor	
N	F/P	P/F	A/P	P/A	A/F	F/A	P/G	A/G	N
1	1.110	.9009	1.1100	.901	1.0000	1.000	0	0	1
2	1.232	.8116	.5839	1.713	.4739	2.110	.812	.474	2
3	1.368	.7312	.4092	2.444	.2992	3.342	2.274	.931	3
4	1.518	.6587	.3223	3.102	.2123	4.710	4.250	1.370	4
5	1.685	.5935	.2706	3.696	.1606	6.228	6.624	1.792	5
6	1.870	.5346	.2364	4.231	.1264	7.913	9.297	2.198	6
7	2.076	.4817	.2122	4.712	.1022	9.783	12.187	2.586	7
8	2.305	.4339	.1943	5.146	.0843	11.859	15.225	2.958	8
9	2.558	.3909	.1806	5.537	.0706	14.164	18.352	3.314	9
10	2.839	.3522	.1698	5.889	.0598	16.722	21.522	3.654	10
11	3.152	.3173	.1611	6.207	.0511	19.561	24.695	3.979	11
12	3.498	.2858	.1540	6.492	.0440	22.713	27.839	4.288	12
13	3.883	.2575	.1482	6.750	.0382	26.212	30.929	4.582	13
14	4.310	.2320	.1432	6.982	.0332	30.095	33.945	4.862	14
15	4.785	.2090	.1391	7.191	.0291	34.405	36.871	5.127	15
16	5.311	.1883	.1355	7.379	.0255	39.190	39.695	5.379	16
17	5.895	.1696	.1325	7.549	.0225	44.501	42.409	5.618	17
18	6.544	.1528	.1298	7.702	.0198	50.396	45.007	5.844	18
19	7.263	.1377	.1276	7.839	.0176	56.939	47.486	6.057	19
20	8.062	.1240	.1256	7.963	.0156	64.203	49.842	6.259	20
21	8.949	.1117	.1238	8.075	.0138	72.265	52.077	6.449	21
22	9.934	.1007	.1223	8.176	.0123	81.214	54.191	6.628	22
23	11.026	.0907	.1210	8.266	.0110	91.148	56.186	6.797	23
24	12.239	.0817	.1198	8.348	.00979	102.2	58.066	6.956	24
25	13.585	.0736	.1187	8.422	.00874	114.4	59.832	7.104	25
26	15.080	.0663	.1178	8.488	.00781	128.0	61.490	7.244	26
27	16.739	.0597	.1170	8.548	.00699	143.1	63.043	7.375	27
28	18.580	.0538	.1163	8.602	.00626	159.8	64.497	7.498	28
29	20.624	.0485	.1156	8.650	.00561	178.4	65.854	7.613	29
30	22.892	.0437	.1150	8.694	.00502	199.0	67.121	7.721	30
31	25.410	.0394	.1145	8.733	.00451	221.9	68.302	7.821	31
32	28.206	.0355	.1140	8.769	.00404	247.3	69.401	7.915	32
33	31.308	.0319	.1136	8.801	.00363	275.5	70.423	8.002	33
34	34.752	.0288	.1133	8.829	.00326	306.8	71.372	8.084	34
35	38.575	.0259	.1129	8.855	.00293	341.6	72.254	8.159	35
40	65.001	.0154	.1117	8.951	.00172	581.8	75.779	8.466	40
45	109.5	.00913	.1110	9.008	.00101	986.6	78.155	8.676	45
50	184.6	.00542	.1106	9.042	.000599	1669	79.734	8.819	50
55	311.0	.00322	.1104	9.062	.000355	2818	80.771	8.913	55
60	524.1	.00191	.1102	9.074	.000210	4755	81.446	8.976	60
65	883.1	.00113	.1101	9.081	.000125	8019	81.882	9.017	65
70	1488	.000672	.1101	9.085	.0000740	13518	82.161	9.044	70
75	2507	.000399	.1100	9.087	.0000439	22785	82.340	9.061	75
80	4225	.000237	.1100	9.089	.0000260	38401	82.453	9.072	80
85	7120	.000140	.1100	9.090	.0000155	64714	82.524	9.079	85
90	11997	.000083	.1100	9.090	.0000092	109053	82.570	9.083	90
95	20215	.000049	.1100	9.090	.0000054	183768	82.598	9.086	95
100	34064	.000029	.1100	9.091	.0000032	309665	82.616	9.088	100
∞	∞	0	.11	9.091	0	∞	82.645	9.091	∞

	Single Payment		Uniform Payment Series				Arithmetic Gradient		
	Compound Amount Factor	Present Worth Factor	Capital Recovery Factor	Present Worth Factor	Sinking Fund Factor	Compound Amount Factor	Present Worth Factor	Uniform Payment Factor	
N	F/P	P/F	A/P	P/A	A/F	F/A	P/G	A/G	N
1	1.120	.8929	1.1200	.893	1.0000	1.000	0	0	1
2	1.254	.7972	.5917	1.690	.4717	2.120	.797	.472	2
3	1.405	.7118	.4163	2.402	.2963	3.374	2.221	.925	3
4	1.574	.6355	.3292	3.037	.2092	4.779	4.127	1.359	4
5	1.762	.5674	.2774	3.605	.1574	6.353	6.397	1.775	5
6	1.974	.5066	.2432	4.111	.1232	8.115	8.930	2.172	6
7	2.211	.4523	.2191	4.564	.0991	10.089	11.644	2.551	7
8	2.476	.4039	.2013	4.968	.0813	12.300	14.471	2.913	8
9	2.773	.3606	.1877	5.328	.0677	14.776	17.356	3.257	9
10	3.106	.3220	.1770	5.650	.0570	17.549	20.254	3.585	10
11	3.479	.2875	.1684	5.938	.0484	20.655	23.129	3.895	11
12	3.896	.2567	.1614	6.194	.0414	24.133	25.952	4.190	12
13	4.363	.2292	.1557	6.424	.0357	28.029	28.702	4.468	13
14	4.887	.2046	.1509	6.628	.0309	32.393	31.362	4.732	14
15	5.474	.1827	.1468	6.811	.0268	37.280	33.920	4.980	15
16	6.130	.1631	.1434	6.974	.0234	42.753	36.367	5.215	16
17	6.866	.1456	.1405	7.120	.0205	48.884	38.697	5.435	17
18	7.690	.1300	.1379	7.250	.0179	55.750	40.908	5.643	18
19	8.613	.1161	.1358	7.366	.0158	63.440	42.998	5.838	19
20	9.646	.1037	.1339	7.469	.0139	72.052	44.968	6.020	20
21	10.804	.0926	.1322	7.562	.0122	81.699	46.819	6.191	21
22	12.100	.0826	.1308	7.645	.0108	92.503	48.554	6.351	22
23	13.552	.0738	.1296	7.718	.0096	104.603	50.178	6.501	23
24	15.179	.0659	.1285	7.784	.0085	118.155	51.693	6.641	24
25	17.000	.0588	.1275	7.843	.0075	133.334	53.105	6.771	25
26	19.040	.0525	.1267	7.896	.00665	150.3	54.418	6.892	26
27	21.325	.0469	.1259	7.943	.00590	169.4	55.637	7.005	27
28	23.884	.0419	.1252	7.984	.00524	190.7	56.767	7.110	28
29	26.750	.0374	.1247	8.022	.00466	214.6	57.814	7.207	29
30	29.960	.0334	.1241	8.055	.00414	241.3	58.782	7.297	30
31	33.555	.0298	.1237	8.085	.00369	271.3	59.676	7.381	31
32	37.582	.0266	.1233	8.112	.00328	304.8	60.501	7.459	32
33	42.092	.0238	.1229	8.135	.00292	342.4	61.261	7.530	33
34	47.143	.0212	.1226	8.157	.00260	384.5	61.961	7.596	34
35	52.800	.0189	.1223	8.176	.00232	431.7	62.605	7.658	35
40	93.051	.0107	.1213	8.244	.00130	767.1	65.116	7.899	40
45	164.0	.00610	.1207	8.283	.00074	1358	66.734	8.057	45
50	289.0	.00346	.1204	8.304	.00042	2400	67.762	8.160	50
55	509.3	.00196	.1202	8.317	.00024	4236	68.408	8.225	55
60	897.6	.00111	.1201	8.324	.00013	7472	68.810	8.266	60
65	1582	.00063	.1201	8.328	.000076	13174	69.058	8.292	65
70	2788	.00036	.1200	8.330	.000043	23223	69.210	8.308	70
75	4913	.00020	.1200	8.332	.000024	40934	69.303	8.318	75
80	8658	.00012	.1200	8.332	.000014	72146	69.359	8.324	80
85	15259	.000066	.1200	8.333	.0000079	127152	69.393	8.328	85
90	26892	.000037	.1200	8.333	.0000045	224091	69.414	8.330	90
95	47393	.000021	.1200	8.333	.0000025	394931	69.426	8.331	95
100	83522	.000012	.1200	8.333	.0000014	696011	69.434	8.332	100
∞	∞	0	.12	8.333	0	∞	69.444	8.333	∞

13%   **End-of-Period Compound Interest Factors**   **13%**

	Single Payment		Uniform Payment Series				Arithmetic Gradient		
N	Compound Amount Factor F/P	Present Worth Factor P/F	Capital Recovery Factor A/P	Present Worth Factor P/A	Sinking Fund Factor A/F	Compound Amount Factor F/A	Present Worth Factor P/G	Uniform Payment Factor A/G	N
1	1.130	.8850	1.1300	.885	1.0000	1.000	0	0	1
2	1.277	.7831	.5995	1.668	.4695	2.130	.783	.469	2
3	1.443	.6931	.4235	2.361	.2935	3.407	2.169	.919	3
4	1.630	.6133	.3362	2.974	.2062	4.850	4.009	1.348	4
5	1.842	.5428	.2843	3.517	.1543	6.480	6.180	1.757	5
6	2.082	.4803	.2502	3.998	.1202	8.323	8.582	2.147	6
7	2.353	.4251	.2261	4.423	.0961	10.4051	1.132	2.517	7
8	2.658	.3762	.2084	4.799	.0784	12.7571	3.765	2.869	8
9	3.004	.3329	.1949	5.132	.0649	15.4161	6.428	3.201	9
10	3.395	.2946	.1843	5.426	.0543	18.420	19.080	3.516	10
11	3.836	.2607	.1758	5.687	.0458	21.814	21.687	3.813	11
12	4.335	.2307	.1690	5.918	.0390	25.650	24.224	4.094	12
13	4.898	.2042	.1634	6.122	.0334	29.985	26.674	4.357	13
14	5.535	.1807	.1587	6.302	.0287	34.883	29.023	4.605	14
15	6.254	.1599	.1547	6.462	.0247	40.417	31.262	4.837	15
16	7.067	.1415	.1514	6.604	.0214	46.672	33.384	5.055	16
17	7.986	.1252	.1486	6.729	.0186	53.739	35.388	5.259	17
18	9.024	.1108	.1462	6.840	.0162	61.725	37.271	5.449	18
19	10.197	.0981	.1441	6.938	.0141	70.749	39.037	5.627	19
20	11.523	.0868	.1424	7.025	.0124	80.947	40.685	5.792	20
21	13.021	.0768	.1408	7.102	.0108	92.470	42.221	5.945	21
22	14.714	.0680	.1395	7.170	.00948	105.5	43.649	6.088	22
23	16.627	.0601	.1383	7.230	.00832	120.2	44.972	6.220	23
24	18.788	.0532	.1373	7.283	.00731	136.8	46.196	6.343	24
25	21.231	.0471	.1364	7.330	.00643	155.6	47.326	6.457	25
26	23.991	.0417	.1357	7.372	.00565	176.9	48.369	6.561	26
27	27.109	.0369	.1350	7.409	.00498	200.8	49.328	6.658	27
28	30.633	.0326	.1344	7.441	.00439	227.9	50.209	6.747	28
29	34.616	.0289	.1339	7.470	.00387	258.6	51.018	6.830	29
30	39.116	.0256	.1334	7.496	.00341	293.2	51.759	6.905	30
31	44.201	.0226	.1330	7.518	.00301	332.3	52.438	6.975	31
32	49.947	.0200	.1327	7.538	.00266	376.5	53.059	7.039	32
33	56.440	.0177	.1323	7.556	.00234	426.5	53.626	7.097	33
34	63.777	.0157	.1321	7.572	.00207	482.9	54.143	7.151	34
35	72.069	.0139	.1318	7.586	.00183	546.7	54.615	7.200	35
40	132.8	.00753	.1310	7.634	.000986	1014	56.409	7.389	40
45	244.6	.00409	.1305	7.661	.000534	1874	57.515	7.508	45
50	450.7	.00222	.1303	7.675	.000289	3460	58.187	7.581	50
55	830.5	.00120	.1302	7.683	.000157	6380	58.591	7.626	55
60	1530	.00065	.1301	7.687	.000085	11762	58.831	7.653	60
∞	∞	0	.13	7.692	0	∞	59.172	7.692	∞

14% **End-of-Period Compound Interest Factors** 14%

	Single Payment		Uniform Payment Series				Arithmetic Gradient		
	Compound Amount Factor	Present Worth Factor	Capital Recovery Factor	Present Worth Factor	Sinking Fund Factor	Compound Amount Factor	Present Worth Factor	Uniform Payment Factor	
N	F/P	P/F	A/P	P/A	A/F	F/A	P/G	A/G	N
1	1.140	.8772	1.1400	.877	1.0000	1.000	0	0	1
2	1.300	.7695	.6073	1.647	.4673	2.140	.769	.467	2
3	1.482	.6750	.4307	2.322	.2907	3.440	2.119	.913	3
4	1.689	.5921	.3432	2.914	.2032	4.921	3.896	1.337	4
5	1.925	.5194	.2913	3.433	.1513	6.610	5.973	1.740	5
6	2.195	.4556	.2572	3.889	.1172	8.536	8.251	2.122	6
7	2.502	.3996	.2332	4.288	.0932	10.730	10.649	2.483	7
8	2.853	.3506	.2156	4.639	.0756	13.233	13.103	2.825	8
9	3.252	.3075	.2022	4.946	.0622	16.085	15.563	3.146	9
10	3.707	.2697	.1917	5.216	.0517	19.337	17.991	3.449	10
11	4.226	.2366	.1834	5.453	.0434	23.045	20.357	3.733	11
12	4.818	.2076	.1767	5.660	.0367	27.271	22.640	4.000	12
13	5.492	.1821	.1712	5.842	.0312	32.089	24.825	4.249	13
14	6.261	.1597	.1666	6.002	.0266	37.581	26.901	4.482	14
15	7.138	.1401	.1628	6.142	.0228	43.842	28.862	4.699	15
16	8.137	.1229	.1596	6.265	.0196	50.980	30.706	4.901	16
17	9.276	.1078	.1569	6.373	.0169	59.118	32.430	5.089	17
18	10.575	.0946	.1546	6.467	.0146	68.394	34.038	5.263	18
19	12.056	.0829	.1527	6.550	.0127	78.969	35.531	5.424	19
20	13.743	.0728	.1510	6.623	.0110	91.025	36.914	5.573	20
21	15.668	.0638	.1495	6.687	.00954	104.8	38.190	5.711	21
22	17.861	.0560	.1483	6.743	.00830	120.4	39.366	5.838	22
23	20.362	.0491	.1472	6.792	.00723	138.3	40.446	5.955	23
24	23.212	.0431	.1463	6.835	.00630	158.7	41.437	6.062	24
25	26.462	.0378	.1455	6.873	.00550	181.9	42.344	6.161	25
26	30.167	.0331	.1448	6.906	.00480	208.3	43.173	6.251	26
27	34.390	.0291	.1442	6.935	.00419	238.5	43.929	6.334	27
28	39.204	.0255	.1437	6.961	.00366	272.9	44.618	6.410	28
29	44.693	.0224	.1432	6.983	.00320	312.1	45.244	6.479	29
30	50.950	.0196	.1428	7.003	.00280	356.8	45.813	6.542	30
31	58.083	.0172	.1425	7.020	.00245	407.7	46.330	6.600	31
32	66.215	.0151	.1421	7.035	.00215	465.8	46.798	6.652	32
33	75.485	.0132	.1419	7.048	.00188	532.0	47.222	6.700	33
34	86.053	.0116	.1416	7.060	.00165	607.5	47.605	6.743	34
35	98.100	.0102	.1414	7.070	.00144	693.6	47.952	6.782	35
40	188.9	.00529	.1407	7.105	.000745	1342	49.238	6.930	40
45	363.7	.00275	.1404	7.123	.000386	2591	49.996	7.019	45
50	700.2	.00143	.1402	7.133	.000200	4995	50.438	7.071	50
55	1348	.00074	.1401	7.138	.000104	9623	50.691	7.102	55
60	2596	.00039	.1401	7.140	.000054	18535	50.836	7.120	60
∞	∞	0	.14	7.143	0	∞	51.020	7.143	∞

15%   **End-of-Period Compound Interest Factors**   **15%**

	Single Payment		Uniform Payment Series				Arithmetic Gradient		
	Compound Amount Factor	Present Worth Factor	Capital Recovery Factor	Present Worth Factor	Sinking Fund Factor	Compound Amount Factor	Present Worth Factor	Uniform Payment Factor	
N	F/P	P/F	A/P	P/A	A/F	F/A	P/G	A/G	N
1	1.150	.8696	1.1500	.870	1.0000	1.000	0	0	1
2	1.323	.7561	.6151	1.626	.4651	2.150	.756	.465	2
3	1.521	.6575	.4380	2.283	.2880	3.473	2.071	.907	3
4	1.749	.5718	.3503	2.855	.2003	4.993	3.786	1.326	4
5	2.011	.4972	.2983	3.352	.1483	6.742	5.775	1.723	5
6	2.313	.4323	.2642	3.784	.1142	8.754	7.937	2.097	6
7	2.660	.3759	.2404	4.160	.0904	11.067	10.192	2.450	7
8	3.059	.3269	.2229	4.487	.0729	13.727	12.481	2.781	8
9	3.518	.2843	.2096	4.772	.0596	16.786	14.755	3.092	9
10	4.046	.2472	.1993	5.019	.0493	20.304	16.979	3.383	10
11	4.652	.2149	.1911	5.234	.0411	24.349	19.129	3.655	11
12	5.350	.1869	.1845	5.421	.0345	29.002	21.185	3.908	12
13	6.153	.1625	.1791	5.583	.0291	34.352	23.135	4.144	13
14	7.076	.1413	.1747	5.724	.0247	40.505	24.972	4.362	14
15	8.137	.1229	.1710	5.847	.0210	47.580	26.693	4.565	15
16	9.358	.1069	.1679	5.954	.0179	55.717	28.296	4.752	16
17	10.761	.0929	.1654	6.047	.0154	65.075	29.783	4.925	17
18	12.375	.0808	.1632	6.128	.0132	75.836	31.156	5.084	18
19	14.232	.0703	.1613	6.198	.0113	88.212	32.421	5.231	19
20	16.367	.0611	.1598	6.259	.00976	102.4	33.582	5.365	20
21	18.822	.0531	.1584	6.312	.00842	118.8	34.645	5.488	21
22	21.645	.0462	.1573	6.359	.00727	137.6	35.615	5.601	22
23	24.891	.0402	.1563	6.399	.00628	159.3	36.499	5.704	23
24	28.625	.0349	.1554	6.434	.00543	184.2	37.302	5.798	24
25	32.919	.0304	.1547	6.464	.00470	212.8	38.031	5.883	25
26	37.857	.0264	.1541	6.491	.00407	245.7	38.692	5.961	26
27	43.535	.0230	.1535	6.514	.00353	283.6	39.289	6.032	27
28	50.066	.0200	.1531	6.534	.00306	327.1	39.828	6.096	28
29	57.575	.0174	.1527	6.551	.00265	377.2	40.315	6.154	29
30	66.212	.0151	.1523	6.566	.00230	434.7	40.753	6.207	30
31	76.144	.0131	.1520	6.579	.00200	501.0	41.147	6.254	31
32	87.565	.0114	.1517	6.591	.00173	577.1	41.501	6.297	32
33	100.7	.00993	.1515	6.600	.00150	664.7	41.818	6.336	33
34	115.8	.00864	.1513	6.609	.00131	765.4	42.103	6.371	34
35	133.2	.00751	.1511	6.617	.00113	881.2	42.359	6.402	35
40	267.9	.00373	.1506	6.642	.000562	1779	43.283	6.517	40
45	538.8	.00186	.1503	6.654	.000279	3585	43.805	6.583	45
50	1084	.00092	.1501	6.661	.000139	7218	44.096	6.620	50
55	2180	.00046	.1501	6.664	.000069	14524	44.256	6.641	55
60	4384	.00023	.1500	6.665	.000034	29220	44.343	6.653	60
∞	∞	0	.15	7.667	0	∞	44.444	6.667	∞

16% — End-of-Period Compound Interest Factors — 16%

	Single Payment		Uniform Payment Series				Arithmetic Gradient		
	Compound Amount Factor	Present Worth Factor	Capital Recovery Factor	Present Worth Factor	Sinking Fund Factor	Compound Amount Factor	Present Worth Factor	Uniform Payment Factor	
N	F/P	P/F	A/P	P/A	A/F	F/A	P/G	A/G	N
1	1.160	.8621	1.1600	.862	1.0000	1.000	0	0	1
2	1.346	.7432	.6230	1.605	.4630	2.160	.743	.463	2
3	1.561	.6407	.4453	2.246	.2853	3.506	2.024	.901	3
4	1.811	.5523	.3574	2.798	.1974	5.066	3.681	1.316	4
5	2.100	.4761	.3054	3.274	.1454	6.877	5.586	1.706	5
6	2.436	.4104	.2714	3.685	.1114	8.977	7.638	2.073	6
7	2.826	.3538	.2476	4.039	.0876	11.414	9.761	2.417	7
8	3.278	.3050	.2302	4.344	.0702	14.240	11.896	2.739	8
9	3.803	.2630	.2171	4.607	.0571	17.519	14.000	3.039	9
10	4.411	.2267	.2069	4.833	.0469	21.321	16.040	3.319	10
11	5.117	.1954	.1989	5.029	.0389	25.733	17.994	3.578	11
12	5.936	.1685	.1924	5.197	.0324	30.850	19.847	3.819	12
13	6.886	.1452	.1872	5.342	.0272	36.786	21.590	4.041	13
14	7.988	.1252	.1829	5.468	.0229	43.672	23.217	4.246	14
15	9.266	.1079	.1794	5.575	.0194	51.660	24.728	4.435	15
16	10.748	.0930	.1764	5.668	.0164	60.925	26.124	4.609	16
17	12.468	.0802	.1740	5.749	.0140	71.673	27.407	4.768	17
18	14.463	.0691	.1719	5.818	.0119	84.141	28.583	4.913	18
19	16.777	.0596	.1701	5.877	.0101	98.603	29.656	5.046	19
20	19.461	.0514	.1687	5.929	.00867	115.4	30.632	5.167	20
21	22.574	.0443	.1674	5.973	.00742	134.8	31.518	5.277	21
22	26.186	.0382	.1664	6.011	.00635	157.4	32.320	5.377	22
23	30.376	.0329	.1654	6.044	.00545	183.6	33.044	5.467	23
24	35.236	.0284	.1647	6.073	.00467	214.0	33.697	5.549	24
25	40.874	.0245	.1640	6.097	.00401	249.2	34.284	5.623	25
26	47.414	.0211	.1634	6.118	.00345	290.1	34.811	5.690	26
27	55.000	.0182	.1630	6.136	.00296	337.5	35.284	5.750	27
28	63.800	.0157	.1625	6.152	.00255	392.5	35.707	5.804	28
29	74.009	.0135	.1622	6.166	.00219	456.3	36.086	5.853	29
30	85.850	.0116	.1619	6.177	.00189	530.3	36.423	5.896	30
31	99.586	.0100	.1616	6.187	.00162	616.2	36.725	5.936	31
32	115.5	.00866	.1614	6.196	.00140	715.7	36.993	5.971	32
33	134.0	.00746	.1612	6.203	.00120	831.3	37.232	6.002	33
34	155.4	.00643	.1610	6.210	.00104	965.3	37.444	6.030	34
35	180.3	.00555	.1609	6.215	.000892	1121	37.633	6.055	35
40	378.7	.00264	.1604	6.233	.000424	2361	38.299	6.144	40
45	795.4	.00126	.1602	6.242	.000201	4965	38.660	6.193	45
50	1671	.00060	.1601	6.246	.000096	10436	38.852	6.220	50
55	3509	.00028	.1600	6.248	.000046	21925	38.953	6.234	55
60	7370	.00014	.1600	6.249	.000022	46058	39.006	6.242	60
∞	∞	0	.16	6.250	0	∞	39.063	6.250	∞

17%   **End-of-Period Compound Interest Factors**   **17%**

	Single Payment		Uniform Payment Series				Arithmetic Gradient		
N	Compound Amount Factor F/P	Present Worth Factor P/F	Capital Recovery Factor A/P	Present Worth Factor P/A	Sinking Fund Factor A/F	Compound Amount Factor F/A	Present Worth Factor P/G	Uniform Payment Factor A/G	N
1	1.170	.8547	1.1700	.855	1.0000	1.000	0	0	1
2	1.369	.7305	.6308	1.585	.4608	2.170	.731	.461	2
3	1.602	.6244	.4526	2.210	.2826	3.539	1.979	.896	3
4	1.874	.5337	.3645	2.743	.1945	5.141	3.580	1.305	4
5	2.192	.4561	.3126	3.199	.1426	7.014	5.405	1.689	5
6	2.565	.3898	.2786	3.589	.1086	9.207	7.354	2.049	6
7	3.001	.3332	.2549	3.922	.0849	11.772	9.353	2.385	7
8	3.511	.2848	.2377	4.207	.0677	14.773	11.346	2.697	8
9	4.108	.2434	.2247	4.451	.0547	18.285	13.294	2.987	9
10	4.807	.2080	.2147	4.659	.0447	22.393	15.166	3.255	10
11	5.624	.1778	.2068	4.836	.0368	27.200	16.944	3.503	11
12	6.580	.1520	.2005	4.988	.0305	32.824	18.616	3.732	12
13	7.699	.1299	.1954	5.118	.0254	39.404	20.175	3.942	13
14	9.007	.1110	.1912	5.229	.0212	47.103	21.618	4.134	14
15	10.539	.0949	.1878	5.324	.0178	56.110	22.946	4.310	15
16	12.330	.0811	.1850	5.405	.0150	66.649	24.163	4.470	16
17	14.426	.0693	.1827	5.475	.0127	78.979	25.272	4.616	17
18	16.879	.0592	.1807	5.534	.0107	93.406	26.279	4.749	18
19	19.748	.0506	.1791	5.584	.00907	110.3	27.190	4.869	19
20	23.106	.0433	.1777	5.628	.00769	130.0	28.013	4.978	20
21	27.034	.0370	.1765	5.665	.00653	153.1	28.753	5.076	21
22	31.629	.0316	.1756	5.696	.00555	180.2	29.417	5.164	22
23	37.006	.0270	.1747	5.723	.00472	211.8	30.011	5.244	23
24	43.297	.0231	.1740	5.746	.00402	248.8	30.542	5.315	24
25	50.658	.0197	.1734	5.766	.00342	292.1	31.016	5.379	25
26	59.270	.0169	.1729	5.783	.00292	342.8	31.438	5.436	26
27	69.345	.0144	.1725	5.798	.00249	402.0	31.813	5.487	27
28	81.134	.0123	.1721	5.810	.00212	471.4	32.146	5.533	28
29	94.927	.0105	.1718	5.820	.00181	552.5	32.441	5.574	29
30	111.1	.00900	.1715	5.829	.00154	647.4	32.702	5.610	30
31	129.9	.00770	.1713	5.837	.00132	758.5	32.932	5.642	31
32	152.0	.00658	.1711	5.844	.00113	888.4	33.136	5.670	32
33	177.9	.00562	.1710	5.849	.000961	1040	33.316	5.696	33
34	208.1	.00480	.1708	5.854	.000821	1218	33.475	5.718	34
35	243.5	.00411	.1707	5.858	.000701	1426	33.614	5.738	35
40	533.9	.00187	.1703	5.871	.000319	3135	34.097	5.807	40
45	1170	.000854	.1701	5.877	.000145	6879	34.346	5.844	45
50	2566	.000390	.1701	5.880	.000066	15090	34.474	5.863	50
55	5626	.000178	.1700	5.881	.000030	33090	34.538	5.873	55
60	12335	.000081	.1700	5.882	.000014	72555	34.571	5.877	60
∞	∞	0	.17	5.882	0	∞	34.602	5.882	∞

18%   **End-of-Period Compound Interest Factors**   18%

	Single Payment		Uniform Payment Series				Arithmetic Gradient		
N	Compound Amount Factor F/P	Present Worth Factor P/F	Capital Recovery Factor A/P	Present Worth Factor P/A	Sinking Fund Factor A/F	Compound Amount Factor F/A	Present Worth Factor P/G	Uniform Payment Factor A/G	N
1	1.180	.8475	1.1800	.847	1.0000	1.000	0	0	1
2	1.392	.7182	.6387	1.566	.4587	2.180	.718	.459	2
3	1.643	.6086	.4599	2.174	.2799	3.572	1.935	.890	3
4	1.939	.5158	.3717	2.690	.1917	5.215	3.483	1.295	4
5	2.288	.4371	.3198	3.127	.1398	7.154	5.231	1.673	5
6	2.700	.3704	.2859	3.498	.1059	9.442	7.083	2.025	6
7	3.185	.3139	.2624	3.812	.0824	12.142	8.967	2.353	7
8	3.759	.2660	.2452	4.078	.0652	15.327	10.829	2.656	8
9	4.435	.2255	.2324	4.303	.0524	19.086	12.633	2.936	9
10	5.234	.1911	.2225	4.494	.0425	23.521	14.352	3.194	10
11	6.176	.1619	.2148	4.656	.0348	28.755	15.972	3.430	11
12	7.288	.1372	.2086	4.793	.0286	34.931	17.481	3.647	12
13	8.599	.1163	.2037	4.910	.0237	42.219	18.877	3.845	13
14	10.147	.0985	.1997	5.008	.0197	50.818	20.158	4.025	14
15	11.974	.0835	.1964	5.092	.0164	60.965	21.327	4.189	15
16	14.129	.0708	.1937	5.162	.0137	72.939	22.389	4.337	16
17	16.672	.0600	.1915	5.222	.0115	87.068	23.348	4.471	17
18	19.673	.0508	.1896	5.273	.00964	103.7	24.212	4.592	18
19	23.214	.0431	.1881	5.316	.00810	123.4	24.988	4.700	19
20	27.393	.0365	.1868	5.353	.00682	146.6	25.681	4.798	20
21	32.324	.0309	.1857	5.384	.00575	174.0	26.300	4.885	21
22	38.142	.0262	.1848	5.410	.00485	206.3	26.851	4.963	22
23	45.008	.0222	.1841	5.432	.00409	244.5	27.339	5.033	23
24	53.109	.0188	.1835	5.451	.00345	289.5	27.772	5.095	24
25	62.669	.0160	.1829	5.467	.00292	342.6	28.155	5.150	25
26	73.949	.0135	.1825	5.480	.00247	405.3	28.494	5.199	26
27	87.260	.0115	.1821	5.492	.00209	479.2	28.791	5.243	27
28	103.0	.00971	.1818	5.502	.00177	566.5	29.054	5.281	28
29	121.5	.00823	.1815	5.510	.00149	669.4	29.284	5.315	29
30	143.4	.00697	.1813	5.517	.00126	790.9	29.486	5.345	30
31	169.2	.00591	.1811	5.523	.00107	934.3	29.664	5.371	31
32	199.6	.00501	.1809	5.528	.000906	1103	29.819	5.394	32
33	235.6	.00425	.1808	5.532	.000767	1303	29.955	5.415	33
34	278.0	.00360	.1806	5.536	.000650	1539	30.074	5.433	34
35	328.0	.00305	.1806	5.539	.000550	1817	30.177	5.449	35
40	750.4	.00133	.1802	5.548	.000240	4163	30.527	5.502	40
45	1717	.000583	.1801	5.552	.000105	9532	30.701	5.529	45
50	3927	.000255	.1800	5.554	.0000458	21813	30.786	5.543	50
55	8985	.000111	.1800	5.555	.0000200	49910	30.827	5.549	55
60	20555	.000049	.1800	5.555	.0000088	114190	30.846	5.553	60
∞	∞	0	.18	5.556	0	∞	30.864	5.556	∞

19% End-of-Period Compound Interest Factors 19%

	Single Payment		Uniform Payment Series				Arithmetic Gradient		
N	Compound Amount Factor F/P	Present Worth Factor P/F	Capital Recovery Factor A/P	Present Worth Factor P/A	Sinking Fund Factor A/F	Compound Amount Factor F/A	Present Worth Factor P/G	Uniform Payment Factor A/G	N
1	1.190	.8403	1.1900	.840	1.0000	1.000	0	0	1
2	1.416	.7062	.6466	1.547	.4566	2.190	.706	.457	2
3	1.685	.5934	.4673	2.140	.2773	3.606	1.893	.885	3
4	2.005	.4987	.3790	2.639	.1890	5.291	3.389	1.284	4
5	2.386	.4190	.3271	3.058	.1371	7.297	5.065	1.657	5
6	2.840	.3521	.2933	3.410	.1033	9.683	6.826	2.002	6
7	3.379	.2959	.2699	3.706	.0799	12.523	8.601	2.321	7
8	4.021	.2487	.2529	3.954	.0629	15.902	10.342	2.615	8
9	4.785	.2090	.2402	4.163	.0502	19.923	12.014	2.886	9
10	5.695	.1756	.2305	4.339	.0405	24.709	13.594	3.133	10
11	6.777	.1476	.2229	4.486	.0329	30.404	15.070	3.359	11
12	8.064	.1240	.2169	4.611	.0269	37.180	16.434	3.564	12
13	9.596	.1042	.2121	4.715	.0221	45.244	17.684	3.751	13
14	11.420	.0876	.2082	4.802	.0182	54.841	18.823	3.920	14
15	13.590	.0736	.2051	4.876	.0151	66.261	19.853	4.072	15
16	16.172	.0618	.2025	4.938	.0125	79.850	20.781	4.209	16
17	19.244	.0520	.2004	4.990	.0104	96.022	21.612	4.331	17
18	22.901	.0437	.1987	5.033	.00868	115.3	22.354	4.441	18
19	27.252	.0367	.1972	5.070	.00724	138.2	23.015	4.539	19
20	32.429	.0308	.1960	5.101	.00605	165.4	23.601	4.627	20
21	38.591	.0259	.1951	5.127	.00505	197.8	24.119	4.705	21
22	45.923	.0218	.1942	5.149	.00423	236.4	24.576	4.773	22
23	54.649	.0183	.1935	5.167	.00354	282.4	24.979	4.834	23
24	65.032	.0154	.1930	5.182	.00297	337.0	25.333	4.888	24
25	77.388	.0129	.1925	5.195	.00249	402.0	25.643	4.936	25
26	92.092	.0109	.1921	5.206	.00209	479.4	25.914	4.978	26
27	109.6	.00912	.1917	5.215	.00175	571.5	26.151	5.015	27
28	130.4	.00767	.1915	5.223	.00147	681.1	26.358	5.047	28
29	155.2	.00644	.1912	5.229	.00123	811.5	26.539	5.075	29
30	184.7	.00541	.1910	5.235	.00103	966.7	26.696	5.100	30
31	219.8	.00455	.1909	5.239	.000869	1151	26.832	5.121	31
32	261.5	.00382	.1907	5.243	.000729	1371	26.951	5.140	32
33	311.2	.00321	.1906	5.246	.000612	1633	27.054	5.157	33
34	370.3	.00270	.1905	5.249	.000514	1944	27.143	5.171	34
35	440.7	.00227	.1904	5.251	.000432	2314	27.220	5.184	35
40	1052	.000951	.1902	5.258	.000181	5530	27.474	5.225	40
45	2510	.000398	.1901	5.261	.0000757	13203	27.595	5.245	45
50	5989	.000167	.1900	5.262	.0000317	31515	27.652	5.255	50
55	14292	.000070	.1900	5.263	.0000133	75214	27.679	5.259	55
60	34105	.000029	.1900	5.263	.0000056	179495	27.691	5.261	60
∞	∞	0	.19	5.263	0	∞	27.701	5.263	∞

20% **End-of-Period Compound Interest Factors** 20%

	Single Payment		Uniform Payment Series				Arithmetic Gradient		
	Compound Amount Factor	Present Worth Factor	Capital Recovery Factor	Present Worth Factor	Sinking Fund Factor	Compound Amount Factor	Present Worth Factor	Uniform Payment Factor	
N	F/P	P/F	A/P	P/A	A/F	F/A	P/G	A/G	N
1	1.200	.8333	1.2000	.833	1.0000	1.000	0	0	1
2	1.440	.6944	.6545	1.528	.4545	2.200	.694	.455	2
3	1.728	.5787	.4747	2.106	.2747	3.640	1.852	.879	3
4	2.074	.4823	.3863	2.589	.1863	5.368	3.299	1.274	4
5	2.488	.4019	.3344	2.991	.1344	7.442	4.906	1.641	5
6	2.986	.3349	.3007	3.326	.1007	9.930	6.581	1.979	6
7	3.583	.2791	.2774	3.605	.0774	12.916	8.255	2.290	7
8	4.300	.2326	.2606	3.837	.0606	16.499	9.883	2.576	8
9	5.160	.1938	.2481	4.031	.0481	20.799	11.434	2.836	9
10	6.192	.1615	.2385	4.192	.0385	25.959	12.887	3.074	10
11	7.430	.1346	.2311	4.327	.0311	32.150	14.233	3.289	11
12	8.916	.1122	.2253	4.439	.0253	39.581	15.467	3.484	12
13	10.699	.0935	.2206	4.533	.0206	48.497	16.588	3.660	13
14	12.839	.0779	.2169	4.611	.0169	59.196	17.601	3.817	14
15	15.407	.0649	.2139	4.675	.0139	72.035	18.509	3.959	15
16	18.488	.0541	.2114	4.730	.0114	87.442	19.321	4.085	16
17	22.186	.0451	.2094	4.775	.00944	105.9	20.042	4.198	17
18	26.623	.0376	.2078	4.812	.00781	128.1	20.680	4.298	18
19	31.948	.0313	.2065	4.843	.00646	154.7	21.244	4.386	19
20	38.338	.0261	.2054	4.870	.00536	186.7	21.739	4.464	20
21	46.005	.0217	.2044	4.891	.00444	225.0	22.174	4.533	21
22	55.206	.0181	.2037	4.909	.00369	271.0	22.555	4.594	22
23	66.247	.0151	.2031	4.925	.00307	326.2	22.887	4.647	23
24	79.497	.0126	.2025	4.937	.00255	392.5	23.176	4.694	24
25	95.396	.0105	.2021	4.948	.00212	472.0	23.428	4.735	25
26	114.5	.00874.	.2018	4.956	.00176	567.4	23.646	4.771	26
27	137.4	.00728	.2015	4.964	.00147	681.9	23.835	4.802	27
28	164.8	.00607	.2012	4.970	.00122	819.2	23.999	4.829	28
29	197.8	.00506	.2010	4.975	.00102	984.1	24.141	4.853	29
30	237.4	.00421	.2008	4.979	.000846	1182	24.263	4.873	30
31	284.9	.00351	.2007	4.982	.000705	1419	24.368	4.891	31
32	341.8	.00293	.2006	4.985	.000587	1704	24.459	4.906	32
33	410.2	.00244	.2005	4.988	.000489	2046	24.537	4.919	33
34	492.2	.00203	.2004	4.990	.000407	2456	24.604	4.931	34
35	590.7	.00169	.2003	4.992	.000339	2948	24.661	4.941	35
40	1470	.000680	.2001	4.997	.000136	7344	24.847	4.973	40
45	3657	.000273	.2001	4.999	.0000547	18281	24.932	4.988	45
50	9100	.000110	.2000	4.999	.0000220	45497	24.970	4.995	50
55	22645	.000044	.2000	5.000	.0000088	113219	24.987	4.998	55
60	56348	.000018	.2000	5.000	.0000035	281733	24.994	4.999	60
∞	∞	0	.2	5	0	∞	25	5	∞

25%   **End-of-Period Compound Interest Factors**   **25%**

	Single Payment		Uniform Payment Series				Arithmetic Gradient		
N	Compound Amount Factor F/P	Present Worth Factor P/F	Capital Recovery Factor A/P	Present Worth Factor P/A	Sinking Fund Factor A/F	Compound Amount Factor F/A	Present Worth Factor P/G	Uniform Payment Factor A/G	N
1	1.250	.8000	1.2500	.800	1.0000	1.000	0	0	1
2	1.563	.6400	.6944	1.440	.4444	2.250	.640	.444	2
3	1.953	.5120	.5123	1.952	.2623	3.813	1.664	.852	3
4	2.441	.4096	.4234	2.362	.1734	5.766	2.893	1.225	4
5	3.052	.3277	.3718	2.689	.1218	8.207	4.204	1.563	5
6	3.815	.2621	.3388	2.951	.0888	11.259	5.514	1.868	6
7	4.768	.2097	.3163	3.161	.0663	15.073	6.773	2.142	7
8	5.960	.1678	.3004	3.329	.0504	19.842	7.947	2.387	8
9	7.451	.1342	.2888	3.463	.0388	25.802	9.021	2.605	9
10	9.313	.1074	.2801	3.571	.0301	33.253	9.987	2.797	10
11	11.642	.0859	.2735	3.656	.0235	42.566	10.846	2.966	11
12	14.552	.0687	.2684	3.725	.0184	54.208	11.602	3.115	12
13	18.190	.0550	.2645	3.780	.0145	68.760	12.262	3.244	13
14	22.737	.0440	.2615	3.824	.0115	86.949	12.833	3.356	14
15	28.422	.0352	.2591	3.859	.0091	109.687	13.326	3.453	15
16	35.527	.0281	.2572	3.887	.0072	138.109	13.748	3.537	16
17	44.409	.0225	.2558	3.910	.00576	173.6	14.108	3.608	17
18	55.511	.0180	.2546	3.928	.00459	218.0	14.415	3.670	18
19	69.389	.0144	.2537	3.942	.00366	273.6	14.674	3.722	19
20	86.736	.0115	.2529	3.954	.00292	342.9	14.893	3.767	20
21	108.420	.0092	.2523	3.963	.00233	429.7	15.078	3.805	21
22	135.525	.0074	.2519	3.970	.00186	538.1	15.233	3.836	22
23	169.407	.0059	.2515	3.976	.00148	673.6	15.362	3.863	23
24	211.758	.0047	.2512	3.981	.00119	843.0	15.471	3.886	24
25	264.698	.0038	.2509	3.985	.000948	1055	15.562	3.905	25
26	330.9	.00302	.2508	3.988	.000758	1319	15.637	3.921	26
27	413.6	.00242	.2506	3.990	.000606	1650	15.700	3.935	27
28	517.0	.00193	.2505	3.992	.000485	2064	15.752	3.946	28
29	646.2	.00155	.2504	3.994	.000387	2581	15.796	3.955	29
30	807.8	.00124	.2503	3.995	.000310	3227	15.832	3.963	30
31	1010	.000990	.2502	3.996	.000248	4035	15.861	3.969	31
32	1262	.000792	.2502	3.997	.000198	5045	15.886	3.975	32
33	1578	.000634	.2502	3.997	.000159	6307	15.906	3.979	33
34	1972	.000507	.2501	3.998	.000127	7885	15.923	3.983	34
35	2465	.000406	.2501	3.998	.000101	9857	15.937	3.986	35
40	7523	.000133	.2500	3.999	.0000332	30089	15.977	3.995	40
45	22959	.000044	.2500	4.000	.0000109	91831	15.991	3.998	45
50	70065	.000014	.2500	4.000	.0000036	280256	15.997	3.999	50
∞	∞	0	.25	4	0	∞	16	4	∞

30%   **End-of-Period Compound Interest Factors**   **30%**

	Single Payment		Uniform Payment Series				Arithmetic Gradient		
N	Compound Amount Factor F/P	Present Worth Factor P/F	Capital Recovery Factor A/P	Present Worth Factor P/A	Sinking Fund Factor A/F	Compound Amount Factor F/A	Present Worth Factor P/G	Uniform Payment Factor A/G	N
1	1.300	.7692	1.3000	.769	1.0000	1.000	0	0	1
2	1.690	.5917	.7348	1.361	.4348	2.300	.592	.435	2
3	2.197	.4552	.5506	1.816	.2506	3.990	1.502	.827	3
4	2.856	.3501	.4616	2.166	.1616	6.187	2.552	1.178	4
5	3.713	.2693	.4106	2.436	.1106	9.043	3.630	1.490	5
6	4.827	.2072	.3784	2.643	.0784	12.756	4.666	1.765	6
7	6.275	.1594	.3569	2.802	.0569	17.583	5.622	2.006	7
8	8.157	.1226	.3419	2.925	.0419	23.858	6.480	2.216	8
9	10.604	.0943	.3312	3.019	.0312	32.015	7.234	2.396	9
10	13.786	.0725	.3235	3.092	.0235	42.619	7.887	2.551	10
11	17.922	.0558	.3177	3.147	.0177	56.405	8.445	2.683	11
12	23.298	.0429	.3135	3.190	.0135	74.327	8.917	2.795	12
13	30.288	.0330	.3102	3.223	.0102	97.625	9.314	2.889	13
14	39.374	.0254	.3078	3.249	.00782	127.9	9.644	2.969	14
15	51.186	.0195	.3060	3.268	.00598	167.3	9.917	3.034	15
16	66.542	.0150	.3046	3.283	.00458	218.5	10.143	3.089	16
17	86.504	.0116	.3035	3.295	.00351	285.0	10.328	3.135	17
18	112.5	.00889	.3027	3.304	.00269	371.5	10.479	3.172	18
19	146.2	.00684	.3021	3.311	.00207	484.0	10.602	3.202	19
20	190.0	.00526	.3016	3.316	.00159	630.2	10.702	3.228	20
21	247.1	.00405	.3012	3.320	.00122	820.2	10.783	3.248	21
22	321.2	.00311	.3009	3.323	.000937	1067	10.848	3.265	22
23	417.5	.00239	.3007	3.325	.000720	1388	10.901	3.278	23
24	542.8	.00184	.3006	3.327	.000554	1806	10.943	3.289	24
25	705.6	.00142	.3004	3.329	.000426	2349	10.977	3.298	25
26	917.3	.00109	.3003	3.330	.000327	3054	11.005	3.305	26
27	1193	.000839	.3003	3.331	.000252	3972	11.026	3.311	27
28	1550	.000645	.3002	3.331	.000194	5164	11.044	3.315	28
29	2015	.000496	.3001	3.332	.000149	6715	11.058	3.319	29
30	2620	.000382	.3001	3.332	.000115	8730	11.069	3.322	30
∞	∞	0	.3	3.333	0	∞	11.111	3.333	∞

35%   **End-of-Period Compound Interest Factors**   **35%**

	Single Payment		Uniform Payment Series				Arithmetic Gradient		
N	Compound Amount Factor F/P	Present Worth Factor P/F	Capital Recovery Factor A/P	Present Worth Factor P/A	Sinking Fund Factor A/F	Compound Amount Factor F/A	Present Worth Factor P/G	Uniform Payment Factor A/G	N
1	1.350	.7407	1.3500	.741	1.0000	1.000	0	0	1
2	1.823	.5487	.7755	1.289	.4255	2.350	.549	.426	2
3	2.460	.4064	.5897	1.696	.2397	4.173	1.362	.803	3
4	3.322	.3011	.5008	1.997	.1508	6.633	2.265	1.134	4
5	4.484	.2230	.4505	2.220	.1005	9.954	3.157	1.422	5
6	6.053	.1652	.4193	2.385	.0693	14.438	3.983	1.670	6
7	8.172	.1224	.3988	2.508	.0488	20.492	4.717	1.881	7
8	11.032	.0906	.3849	2.598	.0349	28.664	5.352	2.060	8
9	14.894	.0671	.3752	2.665	.0252	39.696	5.889	2.209	9
10	20.107	.0497	.3683	2.715	.0183	54.590	6.336	2.334	10
11	27.144	.0368	.3634	2.752	.0134	74.697	6.705	2.436	11
12	36.644	.0273	.3598	2.779	.00982	101.8	7.005	2.520	12
13	49.470	.0202	.3572	2.799	.00722	138.5	7.247	2.589	13
14	66.784	.0150	.3553	2.814	.00532	188.0	7.442	2.644	14
15	90.158	.0111	.3539	2.825	.00393	254.7	7.597	2.689	15
16	121.7	.00822	.3529	2.834	.00290	344.9	7.721	2.725	16
17	164.3	.00609	.3521	2.840	.00214	466.6	7.818	2.753	17
18	221.8	.00451	.3516	2.844	.00158	630.9	7.895	2.776	18
19	299.5	.00334	.3512	2.848	.00117	852.7	7.955	2.793	19
20	404.3	.00247	.3509	2.850	.000868	1152	8.002	2.808	20
21	545.8	.00183	.3506	2.852	.000642	1556	8.038	2.819	21
22	736.8	.00136	.3505	2.853	.000476	2102	8.067	2.827	22
23	994.7	.00101	.3504	2.854	.000352	2839	8.089	2.834	23
24	1343	.000745	.3503	2.855	.000261	3834	8.106	2.839	24
25	1813	.000552	.3502	2.856	.000193	5177	8.119	2.843	25
26	2447	.000409	.3501	2.856	.000143	6989	8.130	2.847	26
27	3304	.000303	.3501	2.856	.000106	9437	8.137	2.849	27
28	4460	.000224	.3501	2.857	.000078	12740	8.143	2.851	28
29	6021	.000166	.3501	2.857	.000058	17200	8.148	2.852	29
30	8129	.000123	.3500	2.857	.000043	23222	8.152	2.853	30
∞	∞	0	.35	2.857	0	∞	8.163	2.857	∞

40%   **End-of-Period Compound Interest Factors**   **40%**

	Single Payment		Uniform Payment Series				Arithmetic Gradient		
	Compound Amount Factor	Present Worth Factor	Capital Recovery Factor	Present Worth Factor	Sinking Fund Factor	Compound Amount Factor	Present Worth Factor	Uniform Payment Factor	
N	*F/P*	*P/F*	*A/P*	*P/A*	*A/F*	*F/A*	*P/G*	*A/G*	*N*
1	1.400	.7143	1.4000	.714	1.0000	1.000	0	0	1
2	1.960	.5102	.8167	1.224	.4167	2.400	.510	.417	2
3	2.744	.3644	.6294	1.589	.2294	4.360	1.239	.780	3
4	3.842	.2603	.5408	1.849	.1408	7.104	2.020	1.092	4
5	5.378	.1859	.4914	2.035	.0914	10.946	2.764	1.358	5
6	7.530	.1328	.4613	2.168	.0613	16.324	3.428	1.581	6
7	10.541	.0949	.4419	2.263	.0419	23.853	3.997	1.766	7
8	14.758	.0678	.4291	2.331	.0291	34.395	4.471	1.919	8
9	20.661	.0484	.4203	2.379	.0203	49.153	4.858	2.042	9
10	28.925	.0346	.4143	2.414	.0143	69.814	5.170	2.142	10
11	40.496	.0247	.4101	2.438	.0101	98.739	5.417	2.221	11
12	56.694	.0176	.4072	2.456	.00718	139.2	5.611	2.285	12
13	79.371	.0126	.4051	2.469	.00510	195.9	5.762	2.334	13
14	111.1	.00900	.4036	2.478	.00363	275.3	5.879	2.373	14
15	155.6	.00643	.4026	2.484	.00259	386.4	5.969	2.403	15
16	217.8	.00459	.4018	2.489	.00185	542.0	6.038	2.426	16
17	304.9	.00328	.4013	2.492	.00132	759.8	6.090	2.444	17
18	426.9	.00234	.4009	2.494	.000939	1065	6.130	2.458	18
19	597.6	.00167	.4007	2.496	.000670	1492	6.160	2.468	19
20	836.7	.00120	.4005	2.497	.000479	2089	6.183	2.476	20
21	1171	.000854	.4003	2.498	.000342	2926	6.200	2.482	21
22	1640	.000610	.4002	2.498	.000244	4097	6.213	2.487	22
23	2296	.000436	.4002	2.499	.000174	5737	6.222	2.490	23
24	3214	.000311	.4001	2.499	.000124	8033	6.229	2.493	24
25	4500	.000222	.4001	2.499	.000089	11247	6.235	2.494	25
26	6300	.000159	.4001	2.500	.000064	15747	6.239	2.496	26
27	8820	.000113	.4000	2.500	.000045	22047	6.242	2.497	27
28	12348	.000081	.4000	2.500	.000032	30867	6.244	2.498	28
29	17287	.000058	.4000	2.500	.000023	43214	6.245	2.498	29
30	24201	.000041	.4000	2.500	.000017	60501	6.247	2.499	30
∞	∞	0	.4	2.5	0	∞	6.25	2.5	∞

50%   **End-of-Period Compound Interest Factors**   50%

	Single Payment		Uniform Payment Series				Arithmetic Gradient		
	Compound Amount Factor	Present Worth Factor	Capital Recovery Factor	Present Worth Factor	Sinking Fund Factor	Compound Amount Factor	Present Worth Factor	Uniform Payment Factor	
N	F/P	P/F	A/P	P/A	A/F	F/A	P/G	A/G	N
1	1.500	.6667	1.5000	.667	1.0000	1.000	0	0	1
2	2.250	.4444	.9000	1.111	.4000	2.500	.444	.400	2
3	3.375	.2963	.7105	1.407	.2105	4.750	1.037	.737	3
4	5.063	.1975	.6231	1.605	.1231	8.125	1.630	1.015	4
5	7.594	.1317	.5758	1.737	.0758	13.188	2.156	1.242	5
6	11.391	.0878	.5481	1.824	.0481	20.781	2.595	1.423	6
7	17.086	.0585	.5311	1.883	.0311	32.172	2.947	1.565	7
8	25.629	.0390	.5203	1.922	.0203	49.258	3.220	1.675	8
9	38.443	.0260	.5134	1.948	.0134	74.887	3.428	1.760	9
10	57.665	.0173	.5088	1.965	.00882	113.3	3.584	1.824	10
11	86.498	.0116	.5058	1.977	.00585	171.0	3.699	1.871	11
12	129.7	.00771	.5039	1.985	.00388	257.5	3.784	1.907	12
13	194.6	.00514	.5026	1.990	.00258	387.2	3.846	1.933	13
14	291.9	.00343	.5017	1.993	.00172	581.9	3.890	1.952	14
15	437.9	.00228	.5011	1.995	.00114	873.8	3.922	1.966	15
16	656.8	.00152	.5008	1.997	.000762	1312	3.945	1.976	16
17	985.3	.00101	.5005	1.998	.000508	1969	3.961	1.983	17
18	1478	.000677	.5003	1.999	.000339	2954	3.973	1.988	18
19	2217	.000451	.5002	1.999	.000226	4432	3.981	1.991	19
20	3325	.000301	.5002	1.999	.000150	6649	3.987	1.994	20
21	4988	.000200	.5001	2.000	.000100	9974	3.991	1.996	21
22	7482	.000134	.5001	2.000	.0000668	14962	3.994	1.997	22
23	11223	.0000891	.5000	2.000	.0000446	22443	3.996	1.998	23
24	16834	.0000594	.5000	2.000	.0000297	33666	3.997	1.999	24
25	25251	.0000396	.5000	2.000	.0000198	50500	3.998	1.999	25
∞	∞	0	.5	2	0	∞	4	2	∞

End-of-Period Compound Interest Factors

	Single Payment		Uniform Payment Series				Arithmetic Gradient		
	Compound Amount Factor	Present Worth Factor	Capital Recovery Factor	Present Worth Factor	Sinking Fund Factor	Compound Amount Factor	Present Worth Factor	Uniform Payment Factor	
N	F/P	P/F	A/P	P/A	A/F	F/A	P/G	A/G	N
1	1.600	.6250	1.6000	.625	1.0000	1.000	0	0	1
2	2.560	.3906	.9846	1.016	.3846	2.600	.391	.385	2
3	4.096	.2441	.7938	1.260	.1938	5.160	.879	.698	3
4	6.554	.1526	.7080	1.412	.1080	9.256	1.337	.946	4
5	10.486	.0954	.6633	1.508	.0633	15.810	1.718	1.140	5
6	16.777	.0596	.6380	1.567	.0380	26.295	2.016	1.286	6
7	26.844	.0373	.6232	1.605	.0232	43.073	2.240	1.396	7
8	42.950	.0233	.6143	1.628	.0143	69.916	2.403	1.476	8
9	68.719	.0146	.6089	1.642	.00886	112.9	2.519	1.534	9
10	110.0	.00909	.6055	1.652	.00551	181.6	2.601	1.575	10
11	175.9	.00568	.6034	1.657	.00343	291.5	2.658	1.604	11
12	281.5	.00355	.6021	1.661	.00214	467.5	2.697	1.624	12
13	450.4	.00222	.6013	1.663	.00134	748.9	2.724	1.638	13
14	720.6	.00139	.6008	1.664	.000834	1199	2.742	1.647	14
15	1153	.000867	.6005	1.665	.000521	1920	2.754	1.654	15
16	1845	.000542	.6003	1.666	.000325	3073	2.762	1.658	16
17	2951	.000339	.6002	1.666	.000203	4917	2.767	1.661	17
18	4722	.000212	.6001	1.666	.000127	7869	2.771	1.663	18
19	7556	.000132	.6001	1.666	.000079	12591	2.773	1.664	19
20	12089	.000083	.6000	1.667	.000050	20147	2.775	1.665	20
∞	∞	0	.6	1.667	0	∞	2.778	1.667	∞

N	F/P	P/F	A/P	P/A	A/F	F/A	P/G	A/G	N
1	1.700	.5882	1.7000	.588	1.0000	1.000	0	0	1
2	2.890	.3460	1.0704	.934	.3704	2.700	.346	.370	2
3	4.913	.2035	.8789	1.138	.1789	5.590	.753	.662	3
4	8.352	.1197	.7952	1.258	.0952	10.503	1.112	.885	4
5	14.199	.0704	.7530	1.328	.0530	18.855	1.394	1.050	5
6	24.138	.0414	.7303	1.369	.0303	33.054	1.601	1.169	6
7	41.034	.0244	.7175	1.394	.0175	57.191	1.747	1.254	7
8	69.758	.0143	.7102	1.408	.0102	98.225	1.848	1.312	8
9	118.6	.00843	.7060	1.417	.00595	168.0	1.915	1.352	9
10	201.6	.00496	.7035	1.421	.00349	286.6	1.960	1.379	10
11	342.7	.00292	.7020	1.424	.00205	488.2	1.989	1.396	11
12	582.6	.00172	.7012	1.426	.00120	830.9	2.008	1.408	12
13	990.5	.00101	.7007	1.427	.000707	1414	2.020	1.415	13
14	1684	.000594	.7004	1.428	.000416	2404	2.028	1.420	14
15	2862	.000349	.7002	1.428	.000245	4088	2.033	1.423	15
16	4866	.000206	.7001	1.428	.000144	6950	2.036	1.425	16
17	8272	.000121	.7001	1.428	.000085	11816	2.038	1.427	17
18	14063	.000071	.7000	1.428	.000050	20089	2.039	1.427	18
19	23907	.000042	.7000	1.429	.000029	34152	2.040	1.428	19
20	40642	.000025	.7000	1.429	.000017	58059	2.040	1.428	20
∞	∞	0	.7	1.429	0	∞	2.041	1.429	∞

	Single Payment		Uniform Payment Series				Arithmetic Gradient		
N	Compound Amount Factor F/P	Present Worth Factor P/F	Capital Recovery Factor A/P	Present Worth Factor P/A	Sinking Fund Factor A/F	Compound Amount Factor F/A	Present Worth Factor P/G	Uniform Payment Factor A/G	N
1	1.800	.5556	1.8000	.556	1.0000	1.000	0	0	1
2	3.240	.3086	1.1571	.864	.3571	2.800	.309	.357	2
3	5.832	.1715	.9656	1.036	.1656	6.040	.652	.629	3
4	10.498	.0953	.8842	1.131	.0842	11.872	.937	.829	4
5	18.896	.0529	.8447	1.184	.0447	22.370	1.149	.971	5
6	34.012	.0294	.8242	1.213	.0242	41.265	1.296	1.068	6
7	61.222	.0163	.8133	1.230	.0133	75.278	1.394	1.134	7
8	110.2	.00907	.8073	1.239	.00733	136.5	1.458	1.177	8
9	198.4	.00504	.8041	1.244	.00405	246.7	1.498	1.204	9
10	357.0	.00280	.8022	1.246	.00225	445.1	1.523	1.222	10
11	642.7	.00156	.8012	1.248	.00125	802.1	1.539	1.233	11
12	1157	.000864	.8007	1.249	.000692	1445	1.548	1.240	12
13	2082	.000480	.8004	1.249	.000384	2602	1.554	1.244	13
14	3748	.000267	.8002	1.250	.000213	4684	1.557	1.246	14
15	6747	.000148	.8001	1.250	.000119	8432	1.559	1.248	15
∞	∞	0	.8	1.25	0	∞	1.562	1.25	∞

90%   90%

N	F/P	P/F	A/P	P/A	A/F	F/A	P/G	A/G	N
1	1.900	.5263	1.9000	.526	1.0000	1.000	0	0	1
2	3.610	.2770	1.2448	.803	.3448	2.900	.277	.345	2
3	6.859	.1458	1.0536	.949	.1536	6.510	.569	.599	3
4	13.032	.0767	.9748	1.026	.0748	13.369	.799	.779	4
5	24.761	.0404	.9379	1.066	.0379	26.401	.960	.901	5
6	47.046	.0213	.9195	1.087	.0195	51.162	1.067	.981	6
7	89.387	.0112	.9102	1.099	.0102	98.208	1.134	1.032	7
8	169.8	.00589	.9053	1.105	.00533	187.6	1.175	1.064	8
9	322.7	.00310	.9028	1.108	.00280	357.4	1.200	1.083	9
10	613.1	.00163	.9015	1.109	.00147	680.1	1.214	1.095	10
∞	∞	0	.9	1.111	0	∞	1.235	1.111	∞

100%   100%

N	F/P	P/F	A/P	P/A	A/F	F/A	P/G	A/G	N
1	2	.5000	2.0000	.500	1.0000	1.000	0	0	1
2	4	.2500	1.3333	.750	.3333	3.000	.250	.333	2
3	8	.1250	1.1429	.875	.1429	7.000	.500	.571	3
4	16	.0625	1.0667	.938	.0667	15.000	.688	.733	4
5	32	.0313	1.0323	.969	.0323	31.000	.813	.839	5
6	64	.0156	1.0159	.984	.0159	63.000	.891	.905	6
7	128	.00781	1.0079	.992	.00787	127.0	.938	.945	7
8	256	.00391	1.0039	.996	.00392	255.0	.965	.969	8
9	512	.00195	1.0020	.998	.00196	511.0	.980	.982	9
10	1024	.00098	1.0010	.999	.00098	1023	.989	.990	10
∞	∞	0	1	1	0	∞	1	1	∞

INDEX

Exhibit 12.7 Recovery periods for MACRS

Recovery Period	Description of Assets Included
3-Year	Tractors for over-the-road tractor/trailer use and special tools such as dies and jigs; ADR < 4 years
5-Year	Cars, buses, trucks, computers, office machinery, construction equipment, and R&D equipment; 4 years ≤ ADR < 10 years
7-Year	Office furniture, most manufacturing equipment, mining equipment, and items not otherwise classified; 10 years ≤ ADR < 16 years
10-Year	Marine vessels, petroleum refining equipment, single-purpose agricultural structures, trees and vines that bear nuts or fruits; 16 years ≤ ADR < 20 years
15-Year	Roads, shrubbery, wharves, steam and electric generation and distribution systems, and municipal wastewater treatment facilities; 20 years ≤ ADR < 25 years
20-Year	Farm buildings and municipal sewers; ADR ≥ 25 years
27.5-Year	Residential rental property
31.5-Year	Nonresidential real property purchased on or before 5/12/93
39-Year	Nonresidential real property purchased on or after 5/13/93

Note: This table is used to find the recovery period for *all* items under MACRS.

Exhibit 12.9 MACRS percentages

Year	Recovery Period					
	3-year	5-year	7-year	10-year	15-year	20-year
1	33.33%	20.00%	14.29%	10.00%	5.00%	3.750%
2	44.45	32.00	24.49	18.00	9.50	7.219
3	14.81	19.20	17.49	14.40	8.55	6.677
4	7.41	11.52	12.49	11.52	7.70	6.177
5		11.52	8.93	9.22	6.93	5.713
6		5.76	8.92	7.37	6.23	5.285
7			8.93	6.55	5.90	4.888
8			4.46	6.55	5.90	4.522
9				6.56	5.91	4.462
10				6.55	5.90	4.461
11				3.28	5.91	4.462
12					5.90	4.461
13					5.91	4.462
14					5.90	4.461
15					5.91	4.462
16					2.95	4.461
17						4.462
18						4.461
19						4.462
20						4.461
21						2.231